Water Resources

D0164998

Now in its second edition, *Water Resources: An Integrated Approach* provides students with a comprehensive overview of natural processes associated with water and the modifications of these processes by humans through climate change and land management, water-related health issues, engineering approaches to water and socio-economic processes of huge importance to water resources. The book contains chapters written by 24 specialist contributors, providing expert depth of coverage to topics.

The text introduces the basic properties of water and its importance to society and the nature of the different regional imbalances between water resource availability and demand. It guides the reader through the changing water cycle impacted by climate and land management, water flows in river basins, surface water quality, groundwater and aquatic ecosystems, and covers the role of water in human health and associated hazards before turning to engineering solutions to water and wastewater treatment and reuse. The book deals with physical and social management strategies required for water resource planning, the economics of water and treatment of issues associated with conflict over water. The concept of virtual water is covered before the text concludes with a chapter considering the challenges of predicting future water issues in a rapidly changing world and where environmental systems can behave in a non-linear way. The need to work across disciplines to address challenges that are connected at both local and global scales is highlighted. *Water Resources* also includes global examples from both the developing and developed world. There are 58 case study boxes. Each chapter is supplemented with these case studies and with reflective questions, project ideas and further reading, as well as links to a glossary of terms. The book is richly illustrated throughout with over 160 full-colour diagrams and photographs.

The text provides a novel interdisciplinary approach to water in a changing world, from an environmental change perspective and interrelated social, political and economic dimensions. It will be an indispensable guide to undergraduates studying water resources and management, geography of water, and water in the environment.

Joseph Holden holds the Chair of Physical Geography at the University of Leeds and is research dean in the Faculty of Environment. He is director of water@leeds, one of the world's largest university-based interdisciplinary water research centres, and director of the UK Natural Environment Research Council's Integrated Catchment Solutions Programme.

Water Resources

An Integrated Approach

Second edition

Edited by
Joseph Holden

Routledge
Taylor & Francis Group

LONDON AND NEW YORK

Second edition published 2020
by Routledge
2 Park Square, Milton Park, Abingdon, Oxon, OX14 4RN

and by Routledge
52 Vanderbilt Avenue, New York, NY 10017

Routledge is an imprint of the Taylor & Francis Group, an informa business

First edition published by Routledge 2014

British Library Cataloguing-in-Publication Data
A catalogue record for this book is available from the British Library

Library of Congress Cataloging-in-Publication Data
Names: Holden, Joseph, 1975- editor of compilation.
Title: Water resources : an integrated approach / Joseph Holden.
Identifiers: LCCN 2019028081
Subjects: LCSH: Water supply. | Water resources development. | Water security.
Classification: LCC TD346 .W28 2020 | DDC 333.91--dc23
LC record available at https://lccn.loc.gov/2019028081

ISBN: 978-1-138-32921-8 (hbk)
ISBN: 978-1-138-32922-5 (pbk)
ISBN: 978-0-429-44827-0 (ebk)

Typeset in Minion
by Swales & Willis, Exeter, Devon, UK

Contents

Contributors

Dr Anamika Barua, Indian Institute of Technology, Guwahati, 781039, India.

Dr Nesha C. Beharry-Borg, visiting researcher, water@leeds, School of Earth and Environment, University of Leeds, Leeds LS2 9JT, UK.

Professor Lee E. Brown, water@leeds, School of Geography, University of Leeds, Leeds LS2 9JT, UK.

Professor Pippa J. Chapman, water@leeds, School of Geography, University of Leeds, Leeds LS2 9JT, UK.

Dr Frances Drake, formerly water@leeds, School of Geography, University of Leeds, Leeds LS2 9JT, UK.

Dr Alison M. Dunn, water@leeds, School of Biology, University of Leeds, Leeds LS2 9JT, UK.

Professor Dabo Guan, Water Security Research Centre, School of International Development, University of East Anglia, Norwich NR4 7TJ, UK.

Dr Christopher Hassall, water@leeds, School of Biology, University of Leeds, Leeds LS2 9JT, UK.

Professor Joseph Holden, water@leeds, School of Geography, University of Leeds, Leeds LS2 9JT, UK.

Dr Nigel J. Horan, formerly water@leeds, School of Civil Engineering, University of Leeds, Leeds LS2 9JT, UK.

Miss Yixin Hu, Water Security Research Centre, School of International Development, University of East Anglia, Norwich NR4 7TJ, UK.

Dr Paul Kay, water@leeds, School of Geography, University of Leeds, Leeds LS2 9JT, UK.

Professor Adrian T. McDonald, formerly water@leeds, School of Geography, University of Leeds, Leeds LS2 9JT, UK.

Dr Gordon Mitchell, water@leeds, School of Geography, University of Leeds, Leeds LS2 9JT, UK.

Dr Colin S. Pitts, water@leeds, School of Earth and Environment, University of Leeds, Leeds LS2 9JT, UK.

Dr Sumit Sinha, water@leeds, School of Earth and Environment, University of Leeds, Leeds LS2 9JT, UK.

Professor Stephen Sitch, Geography, College of Life and Environmental Sciences, University of Exeter, Exeter EX4 4RJ, UK.

Dr Rebecca J. Slack, PlantNetwork and formerly water@leeds, School of Geography, University of Leeds, Leeds LS2 9JT, UK.

Dr Sonja S. Teelucksingh, Global Environment Facility, N8-800, 1818 H Street NW, Washington, DC, 20433, USA.

Professor Martin Tillotson, water@leeds, School of Civil Engineering, University of Leeds, Leeds LS2 9JT, UK.

Mr Kitriphar Tongper, Lady Keane College, Shillong, Meghalaya, 793001, India.

Dr L. Jared West, water@leeds, School of Earth and Environment, University of Leeds, Leeds LS2 9JT, UK.

Miss Zhao Zeng, Water Security Research Centre, School of International Development, University of East Anglia, Norwich NR4 7TJ, UK.

Dr Xu Zhao, Key Laboratory of Integrated Regulation and Resource Development on Shallow Lakes, College of Environment, Hohai University, Nanjing, 210098, People's Republic of China.

Preface

A team of 24 authors have contributed their knowledge and understanding to the second edition of this book, bringing their experiences of water research from around the world. These authors also bring a unique interdisciplinary approach to the subject of water resources. Water scarcity, climate change, pollution and degradation of aquatic ecosystems, lack of adequate sanitation and water-related risk to human health, the need to reduce waste and recover resources and energy from sewage, and the need to understand current and future water demand, including how trade and economics affect this demand and can move virtual water around the globe, are all major international challenges for water. We need to plan for water resources in a cooperative way, changing behaviour, rethinking economics to consider wider water-related ecosystem services to society, and reducing conflict by looking for co-benefits and ensuring social justice, considering whole systems approaches that integrate within and between river basins.

Given the above, water resource experts of the future will need a broad understanding of issues which cover a wide range of disciplines. Hence this book brings together coverage of topics from many disciplines. Each chapter contains boxed features providing more depth or case study examples, often from many different regions of the world. As you read the book you will notice some words that are typeset in bold within the text. These words are highlighted the first time they appear in each chapter and can be found in the glossary with an explanation of their meaning. Each chapter also contains reflective questions, annotated further reading and some project ideas to aid learning.

The book's interdisciplinary nature reflects my own role as director of water@leeds, which was founded in 2009 at the University of Leeds and is now one of the largest interdisciplinary water research groups in any university. water@leeds members work together as scientists, social scientists, engineers and humanities experts to tackle water challenges facing the world. Research undertaken by water@leeds is often in partnership with government bodies, NGOs, industry and practitioners to ensure that it is applicable to societal needs. The water challenges we face are crucial to the survival of the human race, stability and peace, and to the sustainability of the Earth's ecosystems. It is therefore vital that around the world we improve understanding of water resource issues from an interdisciplinary perspective. This book is part of that mission and I hope that you will feel inspired to join us in trying to make a real difference to people's lives through developing a shared understanding of water resource issues.

Joseph Holden, water@leeds, UK.

Acknowledgements

The editor wishes to formally thank the chapter authors for producing such great learning resources. Alex Santos, University of Leeds, is thanked for producing new figures included in this second edition. Alison Manson, formerly University of Leeds, is acknowledged for her work in producing some of the drawings that have been retained from the first edition. Chris Tena of water@leeds is thanked for obtaining figure permissions. Kathryn McKendrick-Smith is thanked for helping to compile some of the glossary. Thanks are also given to the editorial production team at Routledge for their support in producing this volume. Where not originals, figure and table source acknowledgements are provided where known. Every effort has been made to contact copyright holders for their permission to reprint material in this book. The publishers would be grateful to hear from any copyright holder who is not here acknowledged and will undertake to rectify any errors or omissions in future editions of this book.

CHAPTER ONE

Water basics

Joseph Holden

LEARNING OUTCOMES

After reading this chapter you should be able to:

- describe the nature of water;
- describe some of the roles of water in historic and contemporary society;
- outline some of the major global water resource issues.

1.1 INTRODUCTION

Water is a very common substance on Earth, covering 73% of the Earth's surface in oceans, rivers and lakes. However, water is also a special substance. It is one of only a handful of substances that expands when it freezes. Water is one of the few substances found in solid (ice), liquid and gas (water vapour) form within our natural environment. It is the best naturally occurring solvent, meaning that it acts to dissolve and carry more different types of material within it than anything else, being good for cleaning and helping to sustain life in plants by providing them with nutrients. Indeed, water in solid, liquid and gaseous forms has been very important in shaping the landscapes on Earth through its role in **weathering**, erosion and transport of materials.

Water is also fundamental to life on Earth. Humans, for example, are around 70% water by mass and we quickly dehydrate if we do not drink; people will die within a few days without water. Water can also be harnessed for other purposes, such as being a cooling agent for industry and power generation, and it thus serves a wide range of economic functions. Water also takes on a cultural and religious significance for many societies in terms of rituals, blessings, emotions formed around scenic views of waterscapes, leisure activities and physical and mental well-being. Water was fundamental to the development of agriculture and, hence, civilisations.

While global freshwater resources are many times more than required by humans today, because these resources, and human populations, are unevenly distributed there are many regions which do not have adequate fresh-water supplies to meet today's demand. Projecting into the future, there will also be increasing demands for freshwater and it is important to assess these needs to understand the scale of the water challenge that lies ahead.

Accordingly, this chapter sets the scene for the rest of the book. The chapter outlines some of the fundamental properties of water, introduces the role of water management in societal development and culture, and then outlines some of the major issues around global freshwater resource availability. With these things in mind, the reader should be able to see why an interdisciplinary approach is required when considering water resources and how important it is to gain a broad understanding across the rest of the chapters in this book.

1.2 WHAT IS WATER?

Water is not an element in itself. It is a compound made from two other elements: hydrogen and oxygen. Each water molecule contains two hydrogen atoms and one oxygen atom (H_2O), which are strongly bonded together. Each hydrogen atom shares a pair of electrons with the oxygen atom. Such pairing is called a covalent bond. All of the electrons position themselves as far apart from each other as they can, as they are all negatively charged and hence repel one another. However, the electrons that are not bonded to the hydrogen atom are closer to the oxygen atom than the ones being shared with the hydrogen atom and therefore the repulsion forces are slightly stronger for the unpaired electrons. This stronger force pushes the hydrogen atoms slightly closer together so that the two hydrogen atoms rest apart at 104.5° (Figure 1.1). While the overall charge of the water molecule is neutral, the electron

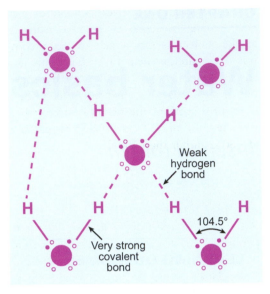

Figure 1.1 A schematic diagram showing water molecules and the strong covalent bonds and weak hydrogen bonds.

arrangement shown in Figure 1.1 indicates that more of the negative (electron) charge is concentrated towards the oxygen side of the molecule. This means that the slightly positive hydrogen side in one water molecule is attracted to the slightly negative concentration on the oxygen side of another water molecule, forming a 'hydrogen bond'. Hydrogen bonds are between 10 and 50 times weaker than covalent bonds.

The strong covalent bonds and the weak hydrogen bonds are both important properties of water, and the latter, in particular, makes water unusual in its nature. The covalent bonds result in it being very difficult to force apart the hydrogen and oxygen atoms within water molecules, which is why water is so prevalent on the planet and has not disappeared. The hydrogen bonds result in water molecules themselves being able to move between one another relatively easily, meaning that water is highly mobile, enabling it to flow rapidly and allowing materials (e.g. fish, people, ships, sediment) to pass through it easily. The hydrogen bonds also enable water to act as an excellent solvent, as the molecules can attach to other compounds, enabling them to be taken up into solution (see below). Finally, the hydrogen bonds mean that water is present as a liquid over most parts of the planet's surface. If it were not for these bonds, water would boil at −80°C rather than 100°C.

1.2.1 Physical properties of water

1.2.1.1 Thermal properties

Water is at its most dense at 4°C. This temperature–density property is highly unusual. At this temperature water is seen to be at a standard volume of 1 g per cm^3 or 1 kg per litre. Water at 4°C will tend to sink towards the bottom of water bodies and there will be thermal stratification. Upon freezing, water expands by about 9% (~0.91 g cm^{-3}) and the hydrogen bonds form a strong structure. Thus, ice, which is less dense than water, floats upon the surface of water. This is important because, if ice were denser than water, it would sink, thereby crushing the creatures living in water bodies or leaving them stranded at the top of the water body. Indeed, lakes would freeze solid quite quickly during winter if it were not for the special properties of water. The ice at the surface insulates the lower liquid water and also reflects more of the Sun's energy, thereby reducing the chance of the lake freezing solid.

Normally, as liquids are heated, the molecules gain energy and bounce around more, thereby taking up more space. However, this does not happen until >4°C for water. As water warms between 0 and 4°C, the hydrogen bonds bend or break frequently, causing the water molecules to pack more tightly. The expansion of water at temperatures greater than 4°C is very small in comparison to that caused by

freezing. For example, in warming from 4°C to 10°C, the expansion of water is less than 0.03%. Nevertheless, this thermal expansion is still important for sea-level rise under climate change. Between 1970 and 2010 the IPCC estimates that thermal expansion of water in the oceans contributed 3.2 cm of global mean sea-level rise out of a total of 8 cm rise (Church et al., 2013). It is estimated that the sea level between 2081 and 2100 will be 40 to 65 cm higher than 1986–2005 levels, with about 40% of this increase being due to thermal expansion. Most of the rest of the contribution to sea-level rise will come from melting of ice on the land surface (see Box 13.2 in Chapter 13). Note that melting of sea ice does not directly affect global sea levels, since the ice is already part of the water body and sea levels are already displaced by the mass of ice sitting in the water. It should also be noted that seawater does not behave like freshwater, due to its high salt content, so that the density is not at a maximum at 4°C and instead it behaves more like a normal liquid with maximum density at its freezing point (Berner and Berner, 1987).

To raise the temperature of 1 kg of water by 1°C requires 4200 joules of energy (the specific heat of water). This is the highest of any liquid except ammonia. Thus heating water for cooking is energy intensive. Cooling water by 1°C would release the same amount of energy. This is why water takes a long time to heat up and cool down and why coastal areas have a moderate climate compared to inland areas (McClatchey, 2017). Coastal areas often have seawater temperatures warmer than air temperature during winter, as the sea cools slowly from the previous summer's heat. During summer, water temperatures can be a lot lower than air temperature, as the water warms only very slowly.

Energy is also used or released when water changes phase between solid, liquid and gas forms. It takes a lot of energy to break the hydrogen bonds in water to form water vapour. In fact, it takes 2.3 million joules to evaporate 1 kg of water without any change in temperature. This energy used to turn a liquid into a gas is known as latent heat of vaporisation, and for water it is one of the highest known because the molecule has to escape both the normal molecular attraction forces of the liquid and the hydrogen bonds. Water vapour may travel large distances as a gas before it condenses. When it condenses (e.g. as droplets in a cloud), the same amount of energy is released (e.g. into the atmosphere). Thus, energy can be moved around the planet and away from the Tropics by water in the atmosphere (see Chapter 2). To put this into context, it takes six times more energy to evaporate water from liquid to gas (at any temperature, even when the liquid is already at 100°C) than it takes to raise the temperature of water from 0°C to 100°C.

While the boiling point of water at normal sea-level atmospheric pressure is 100°C, water vapour is still present all around us. Evaporation of water occurs at the surface of water bodies where sufficient energy has been imparted to break the hydrogen bonds and allow the water vapour to escape. People often think of white clouds or white trails from kettles as being water vapour, but in fact water vapour is invisible. Clouds and kettle trails are liquid forms of water where condensation has occurred. Evaporation is a crucial part of the water cycle and invisible water vapour is a very powerful greenhouse gas. Short-wave radiation comes from the Sun and hits the Earth's surface. It then radiates back up into the atmosphere from the Earth as long-wave radiation, which gets absorbed by several gases, thereby warming the air from below. Water vapour is good at absorbing radiation released from the Earth's surface. This absorption of radiation means that radiation is kept in the atmosphere rather than being allowed to escape back out into space. Greenhouse gases have made our planet hospitable, but changes in the concentrations of water vapour and other greenhouse gases in our atmosphere alter the energy balance and can lead to further warming. Chapter 2 provides more detail on this process and the impacts that water vapour feedbacks may have on climate change. Note that most water vapour in the atmosphere is produced from transfers from the oceans via evaporation of surface water.

1.2.1.2 Surface tension

Surface tension is common to liquids (i.e. it is not unique to water). Within the main volume of a liquid, molecules are attracted in all directions by their neighbours. However, at the surface of a liquid there are no molecules outside and so the molecules are only pulled inward. This has the effect of forcing the liquid to adopt the minimum surface area, which is a spherical shape. Without gravity, liquid would form a sphere. The force pulling the liquid in towards itself is greater than the gravitational force of some objects which might rest on the water surface (Figure 1.2). Invertebrates such as pondskaters can be seen skimming across the water surface without sinking (Figure 1.2a). The surface tension of water is the greatest of all liquids that are not metals.

For liquids contacting solid matter, the behaviour will be a result of the balance between surface tension (cohesive forces) and any attractive forces between the molecules of the liquid and those of the solid

Figure 1.2 a) Pond skaters are invertebrates that move across the water surface using the surface tension of water to maintain their position; b) a paperclip on the surface of a cup of water.

matter (adhesive forces). For many solid materials, water is more strongly attracted to itself and so water forms droplets on the surface. However, for other materials the water is attracted to the material more than to itself and here it will spread out and wet the surface. In other words, if the adhesive force between the liquid and solid exceeds that of the liquid's cohesive force, the liquid wets the surface. Adding a **detergent** to water makes the water reduce its surface tension and therefore allows a substance to be wetted by spreading water.

The attractive forces of liquids result in a phenomenon known as **capillary action**, whereby liquids flow within narrow gaps between solid material, no matter which way gravity acts. In other words, capillary action means that water can flow upward. This is vital for plant growth. As plants lose water from their leaves via **transpiration**, new water can be drawn up from the soil. Capillary action can be seen if you hold some tissue above some water and dip the end of the tissue into the water – some water rises up the tissue. If the diameter of the pore (e.g. glass tube, soil or rock pore

space, tissue pore space) is sufficiently small, then surface tension combined with adhesive forces between the liquid and solid acts to force the water into the pore. The narrower the pore space (e.g. a narrow glass tube; **Figure 1.3**), the greater the rise of liquid. The surface of the liquid near the edge of solid matter will curve upward if it is wetting that surface (**Figure 1.3**). Surface tension then acts as a force to pull the liquid upward as it tries to counteract the curved edge which is in the opposite direction to the natural spherical shape which the cohesive forces naturally pull towards. Chapter 3 provides more information on this topic.

1.2.2 Chemical properties of water

Water is an excellent solvent. Pure water does not readily occur in nature because water so quickly dissolves other substances that it comes into contact with. A substance will dissolve in water if its molecules can interact with charged water molecules. The positive (hydrogen side) and negative (oxygen side) charges around

Figure 1.4 When dissolved in water, each of the sodium and chloride ions is hydrated as it is surrounded by water molecules.

Figure 1.3 Capillary action shown in a series of narrow glass tubes. The water has had dye added to aid vision. The curved meniscus can be seen where water contacts the solid material it is wetting. Surface tension tries to counteract this attraction between the water and solid material by pulling upward in an attempt to move itself back towards a spherical shape where the surface area is minimised. This results in water moving up the narrow glass tubes against the force of gravity. The narrower the tube, the greater the capillary rise.

water molecules are stronger than the charges that hold many other molecules together. For example, a salt such as sodium (Na) chloride (Cl) is torn apart by water in that the Na^+ **ion** (an ion is a charged atom or molecule, due to an unequal number of protons (positive charge) and electrons (negative charge)) is attracted to the negative charge and the Cl^- is attracted to the positive charge on the water molecule. This breaks down the solid crystals and the salt dissolves into the water. Water tends to reduce the electrical attraction between oppositely charged ions 80-fold (Ghali, 2010). The fairly small size of water molecules allows

many water molecules to surround one molecule of a solute (**Figure 1.4**), enabling the ion to be hydrated. Substances such as sugars and alcohols are water soluble not because they are ionic, like salts, but because they have polar molecules (rather like water) with some weak electrostatic charge on parts of their molecules, which attracts opposite charges from parts of water molecules. Substances that are electrically neutral (e.g. oil) dissolve poorly in water. Chapter 4 describes in more detail the chemical properties of water and how water quality is affected by flowpaths and human intervention.

REFLECTIVE QUESTIONS

Why is sweating a really good way of reducing your body temperature when you feel too hot?

Why does water often move upward in soils in defiance of gravity?

1.3 WATER AND ITS IMPORTANCE FOR LIFE

Every known form of life relies on water. Life is maintained in rivers, lakes and oceans by the ability of water to carry food and oxygen to organisms and to remove waste. In either photosynthesis (sunlight driven) or chemosynthesis (chemical energy driven, such as around deep ocean hydrothermal vents), water is combined with carbon dioxide (or methane for some chemosynthesis) to create carbohydrates and release oxygen:

$$6CO_{2(gas)} + 12H_2O_{(liquid)}$$
$$+ \text{photons} \rightarrow C_6H_{12}O_{6(glucose)}$$
$$+ 6O_{2(gas)} + 6H_2O_{(liquid)}$$

carbon dioxide + water + light energy

$\rightarrow$ glucose + oxygen + water

Leaves use the Sun's energy to enable chemical reactions resulting in carbon dioxide being absorbed from the atmosphere, mixed with water and converted into simple sugars, such as glucose (see also Box 2.4 in Chapter 2). Glucose is made into compounds for storage such as starch, and compounds for structure such as cellulose and lignin (which make up cell walls for leaves and stems). These carbohydrates can then be consumed by higher parts of the food chain as the basis of life on Earth.

Plants are able to obtain water from the soil through capillary action. As water is transpired through leaves, new water containing essential dissolved nutrients to allow the plant to grow is drawn up from below. For land plants, water is additionally important because they are mechanically supported by the water pressure inside them. Without water, plants wilt. Only a few plants can survive drying, by switching off the photosynthetic process when drought comes and then restarting it when water is abundant. Some plants cope with drought by shedding leaves or are adapted to drought by having thick, waxy coatings on their leaf surfaces, or small, thick leaves and special water storage areas.

All life is made up from cells. Some species (e.g. bacteria) have one cell while others (e.g. humans) have hundreds of billions. Cells are the building blocks for life. They provide structure for the organism, take in nutrients from food, convert those nutrients into energy and carry out specialised processes. Cells also contain the blueprint for the living thing and information from its hereditary past and can make copies of themselves. The major biochemical processes that operate within cells rely on water, enabling cell structure, function and division.

Mammals are typically 50–80% water by mass. The human brain is 75% water. Human function is all about the flow of water through our system. Water is important to transport chemical signals and nutrients to vital organs of the body and is necessary for our body to digest and absorb vitamins and nutrients and flush out wastes. We generally feel thirsty at about 2–3% water loss. When a human suffers 5–6% water loss they will feel unwell and may have headache, nausea or tingling in the limbs. At dehydration levels greater than this the skin will shrivel, vision dims, delirium may occur, muscle control will be lost and, normally, dehydration beyond 15% of normal body water is fatal.

Drinking water directly is not the only way to obtain water. Food contains water within it and typically one-fifth of human water intake comes from eating food. Typically a lettuce is 95% water and an apple 84%. Water is also essential in the production of food for humans. Some foodstuffs are low in their water requirements, while others have heavy water requirements (Table 1.1), although water use for a particular crop type will depend on soil type and climate. Meat production requires six to 20 times more water than cereal production. Cereals are the most important source of total food consumption worldwide, accounting for over half of the diet of most developing countries. However, a large proportion of cereals are used for animal feed. As countries develop they also tend to demand more food that uses water more intensively (e.g. meat products) and so, as the world's population both expands and develops, more water will be required for agriculture.

Table 1.1 Typical water use to produce some common human foodstuffs

Product	Water use, m³ per kg of food/drink
Fresh beef	15
Fresh lamb	10
Fresh poultry	6
Eggs	3.4
Rice	1.5
Cereals	1.5
Bread	1.3
Coffee	1.2
Citrus fruits	1.0
Pulses, roots and tubers	1.0
Milk	1.0
Wine	0.9
Beer	0.3
Potatoes	0.3

REFLECTIVE QUESTION

What do water and carbon dioxide combine to produce as part of photosynthesis?

1.4 WATER'S ROLE IN SOCIETY

1.4.1 The development of civilisations

Gathering foods and hunting animals originally depended on the natural distribution of plants and animals. However, around 13,000 years ago agriculture developed. Agriculture involves cultivating plants and herding animals. The animals and plants in most cases underwent domestication, indicating that humans have taken control of the reproduction of the individual plants and animals concerned. Almost all domestication occurred over the last 13,000 years (Blockley and Pinhasi, 2011). Domestication of goats, cattle, sheep, horses and so on provided the ability to convert inedible plants (e.g. grass) into food which humans could eat, such as meat and milk. It also provided materials for clothing and the ability to travel more quickly, and provided power (e.g. horses). Most of the earliest sites of domestication were in the 'fertile crescent' of Mesopotamia. Here, around 12,000 years ago, hunter-gatherers foraged for the large seeds of certain wild grasses. They must have realised that planting the seed led to a more predictable crop the following year. They favoured the planting of seeds from those with the largest seeds, and with non-shattering seed heads as these were easier to gather. While most grasses drop their seed to the ground when ripe, a few mutants in any population will retain the seeds on the head. It was these seeds that remained on the plant that were selectively collected and later sown by humans. Consequently, over time, the average seed size of the wheat and barley that they grew increased and, as the seeds remained on the seed heads, these new strains could not reproduce without human help. The new varieties of cereals were joined by other crops such as peas, lentils and beans and by domesticated animals, notably goats in the first instance. Elsewhere in the world, people also began domesticating plants and animals.

As a result of agriculture, populations of farming communities grew rapidly because food supplies were more reliable and there was no longer a need to be nomadic. The overproduction of food by farming communities meant that they needed storage containers, and hence this encouraged technological innovation. It also meant that such food could be traded with others, who could in turn specialise in other activities rather than have to gather and hunt for food. These specialised activities might be woodwork, pottery, metalwork and so on. Thus we have the start of trading and an economy, such that some people accumulate more goods than others. This encouraged stratified societies whereby some people became powerful. The powerful were then able to organise others to contribute to large-scale projects such as **irrigation** schemes and developing monuments (Hassan, 2012).

For some groups, locating settlements close to permanent springs and streams meant that they could enjoy more stability for their farming and be less reliant on rainfall variations from year to year. Thus, those farming communities which settled around water resources had an upper hand over others. As well as having a natural water supply, humans living around watercourses realised that they could divert the flow by building channels across the landscape to move water from the river into farmland. Some of the most famous early civilisations developed because they controlled that water supply. From around 10,000 years ago, groups of farmers came together on the floodplains of the Tigris and Euphrates rivers and diverted the courses of rivers and embanked and arranged fields so that water could be retained and made to flow from one field to another. The combination of a hot climate, fertile floodplain soils and a carefully controlled water supply made these farming techniques very productive, and empires were built on the surpluses that they produced. In the long run, however, problems could emerge. Disputes over water rights, and the political difficulty of building and maintaining such large shared resources, could lead to conflict (Hassan, 2012). On the whole, though, the transformation of Mesopotamia from semi-desert to a breadbasket was an outstanding example of the landscape engineering that allowed complex civilisations to flourish for many millennia (Lawson, 2017).

Civilisations such as the Roman Empire, Egyptian rule and the Omayyad Dynasty were all founded on their access to water. The rise of state societies with kings to organise, defend and rule also had cultural and religious significance. Leaders of civilisations were often seen as representatives of gods or God, or in some cases as gods in themselves. Social practices and rituals focused on water are common in history. During the Pharaonic period in Egypt, the River Nile was worshipped and floods were seen as being sent from the gods. The river levels were also used to determine the level of tax that should be levied on farmers, because the higher the flood, the more land was under agricultural production. The development of cities became possible because of agriculture and the use of water.

Water was used for hygiene as well as drinking water and food supply. Water was also culturally important to early cities. Baths in ancient Rome were socially very important, and enormous volumes of water were channelled into the city in spectacular fashion, while excess water was used to flush out sewage. Water was used in architectural designs for streets and gardens. The Romans also drained wetland to make it agriculturally more productive or to reduce the threat from waterborne disease.

1.4.2 Contemporary society

As with the development of civilisations, control of water resources today provides an economic advantage (Chapter 10) that enables food production, industry, navigation and sustenance for the population. Careful management of water also provides health benefits, through disease control, for example (Chapter 7).

In many religions, water is of great importance. Water is used in ritual washing in Hinduism, Islam, Shinto, Taoism and many other religions. In Shinto, waterfalls are sacred and an individual can become purified by standing under one. For Jews, water is used in ritual washing for several different purposes. Muslims wash with water before performing ritual prayers or handling the Qur'an. Buddhists use water at funeral services. Water is also important to Christian baptism and for blessings or cleansing from evil. In Sikhism, water is used to make Amrit, ready for use in an initiation into the Khalsa through Amrit Sanskar. Some water bodies such as the Ganges River (Hinduism) and the Well of Zamzam (Islam) are sacred.

Water also has wider cultural significance (e.g. see Box 1.1). Water can represent different things to different people or groups of people. For example, for some North American natives, water represents mobility, providing a highway in ancestral tribal cultures where tribes moved between different camps following seasonal patterns of resource availability. Water features in art such as music, paintings, writing and poetry. It is used aesthetically in the layout of cityscapes, gardens (ponds, fountains, etc.) or landscaping. Many sports focus on water, such as swimming, rowing and sailing.

In the developed world, the control of water and its use for industrial, navigational or flood control purposes has been partly replaced by recognition that the wider ecosystem and environment is a user of water. Hence water is seen to be useful for environmental conservation. Rather than drain wetlands, there is a move to preserve or restore them. For example, large parts of the Florida Everglades were drained and cut through by canals in the late nineteenth and early twentieth centuries. Pollution from urban areas and agriculture also threatened the ecology. Since the 1970s there has been conservation protection status on part of the Everglades, but in 2000 a comprehensive package of restoration, to take place over 30 years and estimated to cost more than $12 billion, was agreed. This is thought to be the world's largest ecosystem restoration. Actions include rewetting large areas, infilling drains and canals and reuse of wastewater to safeguard water quality. In many other parts of the world, rivers that had previously been straightened and have become rather dull and monotonous, with low ecological and geomorphological diversity, are now being rehabilitated through the installation of meander bends, removal of concrete bank linings and addition of woody debris (Palmer and Bernhardt, 2006).

Thailand's water culture

Water is central to Thai culture. Both the rivers, and the associated canal networks that were built for transport and agriculture, are crucial to the functioning and economy of the country. The capital of Thailand is Bangkok, which originally formed around a network of canals on which locals built homes on stilts above the water or lived in boats. Shops, entertainment and eating establishments were on boats and locals would travel between them on rowing boats. Even today there is a bustling water scene in Bangkok and elsewhere in the country there are famous floating markets for food and where artists and craftspeople display their creative products (Figure 1.5).

Water is used in Thai wedding ceremonies, with sprinkling on the wrists to bless the bride and groom. The key Thai festivals during the year all involve water. The April Songkran New Year celebration, for example, involves all towns and villages having water parties and mock water battles, splashing water on citizens of all ages. Loi Krathong celebrations, usually held in November, are a thanksgiving to the water gods. People release boats made from banana branches and leaves, with incense and candles. The spectacular Royal Barge Procession is of religious and royal significance and has been performed for around 700 years. It involves 52 gilded barges being rowed down the Chao Phraya River by >2000 oarsman.

Figure 1.5 Bustling Damnoen Saduak floating market, Thailand.
Source: Shutterstock/pierdest.

Continued

Despite the important cultural significance of water and waterways in Thailand, the rivers have suffered from pollution, channelisation and encroachment. However, a combination of government strategic initiatives (Apipattanavis *et al.*, 2018) and civil society movements now use local knowledge and engagement to protect and restore the water quality and habitat in watercourses. Education programmes and dialogue with local communities, combining local/religious beliefs, traditional respect for rivers, and modern conservation approaches, is enabling communities to take responsibility for improving their local water environment (Jompakdee, 2003).

The EU Water Framework Directive requires that water bodies must be in good ecological status and efforts must be undertaken by all EU member states to improve the status of water bodies. In other words, legislation is in place to protect water bodies for their own sake as well as for the sake of society, which enjoys and relies on water bodies for the wider benefits it receives from them. Further information about the EU Water Framework Directive is provided in Chapters 4 and 6. However, there is still a tension between rehabilitating rivers or having 'natural-looking' water landscapes and societal expectations around issues such as flood management, where flood defences are still expected to be in place (see Chapter 3).

REFLECTIVE QUESTIONS

How did control of water resources enable civilisations to develop?

Can you list some cultural water uses that you are familiar with in your home region?

1.5 GLOBAL WATER RESOURCES

1.5.1 The amount of water available

Near, and on, the Earth's surface is a huge amount of water – about 1388 million cubic kilometres. However, approximately 97% of this amount is saline water and only 3% is freshwater (Figure 1.6), although as Chapter 2 notes there are large uncertainties as to the exact values of the stores. The greater portion of this freshwater (between 48 and 69%) is in the form of ice and permanent snow cover in the Antarctic, Greenland, other parts of the Arctic and the mountainous regions. While 30% of freshwater exists as groundwater, <1% of the total amount of freshwater on the Earth is concentrated in lakes, reservoirs and river systems. Hence <1% of the total available water is most easily accessible for our economic needs, and at the same time it is vital for land-based aquatic ecosystems.

The mean volume of the renewable global water resource is estimated at 42,700 km^3 per year, and this is extremely variable in space and time (Shiklomanov, 1998). At the continental level, the Americas have the largest share of the world's total freshwater

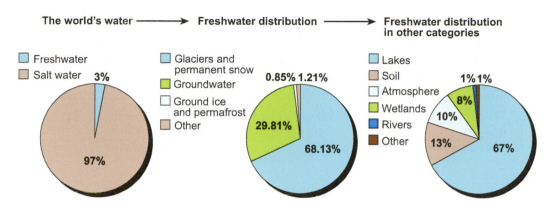

Figure 1.6 Categories of world water resources.

resources, with 45%, followed by Asia, with 28%, Europe, with 16%, and Africa, with 9%. A country-level analysis shows that for the ten countries with the most renewable water resource some such as Brazil or Columbia only withdraw (use) a tiny fraction of that resource, whereas others such as India or USA withdraw a much larger fraction (Table 1.2). Because countries are of different sizes and have different population levels, rather than examine total resource by country, it may be more meaningful to examine the amount of renewable water resource per person. Table 1.2 therefore also provides data for the top ten countries with the most and top ten countries with the least water resource per inhabitant. The countries with the most resource per person are very variable in characteristics ranging from high latitude countries such as Iceland or Norway to tropical countries such as Guyana or Papua New Guinea. However, the countries with the least resource per inhabitant are concentrated in the Middle East. As Table 1.2 shows, many of these countries are annually withdrawing water many times greater than their available annual renewable freshwater resource. This extra water is often sourced from deep groundwater that has taken centuries or longer to accumulate (and hence is not considered 'renewable') and there are serious questions about the longer-term sustainability of such water use.

Tables or maps of countries showing water availability data can mask some very strong within-country variability. Peru, for example, is listed in Table 1.2 as being in the top ten water-rich countries in the world by total annual available freshwater resource, with only 0.1% of its water withdrawn for use each year. However, most of its inhabitants live in desert climate conditions. For example, Lima, the capital of Peru, where around one-third of the country's inhabitants live, only receives an average of 16 mm of rainfall per year. Thus, water must be supplied from the water-rich Andes to lowland, coastal desert regions in the country. Centuries of water diversion and management enable such arrangements, but the infrastructure is under severe stress due to population growth and many of the poorest inhabitants struggle to afford clean water.

Falkenmark (1986) proposed thresholds of 1000 and 500 m^3 of freshwater available per inhabitant to correspond respectively to *water stress* and *water scarcity* levels.

Table 1.2 Water resource values for countries with plentiful or scarce water resources

Water resource measure	Total resources (km³ yr⁻¹)	Resource per inhabitant (m³ yr⁻¹)	% of renewable resource withdrawn
Ten countries with most renewable resource			
Brazil	8647	41,865	0.7
Russia	4525	31,561	1.5
USA	3069	9802	15.6
Canada	2902	84,483	1.4
China	2840	2060	19.5
Columbia	2360	45,432	0.6
Indonesia	2019	8332	5.6
India	1911	1539	33.9
Peru	1880	65,068	0.1
DR Congo	1283	18,935	0.1
Ten countries with least renewable resource			
Kuwait	0.02	7	2075
Malta	0.05	121	71
Qatar	0.06	31	381
Bahrain	0.1	88	206
United Arab Emirates	0.2	19	1867
Djibouti	0.3	331	6
Cape Verde	0.3	599	7
Singapore	0.6	116	32
Libya	0.7	109	610
Gaza	0.8	202	50
Ten countries with most renewable resource per inhabitant			
Iceland	170	524,691	0.1
Guyana	241	318,783	0.7
Suriname	122	230,624	0.6
Congo	832	200,966	<0.1
Papua New Guinea	801	114,200	<0.1
Gabon	164	106,910	0.1
Bhutan	78	105,691	0.4
Norway	382	77,563	0.8
New Zealand	327	74,066	1.4
Peru	1880	65,068	1.0
Ten countries with least renewable resource per inhabitant			
Kuwait	0.02	7	2075
United Arab Emirates	0.2	19	1867

(Continued)

Table 1.2 (Cont.)

Water resource measure	Total resources (km³ yr⁻¹)	Resource per inhabitant (m³ yr⁻¹)	% of renewable resource withdrawn
Qatar	0.06	31	381
Saudi Arabia	2.4	85	934
Yemen	2.1	85	169
Bahrain	0.1	88	206
Libya	0.7	109	610
Singapore	0.6	116	32
Malta	0.05	121	71
Jordan	0.94	148	90.5
Ten countries with smallest % of resource withdrawn			
DR Congo	1283	18,935	0.1
Equatorial Guinea	26	36,111	0.1
Guinea	226	22,109	0.2
Cameroon	286	14,254	0.3
Uganda	66	1913	0.5
Nicaragua	197	33,492	0.7
Brazil	8647	41,865	0.7
Guyana	241	318,783	0.7
Norway	382	77,563	0.8
Comoros	1.2	1592	0.8
Ten countries with largest % of resource withdrawn			
Kuwait	0.02	7	2075
United Arab Emirates	0.2	19	1867
Saudi Arabia	2.4	85	934
Libya	0.7	109	610
Qatar	0.06	31	381
Bahrain	0.1	88	206
Yemen	2.1	85	169
Jordan	0.9	148	90
Malta	0.05	121	71
Gaza	0.8	202	50

Source: Data from FAO Aquastat database (2016).

The OECD has used less severe thresholds to define stress (<1700 m³) and scarcity (<1000 m³) but it does define **absolute water scarcity** as freshwater <500 m³ per person per year (see Section 9.4 in Chapter 9). There are currently 28 countries with absolute water scarcity (e.g. see those listed in **Table 1.2**) and 45 countries with less than 1000 m³ per person per year. Transfer of water between countries can be important, but may be the subject of much dispute. Table 1.2 lists total available freshwater for each country, but this includes internally available freshwater plus freshwater that is transported into the country from a neighbouring country (e.g. by river flow). There are 44 countries which depend on other countries for over 50% of their renewable water resources, and so this increases the risk of conflict over water (Chapter 11). For example, an upstream country may build new dams which reduces water resources flowing downstream into the neighbouring country, or an upstream country may pollute the water resource.

1.5.2 Increasing demand for water use and consumption

The world's population is approaching 8 billion and is growing by about 82 million people per year. The associated water demand is increasing by around 2% per year (i.e. nearly double the rate of population growth) to reach around 6350 km^3 yr^{-1} by 2030. This 2030 value is almost double the global water demand that occurred in 1980. Putting this 2030 projection against the total global renewable freshwater resource of ~43,000 km^3 yr^{-1} it seems, at first glance, that there is still plenty of water to meet demand. However, this is not true because of the uneven distribution of water resources and people. While there are many countries with adequate water resources (e.g. Table 1.2), there is also a long list of countries where water resources are insufficient to meet a growing demand and in many of these countries non-sustainable water resources (e.g. ancient groundwater) are being depleted. An estimated 90% of the 2 billion people who

are expected to be added to the population by 2050 will be in developing countries, many in regions where the current population does not have sustainable access to safe drinking water and adequate sanitation. Good sanitation is important for health (see Chapter 7), yet 4.5 billion people today have to use hygienically unsafe toilets or have to defecate on open land (WHO and UNICEF, 2017). Poor sanitation causes a myriad of waterborne diseases, including severe diarrhoea, which kills around 1.5 million children each year. As populations grow, demand for water increases, due to escalating demand from domestic, agricultural and industrial sectors. While the worldwide demand for water is increasing every year, water supply cannot keep pace with the demand. To many, this is seen as a water crisis.

As shown in the photosynthesis equation in Section 1.3, plants need water to grow. While most agriculture is rainfed, the 15% of agriculture that is supplied by irrigation accounts for 69% of global freshwater use (see Figure 9.1 in Chapter 9). Most irrigation systems used around the world today are inefficient. The vast majority of farming irrigation worldwide essentially consists of large water diversion projects of the type described in Section 1.4.1, which initially enabled the development of civilisations several thousand years ago. In these systems, 60 to 70% of the water may be wasted (e.g. lost to evaporation). In addition, some types of agricultural product require more water than others. For example, as shown in Table 1.1, rice requires three times more water than potatoes to produce 1 kg of food, and beef requires ten times more than rice. Some crops produced by irrigated agriculture may be better

BOX 1.2

The reality of India's water crisis

India is a country of 1.3 billion people. Its population is expected to reach 1.6 billion by 2050. While India is not one of the ten countries with least available freshwater resource per person (Table 1.2), it has a major water crisis. Today, 600 million people face water scarcity, 75% of households have no drinking water on their premises and 84% of rural households have no piped water access. It is estimated that 200,000 people die each year in India due to inadequate access to safe water. The Indian government's own planning department (NITI Aayog, 2018) estimates, that by 2030, 40% of the population will not have access to drinking water and that many major cities will have used up their groundwater reserves.

There are several key reasons for the Indian water crisis, occurring in a country which is in the top ten most water-rich in the world. Water management is devolved to states but these states then compete with each other for the water resources, resulting in inter-state disputes. Agriculture is a major water user in India, with around 80% of freshwater use allocated to irrigation (FAO, 2016). The type of agriculture adopted is not particularly water efficient (e.g. rice, sugarcane) and this cripples water availability elsewhere, and there has been widespread over-abstraction of groundwater by farmers (Fishman *et al.*, 2011). Groundwater abstraction has not been controlled and groundwater has been treated as if it were a freely available resource. Climate change has meant many very hot summers with monsoon rains sometimes failing, leading to crop failure. Shortened winters have resulted in reduced snow cover in the Himalayas that feed the northern rivers (Smith and Bookhagen, 2018). Poor infrastructure means that key supply rivers are polluted as sewage is not properly treated and industrial pollution standards are not adequately enforced. Urban systems in India often have very leaky pipe networks (e.g. Delhi loses 40% of its piped water to leaks or theft; Bhatnagar, 2010). Construction has not been managed to take account of natural groundwater recharge areas. Thus recharge becomes less likely due to paving over these former recharge zones.

Some countries with water scarcity have avoided a water crisis through effective integrated water management systems (e.g. Israel). Wastewater can be recycled, irrigation can be more efficient (e.g. using drip or sprinkler irrigation) and more suitable crops selected for water scarce regions (Sharma *et al.*, 2010). Pollution control from industry should be enforced. Water harvesting and groundwater recharge management can be enhanced. India urgently needs to deal with these issues, as the water crisis grows ever greater each year.

produced elsewhere in locations where there is sufficient rainfall. Thus, if we are to avert a global water crisis we should reduce the amount of water consumed by meat production and by irrigated agriculture by changing the types of products

BOX 1.3

Water risk hotspots for agriculture

An assessment of the key global agricultural areas at risk from future water scarcity based on crop projections, population forecasts and climate change modelling (OECD, 2017) revealed three main countries that are crucial to the global outcome: China, India and USA. These countries lead global production of several of the main agricultural commodities but are also exposed to major future risk. Agricultural water risks are especially prevalent in north-east China, south-west USA and north-west India. North-east China is a key cereal-producing region, but is facing drier conditions with climate change, groundwater depletion and competing demands for water (see Chapter 12). North-west India is a large wheat- and rice-producing region, but this production is reliant on groundwater supplies for irrigation, which are being severely depleted. South-west USA is a region with diverse production, including dairy, livestock, fruit and vegetables, but has been facing drier conditions and depleted groundwater and will be subject to water demand from population growth.

The OECD (2017) noted that when agriculture water risks are realised there can be three very significant impact levels:

- a fall in production
- adverse effects on agricultural markets and trade
- concern over food security and socio-economic issues among a larger set of countries.

Projections suggest that, with no water shortages elsewhere, water risks to the three hotspot areas alone would reduce global production in total and increase prices of major field crops such as wheat, maize and cotton, and fruits and vegetables, and would be likely to exacerbate social tensions and conflicts, driving international migration.

we produce under irrigation, using more water-friendly varieties and enhancing the efficiency of irrigation (Molden *et al.*, 2007). Some important agricultural regions of the world may come under future strain due to competing water demands from population growth and climate change (see Box 1.3). However, climate change may open up some new areas of possibility for agriculture such as the northern high latitudes, which may warm sufficiently to allow some crops to grow (e.g. King *et al.*, 2018).

Some countries are undergoing very rapid industrialisation and population growth, which puts their water resources under stress even if the total water resources available within the country are large (e.g. India, Box 1.2). In developing countries, industry is an essential engine of economic growth, and the industrial demand for water is likely to increase. Global annual water use by industry is expected to rise to 1440 km^3 by 2050, double the use that occurred in the year 2000 (Wada and Bierkens, 2014). Much of the increase will be in developing countries now experiencing rapid industrial development. This will further intensify competition between the domestic, industrial and agricultural sectors for water, leading to sectoral water conflicts (see Chapter 11). While solutions such as **desalination** are used in many water-poor countries, they are costly because they require a lot of energy and water has to be pumped from sea level upwards to the distribution network, rather than relying on flow through gravity-fed systems from the uplands. In theory 0.86 kWh of energy is needed to desalinate 1 m^3 of salt water (at around 34,500 parts per million of salt). However, in reality desalination plants tend to use around 20 times this theoretical level, so further work is required to make the process much more efficient, and to use renewable energy sources where possible. Further information on future water demand and measures to deal with rising demand (e.g. wastewater reuse) are discussed in Chapter 8 and Chapter 9 (see especially Sections 9.4 and 9.5).

REFLECTIVE QUESTIONS

What proportion of the Earth's water is considered to be available and renewable freshwater?

If global water demand is currently less than one-seventh of the available global renewable freshwater resource, why do some consider there to be a water crisis?

1.6 SUMMARY

Water is a unique substance, having important physical and chemical properties, brought about by its covalent bonds and its polar hydrogen bonds, which enable it to act as an excellent solvent, to weather and erode the landscape and to regulate climate. Water is necessary for photosynthesis and for the survival of all forms of life. Water has been pivotal to the development of civilisations through its use in developing societal power relations formed by the onset of agriculture and developed over the past 13,000 years. Control of water resources enabled agriculture, which allowed non-farmers to specialise in other activities, some of which could be traded for food. In modern society, as in ancient civilisations, water has economic, religious and symbolic significance. Today the main use of water by humans worldwide, by volume, is in irrigation agriculture and the processing of agricultural products. Hence a focus on agricultural water efficiency is crucial for the future as the world's population grows and places enormous pressure on global water resources. There are already 45 countries that have less than 1000 m^3 of renewable water resource available per person per year. This water is needed for agriculture, industry and domestic use. It is anticipated that many more countries will reach this water 'stress' threshold in the coming years. These include large, populous countries in Asia such as India and China. A total of 44 countries rely on other countries to provide more than half of their water supply. Thus water security for individual nations is of great importance and typically requires a balance of technological, socio-economic and political solutions. Interdisciplinary water research and water solutions are therefore necessary to enable us to reduce the impacts of the global water crisis.

FURTHER READING

FAO. 2017. *Water for sustainable food and agriculture. A report produced for the G20 Presidency of Germany.* FAO; Rome. Available from: www.fao.org/3/a-i7959e.pdf.

A report focusing on water and agriculture, full of useful information and clear points. The report is written as a set of numbered paragraphs highlighting, in clear language, the main take-home points.

Hassan, F.A. 2012. *Water management and early civilizations: from cooperation to conflict.* Report prepared for the UNESCO–Green Cross Inter-national project From Potential Conflict to Co-operation Potential (PCCP): Water for Peace [online]. Available from: http://tinyurl.com/6tebpg5.

An interesting essay on the development of civilisations and water cooperation and water conflict.

Molden, D. (ed.). 2007. *Water for food, water for life.* Earthscan; London and International Water Management Institute; Colombo.

An excellent resource full of useful facts, theory and necessary adaptation strategies for the future.

Olovsson, I. 2017. *Wonders of water: the hydrogen bond in action.* World Scientific Publishing; Singapore.

The special properties of water are explained in the book, which contains lots of nice imagery.

Strang, V. 2015. *Water: nature and culture.* Reaktion Books; London.

Covers some of the cultural aspects of water touched upon in this chapter including material on religion, imagination, societal use of resources, and pollution.

Classic papers

On the properties of water:

Bernal, J.D. and Fowler, R.H. 1933. A theory of water and ionic solution, with particular reference to hydrogen and hydroxyl ions. *Journal of Chemical Physics* 1: 515–549.

Pople, J.A. 1951. Molecular association in liquids. II. A theory of the structure of water. *Proceedings of the Royal Society of London A* 205: 163–178.

On global water resources:

Vörösmarty, C.J., Green, P., Salisbury, J. and Lammers, R.B. 2000. Global water resources: vulnerability from climate change and population growth. *Science* 289: 284–288.

It is interesting to examine this paper from 2000 to see how the forecasts and concerns from this period have been borne out and how they compare to more recent projections (e.g. see next paper).

Wada, Y. and Bierkens, M.F.P. 2014. Sustainability of global water use: past reconstruction and future projections. *Environmental Research Letters* 9: 104003.

A paper that looks backwards as well as forwards to examine changing water resource use. The paper contains some fascinating maps.

PROJECT IDEAS

- Produce a set of PowerPoint slides, suitable for a lay audience, that would explain the key physical and chemical properties of water.

- Write a report, using case studies, showing the key things people have considered when including water features into the design of artificial landscapes such as cities, parks and gardens.

- Look for reports and data on the Internet that report past or predict future water use for different countries over time. Try to partition the relative roles of climate change, population growth and development on changing water use for some selected countries.

REFERENCES

Apipattanavis, S., Ketpratoom, S., Kladkempetch, P. 2018. Water management in Thailand. *Irrigation and Drainage* 67: 113–117.

Bernal, J.D. and Fowler, R.H. 1933. A theory of water and ionic solution, with particular reference to hydrogen and hydroxyl ions. *Journal of Chemical Physics* 1: 515–549.

Berner, E.K. and Berner, R.A. 1987. *Global water cycle: geochemistry and environment.* Prentice Hall; Englewood Cliffs, NJ.

Bhatnagar, G.V. 2010. 40 per cent of water supply gets wasted. *The Hindu*: 6 January, New Delhi.

Blockley, S.P.E. and Pinhasi, R. 2011. A revised chronology for the adoption of agriculture in the Southern Levant and the role of Lateglacial climatic change. *Quaternary Science Reviews* 30: 98–108.

Church, J.A., Clark, P.U., Cazenave, A., Gregory, J. M., Jevrejeva, S., Levermann, A. Merrifield, M. A., Milne, G.A., Nerem, R.S., Nunn, P.D., Payne, A.J., Pfeffer, W.T., Stammer D. and Unnikrishnan, A.S. 2013: Sea level change. In: *Climate change 2013: the physical science basis. Contribution of Working Group I to the fifth assessment report of the Intergovernmental Panel on Climate Change* [Stocker, T.F., Qin, D., Plattner, G.-K., Tignor, M., Allen, S. K., Boschung, J., Nauels, A., Xia, Y., Bex, V. and Midgley, P.M. (eds)]. Cambridge University Press; Cambridge, UK and New York.

Falkenmark, M. 1986. Fresh water – time for a modified approach. *Ambio* 15: 192–200.

FAO. 2016. *AQUASTAT database.* Food and Agriculture Organization of the United Nations [online, accessed April 2019]. Available from: www.fao.org/nr/water/aquastat.

Fishman, R.M., Siegfried, T., Raj, P., Modi, V. and Lall, U. 2011. Over-extraction from shallow bedrock versus deep alluvial aquifers: Reliability versus sustainability considerations for India's groundwater irrigation. *Water Resources Research* 47: W00L05.

Ghali, E. 2010. *Corrosion resistance of aluminium and magnesium alloys: understanding, performance and testing.* John Wiley and Sons; New York.

Hassan, F.A. 2012. *Water management and early civilizations: from cooperation to conflict.* Report prepared for the UNESCO–Green Cross Inter-national project From Potential Conflict to Co-operation Potential (PCCP): Water for Peace [online, accessed April 2019]. Available from: http://tinyurl.com/6tebpg5.

IPCC. 2007. *Climate change 2007: impacts, adaptation and vulnerability. Contribution of Working Group II to the fourth assessment report of the Intergovernmental Panel on Climate Change.* Parry, M.L. *et al.* (eds), Cambridge University Press; Cambridge.

Jompakdee, W. 2003. Rivers in jeopardy and the role of civil society in river restoration: Thai experiences. *CMU Journal* 3: 59–71.

King, M., Altdorff, D., Li, P., Galagedara, L., Holden, J. and Unc, A. 2018. Northward shift of the agricultural climate zone under 21st-century global climate change. *Scientific Reports* 8: Article number 7904.

Lawson, I.T. 2017. The Holocene. *In:* Holden, J. (ed.), *An introduction to physical geography and the environment* (4th edition). Pearson Education; Harlow, 108–136.

McClatchey, J. 2017. Regional and local climates. *In:* Holden, J. (ed.), *An introduction to physical geography and the environment* (4th edition). Pearson Education; Harlow, 229–249.

Molden, D., Oweis, T.Y., Pasquale, S., Kijne, J.W., Hanjra, M.A., Bindraban, P.S., Bouman, B.A. M., Cook, S., Erenstein, O., Farahani, H., Hachum, A., Hoogeveen, J., Mahoo, H., Nangia, V., Peden, D., Sikka, A., Silva, P., Turral, H., Upadhyaya, A. and Zwart, S. 2007. Pathways for increasing agricultural water productivity. *In:* Molden, D. (ed.), *Water for food, water for life: a comprehensive assessment of water management in agriculture.* Earthscan; London and International Water Management Institute; Colombo, 279–310.

NITI Aayog. 2018. *Composite Water Management Index: a tool for water management.* NITI Aayog; Delhi [online, accessed April 2019]. Available from: http://pibphoto.nic.in/documents/rlink/2018/jun/p201861401.pdf.

OECD. 2017. *Water risk hotspots for agriculture. OECD studies on water.* OECD Publishing; Paris. DOI: 10.1787/9789264279551-en.

Olovsson, I. 2017. *Wonders of water: the hydrogen bond in action.* World Scientific Publishing; Singapore.

Palmer, M.A. and Bernhardt, E.S. 2006. Hydroecology and river restoration: ripe for research and synthesis. *Water Resources Research* 42: doi:10.1029/2005 WR004354.

Pople, J.A. 1951. Molecular association in liquids. II. A theory of the structure of water. *Proceedings of the Royal Society of London A* 205: 163–178.

Sharma, B.R., Rao, K.V., Vittal, K.P.R., Ramakrishna, Y.S. and Amarasinghe, U. 2010. Estimating the potential of rainfed agriculture in India: prospects for water productivity improvements. *Agricultural Water Management* 97: 23–30.

Shiklomanov, I. 1998. *World water resources.* UNESCO; Milton Keynes.

Smith, T. and Bookhagen, B. 2018. Changes in seasonal snow water equivalent distribution in High Mountain Asia (1987 to 2009). *Science Advances* 4: e1701550.

Strang, V. 2015. *Water: nature and culture.* Reaktion Books; London.

Vörösmarty, C.J., Green, P., Salisbury, J. and Lammers, R.B. 2000. Global water resources: vulnerability from climate change and population growth. *Science* 289: 284–288.

Wada, Y. and Bierkens, M.F.P. 2014. Sustainability of global water use: past reconstruction and future projections. *Environmental Research Letters* 9: 104003.

WHO and UNICEF. 2017. *Progress on drinking water, sanitation and hygiene: 2017 update and SDG baselines.* World Health Organization; Geneva.

CHAPTER TWO

The changing water cycle

Stephen Sitch and Frances Drake

LEARNING OUTCOMES

After reading this chapter you should be able to:

- understand the global water cycle and its main components;
- describe each component of the water cycle and how it has changed over recent decades, and how it is projected to change during this century;
- appreciate the importance of climate variability and extremes on the water cycle, and for the provision of freshwater;
- describe the main uncertainties in the present-day and future global water cycle.

2.1 INTRODUCTION

This chapter introduces the main components of the global water cycle, describes the main stores and fluxes, and residence times, and presents the underlying physical processes. A discussion on uncertainties in the contemporary water cycle is included. The impacts of climate variability and extremes on the water cycle will cover different time and spatial scales, and the chapter describes future projections but also draws on evidence from the past to illustrate the nature of possible future changes. Specific examples will be provided for North America, the UK and West Africa. The impact of human modification of the water cycle will be discussed, including

a direct role through land use, and indirectly through modification of atmospheric composition and climate as a result of anthropogenic emissions of **greenhouse gases** and **aerosols**. Finally, this chapter outlines major challenges and unknowns in the global water cycle under the spectre of global environmental change.

2.2 THE GLOBAL WATER CYCLE

The global water cycle is shown in **Figure 2.1**. Approximately 97% is saline water in seas and oceans (1,338,000,000 km^3), leaving only 3% as freshwater. Approximately half of the freshwater is stored in glaciers and snow

Figure 2.1 The global hydrological cycle, with estimated flows and storage. The terrestrial water balance does not include Antarctica. Big vertical arrows show total annual precipitation and evapotranspiration over land and ocean (thousand km³ yr⁻¹), which include annual precipitation and evapotranspiration in major landscapes (thousand km³ yr⁻¹) presented by small vertical arrows; brackets indicate area (million km²). The direct groundwater discharge, which is estimated to be about 10% of total river discharge globally, is included in river discharge.

Source: Oki and Kanae (2006). Global hydrological cycles and world water resources. *Science* 313: 1068–1072. Reprinted with permission from AAAS.

(excluding Antarctica, 24,064,000 km³), and a substantial (although highly uncertain) amount in groundwater (23,400,000 km³).

It is important to consider not only the size of each storage zone (or reservoir) but the exchange or flux of water between reservoirs; after all, it is the exchange of water between reservoirs that forms the global water cycle. Both reservoir size and residence time need to be considered. Average residence times can be computed as the reservoir size divided by the flux (**Table 2.1**). For example, water vapour stays in the atmosphere on average for only nine days

and plays an active role in the water cycle. Note that only a small fraction of the total water is in fact 'available' to drive the water cycle.

2.2.1 River discharge

Although several components (e.g. rivers, snowfall) may appear minor in **Figure 2.1**, they in fact play a disproportionately large role in our water security. For example, rivers account for one-half of the world's potable water, extracted directly or from reservoirs (Barnett *et al.*, 2005). Although snowfall

Table 2.1 Average residence time in major water stores

Reservoir	Average residence time
Oceans	1500 years
Glaciers	1000 years
Soil moisture	6 months
Deep groundwater	10,000 years
Shallow groundwater	200 years
Lakes	50 years
Rivers	14 days
Atmosphere	9 days

represents around one tenth of total terrestrial precipitation, snowmelt plays a dominant role in seasonal runoff in most terrestrial regions above 45° north (with the exception of Western Europe), and snow and glacier melt are also important for the overall basin hydrology in several major South East Asian rivers further south (i.e. the Indus and Brahmaputra) (Immerzeel *et al.*, 2010).

Humans have modified rivers changing their geomorphology (e.g. channelisation), hydrology (dams, abstraction, land-use, irrigation), water quality (point and **diffuse pollution**) and ecology. There are now more than 45,000 large dams (and around a million small dams) on river systems, and most large river systems in North America and Eurasia are highly or moderately regulated (Allan and Castillo, 2007). Dams regulate flow regimes, for example in truncating extremes, producing less frequent and smaller floods and fewer very low flow events (Lytle and Poff, 2004). For example, prior to the dam across the Green River in Utah, USA in 1963, the flow regime was characterised by high springtime runoff, associated with snowmelt (elevated flood risk), and very low flows in autumn and winter. The introduction of the dam enabled a more regular flow regime lacking in seasonal patterns. Section 6.5.2 in Chapter 6 provides more detail on the impact of flow regulation on aquatic ecology.

2.2.2 Evaporation, condensation and transpiration

Water is transferred into the atmosphere through evaporation over the ocean (436,500 km^3 per year) and **evapotranspiration** over land (65,500 km^3 per year). Evaporation is the transformation of water from the liquid to gas phase directly above water bodies, from the soil surface, or from plant canopy **interception** (rainfall intercepted and subsequently evaporated off leaves). Imagine a half-filled jar of water where the water is initially separated from the air. If we then remove the separation device within the jar, evaporation will take place into the air within the jar, while **condensation** will also occur. Initially, because the amount of water vapour in the air within the jar is low, the evaporation rate will exceed the condensation rate, but eventually the two will balance. This equilibrium state is called saturation (Aguado and Burt, 2015). The total pressure exerted by the atmosphere is the sum of the pressures exerted by each component gas of the atmosphere. These pressures are called the **partial pressures** of the individual gases. At any given temperature there is a limit to the density of water in the air and thus a limit on the partial pressure, often referred to in meteorological textbooks as the vapour pressure (e) that water vapour can exert. This limit is known as the **saturation vapour pressure** (e_s). If you try to add more water vapour to an atmosphere that is already saturated at that temperature, then the condensation rate exceeds the evaporation rate. Hence more water droplets will appear in the atmosphere and be able to form cloud water droplets. On the large spatial scale ($10 + km^2$), evaporation over land is predominantly a radiation-driven process (Priestley and Taylor, 1972). Transpiration represents the flux of water from the soil to the atmosphere, passing through plants. Water is lost to the atmosphere as water vapour through small pores in the leaves called stomata.

2.2.3 Linking water and energy

The water cycle is integrally linked to the global energy budget. Water vapour is a greenhouse gas and modulates the radiative cooling of the atmosphere (i.e. the absorption or emission of infrared radiation; Allan, 2011). In fact, water vapour contributes 60% of the natural greenhouse effect (Kiehl and Trenberth, 1997). The tropics receive a surplus and high latitudes a deficit in solar energy over the annual cycle. Energy and its redistribution due to these latitudinal energy imbalances drive atmospheric and oceanic circulation, and thus climate and weather. In other words, warmth from the equator is redistributed polewards. This redistribution is not in a straight northerly or southerly direction as air and water are fluids and the Earth's rotation (via the Coriolis effect) produces circulation effects in them. Approximately 60% of the energy imbalance is redistributed through atmospheric circulation and 40% via ocean circulation (Stevens, 2011).

During evaporation of water, energy is absorbed in the phase change from liquid to gas. During the reverse process, condensation, energy is released in the phase change from gas to liquid. This is known as the latent heat of vaporisation, with 2260 kJ kg^{-1} of energy released or absorbed during a phase change without a change in temperature (see Chapter 1). In this respect, water vapour acts as a global 'air-conditioner', acting to cool the planet's surface (Trenberth *et al.*, 2003), because energy is consumed in evapotranspiration (i.e. changing the bonds between molecules) rather than in raising surface temperatures (i.e. the kinetic energy of molecules). Similarly, we sweat to cool down during hot summer days. As a parcel of air rises in altitude, it cools as it expands (adiabatic cooling). Assuming no change in the total amount of water vapour in the parcel as the air parcel rises and cools, it will eventually reach its dew-point temperature (which represents the cloud base) when the air parcel becomes saturated with water vapour (i.e. 100% relative humidity). At this point, water vapour condenses out, changing from the gaseous to the liquid phase, releasing energy to the atmosphere. The same process is responsible for surface dew on some cold winter mornings. Overnight the surface cools, especially strongly during cloudless nights, where energy can readily escape into space. As surface temperature drops the air may become saturated, and dew be formed. In the atmosphere, condensation leads to cloud water droplets, which themselves are not large enough to form precipitation. There are a number of physical processes which allow the cloud water (or ice) droplets to grow to precipitation size, as discussed below.

2.2.4 Precipitation

The majority of the water evaporated over oceans is precipitated back into the oceans, which is unsurprising given that 71% of the Earth's surface is covered by ocean. The distribution of precipitation also helps to partly explain salinity variations in the ocean. Transportation of water in the atmosphere is intimately linked to general atmospheric circulation. Approximately 10% (45,500 km^3 per year) of the evaporated water over oceans is transported and precipitated over land. An equal amount is returned to the oceans via river runoff (an equal amount is required in a balanced system). This net water vapour transport from ocean to land represents approximately 40% of the total precipitation over land (111,000 km^3 per year), with the remaining 60% of the water for precipitation over land originating from land-surface evapotranspiration (resulting in convective precipitation).

Precipitation refers to all liquid and frozen forms of water: rain, snow, hail, dew, hoar-frost, fog-drip and rime, but only the first two make significant contributions to precipitation totals globally. Precipitation usually occurs in connection with three modes of uplift of air – convective, cyclonic (depression) and orographic (air forced over mountains). Condensation occurs easily in the air because of hygroscopic aerosols, such as dust, smoke, sulfur dioxide and salt. These hygroscopic aerosols are water soluble and this allows condensation to take place before the air is saturated. Small cloud droplets grow faster than large ones. As droplets get larger the growth by condensation lessens. Cloud droplets are typically less than 1 to 50 μm in size and need to grow to 100 μm (i.e. 100 times larger) if they are to fall as precipitation. The processes by which cloud droplets grow to precipitation size depend on the temperature of the cloud. For simplicity, we can distinguish three types of cloud as described below.

2.2.4.1 Warm clouds

In warm clouds, where the temperature is above 0°C (cloud top temperature less than −15°C), all the cloud droplets will be water. To reach precipitation size the cloud droplet would have to continue growing by condensation for several hours, and most precipitating clouds have a lifetime far shorter than this. However, as the droplets grow, different droplet sizes evolve. The larger droplets will fall faster than the small droplets. As the large droplets (greater than 40 μm) fall they will sweep the small droplets up and grow in size. This is the collision–coalescence process. The larger the droplet, the more efficient it is at collecting other droplets it encounters. Even with collision, growth occurs only if the two coalesce. This is more likely to happen if the two droplets vary greatly in size. As they get larger, droplet fall speed increases, sweeping out a larger volume of the cloud. In this case the growth rate increases with the size of droplet and precipitation-size droplets are quickly formed. The thicker the cloud (the longer the droplet has to collect other droplets), the larger the droplet grows.

2.2.4.2 Frozen clouds

If a cloud is below −40°C all the droplets will be frozen. Growth mainly occurs through riming, where supercooled water in the atmosphere is deposited onto the crystal as ice. Growth by this method is relatively rapid. In these 'glaciated clouds' crystal growth can occur by a process analogous to the collision–coalescence process, called aggregation, but growth by this method is slow. These glaciated clouds are so high in the atmosphere that the precipitation rarely reaches the ground.

2.2.4.3 Mixed clouds

In clouds between −40 and 0°C there will be a mixture of frozen and liquid cloud droplets. In these types of cloud the Bergeron–Findeisen process takes place. At temperatures below 0°C the saturation vapour pressure for supercooled water is greater than that for ice (Figure 2.2). A gas will advect from high pressure to low pressure (i.e. there will be a vapour flux along the pressure gradient from close to the supercooled water droplet towards the ice particle). Therefore, in a mixed cloud dominated by supercooled water droplets, the air, which is close to saturation with water, is supersaturated with respect to ice. Ice crystals therefore grow faster than water droplets. They may grow so much that the air becomes sub-saturated with respect to water and the remaining water droplets evaporate. The ice crystals then grow by sublimation, where

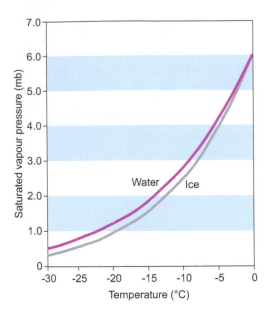

Figure 2.2 Saturated pressure–temperature curves for water and ice.

water vapour is directly deposited onto the ice crystal as a solid, missing the liquid phase. This is the Bergeron–Findeisen process. To see this process in action, look at www.youtube.com/watch?v=-UXpJ3KRO_s. The ice crystals can grow by several means as they fall to the lower, warmer parts of the cloud by aggregation with other ice crystals and collision–coalescence with water droplets.

Clouds do not fall neatly into warm and cold types, and a mixture of both processes may occur within a cloud. The Bergeron–Findeisen process is believed to be the dominant process in mid-latitude precipitating clouds (Jonas, 1994). The type and size of precipitation that falls out of the cloud is determined by the cloud type. However, the precipitation type to reach the ground will be affected by the temperature structure of the air beneath the cloud.

2.2.5 Uncertainties in the global water cycle

Rather surprisingly, there are large uncertainties even in the main components of the global water cycle. For example, based on seven global datasets, mean annual land precipitation estimates vary between 96,000 and 118,000 km^3 per $year^{-1}$ for the period 1979–1999 (Biemans *et al.*, 2009). Assuming no change in water stores, precipitation over land supplies either evapotranspiration or river runoff. Computer models simulate the key ecohydrological processes on land (see e.g. Gerten *et al.*, 2004) and are used to quantify land-based water fluxes. There are large uncertainties in annual evapotranspiration and river runoff of 17% and 20–25%, respectively (Haddeland *et al.*, 2011; Trenberth *et al.*, 2009). Furthermore, there is uncertainty in the surface albedo, which is the percentage of solar radiation reflected back into space. Snow and ice typically have high albedo (up to 90% for fresh snow, compared to only 17% for soil) and reflect a large portion of incoming solar radiation. As the climate warms, spring snowmelt at high latitudes occurs earlier in springtime, and consequently less radiation is reflected back into space, thus further enhancing surface warming. This is known as *snow-ice-albedo feedback* and is the main reason why warming is expected to be most pronounced at high latitudes, where snow is prevalent.

REFLECTIVE QUESTIONS

What are the different processes that can result in rainfall?

What is the seasonality of precipitation and river runoff (snowmelt if applicable) in the region where you live?

2.3 CLIMATE VARIABILITY AND CHANGE

Section 2.2 described the average global water cycle. However, the water cycle varies both spatially and temporally, as we are all well aware when we witness floods and droughts. The Intergovernmental Panel for Climate Change defines climate variability as 'variations in the mean state and other statistics (such as standard deviations, the occurrence of extremes, etc.) of the *climate* on all spatial and temporal scales beyond that of individual weather events'. An extreme weather event is an event that is 'as rare as or rarer than the 10th or 90th percentile of the observed probability density function' (IPCC, 2007a). Climate extreme events occur when 'a pattern of extreme weather persists for some time, such as a season, especially if it yields an average or total that is itself extreme (e.g., drought or heavy rainfall over a season)' (IPCC, 2001). Here we will give examples of climate variability at multiple scales, relevant to the global water cycle, including extreme climate events and variability at multi-annual and multi-decadal scales.

2.3.1 Extremes

The summer heatwave of 2003 across Western Europe was an extreme climate event. Areas experienced elevated temperatures of 3.5°C and below-average precipitation. Across Europe, 30,000 to 50,000 deaths were attributed to this heatwave (Fedoroff *et al.*, 2010), gross primary productivity declined by 30% (Ciais *et al.*, 2005), and crop yields of grains and fruits were reduced by 20–36% (Fedoroff *et al.*, 2010). By studying the conditions surrounding the heatwave we can learn about feedbacks that make extreme events more extreme. Reduced evapotranspiration, due to the dry land-surface conditions, prevented surface cooling, and thus even higher temperatures resulted: a positive

feedback. Teuling *et al.* (2010) showed how initial heating was suppressed more over grasslands than forests because of enhanced evaporation; however, the more conservative use of soil moisture by forests mitigates extreme heat over the longer term. It is also important to consider the effect of soil moisture on precipitation. Figure 2.3 shows the land–atmosphere coupling strength for the northern hemisphere summer. Land–atmosphere coupling is particularly strong across the Great Plains, Sahel and South Asia (Koster *et al.*, 2004). Where such coupling occurs it is likely that when there is low soil moisture there is little supply to the atmosphere and dry conditions are enhanced. Such feedbacks may exacerbate (intensify and/or prolong) extreme conditions. Both soil moisture–temperature and soil moisture–precipitation feedbacks are likely to be more important in Europe under future climate change (Seneviratne *et al.*, 2006).

Extreme events occurred in the summer of 2010 across central Europe, with drought, leading to fires, especially in peatlands, with smoke leading to a deterioration of air quality in places such as Moscow. Concurrently there were catastrophic floods across Pakistan. Both were associated with the same breakdown in the normal flow of the jet stream (high-velocity winds at the top of the lower atmosphere), as explained within Figure 2.4. Patterns in the upper troposphere can be important for determining extreme events and variability from year to year, as described in Box 2.1.

Modelling results provide evidence for a human contribution to recent more extreme events over the northern hemisphere (Mann *et al.*, 2017) including the very hot summer across Europe in 2018, and contributing to more extreme individual flood events (e.g. Schaller *et al.*, 2016). A weakening of the northern hemisphere summer mid-latitude jet stream and associated reduced storm activity since 1979

Figure 2.3 Land–atmosphere coupling strength in June, July and August. The units are dimensionless and just provide a relative picture of where coupling is greatest.
Source: Adapted from Koster *et al.* (2004). Regions of strong coupling between soil moisture and precipitation. *Science* 305: 1138–1140. Reprinted with permission from AAAS.

has been reported with less cloud cover, leading to higher daytime maximum temperatures (Coumou *et al.*, 2015). Blöschl *et al.* (2017) analysed the timing of river floods in Europe over the past five decades and found clear patterns of change in flood timing. They report that warmer temperatures have led to earlier spring snowmelt floods throughout north-eastern Europe; delayed winter storms associated with polar warming have led to later winter floods around the North Sea and some sectors of the Mediterranean coast; and earlier soil moisture maxima have led to earlier winter floods in Western Europe. Box 2.1 provides some further examples of the work being undertaken to tease out human influence on extreme events versus natural variability.

2.3.2 Multi-annual timescales

El Niño Southern Oscillation (ENSO) is important for climate variability and extremes across large parts of the world, through a process called teleconnections (Bjerknes, 1969). ENSO is a modification of the ocean and atmospheric circulation in the equatorial Pacific region, and occurs irregularly at a frequency of three to six years and with varying intensity.

In normal conditions (**Figure 2.6a**), warm water stacks up in the western Pacific, maintained by the easterly (from the east) surface trade winds. The warm water evaporates, producing the large convective rainfall systems over the western Pacific. The waters in the east are cool through evaporation and upwelling of cold,

Figure 2.4 The static jet stream of summer 2010, which led to different extreme events across Europe and Asia. The temperature anomalies shown on the map are from 20–27 July 2010 compared to temperatures for the same dates from 2000 to 2008. The anomalies are based on land-surface temperatures observed by the Moderate Resolution Imaging Spectroradiometer (MODIS) on NASA's Terra satellite. Areas with above-average temperatures appear in red and orange, and areas with below-average temperatures appear in shades of blue.

Source: From NASA/Earth Observatory; NASA image has been altered.

nutrient-rich water from below. During El Niño conditions the easterly winds slow and the body of warm water is shifted eastwards. Rainfall shifts eastwards, leaving the western Pacific region drier than normal, and heavy rainfall occurs over Peru. Upwelling of the cold, nutrient-rich water near the eastern Pacific coast weakens, and leads to drastic reductions in the anchovy populations (and thus harvest); the name, El Niño, is Spanish for 'the boy-child', and the term given by Peruvian fishermen, in reference to the Christ child, as this phenomenon starts around

Christmas time. La Niña is essentially a shift in the ocean–atmosphere circulation in the opposite direction, with stronger trade winds, a greater build-up of warm waters in the western Pacific, with heavier than normal rains in the western Pacific region.

These periodic large-scale changes in ocean–atmosphere circulation have important ramifications for climate and the water cycle across the globe. Typically, global land precipitation is anomalously high and low during La Niña and El Niño periods, respectively, and ENSO is responsible for regional extremes, in terms of both flooding and drought. ENSO teleconnections have been well studied (Moron and Ward, 1998). ENSO has been shown to affect the sub-tropical Pacific basin and eastern and southern Africa. Indonesia, Australia, India and south-east Africa all experience drought (leading to potential crop failure and famine) during El Niño. During El Niño events the Amazon dries, forest fires are more prevalent in Indonesia, and floods and landslides occur across Peru and southern California. Reverse weather conditions occur during La Niña events.

There are very clear changes in the North American seasonal weather pattern associated with ENSO due to displacement of the jet streams, leading to unusual weather outside of the tropics (Coghlan, 2002). The greatest change occurs during the winter months (Figure 2.7). During El Niño the USA experiences wet and windy weather along its west coast and southern states. California was particularly hit by flooding during the 1982/83 El Niño. However, El Niño events are linked to lowered hurricane activity in the Gulf of Mexico (Pielke Jr and Landsea, 1999). The northern states experience mild winters and below-normal precipitation during El Niño. This was particularly marked during the 1997/98 El Niño. The warm weather, paradoxically,

Weather fluctuations or a changing climate?

The average rainfall over the UK in summer (1 June–30 August) is 241 mm. If we look back over the past century we can see large inter-annual variability as shown in **Figure 2.5**. Some summers are extremely wet, as in 2007 (357.8 mm; Mayes, 2008) and 2012 (379.2 mm), which was the wettest summer since 1912 (384.4 mm). In contrast, 2013 (187.0 mm) and 2018 (176.0 mm) were very dry. Much of this variability is due to natural forcings occurring within the atmosphere; however, a challenge facing climatologists is to attribute any change in these extremes to human activity.

If we look at an average surface pressure chart for the mid-latitudes we see areas of low pressure. Travelling depressions or low pressure systems dominate mid-latitude weather and are responsible for most frontal precipitation. Depression systems at mid-latitudes are associated with the convergence of air masses with different characteristics; warm and cold air. The two air masses do not easily mix and the boundary between them is called a front. The warmer, less dense air tends to rise over the cold dense air. As the warm air rises it adiabatically cools, leading to cloud formation, which can then result in precipitation associated with the front. In contrast, high-pressure systems result in dry, calm conditions. In winter this can lead to low cloud or fog but can also provide clear blue skies and cold, frosty weather. In the summer, high pressure leads to dry, bright and warm sunshine (Young, 1994a; Young 1994b).

If we look at an average pressure chart for the **tropopause** (top of the troposphere, which is the lower layer of the Earth's atmosphere) at mid-latitudes we do not see areas of low and high pressure; instead there are undulating waves. These are called **Rossby waves** and cause the troughs and ridges in the fast-flowing upper westerlies (jet streams are areas of maximum velocity) near the tropopause. The formation and direction of depression systems are controlled by these Rossby waves. The formation of depression systems is associated with troughs and high-pressure systems with the ridges of Rossby waves. Rossby waves are anchored by the Rocky Mountains. If the British Isles are stuck under a trough, as in summer 2007, then they can receive continuous depression systems. There is a complex relationship between the upper air flow, jet streams and surface conditions such as temperature, which all affect the position and track of frontal depression systems and, consequently, the precipitation. An excellent video explaining jet streams can be found at http://tinyurl.com/b4nxuvg. Under a ridge, the British Isles receive high-pressure systems. Rossby waves go through a cycle over 20 to 60 days from a smooth zonal flow to undulating waves with pronounced troughs and ridges. The strength of the westerlies is measured by an index. A high index is associated with the strong zonal flow. Some of the warmest summers occur when the Rossby wave index is at its lowest and the waves are most pronounced. At this point cut-off can occur and a blocking anti-cyclone formed. This type of slow-moving high-pressure system leads to exceptionally warm and dry summers when they get stuck over the British Isles, as in summer 2003 or 2006 (Ogi *et al.*, 2005).

The summer of 2012 saw very wet conditions in the UK and drought over large parts of the central USA. In this year the jet stream took up a more southerly path than usual, bringing frontal depression systems over the south of the UK. A useful explanation with diagrams of the jet stream position can be found at http://tinyurl.com/8223wwz. There is some research to

Continued

suggest that the exceptionally wet conditions may also be due to anthropogenic forcing from changes in sea surface temperatures and Arctic sea ice (Dong *et al.*, 2013).

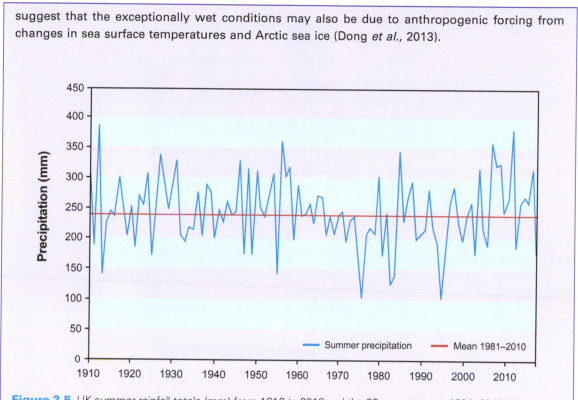

Figure 2.5 UK summer rainfall totals (mm) from 1910 to 2018 and the 30-year average 1981–2010 (241 mm).
Source: Data from www.metoffice.gov.uk/pub/data/weather/uk/climate/datasets/Rainfall/date/UK.txt.

caused ice storms in eastern Canada in January 1998, as the warmer temperatures led to freezing rain rather than the normal snowfall.

The clearest teleconnection is between ENSO and the Indian monsoon. Typically, during El Niño years the monsoon rains in the northern hemisphere summer are weakened and there is less precipitation, particularly in north-western India. As a consequence crops are adversely affected. There are various claims that El Niño affects the climate of northern Africa and Europe. Although climate anomalies can be shown to occur at these locations it is thought that the ENSO influence is small

(Moron and Ward, 1998). The disturbance of upper air flow patterns may well lead to variations in stratospheric humidity and other chemical processes in the stratosphere.

Gergis and Fowler (2009) used proxy data to reproduce ENSO events over the centuries from 1525 and considered what the future might hold. Proxy data are measurements which can be used to reconstruct the climate but in themselves do not measure climate variables such as temperature. A good example of proxy data are tree rings, which can be used in some circumstances to estimate past temperature and rainfall. There are many difficulties with

Figure 2.6 Conditions in the Pacific during a) normal periods and b) El Niño. During normal conditions strong trade winds keep warm water near New Guinea and Australia. During El Niño the trade winds relax along with the temperature gradient of the ocean waters and hence warm surface waters in the west can flow eastwards.

using proxy data to form long-term data series, but used carefully they provide an invaluable record against which to compare present-day climate change. The ENSO reconstructed record revealed that, although extreme ENSO events had occurred throughout the 478-year period, the twentieth century stood out for several reasons. The twentieth century contained around 43% of the most severe ENSOs and also 28% of the long-duration events (Gergis and Fowler, 2009). Furthermore, 30% of extreme ENSOs in the 478-year record have occurred since 1940 (Gergis and Fowler, 2009). The link between surface temperatures and the ENSO cycle means that climate change may well affect the occurrence of extreme and protracted ENSOs. The IPCC (2007a) fourth assessment noted that since 1976–1977 sea surface temperatures in the central and equatorial Pacific have increased and at the same time El Niño events have tended to be more severe and longer. ENSO also influences the global mean temperature because of large heat exchanges between the ocean and atmosphere (IPCC, 2013). Many of the warmest years on record, often occurring since 1990, have had strong El Niño events, helping to increase global mean temperature. However, confidence in climate model projections on whether ENSO will intensify is currently low (IPCC, 2013).

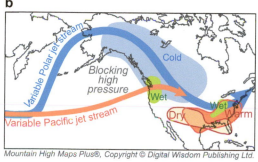

Mountain High Maps Plus®, Copyright © Digital Wisdom Publishing Ltd.

Figure 2.7 Winter weather anomalies during a) El Niño and b) La Niña over North America.

Other examples of variability on multi-annual timescales are drought across western North America between 2000 and 2004, and the famous Great Plains 'Dust Bowl' of the 1930s (see **Box 2.2**). In the latter case, land management practices contributed to the severity of the drought.

2.3.3 Multi-decadal timescales and trends: long-term drought, CO_2 effects, river discharge, glacier melt

An example of multi-decadal variability is the drought occurring in the Sahel (see **Box 2.3**). Climate variations and human activities appear to play important roles in **desertification** of the region. Human activities include the impact of increasing population pressures on land and its management, through to the

role of global dimming, as a result of increasing anthropogenic emissions of scattering aerosols, especially from the industrialised northern hemisphere countries. These aerosols scatter incoming solar radiation, sending a proportion back into space and reducing the incoming radiation reaching the Earth's surface (reducing evapotranspiration). They also modify the temperature structure of the atmosphere, make brighter (more reflective) clouds, reduce the precipitation efficiency of clouds and increase their lifetime (Ramanathan *et al.*, 2001). Global dimming has been hypothesised to explain observed global trends in reduced pan-evaporation (evaporation from a pan filled with water – a widely used meteorological instrument) over the last few decades, despite concurrent increases in global temperature. This further demonstrates the key role of solar radiation in driving evaporation (Roderick and Farquhar, 2002; Roderick *et al.*, 2007). Aerosols impact upon precipitation by forming tiny condensation nuclei, which result in smaller cloud droplets, meaning that fewer droplets are large enough to form rain. The overall effect of aerosols is to weaken the hydrological cycle. Indeed large reductions in precipitation over land and river runoff during 1991 and 1992 have been attributed to the effect of aerosols emitted from the Mount Pinatubo eruption in June 1991 (IPCC, 2007a). However, in some areas brightening has occurred from the 1980s onwards, especially in former industrialised regions where implementation of air-quality measures has reduced air pollution and aerosol emissions. Global evapotranspiration increased between 1982 and 1997, with a small decline between 1998 and 2008 attributed to limited soil moisture supply, as observed from space satellites (Jung *et al.*, 2010).

Over the twentieth century, river discharge increased (Gedney *et al.*, 2006; Peterson *et al.*, 2002). This was mainly due to

BOX 2.2

Dust bowl of 1930s North America

The Dust Bowl of the 1930s is an iconic period in US history, forming the backdrop for the literary classic *The Grapes of Wrath* by John Steinbeck. A period of prolonged drought and dust storms devastated farming in the area known as the Great Plains, particularly in the states of Texas, Oklahoma, Kansas, Colorado and New Mexico (Figure 2.8).

Figure 2.8 Severe conditions during the 1930s Dust Bowl of North America: a) Dust piled up against a farm building, b) a farmer digs in dust, Cimmaron County, Oklahoma.
Source: From Library of Congress, Prints & Photographs Division, FSA-OWI Collection, LC-DIG-fsa-8b38341 DLC (a) and LC-USZ62–131312 DLC (b).

Continued

By the 1930s the Great Plains had become subject to intensive agriculture, and the high price of wheat during the First World War had made the Southern Plains very prosperous. However, in 1931 the rains that had previously made farming possible failed, and by 1932 dust storms had begun to occur. In 1933 there were 38 huge dust storms. These led to hundreds of deaths from 'dust pneumonia', where dust infiltrates the lining of the lungs. It also caused millions of people to leave their homes, with many permanently emigrating to other states, particularly California. There were also severe economic consequences, with bank failures, large falls in land prices and an extension of the Great Depression.

Droughts are a common feature of the mid-latitudes, including those of North America. Observations of rainfall over the twentieth century show a great deal of variability, with long periods of both below- and above-average rainfall over the Great Plains (Garbrecht and Rossel, 2002). It has been suggested that it was anomalously high rainfall prior to the 1930s that encouraged settlers to believe that the climate was suitable for intensive agriculture. However, the observational record shows that during the first part of the twentieth century droughts tended to occur every 20 years. Longer records reconstructed from proxy data show evidence of severe droughts occurring in the Great Plains once or twice a century (Woodhouse and Overpeck, 1998). Particularly intense and prolonged droughts occurred in the thirteenth and sixteenth centuries, with the former lasting for some 38 years (Woodhouse and Overpeck, 1998). It has been suggested that the epic droughts during the Medieval warm period (AD 900–1300) could provide a possible analogue for future droughts in the twenty-first century in western North America (Cook *et al.*, 2004; Woodhouse *et al.*, 2010).

What are the potential causes of these severe droughts, and the Dust Bowl in particular? Drier conditions tend to occur in the south-west and south-east USA during La Niña (**Figure 2.9**). This is when there are lower than average sea surface temperatures in the eastern tropical Pacific. La Niña occurs on an irregular basis, but typically every two to six years. The La Niña at the time of the Dust Bowl, however, was relatively small, given the warmth and intensity of the drought. Also the position of the low rainfall was farther north than might be expected. Typically, La Niña droughts occur in the south-west and Mexico. The drought was also much more widespread, extending as far north as Canada. Climate models have difficulty reproducing the features of the Dust Bowl by simply including La Niña and thus it is likely that there were other contributing factors (Cook *et al.*, 2008). Droughts in North America are also linked to anomalously warm sea surface temperatures in the north Atlantic and these occur on a decadal timescale. By including the sea surface temperature anomalies for the Dust Bowl period in a climate model run, Schubert *et al.* (2004) were able to improve the rainfall results. They found that the tropical Pacific and tropical Atlantic sea surface temperature anomalies were the most important contributing factors. Schubert *et al.* (2004) also pointed out that seasonality is important. In an average year most precipitation falls on the Great Plains during spring and summer. Cold sea surface temperatures in the Pacific in the summer lead to changes on a global scale in the upper troposphere which suppress rainfall in the Great Plains. Warm sea surface temperatures in the Atlantic in summer and autumn produce anomalous cyclonic conditions at the surface which stop moisture from the Gulf of Mexico entering North America, further suppressing rainfall.

Continued

Figure 2.9 Temperature (°C) (top row) and precipitation (mm day^{-1}) (bottom row) anomalies for the Dust Bowl drought from the Climate Research Unit (CRU) and climate model simulations. The left panel shows actual data. The middle and right panels try to reproduce the actual data using model runs to investigate the causes of the Dust Bowl. The middle panel shows sea surface temperature effects only and the right panel shows sea surface temperature, dust aerosol plus crop removal effects combined.

Source: From Cook, B.I., Miller, R.L. and Seager R. 2009. Amplification of the North American 'Dust Bowl' drought through human-induced land degradation. *Proceedings of the National Academy of Sciences* 106: 4997–5001.

A final factor which may help to explain why the Dust Bowl was so intense are the dust storms themselves. Prior to human settlement, the Great Plains had been dominated by grasses and other flora, which were suited to the area and able to withstand prolonged drought. Their roots helped to bind together the topsoil. The farmers planted wheat, which was not drought-tolerant and was unable to hold the soil together. This change in plant cover, poor land-use practice and the low soil moisture content following the lack of rainfall all helped to form dust storms (Cook *et al.*, 2008). Not only did the loss of vegetation lead to an increase in surface albedo, but the dust in the atmosphere blocked out the incoming solar radiation. This meant there was less energy available for evaporation into the atmosphere and therefore the potential for precipitation was reduced (i.e. soil moisture–precipitation feedbacks). By including the effect of dust together with the cold Pacific and warm Atlantic SST anomalies, Cook *et al.* (2008, 2009) were better able to reproduce the Dust Bowl with a climate model, suggesting that all these factors are important (**Figure 2.9**).

increasing global precipitation and physio-logical effects of rising atmospheric CO_2 content reducing plant transpiration (Gedney *et al.*, 2006; Gerten *et al.*, 2008). The role of direct effects of CO_2 on water usage by plants and feedbacks is discussed in detail in **Box 2.4**. Clearly, these effects would not be dominant in areas where vegetation cover was sparse (e.g. Sahel, **Box 2.3**). While increased irrigation has reduced river discharge in many areas (see Chapter 3), Gerten *et al.* (2008) found that land-use change had increased river discharge overall.

Another important consideration for river runoff, aside from changes in

---

BOX 2.3

Sahel drought

The Sahel is a region in sub-Saharan West Africa – an area between the Sahara desert at 18°N and the equatorial forests at 15°N. The Sahel covers several African countries including Senegal, Mauritania, Mali, Burkina Faso, Niger, Nigeria, Chad, Sudan, Somalia, Ethiopia and Eritrea. It has a population of around 50 million inhabitants and a surface area over 5 million km². The vegetation is diverse and contains trees, shrubs and grasslands. In some ways the Sahel acts as a buffer zone from the arid Sahara north of the region. The climate of the area is controlled by the West African monsoon, itself controlled by the seasonal migration of an atmospheric phenomenon known as the **Intertropical Convergence Zone** (ITCZ). The location of the ITCZ broadly follows the seasonal variations in the receipt of the Sun's energy, and is located north of the equator in the northern summer and south in winter. As the ITCZ moves northwards it brings a short period of intense rain between late June and mid-September. Rapid vegetation growth results in a lowering of the surface albedo (as bare soil is light, it has a higher albedo than vegetation). Farming in the Sahel is almost entirely dependent on this rainfall. About 95% of the land in the Sahel is devoted to agriculture and 65% of people depend on agriculture for a livelihood (Druyan, 2011). Traditionally, people in the area maintained a subsistence living by moving to accommodate this climatic variability.

In the twentieth century colonial and postcolonial governments encouraged intensive modern food production in the area. This move accelerated in the 1950s and 1960s, which also coincided with a period of unusually high rainfall in the area (**Figure 2.10**). By the 1970s, however, the Sahel had entered a period of severe drought and there was a 40% decline in rainfall (Biasutti, 2011; **Figure 2.17a**). Drought is associated with either a late onset of the rains, an early finish to the rains or weak rain systems (Druyan, 2011). The Sahel was affected by drought from the 1970s to the 1980s, and although rainfall totals have recovered they are still not consistently back to 1950s levels (**Figure 2.10**). It was thought that such a large drop in rainfall could only be due to the result of human activity. Overgrazing, which led to an increase in albedo and therefore a cooler surface, would mean less evaporation and a decrease in precipitation (Charney, 1975). Note from **Figure 2.3** the strong land–atmosphere coupling over the Sahel, which means there is potential for strong soil moisture–precipitation feedbacks. This represents a positive feedback reinforcing desertification. The blame was placed upon environmental degradation due to traditional grazing methods, suggesting that the people in the area ought to modernise further.

Continued

Figure 2.10 Precipitation difference for June to October (rainy season) from long-term average 1901–2017 in the Sahel region showing the prolonged dry period since the 1970s.

Source: Figure reproduced with kind permission of the Joint Institute for the Study of the Atmosphere and Ocean at the University of Washington, doi:10.6069/H5MW2F2Q.

Whether overgrazing was the dominant cause of precipitation changes has always been open to debate. It is now recognised that the area has a very variable rainfall both spatially and temporally. Rainfall over the Sahel varies year to year and on a decadal timescale. These variations are linked to sea surface temperature anomalies in the Pacific, Indian and Atlantic ocean basins. In the Pacific, the El Niño phase of ENSO leads to drying and La Niña to positive rainfall anomalies. Indian ocean warming and changes in the cross-equatorial temperature gradient in the Atlantic all affect rainfall in the Sahel (Biasutti, 2011). On a decadal timescale the temperature gradient across the northern and southern hemisphere oceans is important, with decreased rainfall when it is colder in the north than in the south (Biasutti, 2011). Land-surface changes due to vegetation dieback are still thought to play a role (Caminade and Terray, 2010).

Several studies suggest that precipitation changes over the Sahel may well have been enhanced through the release of sulfate aerosols from fossil fuel burning in North America and Europe (Held *et al.*, 2005; Ackerley *et al.*, 2011; Booth *et al.*, 2012). Sulfate aerosols have both a direct and an indirect effect on the radiation (and thus energy budget) of the Earth. They reflect more solar radiation away from the Earth, thus cooling the Earth's surface. They also cool the surface by changing the radiative properties of clouds by making them brighter and also increase cloud lifetimes (IPCC AR4). Such cooling will reduce evaporation, leading to less water in the atmosphere and so reducing precipitation. Ackerley *et al.* (2011) performed ensemble climate model runs; this is where a climate model is run many times but each time with small changes to the physics parameters within their known ranges, to see how parameter value assumptions affect the results. They found that a combination of increasing greenhouse gases and sulfate aerosols led to a net decrease in rainfall over the Sahel, mainly by affecting the sea

Continued

surface temperature gradient. Additonally, Haywood *et al.* (2013) showed that three of the driest summers in the Sahel were preceded by large volcanic eruptions in the northern hemisphere. These produced large loads of stratospheric aerosols that cooled northern hemisphere sea surface temperatures thus moving the ITCZ further south and shifting the monsoon circulation and causing drought in the Sahel.

Sahelian climate is very sensitive to factors affecting the Atlantic sea surface temperature gradient as these are strongly coupled to rainfall in the Sahel and it is now well understood that the 1960s–80s drought was affected by both tropospheric and stratospheric aerosol emissions in the northern hemisphere. Furthermore, a deliberate stratospheric aerosol injection has been suggested as an emergency geoengineering strategy to counteract warming of our climate. The geoengineering study by Haywood *et al.* (2013) suggested that if such a strategy were implemented in the northern hemisphere it would bring back the Sahelian drought, whereas if it were concentrated in the southern hemisphere it would induce increased precipitation in the Sahel.

Now, it is generally acknowledged that the Sahel disaster of the 1970s and 1980s was about the implementation of development policies which paid little attention to the environment of the area, combined with external climate factors. There was a failure in recognising that there is a complex interweaving of human socio-economic systems with environmental systems. It is important that we continue to more fully explore the impact of anthropogenic changes of the climate to such vulnerable areas. It is also important that we look to improve water and land management in the Sahel so that millions of people are not regularly pushed into starvation.

In 2007 the African Union agreed an initiative to attempt to reverse desertification. The 'Great Green Wall' is an $8 billion project (www.greatgreenwall.org), that aims to restore 100 million hectares of degraded land and create 10 million jobs in rural areas by 2030, and now involves 20 African countries. The wall will span the width of Africa (8000 km), from Senegal in the west to Djibouti in the east, across Africa's Sahel region. Since its inception, the wall is about 15% complete, with millions of drought-resistant trees planted and degraded lands restored. The hope is that these trees will reduce wind-driven soil erosion, provide shading for crops and thus reduce irrigation, with roots helping to retain soil moisture, and replenish wells. This project is ambitious and targets 15 of the United Nations 17 Sustainable Development Goals (see Figure 13.1 in Chapter 13).

precipitation and CO_2-induced changes in transpiration, relates to changes in snow cover, the timing of seasonal snow, and glacial melt. More than half of the world's **potable water** comes from rivers (Barnett *et al.*, 2005). Glaciers and snow packs act as reservoirs for water storage and are especially important for freshwater supply in snow-dominated regions, supplying water to more than one-sixth of the world's population (Barnett *et al.*, 2005). In South Asia more than 1.4 billion people rely on water from the Indus, Ganges, Brahmaputra, Yangtze and Yellow Rivers, with the Himalayas at their source (Immerzeel *et al.*, 2010). Snow and glacial melt are particularly important for the Indus and Brahmaputra River basins (Immerzeel *et al.*,

2010). Satellite data have been used to estimate an earlier onset of snowmelt by ~0.5 days per year, an earlier finish of spring melt by ~1 day per year, and a shortened length of melting season by ~0.6 days per year over the Pan-Arctic over the last 30 years (Tedesco *et al.*, 2009). Over this period snow cover extent has decreased in May and June by 14% and 46%, respectively, in response to earlier snowmelt (Brown *et al.*, 2010). For the western USA 60% of the climate-related trends of river flow over the second half of the twentieth century have been attributed to human-induced climate change (Barnett *et al.*, 2008).

While there are a few advancing glaciers, for the most apart, right across the globe, glaciers have been in retreat over the last century (**Figures 2.11** and **2.12**). Excluding major ice sheets, for a series of well-studied reference glaciers across 19 regions there was an average retreat of 0.26 m of water equivalent per year between 1981 and 1990, 0.43 m per year for 1991–2000 and 0.76 m per year for 2001–2010 (WGMS, 2018). The rate was 0.9 m of water equivalent per year between 2014 and 2016. This accelerating retreat of glaciers can be seen in **Figure 2.11** with the steepening of the curves in recent years. There was an estimated 80% reduction in the ice extent at the main summit of Mount Kilimanjaro between 1912 and 2003 (IPCC, 2007a), and glacial shrinkage has occurred all over the Himalayas (Singh *et al.*, 2011), the Americas and Europe (IPCC, 2007a). In fact, information on changes in glacial length has been used to reconstruct temperature changes (Oerlemans, 2005). Important for **glacial mass balance** is the consideration of changes in both temperature and precipitation. For example, some of the losses at Kilimanjaro may be attributed to changes in precipitation rather than to direct warming (Kaser *et al.*, 2004).

REFLECTIVE QUESTIONS

What do we think the relative roles were of humans and climate variability on the Great Plains Dust Bowl and the Sahel drought?

What examples of climate variability having an impact on the local water cycle can you think of that have occurred in the last 12 months? What categories of climate variability would you consider them to come under?

Is the rate of glacier loss increasing, stable or decreasing according to **Figure 2.12**?

2.4 IMPACT OF LAND MANAGEMENT ON THE WATER CYCLE

We have already seen how humans affect the global water cycle through changing atmospheric composition (CO_2 and aerosols) and climate. Direct human impacts on water resources include the construction of dams, water extraction, and modification of river channels (Harding *et al.*, 2011), with implications for seasonal water supply, water temperatures and freshwater biodiversity (Lytle and Poff, 2004). In addition, changes in land use and human water use greatly affect the global water cycle and freshwater resources. Land-use and land-cover changes include regional desertification, deforestation (originally in mid-latitudes, Europe and North America, and more recently in tropical regions – see **Box 2.5**), mid-latitude reforestation and afforestation (e.g. during recent decades in China), and agricultural expansion throughout the

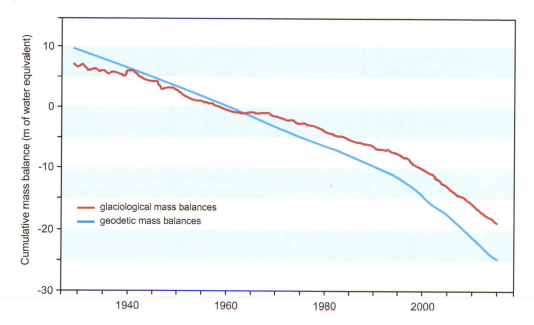

Figure 2.11 Cumulative mass balance for glaciers from 19 major glacial regions around the world compared to 1960 levels. A mass loss of 20 m of water equivalent is the same as 20 tonnes of water lost per m².

Source: WGMS (2018): Fluctuations of Glaciers Database. World Glacier Monitoring Service, Zurich, Switzerland. doi:10.5904/wgms-fog-2018-11.

twentieth century. Humans currently use between 10 and 15% of renewable freshwater, and agriculture accounts for one-third of total land evapotranspiration (Oki and Kanae, 2006). Nevertheless, nearly 80% of the world's population live in areas exposed to either high levels of water scarcity (Chapter 1) or threats to biodiversity (Vörösmarty *et al.*, 2010).

Demand for water outweighs supply in many regions (see section 1.5 in Chapter 1), which can lead to water conflict (see Chapter 11). To help sustain increases in agricultural production over the last 40 years the area of land under irrigation has doubled (Oki and Kanae, 2006), and even tripled across India between 1970 and 1999 (Rodell

et al., 2009). In some regions this has led to potentially unsustainable use of groundwater reserves, for example in parts of South Asia (Rodell *et al.*, 2009). Arid regions, such as Pakistan, are particularly reliant on groundwater for irrigation (Rodell *et al.*, 2009; www.youtube.com/watch?v=o1QsCa7RmmU).

Thus a focus on water-efficiency measures for agriculture is urgently required (see Chapter 1). Future changes in the global water cycle and freshwater resources will depend on population growth. Economic development will impact upon land use and water demand (see also Chapters 1, 11 and 12). Economic development will also indirectly impact upon climate change and atmospheric composition.

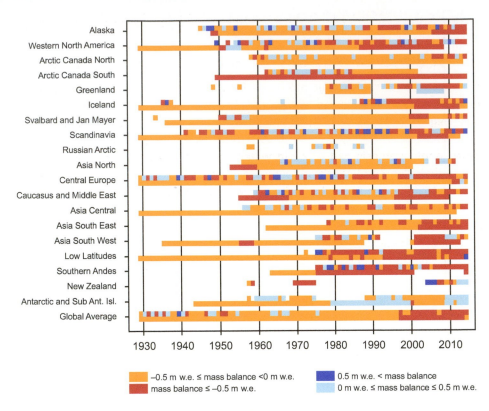

Figure 2.12 Regional glacier mass balances 1930–2015. The upper row for each region represents glacio-logical surveys of annual water balance in metres of water equivalent which is the same as tonnes of water per m². The lower row for each region is the annual geodetic balance, which is based on the measured volume change in glaciers from space or other surveys, and assumes a density of 850 kg m^{-3}. w.e. = water equivalent.

Source: WGMS (2018): Fluctuations of Glaciers Database. World Glacier Monitoring Service, Zurich, Switzerland. doi:10.5904/wgms-fog-2018-11.

REFLECTIVE QUESTIONS

Humans do not directly drink much water each year. So why does population growth pose such a major threat to water resources and the water cycle?

What was the impact of European and US forest clearing up to the middle of the twentieth century and recent reforestation on the regional water cycle?

2.5 IMPACTS OF FUTURE ENVIRONMENTAL CHANGE

2.5.1 Climate change: physical basis

Climate change is often discussed in terms of changes in temperature. Arguably, changes in precipitation will be as, or more, important for many essential ecosystem services, for example freshwater availability and food security, in the coming century. Precipitation must be considered in terms of its season, intensity and frequency rather than annual total alone. For example, rainfall intensity governs the amount of water that percolates

BOX 2.4

Role of direct effects of CO_2 upon water usage by plants

Water provides a medium for most cell functions and is thus vital for plant growth. Small pores, called stomata, regulate the flux of CO_2 going into leaves for photosynthesis (see section 1.3 in Chapter 1 for the photosynthesis equation) and water loss to the atmosphere, known as transpiration (Figure 2.13). There is thus a trade-off for the plants between the need to take up CO_2 for photosynthesis and to avoid water loss. It has been postulated that leaves optimally regulate their degree of stomatal opening or closing (known as stomatal conductance), in order to maximise daily photosynthesis for a given water loss (Cowan and Farquhar, 1977; Barton *et al.*, 2011). This introduces the concept of water-use efficiency; intrinsic water-use efficiency is defined as the ratio of the instantaneous rates of CO_2 assimilation and transpiration through the stomata (Condon *et al.*, 2002). Improvements in plant water-use efficiency would lead to increased crop yields in many dry environments (Condon *et al.*, 2002).

Figure 2.13 Leaf stomata regulate the uptake of carbon dioxide for photosynthesis, and water loss through transpiration.

As it is a greenhouse gas, elevated CO_2 concentrations in the atmosphere will have indirect effects on transpiration and the global water cycle by changing the surface energy balance and temperature. Elevated CO_2 concentrations will also directly affect leaf physiology, for example leaf photosynthesis. CO_2 and O_2 compete as substrates for the photosynthetic enzyme Rubisco (Prentice *et al.*, 2001). O_2 reactions (oxygenation) lead to CO_2 release through photorespiration, and thus reduced efficiency of photosynthesis in C_3 plants (i.e. all tree and many herbaceous species). Elevated atmospheric CO_2 concentrations will increase the rate of carbon-fixing reactions (carboxylation), and decrease oxygenation, and thus, in addition, reduce photorespiration. In free-air-carbon-enrichment (FACE) experiments, elevated CO_2 concentration stimulated photosynthesis by on average 31% (Ainsworth and Rodgers, 2007). There are also direct effects of elevated CO_2 on water usage by plants (Field *et al.*,

Continued

1995). Stomatal conductance decreases on average by just over 20% under elevated CO_2 across a wide range of vegetation types (Medlyn *et al.*, 2001; Ainsworth and Rodgers, 2007). However, these changes may not translate to similarly large changes in regional evapotranspiration or river runoff, due to the net effect of numerous feedbacks, as shown in **Figure 2.14**.

© 1995 Blackwell Science Ltd, Plant, Cell and Environment, 18,

Figure 2.14 Mechanisms connecting changes in leaf conductance to canopy evapotranspiration and soil moisture changes. Positive effects are shown in purple arrows with negative effects in blue arrows. Upward arrows (green) indicate an increase and downward arrows (grey) indicate a decrease in that process (after Field *et al.*, 1995).

Source: Reproduced with permission of John Wiley and Sons based on Field, C.B., Jackson, R.B. and Mooney H.A. 1995. Stomatal responses to increased CO_2: implications from the plant to the global scale. *Plant, Cell and Environment* 18: 1214–1225.

Vapour pressure deficit (VPD) is defined as the difference between the pressure exerted by water vapour in saturated air and the actual water vapour pressure. It is a measure of the 'drying power of air'. The ratio of CO_2 uptake to transpiration rate at the canopy scale appears to be strongly dependent on VPD (Barton *et al.*, 2011). Two negative feedbacks act to reduce the sensitivity of transpiration to leaf conductance. First, a reduction in leaf conductance leads to reductions in transpiration and thus to a drier **boundary layer**, which in turn 'increases the driving gradient for transpiration' (Field *et al.*, 1995), i.e. VPD. Also a reduction in transpiration means a reduction in the latent heat flux. A commensurate increase in **sensible heat** flux is needed to balance net radiation. An increase in sensible heat flux implies an increase in leaf temperature and thus an increase in the VPD between leaf and air, which in turn acts to increase transpiration. However, transpiration does not increase linearly with VPD, owing to the physiological response of stomata to VPD (i.e. stomatal conductance itself decreases with increasing VPD, an example of a positive feedback) (Oren *et al.*, 1999). Accounting for these canopy-scale feedbacks, it has been estimated that for many cropland locations a 40% decrease in stomatal conductance would translate into only a 10% decrease in evapotranspiration (Schulze *et al.*, 1994; Field *et al.*, 1995).

Continued

Other feedbacks which relate to responses of ecosystem processes to reduced stomatal conductance under elevated CO_2 concentrations need to be considered (**Figure 2.14**). For example, a reduction in transpiration will lead to increased soil moisture, which may lead to increases in river runoff (Gedney et al., 2006; Betts et al., 2007). Therefore, the physiological effect of future elevated atmospheric CO_2 concentration would be to increase future fresh-water availability. In water-limited regions increases in soil moisture can lead to reduced plant water stress and potentially to longer growing seasons. This will lead to increased annual **net primary production** (NPP), and increases in leaf area, and consequently to an increase in canopy conductance and transpiration. This represents a negative feedback, whereby the response of the system in growing more leaf area counteracts the initial leaf-level reductions in stomatal conductance and transpiration. Other feedbacks include soil moisture-induced changes in species composition, changes in decomposition rates and nutrient availability for plants, leading to changes in NPP and leaf area. A comprehensive account of feedbacks is given in Field et al. (1995) and references therein.

into soils. Heavy rainfall in wintertime, rather than percolating into soils, is lost as river runoff, and is therefore not available for farmers during the growing season.

With global warming, surface evaporation will increase (absorbing energy in the phase change from liquid to gas), supplying more water vapour into the atmosphere. In addition, the atmosphere can carry 7% more water vapour per degree of temperature increase. Increases in evaporation, together with the ability of the atmosphere to carry more water vapour, lead to an overall intensification of the hydrological cycle. Intensity of heavy rainfall is expected to increase and, while heavy rainfall is local in character, moisture will be gathered from far afield, on average drawing moisture from distances around four times the radius of the precipitating area (Trenberth et al., 2003). An extratropical cyclone with a radius of typically 800 km will draw moisture from 3200 km away, or 30° latitude (Trenberth et al., 2003). This implies less moisture to drive moderate rainfall elsewhere (Allan, 2011). Therefore, an increase in rainfall intensity in one region (e.g. equatorial tropics) implies reduction in duration and/or

frequency elsewhere (e.g. the sub-tropics) (Allan, 2011). Climate change will amplify existing regional differences in precipitation–evaporation, leading to an intensification of the global hydrological cycle (Allen and Ingram, 2002; Held and Soden, 2006).

However, as mentioned earlier, water fluxes and energy are linked, through latent heat exchange during phase change (evaporation at the surface, condensation in the troposphere), and therefore changes in global precipitation are tied to the surface and tropospheric energy budgets (Lambert and Allen, 2009). Water vapour is a powerful greenhouse gas, and clouds affect planetary albedo. Consequently, water vapour, other greenhouse gases and aerosols control precipitation through their effect on the surface and tropospheric energy budget, and the radiative cooling of the atmosphere. In fact, energy provides a stronger constraint to increases in global precipitation than does moisture availability (Lambert and Allen, 2009; Trenberth, 2011). As a consequence an overall more moderate 1–3% increase (rather than 7%) in global precipitation per degree of warming is expected (Held and Soden, 2006; Lambert and Allen, 2009; Lambert et al., 2008).

Rainfall recycling and deforestation

In Section 2.3 the concept of land–atmosphere coupling was introduced and how soil moisture can affect precipitation. In many areas of the world, particularly in continental locations, terrestrial precipitation relies on terrestrial evapotranspiration – rainfall recycling (van der Ent *et al.*, 2010, 2014). Major source regions for continental precipitation are shown in **Figure 2.15a**, and sink regions of this continentally evaporated water in **Figure 2.15b**.

a.

b.

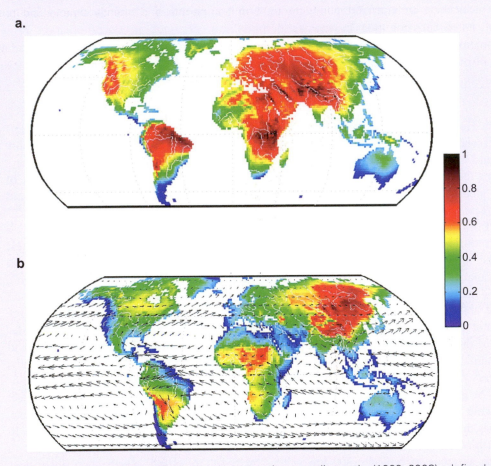

Figure 2.15 Rainfall recycling: a) Continental evaporation recycling ratio (1999–2008), defined as the fraction of evapotranspiration that returns as terrestrial precipitation; b) Continental precipitation recycling ratio (1999–2008), defined as the fraction of precipitation that comes from terrestrial evapotranspiration.
Source: van der Ent *et al.* (2014), under the CC BY 3.0 licence. Modifications were made to the external labelling.

Continued

In South America, moisture from the Amazon forest is important for precipitation over the large La Plata Basin covering southern Paraguay, southern Brazil, Uruguay and Argentina, and major population centres (e.g. Sao Paulo, Buenos Aires) (Lovejoy and Nobre, 2018). It is estimated that over one-third of moisture over the Amazon basin ($3400 km^3 yr^{-1}$) is transported 4000 km further south to distant basins (Nazareno and Laurance, 2015). However, recent studies suggest that Amazon deforestation acts to reduce evapotranspiration and thus diminish the Amazon's water pump (Aragão, 2012; Spracklen *et al.*, 2012), which can lead to amplified forest losses downwind (Zemp *et al.*, 2017), and disrupted moisture flows to distant southern basins (Bovolo *et al.*, 2018 – see video here: https://iopscience.iop.org/article/10.1088/1748-9326/aacf60). This represents a potential tipping point in the Earth system (Boers *et al.*, 2017; Lovejoy and Nobre, 2018). This topic has been of particular relevance given that south-east Brazil is home to 85 million people, and recently experienced its worst drought (2014–15) since 1930 (Escobar, 2015).

An interesting further twist is that elevated future CO_2 levels (see **Box 2.4**) may result in reduced forest evapotranspiration (through enhanced water-use efficiency), which can inadvertently lead to reduced rainfall recycling, and downwind drought (Chadwick *et al.*, 2017).

An increase in atmospheric moisture content further enhances global warming. This is known as *water vapour feedback*, and represents the most important feedback in the climate system. In the absence of feedbacks, the climate would warm by approximately one degree for a doubling of atmospheric CO_2 concentration. The water vapour feedback acting alone approximately doubles the warming due to CO_2 alone (IPCC, 2007a). One of the main uncertainties in climate science relates to the uncertainties in the various feedback mechanisms and their strength (see **Figure 2.16**).

2.5.2 Climate change impacts

With an increase in the water vapour content, rainfall intensity will increase. This will lead to more floods and droughts (Burke *et al.*, 2006). Extreme rainfall events will become more frequent than they are now. However, there may be longer intervals between (non-extreme) individual rainfall events. Climate model projections suggest future increases in precipitation, more intense rainfall, more droughts, and in general wet regions getting wetter and dry regions becoming drier (IPCC 2007a; **Figure 2.17**). This leads to an increase in flood risk, which will pressure physical infrastructure and impact upon water quality. IPCC (2007b) estimated that by the 2080s 20% of the world population will live in areas with increasing river flood risk.

Regionally, the tropics and high latitudes are expected to get wetter, the sub-tropics drier (i.e. enhanced precipitation–evapotranspiration deficit), with a poleward extension of the sub-tropics in both hemispheres (impacting upon the Mediterranean, South Australia, southwest USA). In mid-latitudes, changes in the seasonality of precipitation are expected, with drier summers (likely to impact upon agriculture and wildfire frequency and severity) and wetter winters (implications for flooding). Models project enhancements in the East African monsoon and summer monsoons in South and South East Asia. The main uncertainty in precipitation predictions among climate models is at the boundary of these climate zones.

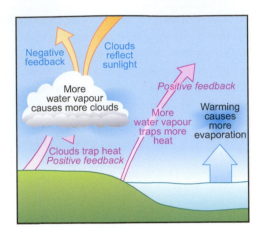

Figure 2.16 Schematic showing example feedbacks of enhanced water vapour in the atmosphere.

Another important consideration is whether the precipitation falls as rain or snow. With warming, the balance shifts to relatively more rain (consequently more river runoff) and earlier snowpack melt, and a general contraction of snow coverage. Snow effectively acts as freshwater storage, reducing seasonality in water input into soils and reducing wintertime runoff losses. This is likely to have significant implications for soil moisture content and freshwater availability in spring and summertime, for example increased requirements for crop irrigation. With global warming, water supplies from glaciers and snow are projected to decrease, reducing water availability in regions dependent on glacier and snowmelt. For South Asia this translates into a future decrease in mean water supply of −8.4%, −17.6%, and −19.6% for the upper Indus, Ganges and Brahmaputra, respectively, despite consideration of the partial offsets from projected increases in monsoon rains (Immerzeel *et al.*, 2010). Accounting for irrigation requirements for crops and population, the Brahmaputra and Indus basins

are most susceptible to future decreased flows, putting an estimated 60 million people at risk of food insecurity (Immerzeel *et al.*, 2010). Warming is likely to decrease **permafrost** (see Chapter 1) extent, challenging infrastructure in affected regions (ACIA, 2004). IPCC AR4 projects permafrost area to decrease by 20–35% by the mid-twenty-first century, and seasonal thaw depth to decrease by 30–50% by 2080.

The combined effects of future increases in precipitation with the competing effects of CO_2 on plant physiology and warming acting to decrease and increase transpiration, respectively, lead to projections of increased river runoff (Betts *et al.*, 2007). The IPCC Assessment Report 4, Working Group II, Chapter 3, 'Freshwater resources and their management', includes the following statements in the executive summary (IPCC, 2007b):

- Future river discharge will increase by 10 to 40% by mid-century at higher latitudes and in some wet tropical areas; decrease by 10 to 30% over some dry regions at mid-latitudes and dry tropics, due to decreases in rainfall and higher rates of evapotranspiration.
- Many semi-arid areas (e.g. the Mediterranean Basin, western United States, southern Africa and north-eastern Brazil) will suffer a decrease in water resources due to climate change.
- The negative impacts of climate change on freshwater systems outweigh its benefits. There will be increased annual runoff in some areas but increased precipitation variability and seasonal runoff shifts on water supply, water quality and flood risk.
- Increased temperatures will further affect the physical, chemical and biological properties of freshwater lakes and rivers, with predominantly adverse impacts on many individual freshwater species, community composition and water quality.

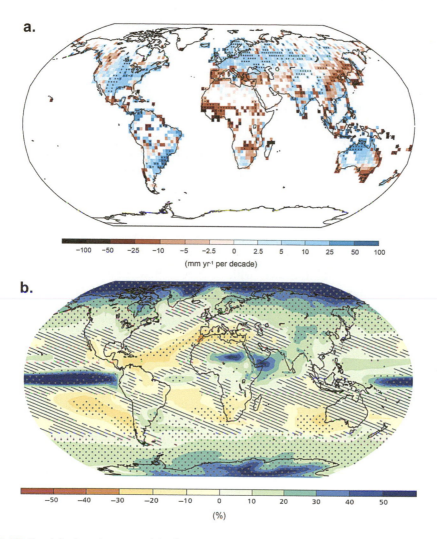

Figure 2.17 Precipitation changes with climate: a) observed change in annual precipitation over land 1951–2010, mm yr^{-1} per decade: black plus signs (+) indicate grid boxes where trends are significant at the 2-tailed 10% significance level; b) % change in average precipitation (1986–2005 to 2081–2100) from 39 global climate models for RCP8.5 future emissions scenario. Hatching indicates regions where the multi-model mean is small compared to natural internal variability (i.e. no change signal). Stippling indicates regions where the multi-model mean is large compared to natural internal variability and where at least 90% of models agree on the sign of change (large change and agreement among models).

Source: a) TFE.1, Figure 2; b) TFE.1, Figure 3 both from Stocker., T.F., Qin, D., Plattner, G.-K., Alexander, L.V., Allen, S.K., Bindoff, N.L., Bréon, F.-M., Church, J.A., Cubasch, U., Emori, S., Forster, P., Friedlingstein, P., Gillett, N., Gregory, M., Hartmann, D.L., Jansen, E., Kirtman, B., Knutti, R., Krishna Kumar, K., Lemke, P., Marotzke, J., Masson-Delmotte, V., Meehl, G.A., Mokhov, I.I., Piao, S., Ramaswamy, V., Randall, D., Rhein, M., Rojas, M., Sabine, C., Shindell, D., Talley, L.D., Vaughan, D.G. and Xie, S.-P. 2013: Technical Summary. In: *Climate Change 2013: The Physical Science Basis. Contribution of Working Group I to the Fifth Assessment Report of the Intergovernmental Panel on Climate Change* [Stocker, T.F., Qin, D., Plattner, G.-K., Tignor, M., Allen, S.K., Boschung, J., Nauels, A., Xia, Y., Bex, V. and Midgley, P.M. (eds)]. Cambridge University Press; Cambridge and New York.

Uncertainties in the magnitude of future climate change relate to uncertainties in future population growth and socio-economic projections, and differences in climate model sensitivity to fossil fuel emissions. The latter relates to uncertainties in the strength of feedbacks in the climate system. Also, although all climate models predict a future increase in global precipitation with warming, there is less consensus on the specific location of precipitation changes, not only the magnitude but even the sign of change differing among models. Projections of future land-use and land-cover changes vary markedly (Sitch *et al.*, 2005) and, together with uncertainty in changes in water demand and climate change, these provide major challenges for future water cycle and freshwater security assessments.

> ### REFLECTIVE QUESTION
>
> What are the major changes to the water cycle, including extreme events, that you would expect to see in the next 50 to 100 years as a result of climate change?

2.6 SUMMARY

The global water cycle is intimately linked to energy balance and transfers across the Earth for a number of reasons, including the latent heat transfer processes associated with evaporation and condensation and the fact that water vapour is an important greenhouse gas. There are large uncertainties in the estimates of the amounts of water being stored and transferred in parts of the global water cycle. There are also large variations between years, between inter-annual periods (e.g. due to ENSO), and between decadal periods (e.g. due to aerosols and land-use change). Future changes in climate, land use and water demand are likely to profoundly affect the global water cycle. There are multiple positive and negative feedbacks operating between factors such as increased CO_2 and plant physiology, aerosols, temperature, precipitation and evapotranspiration, changes in sea surface temperatures and atmospheric circulation patterns. The IPCC projects an intensification of the global water cycle, with future increases in precipitation and evaporation, and more intense rainfall, yet more droughts. Wet regions will get wetter and dry regions will get drier. The impacts of climate change include increases in annual river runoff, with changes in the seasonality of runoff in snow- and glacier-fed river basins, with implications for freshwater supply and food security for human populations and biodiversity.

FURTHER READING

Look at the latest IPCC reports available online for the latest scientific reports on climate change impacts on the water cycle: www.ipcc.ch/publications_and_data/publications_and_data_reports.shtml.

Collins, M., Knutti, R., Arblaster, J., Dufresne, J.-L., Fichefet, T., Friedlingstein, P., Gao, X., Gutowski, W.J., Johns, T., Krinner, G., Shongwe, M., Tebaldi, C., Weaver, A.J. and Wehner, M. 2013. Long-term climate change: projections, commitments and irreversibility. In: *Climate Change 2013: The Physical Science Basis. Contribution of Working Group I to the Fifth Assessment Report of the Intergovernmental Panel on Climate Change* [Stocker, T.F., Qin, D.,

Plattner, G.-K., Tignor, M., Allen, S.K., Boschung, J., Nauels, A., Xia, Y., Bex, V. and Midgley, P.M. (eds)]. Cambridge University Press; Cambridge and New York.

In the above read especially sections 12.4.5 – Change in the Water Cycle, p1074, and 12.4.6 Changes in the Cryosphere, p1087, and How Will the Earth's Water Cycle Change? p1084 in IPCC Working Group 1, Chapter 12, Long-Term Climate Change: Projections, Commitments and Irreversibility.

Seneviratne, S., Corti, T., Davin, E.L., Hirschi, M., Jaeger, E.B., Lehner, I., Orlowsky, B. and Teuling, A.J. 2010. Investigating soil moisture–climate interactions in a changing climate: a review. *Earth-Science Reviews* 99: 125–161.

This document provides a review of soil moisture–climate interactions.

Classic papers

Field, C.B., Jackson, R.B. and Mooney, H.A. 1995. Stomatal responses to increased CO_2: implications from the plant to the global scale. *Plant, Cell and Environment* 18: 1214–1225.

A comprehensive account of feedbacks on the water cycle of changes in plant responses to CO_2 concentrations.

Immerzeel, W.W., van Beek, L.P.H. and Bierkens, M.F.P. 2010. Climate change will affect the Asian water towers. *Science* 328: 1382–1385.

The authors investigate the importance of melt water (snow and glacier) from upstream areas on overall basin hydrology of five major South East Asian (Indus, Ganges, Brahmaputra, Yangtze and Yellow) rivers, which support over 20% of the global population. They then use modelling to project future seasonal and long-term changes in discharge for these basins.

Roderick, M.L. and Farquhar, G.D. 2002. The cause of decreased pan evaporation over the past 50 years. *Science* 298: 1410–1411.

Another example of a feedback effect – here increased cloud cover and aerosols are shown to be linked to reduced measured evaporation.

Stott, P.A., Christidis, N., Otto, F.E., Sun, Y., Vanderlinden, J.P., Van Oldenborgh, G.J., Vautard, R., Von Storch, H., Walton, P., Yiou, P. and Zwiers, F.W. 2016. Attribution of extreme weather and climate-related events. *Wiley Interdisciplinary Reviews: Climate Change* 7: 23–41.

The authors discuss the new science of extreme climate event attribution.

PROJECT IDEAS

- For your local river basin produce a list of all possible impacts and feedback effects of climate change on the water cycle. Use IPCC reports or local climate models for your region to inform your work. What mitigation measures might be adopted in the river basin?

- Produce a list of extreme events that have impacted upon the water cycle in your region over the past 20 years. Investigate the evidence as to the type of climate variability that might have been associated with them.

- What was the impact of the heatwave of summer 2018 on UK water resources? You could use available river flow and precipitation data from stations across the UK.

- Investigate the impact of ENSO 2015 on the tropical water cycle and what differences occurred across continents.

- The Andean region in South America is strongly dependent on water held in the montane forest for freshwater provision and hydroelectricity generation. However, large areas of the Andes have been deforested since the 1950s: a) investigate the impact of deforestation in the Andean region on regional water resources and b) investigate the impact of ENSO on regional water resources.

REFERENCES

ACIA. 2004. *Impacts of a Warming Arctic: Arctic Climate Impact Assessment.* Cambridge University Press; Cambridge.

Ackerley, D., Booth, B.B.B., Knight, S.H.E., Highwood, E.J., Frame, D.J., Allen, M.R. and Rowell, D.P. 2011. Sensitivity of twentieth-century Sahel rainfall to sulfate aerosol and CO_2 forcing. *Journal of Climate* 24: 4999–5014.

Aguado, E. and Burt, J.E. 2015. *Understanding Weather and Climate.* 7th edition. Pearson; Harlow.

Ainsworth, E.A. and Rodgers, A. 2007. The response of photosynthesis and stomatal conductance to rising [CO_2]: mechanisms and environmental interactions. *Plant, Cell and Environment* 30: 258–270.

Allan, R.P. 2011. Human influence on rainfall. *Nature* 470: 344–345.

Allan, J.D. and Castillo, M.M. 2007. *Stream Ecology Structure and Function of Running Waters.* 2nd edition. Chapman and Hall; New York.

Allen, M. and Ingram, W. 2002. Constraints on future changes in climate and the hydrologic cycle. *Nature* 419: 224–232.

Aragão, L.E. 2012. Environmental science: the rainforest's water pump. *Nature* 489: 217–218.

Barnett, T.P., Adam, J.C. and Lettenmaier, D.P. 2005. Potential impacts of a warming climate on water availability in snow-dominated regions. *Nature* 438: 303–309.

Barnett, T.P. *et al.* 2008. Human-induced changes in the hydrology of the western United States. *Science* 319: 1080. doi:10.1126/science.1152538.

Barton, C.V.M. *et al.* 2011. Effects of elevated atmospheric [CO_2] on instantaneous transpiration efficiency at leaf and canopy scales in *Eucalyptus saligna. Global Change Biology* 18: 585–595.

Betts, R.A., Boucher, O., Collins, M., Cox, P.M., Falloon, P.D., Gedney, N., Hemming, D.L., Huntingford, C., Jones, C.D., Sexton, D.M.H. and Webb, M.J. 2007. Projected increase in continental runoff due to plant responses to increasing carbon dioxide. *Nature* 448: 1037–1041.

Biasutti, M. 2011. Atmospheric science: a man-made drought. *Nature Climate Change* 1: 197–198.

Biemans, H., Hutjes, R.W.A., Kabat, P., Strengers, B.J., Gerten, D. and Rost, S. 2009. Effects of precipitation uncertainty on discharge calculations for main river basins. *Journal of Hydrometeorology* 10: 1011–1025.

Bjerknes, J. 1969. Atmospheric teleconnection from the equatorial Pacific. *Monthly Weather Review* 97: 163–172.

Blöschl, G., Hall, J., Parajka, J., Perdigao, R.A.P., Merz, B., Arheimer, B., Aronica, G.T., Bilibashi, A., Bonacci, O., Borga, M., Čanjevac, I., Castellarin, A., Chirico, G.B., Claps, P., Fiala, K., Frolova, N., Gorbachova, L., Gül, A., Hannaford, J., Harrigan, S., Kireeva, M., Kiss, A., Kjeldsen, T.R., Kohnová, S., Koskela, J.J., Ledvinka, O., Macdonald, N., Mavrova-Guirguinova, M., Mediero, L., Merz, R., Molnar, P., Montanari, A., Murphy, C., Osuch, M., Ovcharuk, V., Radevski, I., Rogger, M., Salinas, J.L., Sauquet, E., Šraj, M., Szolgay, J., Viglione, A., Volpi, E., Wilson, D., Zaimi, K. and Živković, N. 2017. Changing climate shifts timing of European floods. *Science* 357: 588–590.

Boers, N., Marwan, N., Barbosa, H.M. and Kurths, J. 2017. A deforestation-induced tipping point for the South American monsoon system. *Scientific Reports* 7: 41489.

Booth, B.B.B., Dunstone, N.J., Halloran, P.R., Andrews, T. and Bellouin, N. 2012. Aerosols implicated as a prime driver of twentieth-century North Atlantic climate variability. *Nature* 484: 228–232.

Bovolo, C.I., Wagner, T., Parkin, G., Hein-Griggs, D., Pereira, R. and Jones, R. 2018. The Guiana Shield rainforests—overlooked guardians of South American climate. *Environmental Research Letters* 13: 074029.

Brown, R., Derksen, C. and Wang, L. 2010. A multi-data set analysis of variability and change in Arctic spring snow cover extent,

1967–2008. *Journal of Geophysical Research* 115: D16111. doi:10.1029/2010JD013975.

Burke, E.J., Brown, S.J. and Christidis, N. 2006. Modeling the recent evolution of global drought and projections for the twenty-first century with the Hadley Centre climate model. *Journal of Hydrometeorology* 7: 1113–1125.

Caminade, C. and Terray, L. 2010. Twentieth century Sahel rainfall variability as simulated by the ARPEGE AGCM, and future changes. *Climate Dynamics* 35 (Special Issue): 75–94.

Chadwick, R., Douville, H. and Skinner, C.B. 2017. Timeslice experiments for understanding regional climate projections: applications to the tropical hydrological cycle and European winter circulation. *Climate Dynamics* 49: 3011–3029.

Charney, J.G. 1975. Dynamics of deserts and drought in Sahel. *Quarterly Journal of the Royal Meteorological Society* 101: 193–202.

Ciais, P., Reichstein, M., Viovy, N., Granier, A., Ogee, J., Allard, V., Aubinet, M., Buchmann, N., Bernhofer, C., Carrara, A., Chevallier, F., De Noblet, N., Friend, A.D., Friedlingstein, P., Grünwald, T., Heinesch, B., Keronen, P., Knohl, A., Krinner, G., Loustau, D., Manca, G., Matteucci, G., Miglietta, F., Ourcival, J.M., Papale, D., Pilegaard, K., Rambal, S., Seufert, G., Soussana, J.F., Sanz, M.J., Schulze, E.D., Vesala, T. and Valentini, R. 2005. Europewide reduction in primary productivity caused by the heat and drought in 2003. *Nature* 437: 529–533.

Coghlan, C. 2002. El Niño – causes, consequences and solutions. *Weather* 57: 209–215.

Collins, M., Knutti, R., Arblaster, J., Dufresne, J.-L., Fichefet, T., Friedlingstein, P., Gao, X., Gutowski, W.J., Johns, T., Krinner, G., Shongwe, M., Tebaldi, C., Weaver A.J. and Wehner, M. 2013. Long-term climate change: projections, commitments and irreversibility. In: *Climate Change 2013: The Physical Science Basis. Contribution of Working Group I to the Fifth Assessment Report of the Intergovernmental Panel on Climate Change* [Stocker, T. F., Qin, D., Plattner, G.-K., Tignor, M., Allen, S.K., Boschung, J., Nauels, A., Xia, Y.,

Bex, V. and Midgley, P.M. (eds)]. Cambridge University Press; Cambridge and New York.

Condon, A.G., Richards, R.A., Rebetzke, G.J. and Farquhar, G.D. 2002. Improving intrinsic water-use efficiency and crop yield. *Crop Science* 42: 122–131.

Cook, B.I., Miller, R.L. and Seager, R. 2008. Dust and sea surface temperature forcing of the 1930s 'Dust Bowl' drought. *Geophysical Research Letters* 35: L08710. doi:10.1029/2008GL033486.

Cook, B.I., Miller, R.L. and Seager, R. 2009. Amplification of the North American 'Dust Bowl' drought through human-induced land degradation. *Proceedings of the National Academy of Sciences* 106: 4997–5001.

Cook, E.R., Woodhouse, C.A., Eakin, C.M., Meko, D.M. and Stahle, D.W. 2004. Long-term aridity changes in the western United States. *Science* 306: 1015–1018.

Coumou, D., Lehmann, J. and Beckmann, J. 2015. The weakening summer circulation in the Northern Hemisphere mid-latitudes. *Science* 348: 324–327.

Cowan, I. and Farquhar, G.D. 1977. Stomatal function in relation to leaf metabolism and environment. *Symposia of the Society for Experimental Biology* 31: 471–505.

Dong, B., Sutton, R. and Woollings, T. 2013. The extreme European summer. *Bulletin of the American Meteorological Society* 94: S28–S32.

Druyan, L.M. 2011. Studies of 21st-century precipitation trends over West Africa. *International Journal of Climatology* 31: 1415–1424.

Escobar, H. 2015. Drought triggers alarms in Brazil's biggest metropolis. *Science* 347: 812.

Fedoroff, N.V., Battisti, D.S., Beachy, R.N., Cooper, P.J., Fischhoff, D.A., Hodges, C.N., Knauf, V.C., Lobell, D., Mazur, B.J., Molden, D., Reynolds, M.P., Ronald, P.C., Rosegrant, M.W., Sanchez, P.A., Vonshak, A. and Zhu, J.K. 2010. Radically rethinking agriculture for the twenty-first century. *Science* 327: 833–834.

Field, C.B., Jackson, R.B. and Mooney, H.A. 1995. Stomatal responses to increased CO_2: implications from the plant to the global scale. *Plant, Cell and Environment* 18: 1214–1225.

Garbrecht, J.D. and Rossel, F.E. 2002. Decade-scale precipitation increase in Great Plains at end of 20th century. *Journal of Hydrologic Engineering* 7: 64–75.

Gedney, N., Cox, P.M., Betts, R.A., Boucher, O., Huntingford, C. and Stott, P.A. 2006. Detection of a direct carbon dioxide effect in continental river runoff records. *Nature* 439: 835–838.

Gergis, J.L. and Fowler, A.M. 2009. A history of ENSO events since AD 1525: implications for future climate change. *Climatic Change* 92: 343–387.

Gerten, D., Schaphoff, S., Haberlandt, U., Lucht, W. and Sitch, S. 2004. Terrestrial vegetation and water balance – hydrological evaluation of a dynamic global vegetation model. *Journal of Hydrology* 286: 249–270.

Gerten, D., Rost, S., von Bloh, W. and Lucht, W. 2008. Causes of change in twentieth century global river discharge. *Geophysical Research Letters* 35: L20405. doi:10.1029/2008GL035258.

Haddeland, I., Clark, D.B., Franssen, W., Ludwig, F., Voß, F., Arnell, N.W., Bertrand, N., Best, M., Folwell, S., Gerten, D., Gomes, S., Gosling, S.N.K., Naota Hanasaki, S., Harding, R., Heinke, J., Kabat, P., Koirala, S., Oki, T., Polcher, J., Stacke, T., Viterbo, P., Weedon, G.P. and Yeh, P. 2011. Multi-model estimate of the global water balance: setup and first results. *Journal of Hydrometeorology* 12: 869–884.

Harding, R., Best, M., Blyth, E., Hagemann, S., Kabat, P., Tallaksen, L.M., Warnaars, T., Wiber, D., Weedon, G.P., Van Lanen, H., Ludwig, F. and Haddeland, I. 2011. WATCH: current knowledge of the terrestrial global water cycle. *Journal of Hydrometeorology* 12: 1149–1156.

Haywood, J.M., Jones, A., Bellouin, N. and Stephenson, D. 2013. Asymmetric forcing from stratospheric aerosols impacts Sahelian rainfall. *Nature Climate Change* 3: 660–665.

Held, I.M. and Soden, B.J. 2006. Robust responses of the hydrological cycle to global warming. *Journal of Climate* 19: 5686–5699.

Held, I.M., Delworth, T.L., Lu, J., Findell, K.L. and Knutson, T.R. 2005. Simulation of Sahel drought in the 20th and 21st centuries. *Proceedings of the National Academy of Science* 102: 17891–17896.

Immerzeel, W.W., van Beek, L.P.H. and Bierkens, M.F.P. 2010. Climate change will affect the Asian water towers. *Science* 328: 1382–1385.

IPCC. 2001. *Climate Change 2001: The scientific basis. Contribution of Working Group I to the Third Assessment Report of the Intergovernmental Panel on Climate Change* [Houghton, J.T. *et al.* (eds)]. Cambridge University Press; Cambridge.

IPCC. 2007a. *Climate Change 2007: The Physical Science Basis. Contribution of Working Group I to the Fourth Assessment Report of the Intergovernmental Panel on Climate Change* [Solomon, S. *et al.* (eds)]. Cambridge University Press; Cambridge.

IPCC. 2007b. *Climate Change 2007: Impacts, Adaptation and Vulnerability. Contribution of Working Group II to the Fourth Assessment Report of the Intergovernmental Panel on Climate Change* [Parry, M.L. *et al.* (eds)]. Cambridge University Press; Cambridge.

IPCC. 2013. *Climate Change 2013: The Physical Science Basis. Contribution of Working Group I to the Fifth Assessment Report of the Intergovernmental Panel on Climate Change* [Stocker, T.F., Qin, D., Plattner, G.-K., Tignor, M., Allen, S.K., Boschung, J., Nauels, A., Xia, Y., Bex, V. and Midgley, P.M. (eds)]. Cambridge University Press; Cambridge and New York.

Jonas, P.R. 1994. Why does it rain? *Weather* 49: 258–260.

Jung, M., Reichstein, M., Ciais, P., Seneviratne, S. I., Sheffield, J., Goulden, M.L., Bonan, G., Cescatti, A., Chen, J., de Jeu, R., Dolman, A. J., Eugster, W., Gerten, D., Gianelle, D., Gobron, N., Heinke, J., Kimball, J., Law, B.E., Montagnani, L., Mu, Q., Mueller, B., Oleson, K., Papale, D., Richardson, A.D., Roupsard, O., Running, S., Tomelleri, E., Viovy, N., Weber, U., Williams, C., Wood, E., Zaehle, S. and Zhang, K. 2010. Recent decline in the global land evapotranspiration trend due to limited moisture supply. *Nature* 467: 951–954.

Kaser, G., Hardy, D.R., Mölg, T., Bradley, R.S. and Hyera, T.M. 2004. Modern glacier retreat on Kilimanjaro as evidence of climate change: observations and facts. *International Journal of Climatology* 24: 329–339.

Kiehl, J.T. and Trenberth, K.E. 1997. Earth's annual global mean energy budget. *Bulletin of the American Meteorological Society* 78: 197–208.

Koster, R.D., Dirmeyer, P.A., Guo, Z., Bonan, G., Chan, E., Cox, P., Gordon, C.T., Kanae, S., Kowalczyk, E., Lawrence, D., Liu, P., Lu, C. H., Malyshev, S., McAvaney, B., Mitchell, K., Mocko, D., Oki, T., Oleson, K., Pitman, A., Sud, Y.C., Taylor, C.M., Verseghy, D., Vasic, R., Xue, Y. and Yamada, T. 2004. Regions of strong coupling between soil moisture and precipitation. *Science* 305: 1138–1140.

Lambert, F.H. and Allen, M.R. 2009. Are changes in global precipitation constrained by the tropospheric energy budget? *Journal of Climate* 22: 499–517.

Lambert, F.H., Stine, A.R., Krakauer, N.Y. and Chiang, J.C.H. 2008. How much will precipitation increase with global warming? *EOS Transactions American Geophysical Union* 89: 193.

Lovejoy, T.E. and Nobre, C. 2018. Amazon tipping point. *Scientific Advances* 4: eaat2340.

Lytle, D.A. and Poff, N.L. 2004. Adaptation to natural flow regimes. *Trends in Ecology and Evolution* 19: doi:10.1016/j.tree.2003.10.002.

Mann, M.E., Rahmstorf, S., Kornhuber, K., Steinman, B.A., Miller, S.K. and Coumou, D. 2017. Influence of anthropogenic climate change on planetary wave resonance and extreme weather events. *Scientific Reports* 7, article number: 54242.

Mayes, J. 2008. Editorial: special issue – the wet summer of 2007 in the UK. *Weather* 63: 251–252.

Medlyn, B.E., Barton, C.V.M., Broadmeadow, M.S.J., Ceulemans, R., De Angelis, P., Forstreuter, M., Freeman, M., Jackson, S.B., Kellomäki, S., Laitat, E., Rey, A., Roberntz, P., Sigurdsson, B.D., Strassemeyer, J., Wang, K., Curtis, P.S. and Jarvis, P.G. 2001. Stomatal conductance of forest species after long-term exposure to elevated CO_2 concentration: a synthesis. *New Phytologist* 149: 247–264.

Min, S-K., Zwiers, F.W. and Hegerl, G.C. 2011. Human contribution to more-intense precipitation extremes. *Nature* 470: 378–381.

Moron, V. and Ward, M.N. 1998. ENSO teleconnections with climate variability in the European and African sectors. *Weather* 53: 287–295.

Nazareno, A.G. and Laurance, W.F. 2015. Brazil's drought: beware deforestation. *Science* 347: 1427–1427.

Oerlemans, J. 2005. Extracting a climate signal from 169 glacier records. *Science* 308: 657–677.

Ogi, M., Yamazaki, K. and Tachibana, Y. 2005. The summer northern annular mode and abnormal summer weather in 2003. *Geophysical Research Letters* 32: L04706. doi:10.1029/2004GL021528.

Oki, T. and Kanae, S. 2006. Global hydrological cycles and world water resources. *Science* 313: 1068–1072.

Oren, R., Sperry, J.S., Katul, G.G., Pataki, D.E., Ewers, B.E., Philipps, N. and Schäfer, K.V.R. 1999. Survey and synthesis of intra- and interspecific variation in stomatal sensitivity to vapour pressure deficit. *Plant, Cell and Environment* 22: 1515–1526.

Pall, P., Allen, M., Aina, T., Stone, D., Stott, P., Nozawa, T., Hilberts, A.G.J. and Lohmann, D. 2011. Anthropogenic greenhouse gas contribution to flood risk in England and Wales in autumn 2000. *Nature* 470: 382–386.

Peterson, B.J., Holmes, R.M., McClelland, J.W., Vörösmarty, C.J., Lammers, R.B., Shiklomanov, A.I., Shiklomanov, I.A. and Rahmstorf, S. 2002. Increasing river discharge to the Arctic Ocean. *Science* 298: 2171–2173.

Pielke Jr, R.A. and Landsea, C.N. 1999. La Niña, El Niño, and Atlantic hurricane damages in the United States. *Bulletin of the American Meteorological Society* 80: 2027–2033.

Prentice, I.C., Farquhar, G.D., Fasham, M.J.R., Goulden, M.L., Heimann, M., Jaramillo, V.J., Kheshgi, H.S., Le Quéré, C., Scholes, R.J. and Wallace, D.W.R. 2001. The carbon cycle and atmospheric carbon dioxide. In: Houghton, J. T. *et al.* (eds) *Climate Change 2001: The Scientific Basis*. Cambridge University Press; Cambridge.

Priestley, C.H.B. and Taylor, R.J. 1972. On the assessment of surface heat flux and evaporation using large-scale parameters. *Monthly Weather Review* 100: 81–92.

Prior, J. and Beswick, M. 2007. The record-breaking heat and sunshine of July 2006. *Weather* 62: 174–182.

Ramanathan, V., Crutzen, P.J., Kiehl, J.T. and Rosenfeld, D. 2001. Aerosols, climate, and the hydrological cycle. *Science* 294: 2119–2124.

Rodell, M., Velicogna, I. and Famiglietti, J.S. 2009. Satellite-based estimates of groundwater depletion in India. *Nature* 460: 999–1002.

Roderick, M.L., Rotstayne, L.D., Farquahar, G.D. and Hobbins, M.T. 2007. On the attribution of changing pan evaporation. *Geophysical Research Letters* 34: L17403. doi:10.1029/2007GL031166.

Schaller, N., Kay, A.L., Lamb, R., Massey, N.R., van Oldenborgh, G.J., Otto, F.E.L., Sparrow, S.N., Vautard, R., Yiou, P., Ashpole, I., Bowery, A., Crooks, S.M., Haustein, K., Huntingford, C., Ingram, W.J., Jones, R.G., Legg, T., Miller, J., Skeggs, J., Wallom, D., Weisheimer, A., Wilson, S., Stott, P.A. and Allen, M.R. 2016. Human influence on climate in the 2014 southern England winter floods and their impacts. *Nature Climate Change* 6: 627–634.

Schubert, S.D., Suarez, M.J., Pegion, P.J., Koster, R.D. and Bacmeister, J.T. 2004. On the cause of the 1930s Dust Bowl. *Science* 303: 1855–1859.

Schulze, E.-D., Kelliher, F.M., Körner, C., Lloyd, J. and Leuning, R. 1994. Relationship among maximum stomatal conductance, ecosystem surface conductance, carbon assimilation rate, and plant nitrogen nutrition: a global ecology scaling exercise. *Annual Review of Ecology and Systematics* 25: 629–660.

Seneviratne, S.I., Lüthi, D., Litschi, M. and Schär, C. 2006. Land–atmosphere coupling and climate change in Europe. *Nature* 443: 205–209.

Seneviratne, S., Corti, T., Davin, E.L., Hirschi, M., Jaeger, E.B., Lehner, I., Orlowsky, B. and Teuling, A.J. 2010. Investigating soil moisture–climate interactions in a changing climate: a review. *Earth-Science Reviews* 99: 125–161.

Singh, S.P., Bassignana-Khadka, I., Karky, B.S. and Sharma, E. 2011. *Climate Change in the Hindu Kush-Himalayas: The State of Current Knowledge*. ICIMOD; Kathmandu.

Sitch, S., Brovkin, V., von Bloh, W., Van Vuuren, D., Eickhout, B. and Ganopolski, A. 2005. Impacts of future land cover on atmospheric CO_2 and climate. *Global Biogeochemical Cycles* 19: GB2013. doi:10.1029/2004GB002311.

Spracklen, D.V., Arnold, S.R. and Taylor, C.M. 2012. Observations of increased tropical rainfall preceded by air passage over forests. *Nature* 489: 282–285.

Stevens, A. 2011. Introduction to the basic drivers of climate. *Nature Education Knowledge* 2: 6.

Stott, P.A., Christidis, N., Otto, F.E., Sun, Y., Vanderlinden, J.P., Van Oldenborgh, G.J., Vautard, R., Von Storch, H., Walton, P., Yiou, P. and Zwiers, F.W. 2016. Attribution of extreme weather and climate-related events. *Wiley Interdisciplinary Reviews: Climate Change* 7: 23–41.

Tedesco, M., Brodzik, M., Armstrong, R., Savoie, M. and Ramage, J. 2009. Pan Arctic terrestrial snowmelt trends (1979–2008) from spaceborne passive microwave data and correlation with the Arctic Oscillation. *Geophysical Research Letters* 36: L21402. doi:10.1029/2009GL039672.

Teuling, A.J. *et al.* 2010. Contrasting response of European forest and grassland energy exchange to heatwaves. *Nature Geosciences* 3: 722–727.

Trenberth, K.E. 2011. Changes in precipitation with climate change. *Climate Research* 47: 123–138.

Trenberth, K.E., Dai, A., Rasmussen, R.M. and Parsons, D.B. 2003. The changing character of precipitation. *Bulletin of the American Meteorological Society* 84: 1205–1217.

Trenberth, K.E., Fasullo, J.T. and Kiehl, J. 2009. Earth's global energy budget. *Bulletin of the American Meteorological Society* 90: 311–323.

van der Ent, R.J., Wang-Erlandsson, L., Keys, P.W. and Savenije, H.H.G. 2014. Contrasting roles of interception and transpiration in the hydrological cycle – Part 2: moisture recycling, *Earth System Dynamics* 5: 471–489.

van der Ent, R.J., Savenije, H.H., Schaefli, B. and Steele-Dunne, S.C. 2010. Origin and fate of atmospheric moisture over continents. *Water Resources Research* 46. doi: 10.1029/2010WR009127.

Vörösmarty, C.J., McIntyre, P.B., Gessner, M.O., Dudgeon, D., Prusevich, A., Green, P., Glidden, S., Bunn, S.E., Sullivan, C.A., Liermann, C.R. and Davies, P.M. 2010. Global threats to human water security and river biodiversity. *Nature* 467: 555–562.

Woodhouse, C.A. and Overpeck, J.T. 1998. 2000 years of drought variability in the central United States. *Bulletin of the American Meteorological Society* 79: 2693–2714.

Woodhouse, C.A., Meko, D.M., MacDonald, G. M., Stahle, D.W. and Cook, E.R. 2010. A 1,200-year perspective of twenty-first century drought in southwestern North America. *Proceedings of the National Academy of Sciences* 107: 21283–21288.

Young, M.V. 1994a. Depressions and anticyclones: part 1. *Weather* 49: 306–311.

Young, M.V. 1994b. Depressions and anticyclones: part 2. Life cycles and weather characteristics. *Weather* 49: 362–370.

Zemp, D.C., Schleussner, C.F., Barbosa, H.M., Hirota, M., Montade, V., Sampaio, G., Staal, A., Wang-Erlandsson, L. and Rammig, A. 2017. Self-amplified Amazon forest loss due to vegetation-atmosphere feedbacks. *Nature Communications*, 8: 14681.

Surface water hydrology

Joseph Holden

3.1 INTRODUCTION

The way water moves across and through a landscape is very important for determining river flow and water quality, and even the evolution of landscapes themselves. It is common to try to understand water movements across landscapes within topographically confined units, often known as **river basins**. It is possible to create a water budget for a river basin, whereby we try to measure the inputs, outputs and stores of water for the system and understand how efficient the system is at turning rainfall into river flow. Depending on the environmental conditions, almost 100% of **precipitation** may reach the river, or as a little as zero. The rate at which water is delivered to the ground surface and the pathways and speed at which water travels across and through a landscape may also be impacted upon by climate change and land management activity, and these factors may affect flood risk and water resources. It is therefore important to understand the flow pathways for water across river basins and how environmental change may impact upon these processes. There are several types of flooding, of which water pouring out over riverbanks is only one, and so management strategies for coping with floods need to be well thought through. In addition, river channels themselves tend to be naturally dynamic, changing course, size and shape through time and migrating across landscapes. Management of river channels has traditionally tried to tie down the channels so that their movements are restricted in order to protect infrastructure or landholdings. However, such modifications impact upon the natural processes within that area and upstream and downstream, resulting in knock-on effects on biodiversity, sediment dynamics and water flows. Hence many river restoration projects are now trying to reverse such negative management impacts and require a more sensitive understanding of channel sediment and water dynamics. This chapter deals with the above topics in order to provide an understanding of surface water hydrology to improve the basis for management decisions.

3.2 HYDROLOGICAL PATHWAYS

3.2.1 Drainage basins

Drainage basins, also known as **watersheds** or **catchments**, are defined by the upslope topographic area which drains to a single point on a river or lake. Hypothetically, if there was no water loss to groundwater or back to the atmosphere, all rainfall or snowmelt in a river basin should flow downslope to that single exit point. A drainage basin can vary in scale from a few hillslopes to the size of the Amazon basin at 6,160,000 km^2. While the Amazon's seemingly vast river basin area amounts to only 2% of the Earth's land mass, the Amazon carries around one-fifth of the world's annual river discharge because the majority of its area rests in a tropical region with plentiful rainfall. A large basin will contain, nested within it, many smaller basins (**Figure 3.1**). According to the Strahler

(1952) definition, a first-order basin is one where the stream has only one main stem. A second-order basin occurs when two first-order basins combine. A third-order basin occurs when two second-order basins combine, and so on. The lower reaches of the Mississippi can be considered to be a tenth-order river and hence sit within a tenth-order river basin, and the lower reaches of the Amazon constitute a twelfth-order river. Drainage basin area is determined using a topographic map (e.g. showing contour lines) that shows the drainage divide where water on one side of the divide flows down into one river and on the other side of the divide it will flow into a different river. Following the exact route of the drainage divide demarcates the boundary of the river basin. Therefore the area of the basin can be calculated from a map. Many software packages (such as ArcGIS) can automatically calculate the basin area for any point on a river if the software contains topographic data and stream layout data. If you are able to explore an online package such as Google Earth, they have settings that allow the user to view the landscape from oblique angles. This means you can better visualise the topography of the landscape and you can explore the nature of different drainage basins all over the world from your computer screen.

3.2.2 Infiltration-excess overland flow

Within a drainage basin, precipitation may land on the ground surface or it may be intercepted by vegetation. Some of the intercepted water may evaporate before it reaches the ground. The water that reaches the land surface may infiltrate (see **infiltration**) into the soil or some of it may pond up and run over the land surface as overland flow. Infiltration is affected by the

Figure 3.1 Nested drainage basins.

Legend:
- 1st order
- 2nd order
- 3rd order
- 4th order
- 5th order
- Contour
- Drainage divide

enter rivers more quickly than water moving below the ground surface. Therefore, if a drainage basin has a large proportion of land with a low **infiltration capacity** (maximum rate at which water will enter the soil under plentiful water supply), then it is likely that the **river discharge** will increase rapidly in response to rainfall.

The infiltration capacity of a soil tends not to be steady (Holden, 2005a). Soils are made up of solid particles between which there are spaces. These pore spaces contain some air and, even in the driest deserts, also contain a small amount of water bound to the edges of soil particles, known as **hygroscopic water**. As water infiltrates and then **percolates** through a soil, the pore spaces can fill with water. Once the pore spaces are full of water, the soil is saturated. As the pore spaces fill up, the infiltration capacity decreases and when the infiltration capacity drops below the rate of rainfall supply, then some of the water will start to pond up on the surface and flow (**Figure 3.2**). This process is called **infiltration-excess overland flow** (also known as Hortonian overland flow). This type of flow is uncommon in many temperate areas except in urban locations, along roads and paths or perhaps along compacted soils in arable fields, created by tractor wheels or overgrazing by animals. Infiltration-excess overland flow may become more common with climate change if rainfall intensities increase due to warming or if surface crusts form after long, dry spells. Infiltration-excess overland flow is more common in semi-arid regions where soil surface crusts have developed and rainfall rates can be rapid. It is also more likely in areas where the ground surface is often frozen. Soils with a deep litter layer, such as those within tropical rainforests, tend to have large infiltration capacities and so, despite high rainfall intensities, infiltration-excess overland flow is restricted. Only

topography, vegetation cover, soil texture, soil structure, the amount and connectivity of soil **pore spaces** and surface compaction. The **infiltration rate**, which is a measure of the volume of water passing into the soil per unit area per unit time, is important because it determines whether all, part or none of the water reaching the ground surface will enter the soil or will move over the land surface as overland flow. Water moving as overland flow generally tends to some parts of a river basin produce infiltration-excess overland flow at any given time and this is known as the **partial contributing area concept** (Betson, 1964). The areas producing infiltration-excess overland flow tend to be the same from storm to storm.

3.2.3 Saturation-excess overland flow

When the soil is saturated, the **water table** (the highest point below which the soil or rock is saturated) is at the surface. Water can therefore flow out of the soil and run out over the surface, producing **saturation-excess overland flow**. If it is raining, then the additional rainwater will find it difficult to enter the soil if it is saturated and so saturation-excess overland flow can be a mix of fresh rainwater and water that has been within the soil for some time. Saturation-excess overland flow is common in shallow soils or at the bottom of hillslopes, where water running through the soil will collect (**Figure 3.3**). In these zones the soil is kept saturated for longer, due to the delivery of water from a large upslope area and low gradient. Water can return to the surface at this point after travelling within the soil, which is why it is sometimes referred to as 'return flow'. The chemistry of saturation-excess overland flow is more affected by the soil properties than is the chemistry of infiltration-excess overland flow, as the former will have had more contact with the surrounding soil (see Chapter 4). Saturation-excess overland flow can occur long after it has stopped raining.

The parts of a hillslope or river basin that produce saturation-excess overland flow will vary during an individual rainfall event. If the system starts off quite dry, then during any following rainfall perhaps only a tiny part of the basin will generate saturation-excess overland flow. However, if rainfall continues for some time, then, as more of the soil becomes saturated, especially in the valley bottoms, a larger area of the landscape will produce saturation-excess overland flow. The source areas for saturation-excess overland flow can also vary markedly with season, with larger areas tending to produce saturation-excess overland flow during cooler, wetter months.

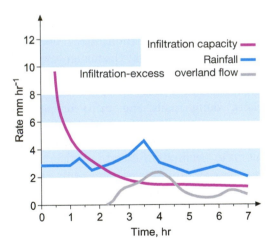

Figure 3.2 An example gentle rainfall event resulting in a reduction in the infiltration capacity of the soil over time and the subsequent production of infiltration-excess overland flow. Note that overland flow is not produced at the immediate point at which rainfall rates exceed infiltration capacity but is very slightly lagged as it takes time for water to pond in small depressions and overflow.

Figure 3.3 Production of saturation-excess overland flow.

This variability in the parts of the landscape that act as source areas for saturation-excess overland flow is known as the 'variable source area concept' (Hewlett and Hibbert, 1967).

3.2.4 Throughflow

Water moving through soils or rocks is called throughflow or interflow. This flow of water through the ground can maintain river flow during dry periods (the baseflow of the river). The pathway that water takes through the ground affects how quickly water can get to a river channel and, therefore, the typical nature of the response of a river to rainfall. Where water moves through the small pore spaces within a rock or soil it is known as matrix flow. Where it moves through larger connected pore spaces such as cracks or animal tunnels it is known as macropore flow. Water moving through macropores tends to travel at faster speeds than matrix flow.

Most soils do not remain saturated for long, as gravity drains some water out of the pore spaces. However, once gravity has played its part, there is still water left within the soil. This is because of the combined attraction of the water molecules to each other and the water to the soil particles. Water that is held in soil against the force of gravity is known as capillary water and it will move within the soil from wet to dry areas. This means that water can move upslope and capillary action is how plants can obtain water from soil. Imagine you hold some tissue paper above a dish of water. When you dip the end of the tissue in the water you can see the water rise up the tissue well above the dish. That is capillary action and demonstrates that the upward 'suction' forces of the pore spaces within the tissue are stronger than downward gravitational forces (see Chapter 1).

The forces holding water within soils are much greater in small pore spaces than larger ones (see Figure 1.3 in Chapter 3, showing how narrow capillary spaces draw water against gravity more strongly than wider capillary spaces), which is why sandy soils are generally drier than clay soils. When water is added to a soil it will tend to fill small pore spaces first, where the 'suction' or 'soil water tension' is greater.

When all of the soil pore spaces are full of water the soil is saturated and there are no suction forces. Instead there will be a positive pore water pressure, caused by the pressure of water from above and there will be a gravitational effect. When the soil is unsaturated there will be forces associated with gravity and negative forces associated with the suction effect. In fact, it is possible to assess how all these forces balance to determine the direction in which water is likely to flow within a hillslope (Hillel, 1982). These effects may be expressed by assigning *potentials* to the soil water. Soil water potential is the energy per unit quantity of water in the soil. The energy state of soil water can be expressed as energy per mass of water (joules per kg), as energy per water volume (joules per m^3, which is also equivalent to pressure measured in pascals) or as energy per unit weight of water (joules per 9810 kg m s^{-2}) (Durner and Or, 2005). If we use the latter, the results simplify down to units of length (m) since one joule is equivalent to 1 kg m^2 s^{-2}. Therefore many hydrologists like to express water potential as a pressure head with units of water column length (m). An example is shown in Figure 3.4.

While there are many forces, when it comes to the dominant hydrological processes in the field they are matric potential (the soil suction effect we described above, which is inversely proportional to the pore space size), submergence potential (positive

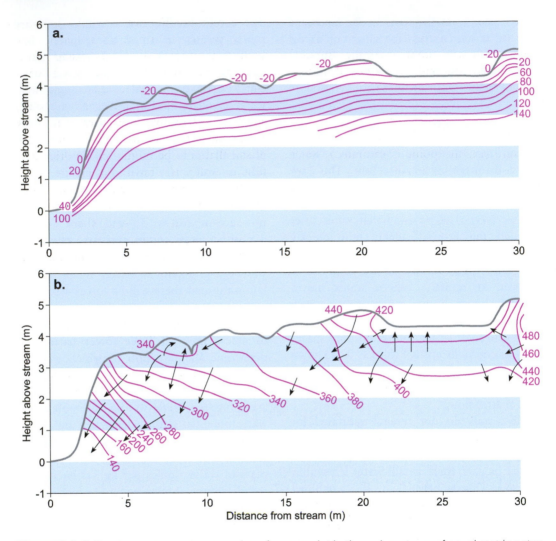

Figure 3.4 Soil water energy status on a slope for one point in time: a) contours of equal matric potential; b) total potential where relative height has been added to matric potential. The arrows show the likely direction of water flow based on the measurements of total potential.

pore water pressure) and elevational potential (the relative effect of gravity). At any point in a soil profile either the soil is saturated or it is not. Therefore, that point in the soil can only either have a matric potential or a submergence potential and not both at the same time. Thus, to work out the **total hydraulic potential** of the soil water all that is needed is a measure of the relative elevation of that point (to determine elevational potential) and either the soil suction (if unsaturated) or the positive pore water pressure (if saturated) and to sum these values. Soil suction will always be negative, and elevation and submergence potentials will always be positive.

The pore water pressure is measured *in situ* using piezometers (**Figure 3.5a**). Piezometers are tubes vertically installed into the ground with a perforated section located at the depth of interest in the soil or rock to allow water through at that point. The rest of the piezometer tube is not permeable. When there is positive pore water pressure (i.e. the point is saturated), water will rise up the piezometer tube. The pore water pressure is equivalent to the elevation of water in the piezometer tube relative to the perforated part of the tube and so merely requires measurement of the water level. Alternatively, the pressure of the water

in the tube can be measured by a pressure sensor housed within the piezometer. Soil suction is measured *in situ* using tensiometers. Tensiometers (**Figure 3.5b**) typically consist of a sealed tube filled with water, a porous ceramic tip and a suction gauge. The tensiometer is inserted into the soil so that the ceramic tip is at the depth of soil that is to be monitored. In drier conditions water is drawn out of the tensiometer through the ceramic tip, which creates a partial vacuum in the tube which can be measured by a gauge (either a visible dial or an electronic sensor). In wetter conditions, water is drawn back into the tube

Figure 3.5 Pore water pressure is measured with the aid of a) piezometers for conditions when there is a positive pore water pressure (the measurement point in the soil is saturated) and b) tensiometers when there is a negative pore water pressure (the measurement point in the soil is not fully saturated).
Source: b) Taco Regensburg, water@leeds, University of Leeds.

and the vacuum is reduced. Combined instruments that automatically measure both pore water pressure and soil suction for the same point (simply a tensiometer and piezometer connected together) using electrical recording sensors are now more common and are useful because the same point is sometimes above the water table and sometimes below the water table.

The total hydraulic potential is also known as the **hydraulic head**. **Figure 3.4** shows a map of hydraulic potential for a cross-section of soil across a hillslope and this provides a way of predicting soil water movement. The rate of change of hydraulic potential with distance (the **hydraulic gradient**) is a key factor that affects the rate of flow. The other key factor that affects the rate of flow is the permeability of the soil. According to Darcy's Law (see Section 5.4.2 in Chapter 5) the discharge (e.g. m^3 s^{-1}) from a cross-sectional area of soil should be equal to the permeability (m s^{-1}) multiplied by the hydraulic gradient (m m^{-1}) multiplied by the cross-sectional area (m^2). Typical permeability values in saturated sandy soils are 0.01 cm s^{-1}, while permeability values of 0.0001 cm s^{-1} are typical of silts and loams, with rates less than 0.00001 cm s^{-1} in clays.

Macropores (**Figure 3.6**) are larger than 0.1 mm in diameter and are also connected conduits for flow. The existence of macropores alone does not mean they will have a strong hydrological effect (Beven and Germann, 1982). A macropore has to be connected to a water supply in order for there to be flow and water must travel some distance along the macropore so that it bypasses part of the soil mass. In some environments, such as crusted sandy soils or peatlands, measurements have suggested that half of the subsurface water movement could be through macropores (Baird, 1997; Holden, 2009; Leonard *et al.*, 2001).

Macropores can be formed by plant roots, shrinkage cracks during hot, dry weather, small landslips or soil animals. Hallam (2018) showed in laboratory experiments that the presence or absence of earthworms and wheat roots both controlled the proportion of infiltrating water that moved through macropores, but that if the earthworm burrows did not reach a more permeable underlying layer then wheat roots can have a stronger impact on the soil hydrology.

Some macropores can enlarge to be several metres in diameter, particularly in erodible soils such as in semi-arid southeast Spain or Arizona (Bryan and Yair, 1982), or in **karst** limestone cave systems. Soil water turbulent flow within these large macropores, often known as natural soil pipes, can rapidly enlarge them until eventually they collapse to form gullies. Macropores can also be enlarged by the dissolving of material into the flowing water. Natural pipe features can sometimes be several hundred metres in length and have been reported to contribute large

Figure 3.6 Side view of soil profile showing a number of macropore pathways which appear like cracks in the soil. Plant roots can also be seen, which can create macropore channels.
Source: Shutterstock/mirbasar.

proportions (occasionally 10 to 50%) of the flow and sediment to rivers in some head-water systems (Holden *et al.*, 2012; Smart *et al.*, 2013; Zhu, 1997). Given this reported importance to flow, macropores and soil pipes have been rather understudied hydrological features (Beven and Germann, 2013).

3.2.5 Water balance

It is possible to calculate a water balance for drainage basin systems. The main input is precipitation (unless there is a large adjacent aquifer flowing into the basin, bypassing the surface topographical river basin boundary). The main outputs are river discharge or evaporation and transpiration. The remainder of the water budget consists of any changes in stores in the soil, rock, vegetation, rivers or lakes. However, measuring each of these components can be challenging, especially since there can be a large amount of spatial variability across a river basin. Therefore, estimates of river basin water budgets are subject to very large errors.

Rainfall is commonly measured using rain gauges and is normally presented in the form of a depth of water (e.g. mm, inches). Rain gauges have to be carefully sited to reduce impacts of turbulence around the gauge orifice affecting the measurements and so they tend either to be buried so that their tops are flush with the ground surface or to have some aerodynamic shape to them or their surrounds. However, as their name suggests, rain gauges are not so good at measuring other forms of precipitation (snow, mist, hail, dew). In eastern Newfoundland, Price (1992) suggested that around half of the water being delivered to the ground surface in summer was in the form of mist caught on vegetation, which then trickled down to the surface, and this was not properly measured by rain gauges.

Heated snow gauges are used in some locations to melt the snow as it arrives so that it can trickle into the measuring device to be recorded at the same time as it falls. Disdrometers are used in some places to measure precipitation drop sizes and consist of either a measurement plate that measures the impact of raindrops or a laser transmitter and a receiver placed at a known horizontal distance apart, which detects droplets passing between the two.

Rainfall can be highly variable across short horizontal distances and with altitude, and so if a river basin of, say, 100 km^2 has one rain gauge that gauge may not be representative of the true inputs to the basin. Thus, networks of gauges are normally installed during monitoring operations, but this still provides only a few values of point rainfall data which have to be interpolated spatially across the basin. Rainfall radar can be used in some locations to record the spatial pattern of rainfall being delivered to a catchment, but the actual amounts of rainfall are usually calibrated using ground rain gauges.

Flowing water which makes it out of a river basin via a river can be measured within the channel. Techniques for measuring river flow are described in Section 3.3. Water outputs in the form of evaporation and transpiration tend to be estimated rather than directly measured, based on data on temperature, humidity, wind and plant types. However, it is possible to set up monitoring of blocks of soil (with growing vegetation left on top) known as lysimeters, in which the soil is weighed *in situ* (i.e. a weighing device is inserted below the block of soil so that the soil is kept in the ground in its natural setting) and where changes in the weight are ascribed to changes in the soil water content. Water draining down through the soil is caught and weighed and so the remaining weight change of the soil is assumed to be due to evapotranspiration. Evaporation rates from open water bodies can be estimated by using evaporation pans, which are containers of water with open tops in which the water level is measured through time. Water inputs to the pan are known and the only output is by evaporation, and so changes in water level can easily be ascribed to evaporation rates.

Changes in water storage within soils and rock can be determined by measuring moisture content (there are dozens of methods – see Chapter 6 in Shaw *et al.* (2011) for a description of methods) or by the groundwater techniques described in Chapter 5. However, these processes are spatially highly variable and most soil water storage changes are simply assumed, based on the difference between measured inputs and outputs from the river basin.

Using a simple form of the water balance, it is possible to examine the water efficiency of a river basin. Such a measure is also known as the runoff efficiency or the rainfall to runoff ratio and tells us what proportion of the input precipitation is produced as an output in the form of river discharge. Over annual cycles or longer, the change in storage is sometimes thought to be zero (unless heavily modified by humans, such as through water abstraction, or unusual weather conditions, such as a drought), and so the simplest form of the water budget can be used to provide the evapotranspiration from the basin as: precipitation – river discharge = evapotranspiration. **Box 3.1** provides a short example of a runoff efficiency calculation.

REFLECTIVE QUESTIONS

What is a fifth-order drainage basin?

What are the main processes of water movement through and across the landscape to reach rivers and lakes?

What would you need to measure to determine the directions of water flow at different points within a hillslope cross-section of soil?

3.3 RIVER FLOW

3.3.1 Hydrographs and river regimes

Rivers are supplied by throughflow (including groundwater) and overland flow. The relative proportions of the different types of flow, along with the soil types, topography, size and type of drainage network, can determine how quickly river flow varies during rainfall events or seasonally. River flows can change during individual rainfall or snowmelt events, or remain fairly stable, depending on the nature of the river basin. There is a delay (lag) between the precipitation occurring and the peak discharge of a river.

Where infiltration-excess overland flow dominates the **runoff** response, the hydrograph (graph showing discharge through time) is likely to have a short lag time and high peak flow (**Figure 3.7a**). The storm hydrograph showing river discharge through time in response to a rainfall or melt event will therefore be quite steep in shape, rising quickly from low flow to the peak flow. Urbanisation increases flood risk, as it reduces the infiltration capacity of the surface through construction, leading to rapid water flow to the river and resulting in steeply rising hydrographs with sharp peaks. If matrix throughflow dominates flow pathways within the river basin, the river may rise and fall very slowly in response to precipitation and the peak may be small. However, since throughflow contributes to saturation-excess overland flow, throughflow

---

BOX 3.1

Calculating the hydrological efficiency of a river basin

Here we take a simple worked example to illustrate how runoff efficiency can be calculated. Assume a river basin area of 80 km^2, precipitation of 1600 mm in a year and that there was a river gauging station which recorded a mean discharge during the year of 2.28 m^3 s^{-1}. What we need to do is put everything into the same units. One way of doing this is to put everything into units of length (depth of water, just like rainfall). Multiplying 2.28 m^3 by the number of seconds in a year gives us 72,000,000 m^3 of water which was discharged by the river. So imagine that you spread all of the 72,000,000 m^3 of water across the river basin evenly; then you would have a depth of water equivalent to: 72,000,000 m^3 divided by the area of the basin, which is 80,000,000 m^2, which gives us 0.9 m or 900 mm. If we compare this depth of water produced as river discharge from the basin to the precipitation that fell on the basin (1600 mm), then we can see that this equates to an efficiency of 56.25%. In other words, 0.5625 of the precipitation that fell on the basin was released from the river basin outlet as river discharge. The remaining 43.75% of water was either lost from the catchment as evapotranspiration or resulted in a change of storage in lakes or groundwater. Similar calculations can be performed for individual storm events.

---

can, in many circumstances, still lead to rapid and large flood peaks. Even without saturation-excess overland flow, a throughflow-dominated catchment may still produce storm discharge if water can get through the soils quickly (e.g. if the soils are very permeable). In some soils, such as peat, only a small amount of infiltration may be needed to cause the water table to rise to the surface (Evans *et al.*, 1999). In other soils there may even be two river discharge peaks caused by one rainfall event (**Figure 3.7b**). This might occur where the first peak is saturation-excess overland flow dominated, or perhaps related to the direct delivery of rainfall into the river channel, and the second peak a little later may be much longer and larger and caused by sub-surface throughflow accumulating at the bottom of hillslopes and in valley bottoms before entering the stream channel (Anderson and Burt, 1978). Throughflow may also

contribute directly to storm hydrographs by a mechanism called piston or **displacement flow**. This is where soil water at the bottom of a slope is rapidly pushed out of the soil by new, freshly infiltrating water entering at the top of a slope. **Figure 3.8** shows how the different processes described in Section 3.2 can combine to create a stormflow response in the river.

Figure 3.9 shows flows in the form of hydrographs over one year for two nearby rivers where the climate is the same. Despite being in the same area, the flows are very different between the rivers. The flows in River 1 appear to be dominated by baseflow and there are no individual storm peaks, unlike for River 2, which appears to be more dominated by overland flow or rapid throughflow (e.g. via macropores). The River 1 basin overlies permeable bedrock, is gently sloping, and has soils that enable good infiltration and little

Figure 3.7 Storm hydrographs: a) where infiltration-excess overland flow dominates and b) where throughflow dominates. Note that (a) also indicates typical hydrograph terminology.

chance of saturation to the surface. Therefore, infiltration-excess or saturation-excess overland flows are a rare occurrence here. For River 2, however, the soils are thin and sit above impermeable bedrock and so there is frequent saturation of the soils and generation of saturation-excess overland flow.

Seasonal variability in river flow is known as the **river regime**, for which there are four major global types:

1 Arid zones, especially in subtropical drylands, tend to experience very occasional but intense rainfall events. Intense rainfall along with little vegetation cover produces infiltration-excess overland flow, rapid runoff and high flood peaks. However, many dryland soils are coarse and sandy, with high infiltration capacities, resulting in little chance of overland flow. Therefore, there is a wide variation of response even if rainfall intensities are very high. In most drylands river flow will stop within a few days of the rainstorm and water often seeps into the riverbeds or is evaporated. River flow is therefore highly intermittent in these systems.

2 Where snow and ice melt dominate, then there can be a major peak of river flow during the late spring or early summer (e.g. the Danube in Hungary or the Mackenzie in North America). River discharge can be extremely low during the winter months on some rivers, even though precipitation may be continuing, as this precipitation is stored on glaciers or snowpacks in winter. There can also be a strong daily change in river discharge due to daily melt cycles of the snow and ice. Night-time discharge immediately downstream from glaciers tends to be much lower than that of the mid-afternoon.

3 In temperate, oceanic areas precipitation occurs all year, perhaps with seasonal maximums. The river flow regime in these areas can change in response to either seasonal changes in groundwater storage and release, or higher evaporation and transpiration rates during the summer months (e.g. Seine, France).

4 Rivers in equatorial areas tend to have a fairly regular regime, while tropical river systems outside of the equatorial areas receive high precipitation during the summer but experience a marked dry season during the winter. Evaporation and transpiration are high at all times so that the streamflow mirrors the seasonal pattern of rainfall. Rivers such as the Brahmaputra or Mekong in South Asia have summer peaks associated with monsoon rains.

Figure 3.8 Mechanisms that lead to stormflow response in a river.

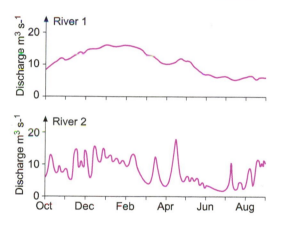

Figure 3.9 Two example annual river flow records showing regimes of each river (after Holden, 2011).

3.3.2 Closed basins

While most rivers around the world drain into the oceans (**exorheic basins**), there are many that do not, and these are often termed 'closed basins' or **endorheic basins**. Almost one-fifth of the Earth's land surface does not drain to an ocean but instead drains to an inland 'sink' (**Figure 3.10**). These tend to be in the interior of continents, in dryland regions, or where surrounding topography such as a mountain range prevents water from reaching the ocean. The largest of these endorheic basins covers much of continental Asia and river water here can end up in the Caspian Sea and Aral Sea. The Great Basin (which is actually several adjacent closed basins) in the USA and much of central Australia (e.g. the Lake Eyre basin) are other examples. The water in lakes into which endorheic rivers drain tends to be very saline because evaporation is a major water loss, leaving behind concentrated salts. For example, the Dead Sea, at the end of the River Jordan, which is located at the lowest point on the surface of the Earth, is so saline that people can lie in the water without sinking. In the centre of many endorheic basins there is no permanent standing water body, but there can often be a white 'salt pan' where once a lake existed, or where the lake occurs only seasonally.

3.3.3 Measuring river discharge

Good measurement of river discharge is important for water resource management,

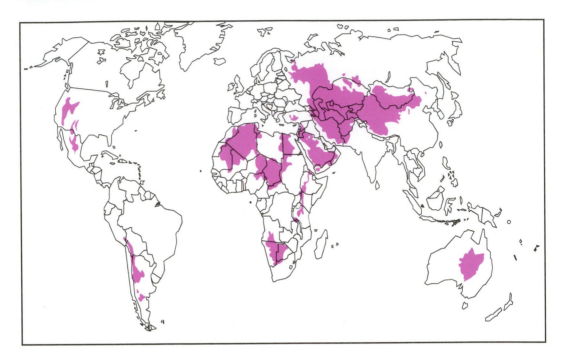

Figure 3.10 The endorheic river basins of the Earth.

flood warnings and monitoring for impacts of environmental change. Rivers act as a point of concentrated water movement where the confinement of water to a channel can assist in measuring the discharge. Typically, gauging stations consist of a method to measure the water levels (such as a pressure sensor) and a method to convert water levels into discharge. To help with this process, many gauging sites use weirs or flumes, although these are not always practical (e.g. on large rivers). The weirs and flumes provide more control over the relationship between water level and discharge. This relationship is known as a rating equation or a stage-discharge equation. Sometimes a design of weir has a standard engineering rating equation, but, even if this is the case, testing in the field is recommended. To derive a stage-discharge equation, so that from any water level at the point in the river being studied we can calculate the discharge, requires us to measure the discharge of the river at a range of water levels.

Various techniques can be used to measure river discharge, including dilution gauging, whereby a known volume and concentration of a tracer, such as salt solution, is added to the river. The more water in the river, the more diluted the salt solution will become. Therefore, measuring the salt's concentration slightly downstream enables us to work out the dilution of the salt and hence what the river discharge is. For a more detailed explanation of salt dilution gauging see the short video at http://tinyurl.com/aopjqxb. Alternatively, the cross-sectional area of the river can be measured and then the river cross-section is split into different segments and the velocity of the water passing each segment is measured. The water velocity may be measured using an impeller meter (typically at 0.6 of the depth from the water surface, as an assumed average of the water velocity in the entire water column; water tends to be slower near the bed, due to friction, and faster near the river surface) or other device (often done by someone standing in the water holding the device). Multiplying the velocity (m s^{-1}) by the area of the segment (m^2) produces the discharge (m^3 s^{-1}) for that segment of the cross-section (**Figure 3.11**). Summing across all segments in the cross-section then yields the total river discharge. This sort of technique can be impractical at very high river flows or where the river is highly vegetated, has lots of boulders or has several shallow channels.

Ultrasonic discharge gauges record the time for acoustic pulses to cross a river at different depths. Devices are placed on opposite banks of the river and the travel time for the pulse is used to calculate the streamflow velocity. A water-level sensor is also needed and the cross-sectional area has to be surveyed in advance for different

water levels. Doppler probes record the change in frequency when acoustic waves are reflected from natural suspended sediment or gas bubbles in river water. The change in frequency is proportional to the velocity of the particle or gas bubble. The Doppler method therefore works well in murky rivers, unlike ultrasonic discharge gauges described above. Sometimes Doppler profilers are deployed from boats or from cables across rivers to provide velocity measurements across the channel. Measuring river discharge can be very costly and is impractical in many very remote places or for large rivers. In fact, the number of gauging stations is reducing worldwide due to cost. However, work is ongoing to refine the use of satellite data to determine river discharge, as described in **Box 3.2**.

3.3.4 Modelling river discharge

Models that can predict how rainfall (either real or simulated) produces river discharge can be very useful for water resource

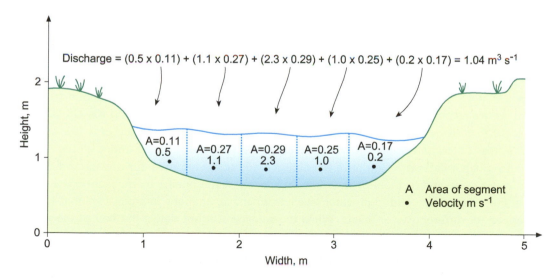

Figure 3.11 A worked velocity-area calculation for river discharge.

planning, flood forecasting, designing flood defence systems and in estimating river flow in basins that have no river gauges. They can also help us understand how river basin systems work, and we can perform landscape-scale experiments using models that we might not be able to do in the real world. Predicting river discharge from rainfall is not as simple as it might at first sound, particularly if the basin is large, because that will most likely mean that the basin has many different types of landscape within it. The river discharge that results from rainfall will vary with antecedent conditions because these affect infiltration rates and saturation conditions, and will vary with the distribution of rainfall across the basin through time. Many hundreds of models have been developed that are used to predict river discharge, and these vary in complexity. However, the quality of data input into models is an important consideration that will affect how good the prediction of discharge is. All models, no matter how complex, are just approximations of reality, and so the choice of model to be used depends on the questions being asked and the type of river basin being studied.

One of the simplest discharge models is known as the unit hydrograph, and was developed by Sherman (1932, 1942). The unit hydrograph is the stormflow that would be produced from an effective rainfall (that amount of rainfall which is released by the river as stormflow) over a unit of time (such as one hour or one day). The model assumes that the rainfall has occurred uniformly through time and across the basin. In essence, the unit hydrograph model requires users to obtain a storm hydrograph, work out the total volume of stormflow (in units of millimetres; i.e. average depth of water if spread across the catchment), and call that the effective rainfall. Then users should divide the hydrograph size (i.e. the y axis) by the effective rainfall so that what we are left with is the size and shape of hydrograph produced by 1 mm of effective rainfall. The hydrograph will now have units of m^3 s^{-1} per mm of rain. This is now the unit hydrograph for the river basin. In order to predict

BOX 3.2

Measuring river flow from space

The Surface Water Ocean Topography (SWOT) mission, to be launched in 2021, supported by NASA and Centre National d'Etudes Spatiales, the Canadian Space Agency, and the UK Space Agency, seeks to directly measure water heights from space with an accuracy of a few centimetres. The mission plans to measure water levels every 100 m across the entire globe every 10 to 11 days. Such a dataset would have the potential to hugely improve our understanding of the global water cycle and to monitor river and lake levels. Additionally, by measuring the topography of the river at low- and high-water levels from space, changes in cross-sectional areas can be derived for different water levels. While the satellite will not detect the bed of the river unless it is dry, the approach will use basic flow laws to estimate discharge for different river water levels. Combining water-level data with cross-sectional information and the slope of the river, all derived from space, could enable river discharge to be estimated for rivers across the globe at 10-day intervals. For further information please see swot.jpl.nasa.gov.

the river discharge for new rainfall events (either real ones or ones that we have made up to test, for example, how big a flood might be) we need to establish how much of the rainfall is effective rainfall. This can be done by using the first storm we derived the unit hydrograph from by subtracting effective rainfall from actual rainfall and averaging the remainder over time to give us an average 'infiltration rate' in mm hr^{-1}. This infiltration rate is really just the average amount of rainfall that does not produce storm discharge and is not a real infiltration rate for the soils in the river basin. It is often called a phi-index. If we use that same phi-index for future rainfall events, we can then determine how much of the rainfall is effective rainfall. If we are then left with, say, 1 mm, 3 mm, 2 mm of effective rainfall over three consecutive hours, then we can establish the predicted hydrograph by multiplying the unit hydrograph by one in hour one and then making a new hydrograph starting one hour later which is three times the size of the unit hydrograph and finally, in the third hour, we have a hydrograph which is twice the unit hydrograph (**Figure 3.12**). The next step is simply to add the three storm hydrographs together to get overall predicted storm discharge from the rainfall event. If we wanted to predict overall river discharge, the final step is to add baseflow.

Clearly, the unit hydrograph model is simple to use, but it also entails a lot of assumptions about uniformity in the river basin and does not deal with spatial processes such as the variable source area concept and issues around antecedent conditions. Models such as the unit hydrograph suffer when tested for different conditions because the same amount of rainfall, even if uniformly distributed across the river basin, can produce a very different storm discharge depending on antecedent conditions (e.g. wet soils or dry

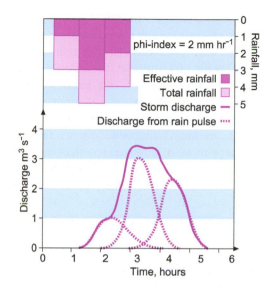

Figure 3.12 Predicted river stormflow discharge using the unit hydrograph model. Each hour of rainfall is associated with its own hydrograph. The first hydrograph produced from the first pulse of rainfall results from an effective rainfall of 1 mm and is therefore the basic unit hydrograph. The second hydrograph results from the second pulse of effective rainfall, which is 3 mm, and the hydrograph is therefore three times the size of the first one. All of the smaller hydrographs resulting from each pulse of effective rainfall are added together to produce the overall storm hydrograph for the catchment predicted to result from the rainfall event shown. Baseflow is not shown but if you wanted to predict total river discharge you would simply add baseflow to total stormflow.

soils). There has therefore been development of a suite of models, such as TOPMODEL, SHE and SWAT, that try to capture elements of flow and saturation processes. Basic overviews of those models, their variants and a range of other models are provided in Shaw *et al.* (2011), with more detail in Beven (2012). In essence, such models use the topography of the river basin and some information about soil types and saturation conditions to move water through and across the landscape to the river system.

River discharge models are typically tested by comparing a predicted river discharge with the observed river discharge. After some fine-tuning to optimise the model to work best (it will never be perfect), it can be applied to situations where we do not yet have river discharge data (e.g. future events, simulated storm rainfall, a change in land management (e.g. see Box 3.3), ungauged river basins). However, users of such models must always be aware that the predictions are just approximate estimates and it is useful for any predictions to be presented along with an estimate of how far we think the prediction might be wrong.

3.3.5 Impacts of land management on river flow

Land-management change can dramatically alter the flows in river systems. Building dams and changing or diverting stream

---

BOX 3.3

Modelling land-cover change impacts on flood peaks in upland peat-covered basins

A model called spatially distributed TOPMODEL (SD-TOPMODEL) was developed by Gao *et al.* (2015). It is based on the traditional TOPMODEL (Beven and Kirkby, 1979), which originally assumed that, aside from topography, the landscape behaved in a uniform way. The new model has a component that takes account of surface vegetation which may change the velocity of overland flow. The spatial distribution combined with the surface roughness component means that for a drainage basin it is possible to test some scenarios of how changing land cover would affect the river flow peak. This is very useful because otherwise it may take many decades of expensive field trials to undertake the same type of analysis. The information is needed to inform land managers about where to most effectively invest resources to change land cover which would result in reduced downstream flood peaks. Such information can ensure resources are efficiently allocated to land-management solutions.

SD-TOPMODEL has been applied to peatland river basins in upland environments. As peatlands tend to be saturated most of the year, saturation-excess overland flow is common. This means that there can be a rapid response to rainfall and river discharge can rise quickly, potentially leading to downstream flooding. Many upland peatlands have been degraded by overgrazing or atmospheric pollution and may have bare areas (**Figure 3.13**). Water velocities across peatlands have been measured for different vegetation types, slopes and water accumulation depths (Holden *et al.*, 2008). Gao *et al.* (2016) ran the model for a number of different scenarios and showed that ensuring there was a rough vegetation cover such as *Sphagnum* moss in the vicinity of watercourses could result in much reduced flood peaks and was more effective than creating this rough surface over other parts of the landscape (**Figure 3.14**). By only modifying 10% of the landscape, large reductions in flood peak could occur and there could be a large delay in the timing of the flood peak. However, modifying 10% of the landscape in other locations in the catchment, such as near the drainage divide or on steep slopes, resulted in very little or no flood reduction. Thus the spatial location of land-cover change is very important for river discharge and so spatially distributed models can be enormously helpful in supporting management decisions.

Continued

Figure 3.13 Upland peatland with a) bare areas and b) *Sphagnum* moss that slows water flow.

Figure 3.14 Modelling discharge for a scenario on Trout Beck catchment, northern England, where the same 10% of the landscape, in concentrated strips next to watercourses, was changed from the current sedge-dominated system to either a bare system or a *Sphagnum* moss-covered system. Twenty millimetres of rainfall occurred over one hour commencing at a time of one hour into the model simulation.
Source: Jihui Gao.

courses alters river regimes. In fact the Colorado River, which is the main river in the south-west of the USA, has been so heavily modified through major dams and abstraction schemes that discharge along its lower course is massively reduced so that it now rarely flows to its natural exit in the Gulf of California.

Major land-management change may alter the water budgets of river basins. For example, deforestation may result in a large reduction in transpiration rates. In 1934 a hydrological monitoring station was set up at Coweeta in the southern Appalachian Mountains to look at how forest management affected river flows. The research showed that the amount of water flowing from a forested basin increases after timber harvest, mainly because of reduced transpiration from trees. The volume of water from storms (efficiency of the river basin) and peak flow increases after deforestation. As the vegetation grows back and trees begin to recover, the peak flows and the basin efficiency start to decline again.

Any management activity that alters the flowpaths for water and the proportion of water moving via that flowpath – infiltration/saturation-excess overland flow, matrix or macropore flow – has the potential to alter the river hydrograph responses to storm events or the regime of the river. Agricultural activity can have a major impact on soil infiltration. Ploughing can often increase the infiltration capacity of the soil, which may then decline during the cropping season until the soil is ploughed once more (e.g. Imeson and Kwaad, 1990). Soils under woody hedgerows can provide important functions on farmland including promoting infiltration and storing more water during storm events (Holden *et al.*, 2019). It has been shown that livestock grazing has profoundly affected soils throughout the world, from arid and semi-arid rangelands to temperate moorlands. A review by Greenwood and McKenzie (2001) suggested that grazing animals exert pressure on the ground comparable to that of agricultural machinery, which leads to soil compaction. Research has generally shown that as grazing pressure increases, vegetation cover declines, soil bulk density increases, water infiltration rates decrease and surface runoff levels increase (Abdel-Magid *et al.*, 1987; Mwendera and Saleem, 1997). Evidence from a UK study showed that alterations of vegetation and soil properties due to grazing greatly increased active source areas for overland flow and discharge during storm events (Meyles *et al.*, 2006).

Field under-drainage and the use of open ditches have been common land-management techniques in agricultural areas (Rogger *et al.*, 2017). These systems tend to be installed where the land is deemed to be too wet for optimum agricultural performance and their aim is to lower the water table. Such drainage often contributes to river regime by increasing base-flows and potentially increasing the flood peak. Drains and ditches act as preferential flowpaths for the fast channelling of water to watercourses. In peatland systems the use of open ditches has lowered the water table and caused organic soils to dry and crack, which has resulted in enhanced flow of water through macropores and the expansion of natural pipe networks (Holden, 2005b). Flow through these pipes can erode the system from below, deliver a lot of carbon to the stream (Holden *et al.*, 2012) and lead to gully development when pipes collapse.

Urban development often reduces the local infiltration capacity to zero, which can increase the propensity for downstream flooding (Perry and Nawaz, 2008). The loss of vegetation cover also reduces transpiration and hence increases the annual and storm efficiency of the river basin. Urban areas tend to have a subsurface drainage system which takes water away from the surface using sinks which pipe it to watercourses. Traditionally these underground drainage systems have served as fast pathways for water movement across the basin and have increased flood peaks. More modern developments use **sustainable urban drainage systems**, which try to minimise the downstream impact on river flows of urban development (see Section 4.5.2.2 in Chapter 4). They often do this by encouraging the use of more permeable surfaces and also by having pipes and conduits that discharge into a series of ponds or storm water-collecting channels, which then drain more slowly into the river network at a later stage. In essence, sustainable urban drainage systems have made space for storm water in their design. However, during intense rainfall events these systems can still be overwhelmed; clearly, the costs and space required for large storage are greater and so a judgement has to be made as to how much space to allocate for sustainable urban drainage, without overly compromising the aesthetics of the landscape or the commercial value of the land. Many sustainable urban drainage systems also seek to clean the water before returning it to rivers (e.g. by using lake or wetland systems to settle out sediment). These systems can be designed to look aesthetically pleasing and perhaps also serve other purposes such as providing recreational and leisure services (e.g. see **Figure 8.14** in Chapter 8).

Predicting the effects of land-management change on river flow is not straightforward because river basins are not uniform and have varying soils, topography and drainage networks. Therefore river basin models are often deployed to investigate the potential impacts on river flow of management interventions (e.g. **Box 3.3**). In fact, the same land management can have a very different impact on river discharge (e.g. the flood peak), depending on where in the river basin that management takes place. Section 3.4 provides further explanation of this issue. These spatial effects, related to how water is routed through the drainage basin and the timing of delivery of water from different parts of the basin, are another reason why using spatial models to help understand the impacts of land-management change on river flows is important. Models can be used to tell us which parts of the river basin might be more suitable for certain types of management change, as the changes in those locations might have a much lower impact on river flow than the same changes in another part of the river basin.

REFLECTIVE QUESTIONS

What are the main types of river regime globally?

What methods could be used to estimate river discharge for an accessible site and for a completely inaccessible site?

For the drainage basin where you live, what key features of the hydrology should be included in a model to predict river discharge in response to rainfall under different future land-cover scenarios?

3.4 FLOODING

Flooding is a natural phenomenon and is to be expected. There is always a flood occurring somewhere in the world and

http://floodobservatory.colorado.edu provides maps of current floods and
a database of flood events. Flooding can
bring benefits to agricultural zones through
supplying nutrients to the soil, but can also
bring devastation when it affects residential
areas. The August 2018 floods in Kerala,
India, killed around 500 people with about
1 million people evacuated. Over 400 died
in East Africa in May and June 2018 directly as the result of flooding, with hundreds of thousands of people displaced.
Many have suffered due to longer-term
health impacts of the flooding. Details
about the wider health impacts of flooding
are provided in Chapter 7.

When rivers overtop their banks and spill
onto surrounding land it is known as fluvial
flooding. Pluvial flooding occurs when there
is heavy rainfall which leads to concentrated
overland flow inundating an area. This is
particularly common in urban areas, due to
impermeable surfaces quickly generating
overland flow. Groundwater flooding can
also occur where saturation-excess overland
flow concentrates. Coastal flooding caused
by very high tides, storm surges or tsu-
namis is also important.

Often the fluvial flood record is analysed
for the number of times the water level in
a river peaked above a given (often critical)
level. Sometimes this water-level value is
not equivalent to a discharge value because
the lower parts of many catchments are
affected by tides or because the channel
has changed dimension (e.g. become silted
up). If a stormflow upstream coincides
with a spring high tide, then the flood risk
may be enhanced. If long-term river-level
records are available, then simple return
period or recurrence interval calculations
are possible. For example, if a water level
on a river greater than 10 m occurred five
times in five years we would say that the
return frequency of the 10 m flood at that

given point is on average once per year. Of
course, we could get a 10 m flood occurring three times in one year, and the
return frequency is just an average value.
Similarly, it is possible to estimate the size
of the one in 100-year flood and to map
its likely extent. These sorts of maps might
be used by the insurance industry and
management agencies to manage flood risk.
Traditionally, flood defences were designed
to deal with a flood of a particular magnitude or recurrence interval. However, infer-
ring what might be expected in the future
from flood events that have happened in
the past may not be reliable, given that
land-management change and climate
change might impact hydrological processes
operating within a given catchment. Hence,
flood-risk mapping often requires some
future climate modelling scenarios. Additionally, there has to be some acceptance
of the risk of flooding and that engineering
design does not rule out all possible flood
events.

Humans have decided to live in low-
lying areas subject to flooding. This is
because these areas tend to be where there
are fertile soils (often made more fertile
by regular flooding) suitable for crops and
where navigation of rivers by boats allows
transport of goods and people. In many
places the reaction to flooding has been
the construction of flood defences and
major river engineering projects. Many
flood solutions have involved building
levees or embankments next to rivers or
straightening the river channel and clearing out the sediment and vegetation to
allow faster, more uniform river flows.
However, in many cases these techniques
have led to worse flooding downstream
because sending the water more quickly
through one part of the river system
simply reduces the lag time downstream
and increases the overall flood peak.

Floodplains, when functioning naturally, act as a temporary store of water. Therefore, if we reduce the space for flood waters upstream we can make flooding worse downstream. However, the demand for building or farming on the floodplains of the world means that fewer and fewer of them are freely available for a river to store its water during wet periods. Flooding in lowland areas is therefore a bigger problem, as the extra water is brought downstream more quickly and in greater quantities than ever before.

Solutions to the flood problem need to be mixed, and include land-management solutions, flood defence solutions and also getting people to accept the risk of flooding and designing coping strategies. Understanding risk means undertaking to identify the consequences (e.g. cost, potential lives lost), the probability of the consequences and the significance of the risk. In the EU, the Floods Directive requires EU Member States to complete flood-risk assessments (topographical maps, descriptions of previous floods, land use, etc.), produce flood-risk maps (for extreme events with low probability, medium events with a return period of at least 100 years and high-probability events) and to produce flood-risk management plans. The latter are meant to set out how local agencies will reduce the impacts of flooding on the economy, human health, the environment and cultural heritage. The plans are not meant to focus on improving hard-engineered protection from flooding but to focus on all aspects, including prevention, preparedness and protection. Dealing with the flood problem means that we need to build resilience into the built environment in flood-prone areas. An example of resilience might be designing buildings that have very flexible ground floor use, such as parking areas for cars. When there is

a flood warning, then the parking area can be evacuated of cars and flooding can be allowed to occur without any damage to residential dwellings on higher levels.

There has been recent significant attention on 'nature-based solutions' or 'natural flood management'. For example, in China, several cities, such as Shanghai and Tianjin, have been selected as 'sponge cities' where urban design is being modified to promote water infiltration, and to slow and trap water flow by collecting rainfall on roof gardens and in green cityscapes. However, in other regions whole catchment approaches are being adopted. For example, for the city of Leeds, UK, peatland restoration, woodland planting and pond creation in the hills and tributaries far away from the city are being adopted as part of the flood alleviation scheme for the city. These significant investments are in addition to river channel engineering solutions and the creation of 'blue corridors' that are designed flexibly near the main river channel so that they can flood without causing significant damage to households, businesses or infrastructure. These river basin-wide approaches are being supported by many organisations working together to optimise resources and integrate solutions together. This work is being supported by the NERC Integrated Catchment Solutions Programme: https://icasp.org.uk, led by water@leeds at the University of Leeds (see **Box 13.3** in Chapter 13).

Changes in flood risk caused by land-management change are often complex to understand and measure. It is not possible to say that if you cover 10% of a catchment with trees to take up more water you will reduce the flood risk by 10%. In fact, activities that reduce flooding in a small area of a catchment may lead to the opposite effect downstream and vice versa, depending on the configuration of the river basin. **Figure 3.15** shows a small tributary catchment

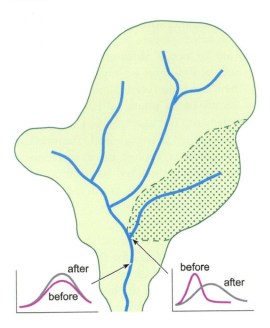

Figure 3.15 A river basin showing what happens to the main channel flood hydrograph after forestry plantation within a tributary basin due to flood-wave synchronisation. The growth of trees both lowers and slows the flood peak from the tributary basin. However, because the flood peak timing from the tributary as it enters the main channel now coincides with the main channel flood peak, this increases the flood peak in the main channel.

within a larger basin. If land management were to introduce a change in the tributary catchment (e.g. a forestry plantation) that resulted in a lower flood peak and delayed lag times from the tributary, this might not necessarily be beneficial. If the peak from the small tributary has been delayed, it may peak at the same time as the discharge peak in the main river channel. Therefore, the tributary will contribute most discharge at the same time as the main river. It is therefore important to understand how all parts of the drainage basin interact and how the whole drainage network operates in order to fully understand the impacts of management change on flood risk.

3.5 RIVER CHANNEL DYNAMICS

3.5.1 Channel networks, planform and long profiles

Rivers are dynamic. They often move position around the landscape through time, change their shape and move sediment, water and dissolved materials. They are an important agent in removing weathered material from the land surface and redistributing it across the landscape or into oceans. This process balances out the mountain building caused by plate tectonics so that, over long time periods, plate tectonics might build mountains but weathering, erosion and removal by river systems or ice masses (see later in this chapter) smooth out the landscape.

River channels are rarely singular in form and tend instead to be part of branching networks. A river basin that has more river channels in it than another basin of exactly the same area is likely to be the more efficiently drained of the two. This means that discharge peaks are likely to be greater during storm events for the systems with a more dense drainage network. The **drainage density** is defined as the length of the stream channel divided by the drainage area. If you look down at the landscape from above, say from an airplane, or using a website that provides aerial images of the

Earth, it is possible to see a wide range of patterns and sizes of rivers. From a great distance you can see the whole drainage network, and these are often different in form depending on topography and geology (Figure 3.16). Therefore, the river regime is often a function of the topography and geology, more so than just being a function of the pattern of the drainage network. Dendritic drainage networks are the most common. Rectangular networks develop where rocks have two directions of jointing at right angles. Joints tend to be eroded more easily, and so streams can develop along the lower-lying eroded joints which meet every now and then at approximately right angles. Parallel drainage networks tend to occur where the slopes are steep and flow in the same direction. The steepness of the slopes means there are few branching tributaries and the channels tend to be straight. Trellised drainage systems are characteristic of folded mountains (such as the Appalachians), where there is a strong structural control on the direction of streams, due to geology. Deranged systems occur in areas where the topography is very irregular and there are lots of low points in which lakes can form.

Individual channels also have distinctive forms, the main ones being braided, meandering and straight (Figure 3.17), but one river may have each of these patterns in different sections along its course. Braided channels consist of lots of individual smaller channels which separate and come together, with the islands in between known as bars. The channels can rapidly move locations, with bars eroding on one side and growing on the other. Braided rivers are common where there is a lot of mobile sediment, such as downstream of a glacier. A drop of water running down the surface of a perfectly smooth piece of glass will take a wavy, looping course. It is therefore unsurprising that most rivers are

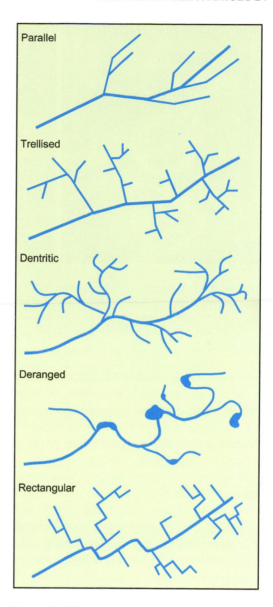

Figure 3.16 The main types of drainage network pattern.

meandering. The meandering pattern is measured by sinuosity, where sinuosity is the ratio of the length of river between two points compared to the length of the straight-line valley between these two points. Straight channels are defined as having a sinuosity of

Figure 3.17 Examples of river channel forms: a) the braided Kennicott River, Alaska; b) a meandering river, and c) a straight river – the River Tara, Montenegro, occurring in a natural steep-sided valley.
Sources: a) Shutterstock/melissanm; b) Shutterstock/Graham Prentice; c) Shutterstock/Kurt.

less than 1.5. A meandering channel refers to a single channel with a number of bends which result in a channel sinuosity in excess of 1.5. Straight rivers are more controlled by human action and natural straight rivers are often unstable and become meandering, unless there is a very strong geological control such as in parallel drainage networks.

Channel slope and discharge exert a control on the planform of river channels. For a certain discharge there is often a critical slope above which channels will meander and then a further threshold above which they will braid. These slope thresholds decrease with increasing discharge. Thus, braided sections are usually found on large rivers or on small rivers with steep slopes. Braided rivers will occur where there is coarse sediment, erodible banks and where the main sediment transport mechanism is **bedload transport** rather than suspended transport.

The slopes of river channels (**long profiles**) from their source to mouth are mainly concave, with progressively lower gradients downstream. The long profile varies with geology and water flow. Other profile shapes also occur when there is interruption by lakes or very resistant rocks (which often result in waterfalls), or through large changes in sea level. If sea levels fall, the whole river may start to erode its bed downward in response. If sea levels rise, the river may deposit more sediment and bed levels will be raised along its length. Thus, future sea-level rise will enhance flood risk inland.

3.5.2 River cross-sections

The shape and size of channel cross-sections adjust to the discharge and sediment load conditions of the river (Leopold and Maddock, 1953). Channels are typically wider and deeper if the discharge is greater. Further downstream, river channel cross-sections tend to get larger, although there is not a perfect relationship between discharge and channel cross-sectional area because larger channels are more efficient at carrying water (less friction around the channel edges per volume of water). Channels with a high percentage of silt or clay in their banks (which are often more characteristic of lowland sections of river), and rivers transporting much of the sediment load in suspension, tend to be narrower and deeper than sand and gravel-bed rivers, where most of the sediment transport occurs via movements of sediment close to the bed and banks. Vegetation can also be important in controlling cross-section shapes by influencing bank resistance through root systems that bind the sediment. Removing bankside vegetation can lead to rapid bank erosion.

Within channel cross-sections there can be large variations in the velocity of water flow. Close to the bed or banks of the river the velocities tend to be slower. On bends of rivers there can be forces exerted that increase the pressure on the outer bank as the water flows by with less pressure on the inner bank of the bend. In addition, a circulatory flow called **helicoidal flow** (corkscrew-type circulation) is superimposed on the downstream flow of water (**Figure 3.18**). On the meander bend there will be an outward flow close to the water surface, a downward flow close to the outer bank of the bend and an inward flow towards the inner bank of the bend. Bank erosion is more likely on the outer bank, as faster water has more chance of picking up sediment. On the inner bank, sediment is more likely to be deposited. This causes further development of the meander (**Figure 3.19**) and means that the river is continuously eroding and depositing sediment as a natural process, and so the exact position of the riverbank will change through time. Within meandering channels, because water velocity is greatest close to the outer bank, sediment size tends to be greatest here and slowly decreases towards the inner bank. When meander bends become too

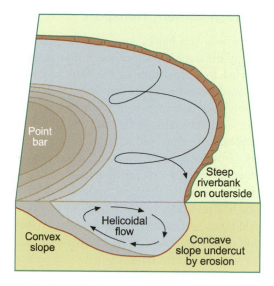

Figure 3.18 Helicoidal flow through a meandering channel.

exaggerated the river will cut through the meander to a direct course downstream for that short section, leaving behind a crescent-shaped section of channel that is cut off from the main river, known as an **oxbow lake** (**Figure 3.19**).

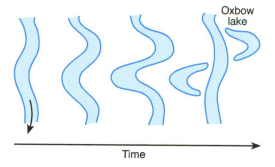

Figure 3.19 River meanders tend to become more exaggerated until eventually the meander is cut through by the river, leaving behind an oxbow lake. Meandering rivers can, over time, follow a course that slowly wanders across the entire flood plain.

3.5.3 Sediment transport

Riverbeds and banks consist of sediments of many different sizes. Sediment that is transported by suspension in the flow is known as wash load or suspended load, while sediment that is transported by forces pushing the sediment along the bed is known as bedload. If some sediment is to be picked up (entrained) from the riverbed or bank by flowing water a critical threshold has to be passed above which the water velocity or shear stress is sufficiently large to overcome frictional forces that resist erosion. At very low water velocities only fine material such as very fine sand or silt will be transported in suspended load (**Figure 3.20**). However, for faster flows larger grains can be transported. It is generally rare for sediment larger than 1 mm in diameter (very coarse sand) to be transported in suspension. Sometimes the smallest diameter sediments can be consolidated and 'stuck' together and so, as shown in **Figure 3.20**, it can take higher-velocity flows to entrain the very finest particles (0.001 or 0.01 mm) into flow in this situation than required for fine sand (0.1 mm). Importantly, the velocity of water required to overcome friction and entrain a particle into flow is generally greater than the velocity that is then required to keep it in suspension (**Figure 3.20**). Some fine sediment grains can be transported in suspension down rivers for hundreds of kilometres before they are deposited. Finer particles are preferentially moved downstream. The characteristics of riverbeds tend to change downstream. Often there are bedrock channels in the upper section of a river network or large boulders and cobbles. There is a sharp decline of bed material size downstream. This pattern is because smaller particles are transported downstream more easily and because of wearing down of larger sediments within the river by colliding and grinding, which causes material to get smaller and more rounded downstream. However, these patterns are not seen everywhere and may be disturbed by local sediment inputs to the river network.

Bedload transport occurs by sliding and rolling along the bed or by saltating (hopping) along or close to the bed. Saltation tends to occur in the water column above the bed within a height of about ten times the diameter of the particle. If the flow velocity does not change, a particle will come to rest only if it becomes lodged against an obstruction or falls into a sheltered area of a larger particle. There has been decades of work trying to predict rates of bedload transport. Because there is a lot of complexity around the different forces to be considered and the irregular patterns of sediment and water velocities along riverbeds, the equations can sometimes get rather daunting for new students. However, Chanson (2004) presents some worked examples for the different formulae if you want to investigate further.

Where erosion exceeds deposition within a particular section of a river, there will be lowering of the riverbed or widening of the banks. If erosion and deposition occur at the same rate, then the river will stay at the same level. Eroding channels may undermine structures such as bridges, while depositing channels may submerge structures such as roads. Stable channels, especially those whose beds are lined with bedrock, are less likely to be a problem to engineering structures, but fluctuations in river channel dimensions and locations caused by flooding or sediment pulses moving down the river can be problematic.

Within rivers there are several erosional and depositional landforms. In bedrock channels potholes can be found, formed by: mechanical wearing and grinding of small particles, enlarging an existing small depression or weakness in the rock; pressure changes, due to bubble collapse in turbulent flow; and chemical

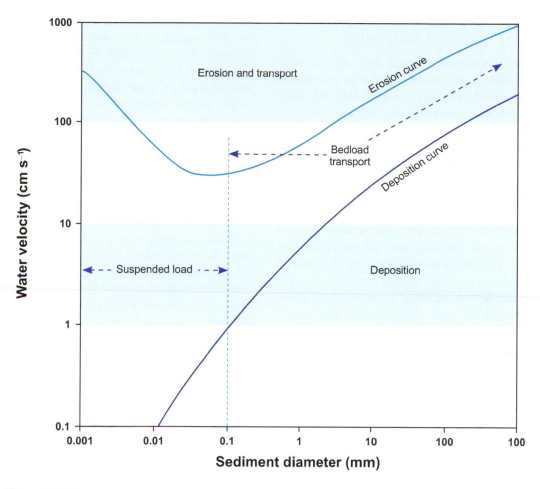

Figure 3.20 The Hjulström curves showing, for a 1 m-deep river, the different flow velocities required for entrainment or deposition of sediment of different sizes.

weathering. Within rivers that have gravels on their bed the most common landforms on the bed are sequences of pools and riffles. Pools are deep sections with relatively slow-flowing water with fine bed material. Riffles are formed by accumulation of coarse sediment with shallow, fast-flowing water. The spacing of pools and riffles is often five to seven times the channel width, but this does vary. The beds of sandy sections of rivers can have small ripples in the sand which are less than four centimetres in height, and then also larger dune features. The size and shape of these dune features change with discharge during rainfall events. The dunes and ripples tend to migrate downstream as sand is carried up the upstream-facing side of the ridge of the feature and then falls down the downstream-facing side. At very high flow velocities, a flat riverbed can be formed or dunes can even migrate upstream, since erosion from the downstream side of the dune allows suspension of material in the water, which occurs faster than it can be replenished from upstream.

3.5.4 Modification of river channels and river restoration

River channels change their slope, cross-section, planform and bed forms in response to environmental change. Humans have modified river channels over the past few thousand years, although changes during the last two centuries have been greatest. Activities such as dam construction, urbanisation, mining, land drainage and deforestation can all impact upon river channels. Faster, more peaky flow from urbanisation or deforestation can accelerate erosion. Dams have a major impact on river flows and sediment dynamics. The Nile now transports only 8% of its natural load of silt below the Aswan Dam, thereby reducing the fertility of the downstream flood-plains and accelerating riverbed and coastal erosion, as the lack of sediment entering the sea no longer replenishes the sediment being eroded by wave action in the **delta**. There are concerns for channel change in many major rivers in the world, including in their deltas. **Box 3.4** provides the example of the Indus.

Humans have directly modified river channels for the purposes of flood control, navigation and protection of property and infrastructure from bank erosion. Many river channels have been straightened, dredged, widened, culverted or lined along their bed or banks, had vegetation or other obstacles removed and had levees constructed. However, rivers are naturally dynamic and move across the landscape. There are many places where river channel modifications were simply reversed by the natural forces of the river. The Mississippi has regained much of its meandering sinuosity since channel straightening took place at the start of the twentieth century. In other places it is clear that many river channel modifications have worked against the natural processes in rivers and have resulted in major problems. For example, where channels are straightened it shortens their length and therefore steepens the slope of the river channel. This speeds up flow and hence increases sediment entrainment from the banks and bed, and so there can be enhanced local erosion. If the bed or banks erode due to faster water velocities in the straightened section, there is a possibility that this will have upstream effects, as the lowered bed elevation in the straightened reach may incise upstream, lowering the elevation of the riverbed upslope. This may also result in enhanced sediment deposition downstream of the straightened section, causing channel instability there. When river channels get choked with sediment they are more liable to suddenly jump to a new course, in a process known as channel **avulsion**.

In addition to the geomorphological impacts of channel modification, there have also been biodiversity impacts. The creation of uniform channels lacking in diversity of flow velocities, sediments, shading or backwaters means that the habitats for many aquatic species have been lost on modified rivers. If the diversity of flow velocities and diversity of physical features within river channels are reduced this inevitably also reduces the diversity of species present.

The realisation that rivers could be valued for their biodiversity and aesthetic qualities and that many channel modifications in the past have not dealt with river channel processes properly has encouraged the practice of river rehabilitation (also known as river restoration). Large sums are being spent on river rehabilitation, which often involves actions such as reinstating meander bends, **riparian** planting, removal of hard banks or beds, bank reprofiling, and the addition of pool and riffle sequences. For an example of a river rehabilitation project see **Box 3.5**. The scale of rehabilitation projects varies from tens of metres to tens of kilometres. One of the major problems that river rehabilitation practitioners face is how to compromise between different stakeholder wishes. Conflicts easily arise between

BOX 3.4

Impacts of river flow and sediment reduction on the Indus delta

The Indus is around 3000 km long; it has its source in Tibet and flows to the Arabian Sea. Over the past 10,000 years the Indus has formed a huge delta of 6200 km^2. However, 60% of the river water is now used for irrigation of agriculture. There has been a dramatic reduction of water and sediment delivery to the lower Indus (**Figure 3.21**). The Indus flowed continuously into its delta until the 1960s and now it may flow to the delta for only two months of the year, during the monsoon season. Delta systems normally have river freshwater flow, which forces itself up against seawater. The lack of water now flowing into the delta means that more salty seawater can move inland up the delta. There has therefore been incursion of seawater into groundwater supplies and damage to farmland by salt deposits, displacing several hundred thousand people (Inam *et al.*, 2007). Mangrove forests in the delta have also been affected by reductions of water flow. This is of vital importance because, of the 1.2 million people living around the forests, 130,000 rely on the mangroves for their livelihoods (Shah, 1999). Prosperous ports such as Keti Bandar have become small villages. Currently the Indus contributes hardly any sediment to the delta and the area of active delta has reduced to one-fifth of its former extent. Coastal erosion via wave action and lack of sediment and flow from inland is resulting in shoreline retreat at a remarkable rate of around 50 m per year.

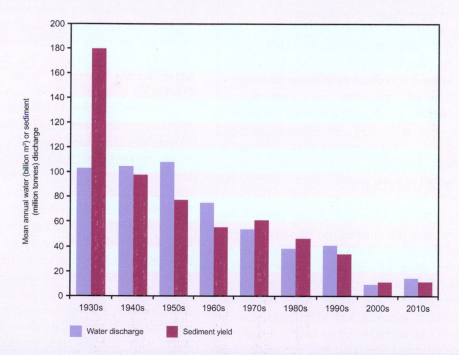

Figure 3.21 Decadal changes in mean annual water and sediment yield in the main channel of the Indus River below Kotri.

Continued

Satellite imagery is being used to detect rates and patterns of change in deltas all over the world (Cazenave *et al.*, 2018), and the most common concern is that many deltas are now subject to enhanced sea erosion due to being starved of freshwater sediment input.

water companies, fisheries, landowners, local residents and environmental organisations. Hence, rehabilitating a degraded river requires the coming together of disciplines such as ecology, hydrology, geomorphology and sociology as well as parties from all sides.

REFLECTIVE QUESTIONS

How might the complete removal of a dam which has become full of sediment over the past 100 years impact upon downstream river channel dynamics?

What are the main types of sediment transport processes in rivers?

If a major flow diversion added significant new water and cobble-rich sediment to an existing meandering river, what might happen to channel cross-sections in the stream?

3.6 SUMMARY

Water reaches rivers and lakes by direct deposition into the channel, or movement across and through the landscape. Overland flow can occur when surface water supply is greater than the infiltration capacity of the ground. Overland flow also occurs when the ground surface is saturated and water returns to the surface as saturation-excess overland flow. Saturation-excess overland flow can occur long after rainfall has stopped, especially at the foot of long, gentle-gradient slopes. Throughflow

occurs through small pores as matrix flow and through large pores as macropore flow. Throughflow can contribute to river stormflow response as well as the baseflow of the river.

There is a wide spectrum of river regimes and responses to rainfall events controlled by climate, topography, drainage networks, geology, soils, vegetation and human management. Fast-responding systems with a rapid rise and fall of the hydrograph tend to be those with steep slopes, impermeable surfaces or quickly saturating soils. It is possible to take measurements to calculate the annual, seasonal or storm efficiency of river basins in delivering river discharge resulting from precipitation. Land management can alter the hydrological efficiency of the system by utilising water for agriculture, industry or domestic use or by changing the land cover and soil characteristics which affect the flowpaths for water across the river basin. Modelling approaches can be useful for understanding how land-management or climate change might influence river regimes in the future.

River flooding is a natural process and should be expected. There are other types of flooding too, including pluvial flooding, groundwater flooding and coastal flooding. Managing flood risk requires more than hard-engineering solutions, particularly as allowing some areas to flood may make the impacts of flooding elsewhere much less severe. Resilience needs to be built into flood management solutions, which include improving perceptions of risk, developing flood warning systems and building landscapes that allow flooding to occur with minimal loss to the economy. Nature-based solutions are becoming increasingly popular including whole-drainage basin

BOX 3.5

The Provo River rehabilitation project, Utah

Many rivers in the USA were straightened for flood control, navigation or redirecting waters for supply systems. However, such changes often led to widespread ecological damage. The middle Provo River, Utah, was a key water diversion route to supply central Utah's growing population with water. Several diversions of water from other catchments into the middle Provo River occurred in the 1940s. To accommodate the extra water the river channel was straightened and protected by dykes. The net result was that the once-rich ecosystem around the river collapsed.

In 1999 work began to rehabilitate a 19 km stretch of the river along the Heber valley at a cost of $45 million. Meanders were excavated to resemble natural historic conditions, and the river was reconnected to old secondary channels. Ponds and other backwater features were also created to enhance habitat for wildlife (Figure 3.22). The dykes were moved back to allow the river to flood more naturally in its floodplain. Public access for anglers was also provided.

Figure 3.22 The re-meandered Provo River in the Heber Valley. The remnants of the old single, straightened channel, which has been infilled, can be seen in the top left of the photo.
Source: Shutterstock/Johnny Adolphson.

The redesigned river channel with its many backwaters and meanders is now associated with abundant wildlife and is one of the most fished streams in Utah with leisure fishing adding value to the local economy. As insect and fish life have been enhanced, so too has local birdlife.

land-management planning to reduce flood risk. However, numerical modelling is often required to understand how changes in land management or channel configuration will affect flood risk, because river basins are not simple systems and a change in management in one location will not have the same impact on flooding as the same management change in another location of the river basin.

River channels can take on a number of planforms with different densities of drainage and different channel patterns, including braided and meandering forms. Modification of river channels, such as channel straightening or building flood protection embankments, can alter the nature of the river regime or storm hydrographs by altering the amount and velocity of water flows through the channel. These changes can also impact upon channel sediment dynamics, with knock-on impacts for river channel dynamics upstream and downstream of the managed river reach, as well as negative impacts on biodiversity. Thus, river rehabilitation schemes are being developed which try to re-engage rivers with their natural processes. One of the major problems for many of the world's largest rivers is that of reduced water and sediment discharge, due to human abstraction from dams or diversion channels for agriculture. In many cases, such as the Colorado or the Indus, this means that those large rivers now rarely flow to their original mouth at the sea. Starving the river mouth of freshwater and sediment can have huge knock-on effects for coastal erosion and intrusions of saline water inland.

FURTHER READING

Arnell, N. 2002. *Hydrology and global environmental change*. Pearson Education; Harlow.

An excellent book full of useful examples with very clear explanation of processes.

Beven, K.J. 2012. *Rainfall-runoff modelling – the primer*. 2nd edition. Wiley-Blackwell; Chichester.

Very useful overview of modelling approaches and challenges.

Gupta, A. (ed.) 2007. *Large rivers: geomorphology and management*. John Wiley and Sons; Chichester.

A great resource with descriptions and research results from many of the world's large rivers. The book includes discussion of potential future changes due to human modification and climate change.

Shaw, E.M., Beven, K.J., Chappell, N.A. and Lamb, R. 2011. *Hydrology in practice*. 4th edition. Spon Press; London.

A very detailed book dealing with hydrological measurements and modelling.

Robinson, M. and Ward, R.C. 2017. *Hydrology: principles and processes*. IWA Publishing; London.

One of the best textbooks on river basin hydrology.

Classic papers

Anderson, M.G. and Burt, T.P. 1978. Role of topography in controlling throughflow generation. *Earth Surface Processes and Landforms* 3: 331–344.

One of the early papers on throughflow mapping.

Brookes, A. 1985. River channelisation, traditional engineering methods, physical consequences and alternative practices. *Progress in Physical Geography* 9: 44–73.

A review article that shows how traditional engineering approaches to channelisation lead to geomorphic problems and environmental degradation.

Leopold, L. and Maddock, T. 1953. The hydraulic geometry of stream channels and some physiographic implications. *Geological Survey Professional Paper 252*. US Department of Interior; Washington, DC.

The paper which identified simple power law relationships between channel width, depth, velocity and discharge.

Rogger, M. *et al.* 2017. Land-use change impacts on floods at the catchment scale – challenges and opportunities for future research. *Water Resources Research* 53: 5209–5219.

A review and future scoping paper which provides an overview about what is and is not known about land-use impacts on flooding.

PROJECT IDEAS

■ Borrow or build a tension-infiltrometer such as the one demonstrated on this video: www. youtube.com/watch?v=ZUxqQq5-oD8. Design and run an experiment to measure steady-state infiltration rates (i.e. when the rate is steady, normally after a few minutes) for two or three different soil types or land-cover types (e.g. field versus hedgerow). Run the experiment at 0, −1, −3, and −6 cm suctions and work out what proportion of the infiltrating water flows through the soil for pore sizes of >3 mm, 3 to 1 mm, 1 to 0.5 mm and <0.5 mm diameter (i.e. a suction of −3 cm in the infiltrometer will mean only pores smaller than 1 mm in diameter will be able to draw water out of it).

■ Find out about a river restoration programme near your location and determine the people and organisations involved, why the restoration is planned or has taken place and what the anticipated benefits might be. Find out if any monitoring is taking place to establish benefits for both the natural environment and local communities.

■ Measure the water efficiency for a small headwater river basin. Measure it also for a slightly larger area (which contains the smaller one) and again at another slightly larger scale. Are there differences in efficiency between scales? Why might this be?

REFERENCES

Abdel-Magid, A.H., Schuman, G.E. and Hart, R. H. 1987. Soil bulk density and water infiltration rates as affected by three grazing systems. *Journal of Range Management* 40: 307–310.

Anderson, M.G. and Burt, T.P. 1978. Role of topography in controlling throughflow generation. *Earth Surface Processes and Landforms* 3: 331–344.

Arnell, N. 2002. *Hydrology and global environmental change.* Pearson Education; Harlow.

Baird, A.J. 1997. Field estimation of macropore functioning and surface hydraulic conductivity in a fen peat. *Hydrological Processes* 11: 287–295.

Betson, R.P. 1964. What is watershed runoff? *Journal of Geophysical Research* 69: 1541–1552.

Beven, K.J. 2012. *Rainfall-runoff modelling – the primer.* 2nd edition. Wiley-Blackwell; Chichester.

Beven, K. and Germann, P. 1982. Macropores and water-flow in soils. *Water Resources Research* 18: 1311–1325.

Beven, K. and Germann, P. 2013. Macropores and water-flow in soils revisited. *Water Resources Research* 49: 3071–3092.

Beven, K. and Kirkby, M.J. 1979. A physically based, variable contributing area model of basin hydrology. *Hydrological Sciences Bulletin* 24: 43–69.

Brookes, A. 1985. River channelisation, traditional engineering methods, physical consequences and alternative practices. *Progress in Physical Geography* 9: 44–73.

Bryan, R.B. and Yair, A. 1982. Perspectives on studies of badland geomorphology. *In:* Bryan, R.B. and Yair, A. (eds) *Badland geomorphology and piping.* GeoBooks; Norwich, 1–12.

Cazenave, A., Le Cozannet, G., Benveniste, J., Woodworth, P.L. and Champollion, N. 2018.

Monitoring coastal zone changes from space. *Eos* 99 (2): 14–15.

Chanson, H. 2004. *The hydraulics of open channel flow: an introduction.* 2nd edition. Butterworth-Heinemann; Oxford.

Durner, W. and Or, D. 2005. Soil water potential measurement. *In:* Anderson, M.G. and McDonnell, J.J. (eds) *Encyclopedia of hydrological sciences.* John Wiley and Sons; New York, 1089–1102.

Evans, M.G., Burt, T.P., Holden, J. and Adamson, J. 1999. Runoff generation and water table fluctuations in blanket peat: evidence from UK data spanning the dry summer of 1995. *Journal of Hydrology* 221: 141–160.

Gao, J., Holden, J. and Kirkby, M.J. 2015. A distributed TOPMODEL for modelling impacts of land-cover change on river flow in upland peatland catchments. *Hydrological Processes* 29: 2867–2879.

Gao, J., Holden, J. and Kirkby, M.J. 2016. The impact of land-cover change on flood peaks in peatland basins. *Water Resources Research* 52: 3477–3492

Greenwood, K.L. and McKenzie, B.M. 2001. Grazing effects on soil physical properties and the consequences for pastures: a review. *Australian Journal of Experimental Agriculture* 41: 1231–1250.

Gupta, A. (ed.) 2007. *Large rivers: geomorphology and management.* John Wiley and Sons; Chichester.

Hallam, J. 2018. Soil hydraulic function: earthworm – plant root interactions. PhD thesis, University of York, UK.

Hewlett, J.D. and Hibbert, A.R. 1967. Factors affecting the response of small watersheds to precipitation in humid areas. *In:* Sopper, W.E. and Lull, H.W. (eds) *Forest hydrology.* Pergamon Press; New York, 275–290.

Hillel, D. 1982. *Introduction to soil physics.* Academic Press; New York.

Holden, J. 2005a. Infiltration, infiltration rate and infiltration capacity. *In:* Lehr, J.H. and Keeley, J. (eds) *Water encyclopedia*, Volume 5. John Wiley and Sons; New York, 212–214.

Holden, J. 2005b. Controls of soil pipe density in blanket peat uplands. *Journal of Geophysical Research* 110: F010002 doi: 10.1029/2004JF000143.

Holden, J. 2009. Flow through macropores of different size classes in blanket peat. *Journal of Hydrology* 364: 342–348.

Holden, J. 2011. *Physical geography: the basics.* Routledge; London.

Holden, J., Kirkby, M.J., Lane, S.N., Milledge, D.J., Brookes, C.J., Holden, V. and McDonald, A. T. 2008. Factors affecting overland flow velocity in peatlands. *Water Resources Research* 44: W06415.

Holden, J., Smart, R.P., Dinsmore, K., Baird, A., Billett, M.F. and Chapman, P.J. 2012. Natural pipes in blanket peatlands: major point sources for the release of carbon to the aquatic system. *Global Change Biology* 18: 3568–3580.

Holden, J., Grayson, R., Berdeni, D., Bird, S., Chapman, P.J., Edmondson, J.L., Firbank, L. G., Helgason, T., Hodson, M.E., Hunt, S., Jones, D.T., Lappage, M., Marshall-Harries E., Nelson, M., Prendergast-Miller, M., Shaw, H., Wade, R. and Leake, J.R. 2019. The role of hedgerows in soil functioning within agricultural landscapes. *Agriculture, Ecosystems and Environment* 273: 1–12.

Imeson, A.C. and Kwaad, F.J.P.M. 1990. The response of tilled soils to wetting by rainfall and the dynamic character of soil erodibility. *In:* Boardman, J., Foster, I.D.L. and Dearing, J.A. (eds) *Soil erosion on agricultural land.* John Wiley and Sons; Chichester, 3–14.

Inam, A., Clift, P.D., Giosan, L., Tabrez, A.R., Tahir, M., Rabbani, M.M. and Danish, M. 2007. The geographic, geological and oceanographic setting of the Indus River. *In:* Gupta, A. (ed.) *Large rivers: geomorphology and management.* John Wiley and Sons; Chichester, 333–346.

Leonard, J., Perrier, E. and de Marsily, G. 2001. A model for simulating the influence of a spatial distribution of large circular macropores on surface runoff. *Water Resources Research* 37: 3217–3225.

Leopold, L. and Maddock, T. 1953. The hydraulic geometry of stream channels and some physiographic implications. *Geological Survey Professional Paper 252*. US Department of Interior; Washington, DC.

Meyles, E.W., Williams, A.G., Ternan, J.L., Anderson, J.M. and Dowd, J.F. 2006. The influences of grazing on vegetation, soil properties and stream discharge in a small Dartmoor catchment, southwest England, UK. *Earth Surface Processes and Landforms* 31: 6221–6631.

Mwendera, E.J. and Saleem, M.A. 1997. Infiltration rates, surface runoff, and soil loss as influenced by grazing pressure in the Ethiopian highlands. *Soil Use and Management* 13: 29–35.

Perry, T. and Nawaz, N.R. 2008. An investigation into the extent and impacts of hard-surfacing of domestic gardens in an area of Leeds, United Kingdom. *Landscape and Urban Planning* 86: 1–13.

Price, J.S. 1992. Blanket bog in Newfoundland: part 1. The occurrence and accumulation of fog-water deposits. *Journal of Hydrology* 135: 87–101.

Robinson, M. and Ward, R.C. 2017. *Hydrology: principles and processes*. IWA Publishing; London.

Rogger, M., Alaoui, A., Agnoletti, M., Bathurst, J., Bodner, G., Borga, M., Chaplot, V., Gaál, L., Gallart, F., Glatzel, G., Hall, J., Holden, J., Holko, L., Horn, R., Kiss, A., Kohnova, S., Leitinger, G., Lennartz, B., Parajka, J., Peth, S., Perdigão, R., Plavcová, L., Robinson, M., Salinas, J.L., Santoro, A., Szolgay, J., Tron, S., van den Akker, J., Viglione, A. and Blöschl, G. 2017. Land-use change impacts on floods at the catchment scale – challenges and opportunities for future research. *Water Resources Research* 53: 5209–5219.

Shah, G.R. 1999. Sociological conditions of Indus Delta Rehabilitation and Replanting of Mangrove Project (IDRRMP) community and the use of mangrove ecosystem. *In: Proceedings of the National Seminar on Mangrove Ecosystem Dynamics of the Indus Delta*. Sindh Forest and Wildlife Department; Karachi, 112–123.

Shaw, E.M., Beven, K.J., Chappell, N.A. and Lamb, R. 2011. *Hydrology in practice*. 4th edition. Spon Press; London.

Sherman, L.K. 1932. Streamflow from rainfall by the unit graph method. *Engineering News Record* 108: 501–505.

Sherman, L.K. 1942. The unit hydrograph method. *In:* Meinzer, O.E. (ed.) *Hydrology*. Dover Publications; Englewood, CO, 514–525.

Smart, R.P., Holden, J., Dinsmore, K., Baird, A. J., Billett, M.F., Chapman, P.J. and Grayson, R. 2013. The dynamics of natural pipe hydrological behaviour in blanket peat. *Hydrological Processes* 27: 1523–1534.

Zhu, T.X. 1997. Deep-seated, complex tunnel systems – a hydrological study in a semi-arid catchment, Loess plateau, China. *Geomorphology* 20: 255–267.

CHAPTER FOUR

Surface water quality

4

*Pippa J. Chapman,
Paul Kay, Gordon
Mitchell and Colin S.
Pitts*

LEARNING OUTCOMES

After reading this chapter you should be able to:

■ understand the natural factors that control surface water chemistry;
■ understand how spatial and temporal patterns in surface water chemistry occur;
■ explain how water use by humans leads to a deterioration in water quality;
■ define the different sources of pollution of surface waters;
■ explain how changes in agricultural practices and urbanisation affect water quality;
■ discuss the causes of acid rain and acid mine drainage, and their impact on water quality;
■ understand the role of national and international policies and legislation in reducing water pollution.

4.1 INTRODUCTION

Surface waters refer to rivers, streams, lakes, ponds and reservoirs. Most of the water held in these zones is stored in lakes, ponds and reservoirs, while less than 1% is held in rivers and streams. However, river networks are important components of the landscape around which many human populations have centred.

As such they can be a key source of drinking water for many major conurbations.

When rain falls on the land it either seeps into the ground to recharge groundwater aquifers (see Chapter 5) or becomes **runoff**, which flows downhill over and through the soil into streams, rivers, ponds and lakes (see Chapter 3). However, surface water bodies do not just receive water from runoff; many receive inputs from groundwater (see **Figure 4.1**), the contribution of which generally increases during periods of low flow. Streams and rivers form where surface water accumulates and flows from land of higher altitude to lower altitude on its journey towards the oceans. Lakes or ponds form where surface runoff accumulates in a flat area, relative to the surrounding land, and the water entering the lake or pond comes in faster than it can escape, whether via outflow in a river, seepage to groundwater, or evaporation. This means that lakes and ponds are standing or very slow-moving bodies of water, while rivers and streams are distinguished by a fast-moving current. While most lakes contain freshwater, some, especially those where water cannot escape via a river, are salty. In fact, some lakes are saltier than the oceans (see also Section 3.3.2 in Chapter 3 on closed basins). The terms 'lakes' and 'ponds', and 'rivers' and 'streams', are often used interchangeably, because in reality there is no obvious distinction between them, although the latter term is typically used when describing a 'smaller' standing body of water or running watercourse, respectively. Reservoirs, also called impoundments, are human-made lakes. However, they can display characteristics of both rivers and lakes because they are created by building a dam across a river and flooding the valley. This damming and flooding creates an artificial lake, filled by the river inflow. Thus, the upstream section of the reservoir has predominantly river-like qualities, meaning there is often still some current, while the area closest to the dam is more lake-like.

The precipitation that falls on the land percolates over and through vegetation and soil, picking up **solutes** (dissolved matter) and sediment along its route to surface waters. This runoff delivers different amounts of solutes and sediment to rivers and lakes depending on the **hydrological pathway** it has taken through the catchment

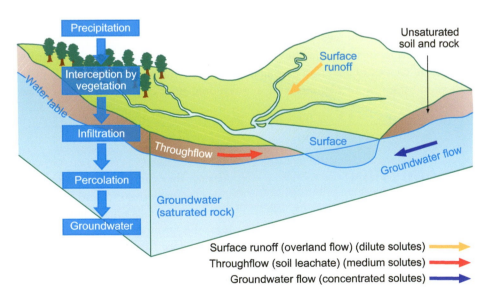

Figure 4.1 The different hydrological pathways that precipitation may take to reach surface waters and their effect on solute concentrations.

and the characteristics of the surrounding landscape (see **Figure 4.1**). Precipitation, which has low solute concentrations, may flow downhill over the soil surface, as **infiltration-excess overland flow** (see Chapter 3), or rapidly through the soil via **macropores** to reach rivers and lakes. As the residence time of this water within the catchment is short, the solute concentration of this runoff is usually very similar to that of precipitation. Alternatively, precipitation may flow through the soil horizons, where the residence time is longer, before reaching surface waters, and solute concentrations increase due to inputs from weathering reactions and microbial activity. Sometimes, when the soil is saturated, **saturation-excess overland flow** occurs (see Chapter 3) and this water can be a mix of fresh precipitation and water which has more solutes from within the soil. Lastly, precipitation may percolate through the soil into the bedrock below, if it is permeable. At a certain depth below the land surface, called the **water table,** the ground becomes saturated with water, whereas the ground above the water table is unsaturated. If a river cuts into this saturated layer, as shown in **Figure 4.1**, then water may flow out of the groundwater into the river. This is why even during dry periods there is usually some water flowing in streams and rivers. In addition, owing to the longer residence time of this pathway, groundwater tends to be enriched in solutes derived from weathering reactions. Surface water chemistry is therefore controlled by geology, vegetation and hydrological processes occurring in the river's basin. Thus, any changes that occur in the catchment also lead to a change in surface water chemistry. River systems also provide a vital linkage between the terrestrial and aquatic ecosystems. Thus streams and rivers have been referred to as 'the environment's circulatory system' (Wetzel, 2001).

Water is essential for the survival of terrestrial life, and without it humans could not survive. In addition to drinking water, humans use water for a wide variety of purposes that both consume and contaminate the water. Since ancient times, surface waters have been used as a convenient disposal route for our waste. However, this use (or abuse) of surface waters for waste disposal conflicts with almost all other uses of water and, most seriously, with the use of surface waters for drinking water. Even today, globally an estimated 80% of all industrial and municipal wastewaters are released to the environment without any prior treatment, resulting in a growing deterioration of overall surface water quality with detrimental impacts on human health, the environment and sustainable development (WWAP, 2017). The main regions that are subject to water quality threats are largely associated with high population densities and areas of economic growth. Since the 1990s, water pollution has increased in almost all rivers in Africa, Asia and Latin America (UNEP, 2016a). The deterioration of water quality is expected to escalate over the next decades, with consequences for human health and ecosystems.

4.2 CHEMICAL COMPOSITION OF SURFACE WATERS

It is important to understand the natural processes that control surface water chemistry before exploring the effect of human activities on water quality. Water is a dynamic chemical system containing gases, mineral and organic substances in both suspension and solution. The distinction between dissolved and suspended substances is based on filtration, usually through a 0.45 μm membrane filter, although other-sized filters are also used. The variety in natural surface water composition is defined not only by the occurrence of the major and minor constituents but also by the different forms and the concentrations and presence of each of them. This section briefly outlines the main constituents of surface water and ways by which surface water chemistry can be quantified and classified.

4.2.1 Dissolved inorganic substances

Mineral substances contained in surface waters in the dissolved state (in the form of ions, complex ions and colloids) are conventionally subdivided into major and minor components. When mineral substances dissolve in water they break down (dissociate) to form ions. For example, NaCl (sodium chloride) added to water produces Na^+ and Cl^- ions. Ions are charged atoms or molecules caused by having an unequal number of protons (positively charged) and electrons (negatively charged); positively and negatively charged ions are normally attracted to each other. Cations are positively charged, as they have more protons than electrons. Anions are negatively charged. Some ions are very soluble, while others have a strong affinity to become attached to suspended sediment.

The major dissolved ions in surface waters, which occur at concentrations exceeding 1 mg L^{-1}, are bicarbonate (HCO_3^-), sulfate (SO_4^{2-}), chloride (Cl^-), calcium (Ca^2), magnesium (Mg^2), sodium (Na) and potassium (K^+). In fact these seven ions along with silica (which occurs as $Si(OH)_4$ at the pH of most natural waters) constitute around 95% of the total inorganic solutes in surface waters. This reflects their relative abundance in the Earth's crust and the fact that they are moderately to very soluble. In contrast, the metals and metalloids, which occur at much lower concentrations (μg L^{-1} and <1 mg L^{-1}), are generally found in or bound to the particulate matter in surface waters. An intermediate position is occupied by ions of hydrogen, compounds of nitrogen, and phosphorus dissolved in water.

The sum of all the dissolved solutes plus silica present in a water body is known as the total dissolved solids (TDS). The TDS can be determined gravimetrically by evaporating a known volume of water and measuring the mass of the residue left. Alternatively, it can be determined by the measurement of electrical conductivity, as the dissolved ions present in the water create the ability for water to conduct an electric current. When correlated with gravimetric measurements of TDS, conductivity provides an approximation of the TDS concentration of a water sample. However, in very low-electrical conductivity waters, where the TDS is small, such as in some peatland stream waters, a normal conductivity meter will not work because an electrical circuit will not be completed. In these circumstances, special conductivity probes are required where the electrodes are placed very close together to measure low-TDS waters (including rainwater). Total dissolved solids in streams and rivers can vary between 50 and 1000 mg L^{-1}, compared with a typical value of less than 20 mg L^{-1} for rainwater (Meybeck et al., 1996). In lakes with outlets, TDS are similar to those of streams and rivers, whereas for lakes without outlets TDS can range from 1000 to 100,000 mg L^{-1} (Meybeck et al., 1996). Most bottles of mineral water will indicate the value of TDS on their label. In the United States, bottled water must contain greater than 250 mg L^{-1} TDS to be labelled as 'mineral water'. If the TDS exceed 1500 mg L^{-1} then a 'high mineral content' label is used. You should find that the higher the TDS, the more distinct the taste of the water.

Some elements exist in a number of different forms while others occur in more than one oxidation state. For example, the most common ionic forms of dissolved inorganic nitrogen in aquatic ecosystems are ammonium (NH_4^+), nitrite (NO_2^-) and nitrate (NO_3^-), while iron can be present as Fe^{2+} (ferrous iron) or Fe^{3+} (ferric iron). The form and oxidation state of an element is mainly controlled by environmental factors, particularly pH and redox potential (see **Box 4.1** for definitions of these terms), and has an important control on the solubility and **toxicity** of an element. Other major elements found in surface waters that are strongly influenced by redox reactions are carbon, nitrogen, sulfur and manganese.

4.2.2 Hydrogen ions (ph)

In the composition of surface waters, hydrogen ions (H^+) take an especially important place even though their absolute concentration as compared with other ions is very low. As outlined in **Box 4.1**, water that has more free hydrogen ions is acidic, whereas water that has more free hydroxyl ions is alkaline and we measure the relative amount of free hydrogen ions using the pH scale which goes from 0 to 14; pH below 7 indicates the water is acidic, and if it is above 7 it is alkaline. In reality the pH of

---

BOX 4.1

pH and redox potential

Whether a water body is acidic, neutral or alkaline is determined by measuring the hydrogen ion concentration in solution. In pure water at 24°C, water ionises (forms ions) to give equal concentrations of hydrogen (H^+) and hydroxide (OH^-) ions: $H_2O \leftrightarrow H^+ + OH^-$. The concentration of both H and OH^- ions is $1 \times 10-7$ (0.000 000 1) moles per litre. To overcome using these very small concentrations of H+ ions, a simpler method of using the negative logarithmic of the hydrogen ion concentration was developed, known as pH: $pH = -\log^{10} [H^+]$. Although the pH scale ranges from 0 to 14, with low values the most acidic and high values the most alkaline, most surface waters have a pH range of 4 to 9. As the pH scale is logarithmic it should be remembered that a change of one unit represents a ten-fold change in H^+ ion concentration.

A redox reaction is one where the oxygen state of the substance is changed. For example, a redox reaction occurs when carbon reacts with oxygen to become carbon dioxide. Oxidation involves the increase of oxidation state of a substance (such as in carbon to carbon dioxide) and is associated in a loss of electrons. Reduction is the reverse case. Redox potential (also known as oxidising or reducing potential or Eh) is a measure (in volts) of the tendency of a substance to acquire electrons compared with a standard. The standard used is hydrogen and its redox potential is set to zero:

$$E^+ + e^- \leftrightarrow \frac{1}{2}H_{2(g)}$$

As the redox potential decreases, the solution is more reduced (i.e. has more electrons to give), and as the redox potential increases, the solution is more oxidised (i.e. will accept more electrons).

most surface waters ranges between 4 and 10.5 (Table 4.1).

If the pH of water is too high or too low, the aquatic organisms living within it will die. The majority of aquatic creatures prefer a pH range of 6.5–9.0, though some can live in water with pH levels outside of this range. As pH levels move away from this range (up or down) this can stress animal systems and reduce hatching and survival rates. The further outside of the optimum pH range a water body is, the higher the mortality rates. The more sensitive a species, the more affected it is by changes in pH. In addition to biological effects, pH controls the solubility of chemicals such as nutrients (phosphorus, nitrogen and carbon) and heavy metals (lead, copper, cadmium, aluminium, etc.) and biological availability (amount that can be used by aquatic life). For example, in addition to affecting how much and what form of phosphorus is most abundant in the water, pH also determines whether aquatic life can use it. In the case of heavy metals, the degree to which they are soluble determines their toxicity. Metals tend to be more toxic at lower pH because they are more soluble.

Excessively high and low pH can also be detrimental for the use of water. High pH causes a bitter taste, water pipes and water-using appliances become encrusted with deposits of minerals, and it depresses the effectiveness of the disinfection of chlorine, thereby causing the need for additional chlorine when pH is high. Low-pH water will corrode or dissolve metals and other substances. Pollution can change a water's pH (see Section 4.5.3), which in turn impacts upon aquatic ecosystems. For instance, water coming out of an abandoned coal mine can have a pH of 2, which is very acidic and would definitely affect fish and macroinvertebrates.

4.2.3 Suspended sediment

The concentration of suspended particles (usually referred to as suspended sediment) can have a large impact on water quality. Suspended sediments include soil particles (clay, silt and sand), algae, plankton, microbes and other substances, and are typically in the size range of 0.004 mm (clay) to 1.0 mm (sand). In rivers, the main sources of suspended sediment include soil erosion, waste discharge, urban runoff and eroding stream banks, whereas in lakes decaying algae, plants and animals and bottom-feeding fish play a major role in controlling suspended sediment concentrations. High concentrations of suspended sediment can block sunlight from reaching submerged vegetation, which slows down photosynthesis. Reduced photosynthesis results in less dissolved oxygen being released into the water by the plants, and in extreme cases can result in the death of the plants. In addition, suspended sediment causes the water temperature to increase, as it absorbs heat from sunlight. This can result in the dissolved oxygen content reducing further because warm water holds less dissolved oxygen than cold water does. If the dissolved oxygen content of a water body becomes very low, fish may die. As suspended sediments settle they can cover the river/lake bottom, especially in slower waters, and smother fish eggs and benthic

Table 4.1 Typical pH values of surface waters

Water body type	pH range
Wetlands and swamps	4–6
Headwater streams	5–7
Major rivers	6.8–7.8
Large lakes	7.3–9.2
Salt lakes	up to 10.5

macroinvertebrates. The greater the amount of suspended sediment in the water, the murkier it looks and the higher the measure of **turbidity**. Turbidity refers to the clarity of the water and how much the particles suspended in water decrease the passage of light through the water. Turbidity is not a measurement of the amount of suspended sediment present, since it measures only the amount of light that is scattered by suspended particles. Measurement of total solids is a more direct measure of the amount of material suspended and dissolved in water. A turbidity meter consists of a light source that illuminates a water sample and a photoelectric cell that measures the intensity of light scattered at a 90° angle by the particles in the sample. It measures turbidity in nephelometric turbidity units, or NTUs. In rivers, turbidity readings are usually less than 10 NTU during periods of low flow (baseflow) but these can increase markedly during storms, when particles from the surrounding land are washed into the river. In addition, during storms, water velocities are faster, which can resuspend material from the stream bed (see Section 3.5.3 in Chapter 3). Chapter 6 describes in more detail the impact of turbidity on aquatic ecosystems.

4.2.4 Dissolved gases

All surface waters contain dissolved gases. Oxygen (O_2) and carbon dioxide (CO_2) are the dissolved gases of primary interest in most aquatic ecosystems because they both influence and reflect biological processes. As a result, they serve as tracers of electron flow (i.e. energy flow) in an ecosystem. Reactions that convert energy into an organic form will reduce CO_2. In the case of photosynthesis, energy is derived from light, and water is the electron donor, with the resultant production of O_2. Conversely, the decomposition of organic material results in the production of CO_2 and the loss of O_2. In general, the balance between carbon reduction and oxidation in lakes and rivers is controlled by light-driven photosynthesis. This, and the physical exchange of gases between water and the atmosphere, results in deep waters having higher CO_2 concentrations and lower dissolved O_2 concentrations than surface waters.

Given that dissolved oxygen (DO) in surface water is used by all forms of aquatic life, it is routinely measured to assess the 'health' of lakes and rivers. Concentrations of DO in surface waters can range between 0 and 18 mg L^{-1}, Concentrations above 9 mg L^{-1} can support abundant fish populations, but, as DO levels in water drop below 5 mg L^{-1}, fish and other aquatic life are put under stress and levels than remain below 1–2 mg L^{-1} for a few hours can result in large fish kills. The concentration of dissolved gases in water in equilibrium with a gaseous phase generally increases with pressure and decreases with temperature. Therefore, DO has both a seasonal and a daily cycle, with higher concentrations observed in the night than the day and in winter than summer.

4.2.5 Organic substances

The predominant source of natural organic substances in surface waters is the large pool of carbon-based compounds found within organic matter. It is matter that has come from the remains of organisms such as plants and animals and their waste products in the environment and can be divided into two components: a) dissolved organic matter (DOM), measured as coloured dissolved organic matter (CDOM) or dissolved organic carbon (DOC), and b) particulate organic matter (POM). They are typically differentiated by that which can pass through a 0.45-micrometre filter (DOM) and that which cannot (POM). Organic matter plays an important role in drinking water and wastewater treatment and recycling and natural aquatic ecosystems. It is an important component of the carbon cycle and energy balance in rivers and lakes and a primary food source in the aquatic food web. DOC is an important component of the acid-base chemistry of many surface waters and forms water-soluble complexes with nutrients and trace metals and therefore controls their mobility and transport from soil to water. In high concentrations, DOC can reduce light penetration in lakes and rivers. A variety of analytical detection methods for organic matter exist. These include, but are not limited to: total and dissolved organic carbon, mass spectrometry, nuclear magnetic resonance (NMR) spectroscopy, infrared (IR) spectroscopy, UV-visible spectroscopy and fluorescence spectroscopy. Each of these methods has its own advantages and limitations.

Humans have increased the input of organic matter to surface waters by increasing wastewater discharge due to urbanisation and intensification of livestock farming. Large quantities of organic matter (organisms and decaying organic waste) lead to higher rates of metabolic processes that demand oxygen. This can result in the development of water zones without oxygen (anaerobic conditions). The transformation of nitrogen to reduced forms under anaerobic conditions, in turn, leads to increased concentrations of ammonium, which is toxic to aquatic life above certain concentrations, depending on water temperature, salinity and pH. The key indicator for the oxygenation status of water bodies is **biochemical oxygen demand (BOD)**, which is the demand for oxygen resulting from organisms in water that consume oxidisable organic matter. Therefore

BOD is a commonly measured feature of water quality in surface waters. It equates to the amount of dissolved oxygen (in milligrams) needed by biological organisms in a water body so that they can break down organic matter in a sample of water (normally one litre) over a set time period (normally five days) for a given temperature (normally 20°C). Typically, if there are a lot of organic compounds present in the water sample the BOD will be very high. Most unpolluted rivers are thought to have a BOD of less than 1 mg L^{-1}, while untreated sewage might have a BOD of 50–500 mg L^{-1} (see Section 8.3 of Chapter 8). If there is a discharge of organic pollutants (e.g. failing septic tank, urban storm water runoff, effluent from a pulp and paper mill) into a river, the dissolved oxygen concentration will quickly drop, which might result in the deaths of many aquatic animals that need to extract oxygen from the water in order to survive. This drop in dissolved oxygen concentration after a pollution incident is often known as a sag curve and is shown in Chapter 8 (**Figure 8.6**). Useful methods for measuring BOD can be found at: http://water.epa.gov/type/rsl/monitoring/vms52.cfm.

Using in-stream BOD as an overall indicator of organic river pollution Wen *et al.* (2017) calculated historical (2000) and future (2050) BOD concentrations in global river networks. Despite significant self-cleaning capacities of rivers, they estimated that the number of people affected by organic pollution (BOD >5 mg L^{-1}) is projected to increase from 1.1 billion in 2000 to 2.5 billion in 2050, with developing countries disproportionately affected. In particular, wastewater treatment in the Indian subcontinent, mid-eastern China, South Korea, Brazil, Mexico, as well as smaller regions in Africa and south-eastern Asia, were insufficient to keep BOD concentrations below 5 mg L^{-1} (Wen *et al.*, 2017), which points to a growing need for affordable wastewater solutions to protect surface water quality. Another variation of the BOD is the sediment oxygen demand (SOD), which is the usage of dissolved oxygen in the overlying water by sediment chemicals and organisms that live in the sediment on the bed of the water body. Such organisms include burrowing fauna, worms, insect larvae, bacteria, protozoa and fungi. To measure SOD, sediment cores are normally extracted and then oxygen use is measured in the laboratory over a certain period of time at a controlled temperature.

The **chemical oxygen demand (COD)** is another measure of pollution by organic compounds, but one which tests the chemical demand for oxygen within the water body without including the biological processing. The COD is a measure of the amount of dissolved oxygen needed to oxidise the organic matter within a litre of water using a standard chemical oxidising agent. COD is measured in milligrams (mg) of oxygen per litre (L) of water and is more commonly used in testing wastewater than surface waters in rivers and lakes (see Chapter 9).

Hundreds of individual organic substances both natural and synthetic can now be analysed and found in freshwaters. A few of them may be found at very low levels in natural waters, such as phenols, hydrocarbons and some polyaromatic hydrocarbons. Most of them are not found in natural conditions (they are xenobiotics) and are derived from pesticides and solvents. In polluted surface waters, phenols and hydrocarbon concentrations are much higher than natural background values.

4.2.6 Temperature

Water temperature exerts a major influence on biological activity. Temperature governs the kinds of organisms that can live in rivers and lakes. Fish, insects, zooplankton, phytoplankton and other aquatic species all have a preferred temperature range. As temperatures get too far above or below this preferred range, the mortality rate increases.

Lakes and reservoirs are characterised by alternating periods of **stratification** and vertical mixing, which has an effect on water chemistry. Stratification in lakes or reservoirs occurs when the water acts as two different bodies with different densities, one floating on the other. It is most commonly caused by temperature differences within the water column, particularly in deep (>10 m) temperate lakes and reservoirs, because of seasonal changes in temperature. It is less common in tropical lakes, where temperature varies little over the year. Stratification results in different processes dominating in the different layers of the lake. In temperate regions during the summer, the surface layer, known as the **epilimnion**, receives more sunlight, is warmer and has an oxidising environment because it is exposed to the atmosphere, than the deeper layer, known as the **hypolimnion**. Owing to the isolation of the hypolimnion from the atmosphere, it is cooler and receives less oxygen, leading to anoxic conditions developing in the lake-bottom sediments. This gradient of dissolved oxygen means that elements that have more than one oxidation state occur in the oxidised state in the epilimnion and the reduced form in the hypolimnion (**Figure 4.2**). Thus, stratification of reservoirs can lead to concentrations of iron and manganese exceeding drinking water quality standards. This is because the reduced forms of these metals are much more soluble than the oxidised forms. Therefore the metals within the sediments on the reservoir bottom will leach out and dissolve in the water, and, if water is abstracted from the

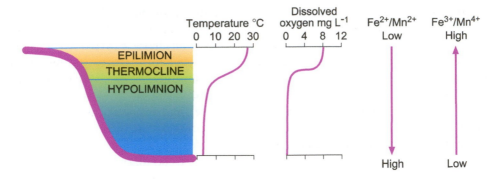

Figure 4.2 Cross-section of a thermally stratified eutrophic lake in summer showing the location of the epilimnion, thermocline and hypolimnion and associated changes in temperature, dissolved oxygen and iron and manganese concentrations with depth.

hypolimnion in the reservoir, high concentrations of dissolved lead and manganese will be found in the abstracted water.

4.2.7 Trophic status

The trophic status of a water body (Table 4.2) can roughly be assessed by using information about the concentration of the limiting nutrient (phosphorus in freshwaters), chlorophyll (an indicator of phytoplankton biomass) and transparency (dependent on both algal biomass and sediment resuspension, expressed as Secchi depth). Oligotrophic lakes are characterised by low nutrient status, high biodiversity and low levels of primary production.

Eutrophication is the process by which waters become enriched with nutrients (phosphorus in freshwaters and nitrogen in marine waters), which disturbs the ecological balance of production between aquatic macrophytes and phytoplankton, shifting systems to the eutrophic end of the spectrum (Table 4.2). The process can occur naturally but it is widely augmented by human activities (Moss, 2010). Enrichment causes increased algal production, turbidity and a change in the structure and functioning of lake ecosystems. Phosphorus is regarded as being the most

Table 4.2 Critical thresholds of classification for trophic status

Trophic status	Mean annual total P concentration ($\mu g\ L^{-1}$)	Mean annual chlorophyll a concentration ($\mu g\ L^{-1}$)	Mean depth of Secchi disc visibility (m)
Ultra-oligotrophic	<4	<2.5	>6.0
Oligotrophic	<10	2.5–8	>3.0
Mesotrophic	10–35	8–25	3–1.5
Eutrophic	35–100	25–75	1.5–0.7
Hypertrophic	>100	>75	<0.7

Source: After Dobson and Frid, 2009. By permission of Oxford University Press: *Ecology of aquatic systems.* 2nd edition.

important nutrient driving eutrophication because it usually limits primary production in freshwaters, due to its relative scarcity (Dillon and Rigler, 1974). The optimum atomic ratio of nitrogen to phosphorus for primary production is 16:1 (the Redfield Ratio), but in most circumstances this ratio is exceeded, implying that phosphorus is the limiting nutrient. Algae are able to take advantage of the nutrients more readily than the macrophytes and a larger phytoplankton biomass leads to increased turbidity. This reduces light penetration and therefore further reduces the extent and presence of rooted macrophytes (Irvine *et al.*, 1989). Blue-green algae, also known as cyanobacteria, are the type of algae that are able to take the greatest advantage of the nutrient enriched conditions. The ecological implications of their dominance in nutrient enriched waters are discussed further in Chapter 6 (Section 6.5.5). In addition, some types of blue-green algae are also able to produce toxins which are harmful to human health (see Section 7.4.1 in Chapter 7). For example, Toledo, Ohio, had its tap water supply shut down for half a million people over several days in 2014 when Lake Erie's water became too toxic to treat. During the last five decades, eutrophication has been a major global threat to the quality of our freshwater resources. For example, approximately 30% of rivers and 40% of lakes in Europe were eutrophic or hypertrophic in 2008–2011 (EC, 2013).

The origins of phosphorus from rural sources are largely due to erosion of soil particles from agriculture and forestry practices. The phosphate (PO_4^{3-}) transported on these sediments are usually bound with metals. However, the main source of phosphorus to freshwaters is through treated sewage effluents and storm drainage, which can carry high concentrations of sodium tripolyphosphate, a compound used as a component of detergents. Agricultural fertilisers are the main source of nitrogen in the form of nitrate (NO_3^-), which is highly soluble in water. Current global estimates of nitrate fertiliser production are around 100 million tonnes, compared to less than 10 million tonnes in 1950. Consequently, the environmental impacts of this escalating use are of increasing concern.

REFLECTIVE QUESTIONS

Why might the depth of a water abstraction point on a reservoir be important for the resultant water quality of the abstracted water?

What are the sources of organic substances in surface waters?

What factors can lead to eutrophication?

4.3 NATURAL SPATIAL VARIATION IN SURFACE WATER CHEMISTRY

In natural conditions, the chemistry of surface waters is controlled by three main processes (Gibbs, 1970; Meybeck *et al.*, 1996; Walling and Webb, 1986): a) atmospheric inputs, b) weathering of bedrock and c) climate (evaporation–crystallisation processes). The main source of atmospheric inputs is from oceanic aerosols (Na and Cl); however, other sources include volcanic emissions (such as SO_2) and the dissolution of soil-derived particulates (such as Ca, HCO_3 and SO_4). The distance from the ocean will determine the extent to which sea-spray rich in Na, Cl, Mg and SO_4 influences surface water chemistry. In coastal areas, where oceanic inputs may dominate over all other sources, Na and Cl are the most abundant solutes in surface waters. These solutes may also dominate in areas where rocks are extremely resistant to weathering, such as in the central Amazon.

In nearly all major rivers of the world, calcium is the predominant cation and bicarbonate the major anion. In fact, 80% of the dissolved load of the major rivers of the world is made up of just four solutes: calcium, bicarbonate, sulfate and silica (Walling and Webb, 1986). The major source of these solutes is the weathering of bedrock. The susceptibility of rocks to weathering varies from low for **plutonic** and **metamorphic** crystalline rocks to high for **sedimentary rocks** such as limestone and chalk. Thus TDS in rivers and streams draining catchments underlain by sedimentary rocks are five times greater than in rivers draining catchments dominated by crystalline rocks and about two and a half times greater than in catchments underlain by volcanic rocks (Meybeck, 1981). In some areas of the world, where rock type is very homogeneous, such as the Canadian and African shields and the Amazon and Congo basins, stream-water chemistry may vary very little over large areas. In contrast, stream-water chemistry can be highly variable in areas where geology varies on the finer scale. Overall, approximately 60% of all the solutes present in surface waters are derived from the weathering of rocks (Walling and Webb, 1986).

The influence of climate on surface water chemistry is very important at the global scale. As well as controlling the rate of weathering, through its influence on temperature and water availability, climate controls the water balance. In areas where evaporation exceeds precipitation, solutes derived from both weathering and atmospheric inputs are gradually concentrated. If evaporation is high enough, some minerals will start to precipitate, such as calcite ($CaCO_3$) and dolomite ($MgCO_3$). The precipitation of minerals results in the surface waters containing lower concentrations of Ca and HCO_3 and higher concentrations of Na, Cl, Mg and SO_4. In contrast, in areas where annual rainfall is high, the concentration of solutes in surface waters is diluted.

It has been proposed that because the above three processes influence the concentration of the major solutes in different ways, the relative abundance of the dominant cation and anion can be used to evaluate which of the processes is most significant in controlling surface water chemistry (**Figure 4.3**). High concentrations of Na and Cl, with low TDS, are observed in wet coastal regions where atmospheric inputs are the dominant source of solutes, and inputs derived from weathering are low. Sodium and Cl can also dominate in semi-arid areas where evaporation-crystallisation processes dominate, resulting in very high TDS. However, Gibbs (1970) found that most surface waters were dominated by Ca and HCO_3 derived from the weathering of bedrock, and plot in the central region of **Figure 4.5**. This was supported by the findings of Meybeck (1981) in his survey of global rivers (representing about 60% of the water discharge to the oceans), where he observed that 97% of rivers were dominated by Ca and bicarbonate and that Na was the dominant cation in only 1.7% of rivers and Cl and SO_4 were dominant in just over 1% of rivers. Thus the classification by Gibbs (1970), although attractive in its simplicity, is of limited value. In addition, other surface waters deviate from the plot in **Figure 4.5**, owing to extreme geological or climatic conditions. For example, very acidic waters are often found in active volcanic regions or in sedimentary rocks rich in metal sulfides.

REFLECTIVE QUESTION

Are TDS in surface waters likely to be greater or smaller in warmer, drier climates than elsewhere?

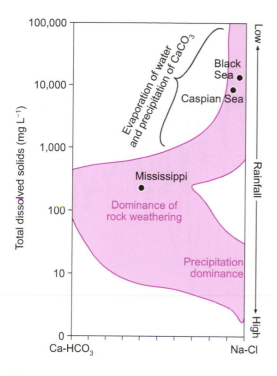

Figure 4.3 The influence of geology and climate on the total dissolved solids (TDS) and major chemical composition of surface waters.

Source: Adapted from Gibbs (1970) Mechanisms controlling world water chemistry. *Science* 170: 1088–1090. Reprinted with permission of AAAS.

4.4 TEMPORAL VARIATIONS IN SURFACE WATER CHEMISTRY

Surface water chemistry naturally varies over time for two main reasons. First, the processes controlling the supply of solutes available for leaching vary over time, and, second, the hydrological processes that generate runoff within a catchment are dynamic and vary with climate and season. The major temporal patterns in surface waters can be considered at the following timescales:

1. during storms (hours) as a result of changes in hydrological flowpaths delivering different sources and amounts of solutes;
2. diurnal (24-hour) variations in, for example, pH and dissolved oxygen, resulting from biological processes and changes in daylight;
3. seasonal variations associated with biological and hydrological cycles;
4. year-to-year trends, usually as a result of anthropogenic activities in the catchment.

Fluctuations in dissolved oxygen concentration and pH are primarily related to photosynthesis and aerobic respiration of aquatic plants. Photosynthesis, which is driven by sunlight, removes carbon dioxide (CO_2) from the water and produces free oxygen, which causes an increase in surface water pH and dissolved oxygen. However, stream-water pH and dissolved oxygen concentrations are also affected by changes in air pressure and temperature. The solubility of gases increases as air pressure increases; this can influence the pH of water, as more CO_2 will dissolve in the water as air pressure rises. In contrast, increases in temperature reduce the solubility of most gases; cold water can hold more dissolved oxygen than warm water. Therefore, the dissolved oxygen content of surface water has both a seasonal and a diurnal cycle, as shown in **Figure 4.4**. This diurnal variation in dissolved oxygen is not generally observed from collecting regular water samples, as this typically occurs in the daytime. However, the use of continuous monitoring probes has enabled these small diurnal variations to be clearly observed (**Figure 4.4**).

During storms, changes in dominant hydrological flowpaths through the catchment (Chapter 3) can lead to large changes in solute concentrations, over relatively short periods of time. For some solutes, an increase in river discharge can lead to an increase in concentration as solutes are

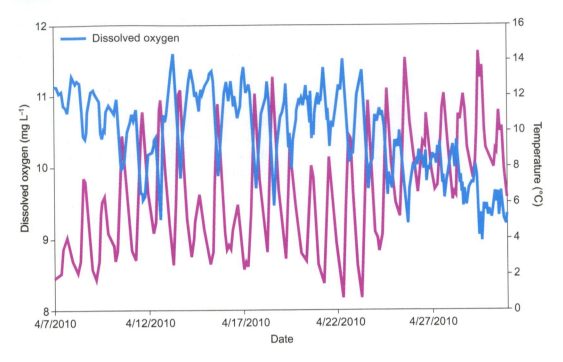

Figure 4.4 Dissolved oxygen concentrations and temperature recorded at 15-minute intervals at a stream in the North Pennines, UK, shows clear diurnal variations.
Source: Data kindly provided by Katie Aspray, University of Leeds.

moved rapidly into the river by overland flow and shallow throughflow. For example, dissolved organic carbon (DOC) concentrations can increase with stream discharge during storms in some catchments (Chapman *et al.*, 1993; Hinton *et al.*, 1997; **Figure 4.5**). This reflects the change in hydrological flowpaths, with more flow through the upper organic surface soil horizon in the catchment where DOC is produced. In contrast, other solutes that are mainly derived from the weathering of bedrock, such as Ca, Mg and Si, and transported to surface waters by deep soil throughflow and groundwater display a negative relationship with discharge during storms (**Figure 4.5**), as their concentrations are diluted by surface runoff.

In general, the higher the annual discharge of a river, the lower the concentration of solutes, and hence TDS, in a river. This inverse relationship can be explained simply by the dilution of the available solutes as surface runoff volumes increase. However, as the volume of runoff increases, so does the total amount of solutes released from the catchment and hence available for transport. This is known as the **solute load** or flux and is determined by multiplying the solute concentration by the discharge at a specific location on a river at a certain time. To calculate the annual load of a specific solute, continuous measurements of discharge and concentration are required. However, while discharge can be measured regularly, solute concentration is usually measured less frequently, owing to the expense of collecting and analysing water samples. This means that solute concentration must be estimated

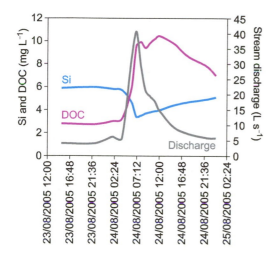

Figure 4.5 Changes in stream discharge and silicon and dissolved organic carbon (DOC) concentrations during a storm event in a headwater stream in north-east Scotland.
Source: Adapted from Stutter *et al.* (2012b).

between sampling periods. Several studies have compared different methods for estimating solute and sediment loads (e.g. Littlewood, 1992; Webb *et al.*, 1997) and have found that regular monitoring programmes tend to over-emphasise baseflow components, which is compounded the longer the time interval between samples.

In catchment studies, the annual dissolved load serves as an integrated measure of all the processes that occur within the river basin that affect stream-water chemistry. While measurement of solute concentration informs us how much of a particular solute is present at a specific point in time and is useful for comparing with water quality standards, solute loads can be used to quantify the amount of solute entering a specific water body, such as a lake or ocean, from terrestrial sources, or the impact of environmental changes on the terrestrial cycle of nutrients. In order to compare solute loads from catchments of

different sizes the load is divided by the area of the catchment and is generally expressed in tonnes per km^2 per year (which is equivalent to $g\ m^{-2}\ yr^{-1}$) (see **Box 4.2**).

Many solutes display strong seasonal patterns (**Figure 4.6**). These patterns can be divided into those that are controlled by seasonal changes in hydrology and those that are mediated by biological processes. In streams and rivers, pH and concentrations of Ca and Mg often display a maximum in summer and minimum in winter, which reflects seasonal changes in the dominant source of water (**Figure 4.6a**). In summer, groundwater enriched with base cations often dominates stream flow, whereas in winter soil water and surface runoff, which is more acidic and contains lower concentrations of base cations, dominates. In natural waters, nitrate (NO_3) and potassium (K) also display strong seasonal patterns, with a summer minima and winter maxima (**Figure 4.6b**). This reflects seasonal changes in the availability of these nutrients for leaching, as plant uptake during the summer reduces the concentration of these nutrients in soil solution. DOC is also strongly mediated by biological processes, being produced via the decomposition of organic matter during summer and washed from the soil during the autumn. Hence DOC displays a maximum in late summer and autumn and a minimum in early spring (**Figure 4.6c**).

Long-term patterns in surface water chemistry are usually driven by climatic variability and anthropogenic activities, such as changes in land management or atmospheric pollution. However, it is often difficult to identify the main driver of change, as many of these factors are changing at the same time or the record of solute concentration (often referred to as a time-series) is of insufficient length. A number of studies have highlighted that caution is needed in

BOX 4.2

Calculating solute loads for Eagle Creek, Indiana, USA

Table 4.3 shows some example data for stream discharge and nitrate-N (NO_3-N) concentrations at hourly intervals for Eagle Creek at Zionsville in Indiana, which were obtained from the United States Geological Survey (USGS) database, which is a public domain source. Data are for one day in January 2013. The drainage area is 274.5 km^2. We would like to determine the total flux of nitrate-N (nitrate-N refers to the amount of nitrogen present in the nitrate ion, so if we want to calculate the total nitrate (NO_3) flux we need to multiply the nitrate-N flux by 4.43, as 10 mg L^{-1} of NO_3-N is equivalent to 44.3 mg L^{-1} of NO_3) from the catchment and, so that we can fairly compare it to data from other sites, to calculate the solute load for the day per unit of catchment area.

Multiplying the discharge value for each hour by the number of seconds in each hour gives us an estimate of total discharge in that hour in m^3. In order to put everything into the same units we can divide the concentration of nitrate by 1000 to express it as mg per m^3 of water (since there are 1000 litres in 1 m^3). If we then assume that the concentration of nitrate-N applies for the whole hour, then, multiplying that concentration by the total discharge for the hour, we end up with a value of mg of nitrate that has moved past the gauging point on the river. Taking the first row of data from Table 4.3 as an example gives us 7848 m^3 of water for that hour (multiply 2.18 m^3 s^{-1} by 3600 s). The value of 3.22 mg L^{-1} of nitrate-N is equivalent to 3220 mg m^{-3} (1000 litres of

Table 4.3 Nitrate-N concentrations and discharge for a 24-hour period 2–3 January 2013 for Eagle Creek, at Zionsville, Indiana

Time	Discharge (m^3 s^{-1})	Nitrate-N (mg L^{-1})
18:00	2.18	3.22
19:00	2.15	3.31
20:00	2.10	3.34
21:00	2.18	3.39
22:00	2.35	3.36
23:00	2.29	3.39
00:00	2.29	3.39
01:00	2.35	3.40
02:00	2.38	3.36
03:00	2.55	3.37
04:00	2.55	3.40
05:00	2.61	3.42
06:00	2.61	3.45
07:00	2.69	3.45
08:00	2.80	3.48
09:00	2.97	3.40
10:00	3.14	3.45
11:00	3.37	3.47
12:00	3.60	3.40
13:00	3.09	3.36
14:00	2.61	3.37
15:00	2.10	3.45
16:00	1.95	3.50
17:00	2.01	3.60

water in one cubic metre). If we multiply 3220 mg m^{-3} of nitrate-N by 7848 m^3 of water we obtain a total flux for the hour of 2,527,560 mg of nitrate-N. This is equivalent to 25.27 kg of nitrate-N. Repeating this for each hour and summing the total provides us with an estimate of 747 kg of nitrate-N for the 24-hour period shown in the table. As the catchment area is 274.5 km^2 we can suggest that for one 24-hour period the nitrate-N flux from Eagle Creek was equivalent to 2.7 kg km^{-2}. Obviously we could repeat this for the whole year quite easily in a spreadsheet to provide a more realistic annual flux estimate.

In many cases we do not have data at such high frequency as is presented in Table 4.3. While we might have good discharge data collected at high frequency (e.g. every 15 minutes), often water quality samples for different solutes are taken once per week or sometimes less frequently, occasionally supplemented by sampling during some storm events. In these instances it is not suitable simply to multiply discharge at the time of sampling by concentration of the solute and then sum the values. That is because we do not have values of concentration for all of the times in between (when the discharge and concentrations might be quite different). Therefore, we need to interpolate values between sample points to produce a value of annual solute load. The methods most commonly used vary depending on what data are available and are described in Walling and Webb (1985).

interpretation of 'long-term' river and lake water chemistry and that misleading conclusions may be drawn from datasets of ten years or less (e.g. Burt *et al.*, 1988; Driscoll *et al.*, 1989). For example, analysis of stream-water NO$_3$ concentrations from the Hubbard Brook Experimental catchment, north-east United States, for the period 1964–74 suggested a strong increasing trend (Likens *et al.*, 1977). However, no long-term trend was observed in the dataset for the period 1964–1987 (Driscoll *et al.*, 1989). This illustrates the importance of maintaining long-term sites and networks of surface water chemistry.

In the 1980s and 1990s, many countries set up atmospheric and stream or lake monitoring networks to determine the impact of emission controls implemented at coal-fired power stations on improving air and water quality. In response to rapid declines in atmospheric sulfate deposition across North America and Europe since the late 1980s (e.g. Jeffries *et al.*, 2003; Fowler *et al.*, 2005), there have been widespread reductions in stream water sulfate and inorganic aluminium concentrations (Burns *et al.*, 2006; Davies *et al.*, 2005; Skjelkvåle *et al.*, 2005; Stoddard *et al.*, 1999; Figure 4.7a), and in some regions an increase in water pH. However, streamwater recovery has been only partial and many waters remain in an acidified state. This lag in surface water acidification recovery (Figure 4.7c) has been attributed to a range of processes, including deposition of N remaining high, depletion of base cations and increase in DOC concentrations (Figure 4.7b). This demonstrates the long-term influence on surface water chemistry of biogeochemical processes occurring within a catchment.

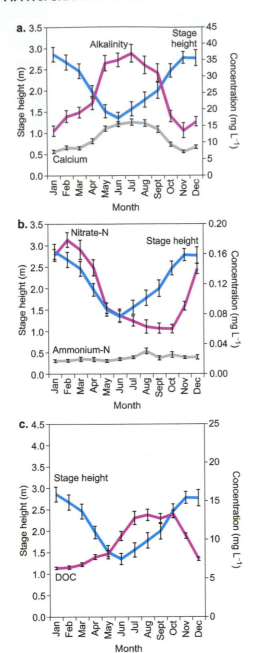

Figure 4.6 Mean monthly concentration in stream water collected weekly from the headwaters of the River Tees, northern England (1993–2010): a) calcium and alkalinity, b) nitrate and ammonium, and c) DOC. Error bars represent the standard error of the means (n = between 72 and 79 per month).

Source: Data kindly provided by the Environmental Change Network, 2012, citation code: ECN:PC7/12.

REFLECTIVE QUESTIONS

Ignoring human influence, what are the key reasons that solute concentrations naturally vary over time in surface waters?

What things do you need to measure if you want do know the total solute load for a particular solute over a given time period?

4.5 SURFACE WATER POLLUTION: CAUSES, CONSEQUENCES AND MANAGEMENT

The industrialisation of a country leads to a shift in the pattern of water use from rural and agricultural to urban and industrial, and the growth of large cities has been accompanied by increasing stress on the aquatic environment, leading to a reduction in the quality of the water (McDonald *et al.*, 2011). Poorer water quality means water pollution.

Water pollution is usually defined as 'where one or more substances have built up in a water body to such an extent that it can be harmful to organisms and plants which live in the water body and to animals and humans that may drink the water'. Although water pollution usually refers to a deterioration in water quality caused by human activities, the same effect may occur naturally. For example, large amounts of sediment can be washed into streams and rivers during rainstorms, and toxic metals find their way into surface waters, where concentrations of metal ores occur. These are natural processes and the environment will be able

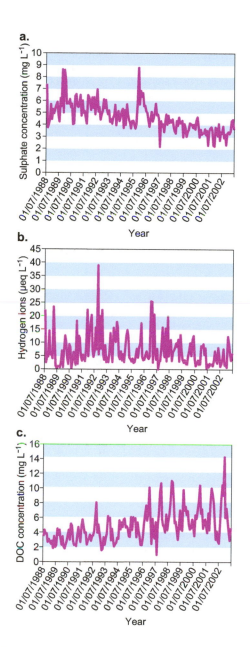

a.

b.

c.

Figure 4.7 Long-term trends in the concentration of a) sulfate (SO$_4$), b) DOC and c) hydrogen (H$^+$) ions, in samples collected from the River Etherow, Peak District, UK, forming part of the UK Upland Waters Monitoring Network (UWMN).

Source: Data kindly provided by CEH Lancaster as part of the UWMN.

to neutralise the effects on water quality over time. Pollution caused by human activities, on the other hand, is usually on a larger scale, can often happen rapidly and can take a much wider variety of forms, as outlined in Table 4.4.

Pollution sources that affect surface water are usually separated into two categories: a) point sources and b) non-point sources or **diffuse pollution**. If pollution comes from a single location, such as a discharge pipe attached to a factory, it is known as **point-source pollution**. Other point sources include sewage treatment plants, industrial discharges or any other type of discharge from a specific location (commonly a pipe) into a stream. By contrast, **non-point sources** are diffuse sources of pollutants that are not as easily identified or measured as point sources. Common examples include:

- leaching of fertilisers and pesticides from agricultural and residential lands;
- sediment from agricultural and forest lands, construction sites, and eroding stream and riverbanks;
- bacteria and nutrients from livestock waste and leaking septic tank systems;
- salt, oil, grease and heavy metals from urban runoff;
- acid drainage from abandoned mines;
- atmospheric deposition.

When point-source pollution enters the aquatic environment, the place most affected is usually the area immediately around the source. However, it is also dependent on the capacity of the receiving water body to dilute the input, and the ecological sensitivity of the receiving water. In contrast, the location most affected by diffuse pollution may be far downstream from the location where the pollutant entered the watercourse, as it is

Table 4.4 The nature, sources, effects and control of some major types of pollutants

Pollutant	Nature	Source	Effects	Control
Bacteria	Disease-causing organisms	Domestic sewage and animal waste	Illness in humans and curtailed recreational use	Treatment of waste (e.g. at sewage treatment works) and chlorination of drinking water
Nutrients	Mainly refers to nitrogen and phosphorous species	Domestic sewage, industrial waste and agricultural land (mainly inorganic fertiliser)	Excessive growth of aquatic plants, leading to eutrophication; high nitrate concentrations in drinking water, potentially toxic	Phosphorus removed at sewage treatment works. Reduce nitrate leaching by controlling fertiliser inputs and good land management
Trace metals		Mining, industrial processes, runoff from roads	Toxic effects on aquatic ecosystems	Difficult
Organic chemicals	Pesticides, detergents, industrial by-products and pharmaceutics	Domestic sewage, industrial waste and farms	Toxic threat to fish and other aquatic fauna; possible impacts on human health	Difficult to control, as not removed by conventional sewage or water treatment works
Natural organic matter	Biodegradable organic material decomposed by aerobic bacteria (requires dissolved oxygen)	Domestic sewage, animal manure, food processing	Causes depletion in dissolved oxygen content, which has an impact on aquatic fauna and flora	Treatment of sewage, containment of sewage, animal slurry and manure and silage effluent
Sediment	Mainly soil material (organic and inorganic); industrial by-products	Land erosion by water and wind, construction sites, some quarrying and mining processes	Removal of fertile soil, filling of reservoirs and lakes, sedimentation of gravel beds; reduced aquatic life	Controlled by soil conservation and best management practices; use of settling ponds; flood control
Heat	Heated water returned to rivers and lakes	Power stations, steel mills and refineries and other industrial cooling units	Reduces dissolved oxygen content of water, with potential harm to aquatic life	Minimised by recirculation and reuse of industrial cooling waters

often the cumulative effect of small amounts of contaminants gathered from a large area that results in an impact. Where pollution enters the environment in one place but has an effect hundreds or even thousands of miles away, this is known as transboundary pollution. An example of this is the emission of pollutants, such as sulfur dioxide (SO_2), into the atmosphere in one country which are then deposited in another country, having an impact on water quality (see Section 4.5.4 on acid rain).

Different sources of pollution respond to storms in contrasting ways. Usually, the concentration of pollutants from non-point sources will increase as flow increases during storm runoff. In contrast, concentrations of pollutants from point sources generally decrease through dilution during storm runoff. Thus, water monitoring programmes that collect water samples at regular time intervals (usually weekly or monthly) are much more likely to identify point sources of pollution which can be treated. In contrast, pollution from diffuse sources is more difficult to treat and remains the major challenge to improving surface water quality globally.

There are two main ways of measuring the quality of water. One is to take water samples and measure the concentrations of different solutes and sediments that they contain. If the concentrations exceed a certain limit, we can regard the water as polluted. Measurements like this are known as chemical indicators of water quality. An alternative method is to examine the fish, insects and other invertebrates that live in the aquatic environment. A widely used system focuses on macroinvertebrate benthic fauna and utilises the Biological Monitoring Working Party (BMWP) scoring system (Armitage *et al.*, 1983) to assess water quality. Measurements like this are called biological indicators of water quality. Often both types of measurements are made in order to classify the quality of a water body. For example, the European Water Framework Directive (WFD) requires Member States to use chemical and biological indicators, along with hydrogeomorphological parameters, to rank the ecological status of rivers (see **Box 6.4** in Chapter 6).

4.5.1 Agriculture

Farm types can be split generally into arable and livestock. Common arable crops include wheat, barley, rice, oats and oilseed rape, as well as vegetables, such as potatoes and other root crops. Livestock farms tend

to focus on one particular type of production, which might be poultry, dairy, beef, sheep or pigs. Some of these farms will operate intensive indoor systems, while others will use more extensive outdoor production techniques.

4.5.1.1 Impacts of agriculture on water quality

Agriculture can have a huge impact on water quality, owing to the vast array of chemicals that are used to increase production. In addition, agriculture is by far the biggest user of available freshwater globally (see **Figure 9.1** in Chapter 9): almost 70% of water withdrawals from rivers and groundwater are for agricultural usage, three times more than 50 years ago. By 2050, the global water demand of agriculture is estimated to increase by a further 19% due to irrigation needs. Approximately 40% of the world's food is currently cultivated in artificially irrigated areas, especially in the densely populated regions of South East Asia. Of the water used in agriculture, only about half of it can be reused, because much of it evaporates or is lost during transit. This means that the discharges of major rivers are reduced, especially during dry seasons, and therefore the impact of pollutants on aquatic ecosystems is exacerbated. Diffuse water pollution from agriculture is a critical problem worldwide, including in developed countries, causing eutrophication, human health problems and increased water treatment costs and reducing the recreational potential of water bodies. Agriculture produces four distinct types of pollution: nutrients, sediments, pesticides and faecal bacteria. We deal with each in turn below.

Nutrients are one of the most well-studied agricultural pollutants and comprise largely of nitrogen and phosphorus, which are applied to land in manure and slurry (**Figure 4.8**) as well as in inorganic fertilisers. Moreover, dung is deposited on fields by grazing livestock. Nitrate is the major form of nitrogen found in surface waters and is very mobile, unlike some other forms, such as ammonium, which is more strongly retained by soils (Hatch *et al.*, 2002). Nitrate leaching is a problem in most agricultural areas and peak concentrations in rivers usually exceed 10 mg N L^{-1}, although the other nitrogen species will be found at much lower concentrations (Neal *et al.*, 2006). Although it is not particularly toxic in the environment, nitrate must be removed from drinking water, which is costly to water utilities and consumers (Pretty *et al.*, 2000). In contrast, phosphate tends to form complexes with soil particles, and concentrations in rivers are thus often associated with levels of suspended solids. Nevertheless, orthophosphate is soluble and is therefore bioavailable to aquatic organisms, which can lead to high growth rates in microorganisms and plants (Jarvie *et al.*, 2007) which at low levels causes a loss of biodiversity, and at high levels can cause eutrophication.

It is thought that the aquatic life in the Gulf of Mexico is severely degraded by agricultural pollution brought down the Mississippi River from the heavily agricultural Midwest of the USA. The United States Geological Survey (USGS) estimates that 70% of the nitrogen and phosphorus that enters the Gulf of Mexico comes from agricultural sources in the USA. This compares to 9–12% from urban sources. Therefore, managing diffuse runoff of excess nutrients from agriculture is regarded as the most prevalent water quality-related challenge globally (OECD, 2017; UNEP, 2016), especially as, in OECD countries, agriculture is the predominant source of nitrate discharged into the environment and a significant source of phosphorus. For example, in New Zealand 75% of nitrate derives from agriculture and in the USA 36% of nitrate and 41% of phosphorus is from agriculture. In England, diffuse pollution from agriculture is believed to contribute 50–60% of nitrate (NO$_3$), 20–30% of phosphate (PO$_4$) and 75% of sediment in rivers (Edwards and Withers, 2008), as well as being a major source of pesticides, pharmaceutical products and faecal and other bacteria.

Sediments are one of the most obvious agricultural pollutants (Edwards and Withers,

Figure 4.8 Slurry being injected into a field by a specialist machine (a) with the deposited slurry shown in (b).

2008), as fields can be left bare once a crop has been harvested, and even when the next crop has been sown it can take months before there is vegetative cover, particularly when a winter crop is planted. Furthermore, overgrazing and trampling of soil by livestock can increase overland flow and erosion. Erosion washes topsoil into water bodies, where concentrations of suspended sediment can reach several hundred mg L^{-1} and causes problems in channels downstream, increasing flood risk, smothering of the benthos and fish breeding grounds and reducing biological diversity in riverbed gravels. Soils can also carry large amounts of phosphate and agrichemicals that bond to clay particles. The degree of erosion varies widely depending on land use, local geology and land management practices.

Pesticides, including herbicides, insecticides and fungicides, are applied to a wide variety of crops in order to improve growth by eliminating competitors and parasites. Residues of these chemicals are frequently transported from crops and soils to waters in runoff (Garrod *et al.*, 2007). Depending on the characteristics of a particular soil and chemical, certain amounts of a substance will be present in the solid and liquid phases. While some chemicals bond very strongly with soils, others tend to be found in the soil water and are much more mobile. The extent to which these substances are present in water will depend largely on their physicochemical properties, which determine their degradation and sorption (how one substance becomes attached to another) characteristics (Kladivko *et al.*, 1991). The degradation rate of a pesticide is usually expressed as its half-life (DT$_{50}$) (i.e. the time taken for the substance to degrade to 50% of its original concentration). Sorption characteristics are expressed as values representing the amount (per kg) of a substance that will attach to soil (abbreviated

to K$_d$). As for phosphorus, sorption of a pesticide to soil does not mean that it will be immobile, as substances can be transported in the particulate phase in overland flow and subsurface drainflow (Øygarden *et al.*, 1997). Concentrations of pesticides detected in water bodies range from nanograms (billionths of a gram) per litre, ng L^{-1}, up to tens of micrograms (millionths of a gram) per litre, µg L^{-1}, although they are commonly measured in streams at peak concentrations below 10 µg L^{-1} (Du Preez *et al.*, 2005).

Faecal bacteria sources include animal manure and slurry stores on farms. The presence of faecal matter in water (identified by the presence of faecal 'indicator organisms') affects drinking water processing, shellfish safety and bathing water quality. In addition, organic pollution from manures and silage effluent leads to significant reductions in the dissolved oxygen content of water bodies (Hooda *et al.*, 2000).

Other important water quality issues resulting from agricultural production include veterinary medicines, which have similar physicochemical properties to pesticides but reach the land following administration to livestock to prevent and treat disease. Residues of these compounds can be found in manure and slurry applied to land and in dung deposited by grazing livestock. Moreover, wash-off can occur following topical application. Antibiotics are the most commonly used group of veterinary medicines and have been measured in rivers at concentrations of up to a few µg L^{-1}. Toxic impacts are unusual in the aquatic environment, although concern exists about the spread of antibiotic resistance (Kay *et al.*, 2004). See Section 7.4 of Chapter 7 for a discussion of the health effects of nutrients, pesticides and veterinary medicines.

4.5.1.2 Agricultural stewardship

Owing to the known impacts that agriculture can have on the aquatic environment, the implementation of stewardship measures is now commonplace and many of the subsidies available to farmers today are offered on this basis rather than being related to production. Specific measures have long been proposed to reduce the impacts of agriculture on water quality. Stewardship measures can be split into three general categories: those that reduce inputs of pollutants; others that limit transport to water bodies; and edge-of-field measures (Kay *et al.*, 2009). Hundreds of measures are available that can be implemented by farmers, but some of the most common include **buffer zones**, wetlands and specific soil management practices. Buffer zones are permanently vegetated areas of land perhaps 5–100 m in width, usually, but not exclusively, located adjacent to a watercourse, where farming practices are restricted with a view to protecting water from agricultural diffuse pollution. Restrictions can include one or more of the following: no fertiliser applied, no pesticides and herbicides applied, no cultivation, no livestock grazing allowed, no farming at all allowed, particular plants or types of plant must be grown/allowed to grow. Such a buffer may provide a biochemical and physical barrier against pollution inputs, such as sediment, nutrients and pesticides, and therefore reduces the connection between the potential pollution source and the receiving water body. Research has shown that, while the impact can be variable, sediment, nutrient and pesticide pollution is often reduced significantly by buffer zones (by up to 100% in some cases). Effects have been found to be site specific, though, and depend on the nature of the pollutant, soil characteristics and hydrology. Evidence suggests that more established vegetation, particularly the presence of trees, will lead to greater pollution reduction. Recently we have started to value the landscape in terms of the provision of a range of interacting **ecosystem services** and goods. For buffer zones these services could include, among others, natural flood management, biodiversity, diffuse pollution mitigation and biofuel production. Therefore we should be revising our methodologies of designing, understanding and evaluating buffer zones to maximise these possible benefits (Stutter *et al.*, 2012a).

There are many examples of agricultural stewardship schemes around the world, including work undertaken by the Environmental Protection Agency in the USA and the Landcare scheme in Australia. Furthermore, many stewardship measures have been implemented in the UK in recent years through the England Catchment Sensitive Farming Delivery Initiative (ECSFDI). This is a scheme set up to encourage farmers to adopt environmentally friendly practices through programmes of advice and the availability of subsidies.

4.5.2 Urbanisation

4.5.2.1 Urban pollution and non-point sources

Many towns and cities develop in close proximity to lakes and rivers. The construction of housing, commercial buildings, roads and other infrastructure reduces the permeability of the surrounding land, and so alters the hydrological characteristics of storm water flowing to those lakes and rivers. Runoff volumes increase and, most obviously, discharges become 'flashy', with a shorter time to a greater peak discharge, resulting in increased bed and bank scour and increased risk of flooding. Loss of in-stream and **riparian** biodiversity also occurs, exacerbated by the low or zero flows that can then occur during dry weather. Urbanisation also has major impacts on water quality. Historically, the impacts of most concern were those associated with sewage disposal and end-of-pipe (e.g. factory) discharges. Investment in sewers and wastewater treatment processes, driven by public health imperatives (see Chapter 9), and greater regulation of commercial point-source discharges have led to dramatic improvements in the quality of urban surface waters in many countries over the last 150 years. However, urban waters often remain unacceptably polluted, due to emissions from non-point sources which originate over an area. These emissions comprise a diverse mix of pollutants that accumulate on built surfaces until they are washed off during storm events (see **Box 4.3** for an example). Impacts may then arise due to sedimentation of rivers (e.g. light loss, gill rot in fish), dissolved oxygen depletion, eutrophication, transmission of infectious pathogens and a wide range of toxic effects from metals, biocides and other persistent organic pollutants.

Table 4.5 indicates the range of urban pollutants involved and their principal sources. While most pesticide pollution is associated with agriculture, some pesticides are often used in gardens in urban areas, which can be problematic (e.g. metaldehyde products are used to kill slugs). **Table 4.6** reveals that the concentration of many of these pollutants in urban storm water exceeds that of untreated sewage or from combined sewage overflows (CSOs), which comprise a mix of untreated sewage and storm water discharged to rivers when storm discharges would otherwise exceed sewer capacities (see **Figure 8.8a** in Chapter 8 for a photo of a CSO). Data limitations mean that it is difficult to understand precisely the role of urban diffuse sources in river degradation, but evidence (see, for example, **Table 4.7**) suggests that in developed countries urban diffuse pollution is now one of the main reasons why rivers fail to reach their water quality objectives. Note that, when CSOs are also considered, urban storm water pollution accounts for more kilometres of poor water quality than any other cause.

Many possible urban water pollutants are not yet being monitored, as the detection techniques are only just becoming available. These pollutants include those derived from pharmaceuticals and personal care products such as steroids, antibiotics, fire retardants, fragrances and flavours, insect repellents, disinfectants, cosmetics, solvents, hair dyes and so on. In many cases it is not known what effects many of these substances might have on aquatic organisms or human health and this is an emerging area of study for water science (**Box 4.4**). However, it is known that many synthetic chemicals act as endocrine disrupters in the aquatic environment. This means that they can cause the sex of some creatures to be altered at the foetal stage. Section 8.3 of Chapter 8 provides more detail on the possible health impacts of these emerging contaminants.

4.5.2.2 Mitigation of urban pollution

The areal nature of diffuse sources means that they are much harder to control than point sources, which are amenable to end-of-pipe treatment. Urban diffuse sources are particularly challenging, as cities are highly heterogeneous in terms of land cover, land use and intensity of activity (e.g. housing density, traffic flow on roads), making effective targeting of mitigation measures more difficult. Such measures are of two fundamental types: structural and non-structural. In the USA both are covered by the term Best Management Practices (BMP), while in the UK the structural measures are known as **sustainable urban drainage systems** (SUDS). The term 'source control' is

Table 4.5 Some urban diffuse pollutants and their sources

Source	Oil and hydrocarbons	PAH	Pesticides	Solvents/paintsanddyes	Sediment	Bacteriaandpathogens	Metals	Organicpollution	Nitrogen	Phosphorus
Residential runoff	2	2	2	1	3	2	2	2	2	2
Industrial and commercial runoff	2/3	2	1	2	2	1	2	1	1	1
Highway runoff	3	3	2	0	3	1	3	1	1	1
Construction industry	0	0	0	0	3	0	1	0	2	1
Rail track runoff	2	2	2	0	2	0	1	1	1	1
Garage/petrol/service stations	3	2	0	0	0	0	1	0	0	0
Road/rail weed control	0	0	3	0	0	0	0	0	1	0
Gardening	0	0	3	0	0	0	0	0	1	1
Pets/birds	0	0	0	0	0	2	0	3	3	3
Misconnections	2	1	1	1	1	1	2	2	2	2
Sewer leaks	0	0	0	0	1	2	1	2	2	2
Septic tanks	0	0	0	0	0	1	1	2	2	2
Litter/waste disposal	1	1	1	1	1	1	1	2	2/3	2
Car/vehicle emissions	2	2	0	0	0	0	2	0	2	2

Source: From Ellis and Mitchell (2006). Urban diffuse pollution: key management issues for the Water Framework Directive. *Journal of the Chartered Institution of Water and Environmental Management* 20: 19–26. Reproduced with kind permission of John Wiley & Sons.
Note: 0, unlikely source; 1, possible source; 2, documented source; 3, significant source.

also commonly used, and recognises that BMPs and SUDS are most effective when applied close to the source of storm water.

The philosophy behind SUDS is that they act to reintroduce a more natural hydrological regime to the urban environment through detention and retention of storm discharges, through a variety of physical devices (Table 4.8). The term 'SUDS triangle' is often used to denote the primary benefits that SUDS offer. First, attenuation of storm discharge leads to reduced flood risk, and enhanced groundwater replenishment and dry-weather flow. Second, this discharge attenuation gives more time for pollutant degradation through natural processes of volatilisation, sedimentation and filtration, precipitation, adsorption, photolysis, microbial degradation and plant uptake. Finally, SUDS that include natural elements in their structure are able to enhance the biodiversity and amenity value of buildings and urban areas. The carbon sequestration function of vegetated SUDS is also now recognised as a valuable additional benefit, given the importance of climate

Table 4.6 Pollutant concentrations in wastewater and storm water across Europe and North America

	Untreated sewage	CSO	Urban storm water
BOD mg L^{-1}	160	60–200	10–250
SS mg L^{-1}	225	100–1100	3000–11 000
Total N mg L^{-1}	35	3–24	3–10
Total P mg L^{-1}	10	1–11	0.2–1.7
Coliforms MPN/100 mL	10^7–10^9	10^5–10^7	10^3–10^8
Cadmium µg L^{-1}	0.15–0.38	1.4–9.6	0.2–63
Copper µg L^{-1}	110–240	80–170	2–1270
Zinc µg L^{-1}	85–180	100–1070	17–3550

Sources: Data from Mitchell, 2001; Novotny, 2002; Rule *et al.*, 2006.

change mitigation. The features and relative advantages of the many different types of sustainable urban drainage devices, as well as a varied range of application case studies, can be found on the UK SUDS practitioners website: www.susdrain.org. There are plenty of photographs of different SUDS devices, and one example is also shown in **Figure 8.14** in Chapter 8.

SUDS differ in performance with respect to the main benefits they offer. While most offer multiple benefits, some perform particularly well with respect to flow attenuation, and others with respect to water quality remediation or ecological performance. This gives rise to a concept known as 'the treatment train', in which SUDS are developed in series, so as to maximise the benefits on offer from source to final discharge (whether watercourse or sewer). Thus a development might have green roofs at the building level, infiltration trenches and swales at the site level and ponds or wetland at the regional (neighbourhood) level. This approach builds robustness into the treatment of storm water, as it increases the range of pollutant remediation processes at work, and, should one SUDS device fail, others are there to treat the runoff.

SUDS do pose particular problems, but these are now better understood and addressed. For example, safety fears with respect to urban ponds are tackled by careful design and edge planting; concerns over maintenance costs have been allayed, as experience shows that whole-lifecycle costs are comparable to or less than those of conventional sewers; and a growing body of evidence on the performance of SUDS (both quality and quantity) means that SUDS designers are better able to develop urban drainage for new sites that meets specified performance criteria.

Although SUDS have received much attention, the importance of non-structural BMPs in urban storm water management must be recognised. These are measures designed to prevent discharge of polluted storm water, and should sit above source control in the BMP hierarchy. These include a very wide range of 'good housekeeping' measures, such as street sweeping, rainwater harvesting and reuse, using bunded tanks for storing hazardous materials, covering material in goods yards that might otherwise be washed to drains during rainfall,

BOX 4.3

De-icing agents and diffuse pollution

In cold regions, highway authorities often undertake preventive measures to avoid road icing in winter. Salt is the most common de-icing agent and typically comprises about 90% sodium chloride, with the remainder made up of insoluble residues such as clay marl and trace amounts of metals. Colwill *et al.* (1984) estimated that de-icing contributed 25% of suspended loadings and 90% of soluble loadings in winter road runoff from a rural motorway in England, although clearly this will vary greatly with the frequency and extent of salt application. The chloride contained in road salt is of concern, owing to its impact on ground and surface waters used for drinking supply and aquatic habitat. The metals contained in de-icer salt, including nickel, lead, zinc and chromium, are toxic in runoff, while the salt itself increases the solubility and mobility of metals previously deposited on road surfaces by vehicles. De-icer salt also contains sodium ferrocyanide, added as an anti-caking agent, which in strong sunlight and acidic conditions can generate toxic cyanide forms, although in surface waters this usually evaporates as hydrogen cyanide.

Monitoring in north-eastern USA from 1991 to 2004 indicated that chloride concentrations exceeded environmental standards in around 2% of the 1300 groundwater wells analysed, while 15 of 100 catchments had winter chloride concentration in excess of aquatic habitat standards (Mullaney *et al.*, 2009). Concentrations were greatest in urban areas, where available data indicated an increasing historical trend in chloride loading. For the urban areas, chloride:bromide ratio analysis indicated that much of the chloride was due to de-icer application, although septic systems, wastewater discharge and saline groundwater plumes from landfills and salt storage areas were also significant sources. The increasing load in chloride, consistent with literature reports of long-term increases in chloride concentrations, was attributed to the expansion of road networks and impervious areas that require de-icing.

Reductions in de-icer application have been achieved through highway design, such as snow fences, from improved weather forecasting and ice prediction that allows more targeted applications in space and time, and by measurement of residual salt loadings, so that less salt is used when topping up. Because salt is corrosive, not only to vehicles but also to critical structures such as bridges, alternative de-icers, most commonly urea and ethylene glycol, may be used on parts of a highway. These de-icing chemicals are also used by the aviation industry on runways and aircraft. However, these de-icers also pose a risk to water quality. For example, urea can mobilise metals, such as copper, which occur in high concentrations on heavily trafficked roads. Urea also hydrolyses to ammonia, which exerts an oxygen demand on water and, in its un-ionised form, is also toxic. In the aviation industry, the use of urea as a de-icer is being replaced by calcium and magnesium acetate-based products.

Table 4.7 Pressures significantly impairing attainment of Water Framework Directive good status in European Union rivers

Pressure type	River length affected (1000s of km)	%
Point sources	374.5	32
Urban wastewater	296.4	
CSO overflows	68.5	
IED plants	66.5	
Non-IED plants	60.9	
Brownfield land	13.6	
Waste disposal sites	8.6	
Aquaculture	7.3	
Mine waters	6.4	
Other	30.1	
Diffuse sources	520.3	44
Agriculture	405.3	
Discharges not connected to sewer	135.3	
Urban runoff, transport and brownfield land	85.7	
Forestry	31.3	
Mining	20.2	
Other	75.5	
Hydromorphology	548.6	47
Atmospheric deposition	252.3	22
Abstraction	171.5	15
Other	229.4	
No significant anthropogenic pressure	248.3	21
Total river length monitored in EU28	1172.5	

Notes: Data relate to the second round of WFD River Basin Management Plans (RBM2), monitoring 2012–18: www.eea.europa.eu/themes/water/water-assessments/pressures-and-impacts-of-water-bodies and summarised in EEA Report No 7/2018. Data do not sum to 100% or total river length owing to multiple pressures exerted on water bodies. IED is Industrial Emission Directive 2010/75/EU, which covers a prescribed list of industrial activity types.

and having dedicated loading areas where a spill will not drain to a watercourse.

Other technical measures exist that improve the operation of the conventional sewer system. These include partial treatment of wastewater in sewers, through the use of aeration, biofilm plates and activated sludge sites (see Chapter 8); real-time control of sewers, so that free capacity can be used to store water in the sewer and release it back to the treatment works after the storm event; and real-time control of discharge to river, to make best use of natural assimilative capacities.

Arguably, the main challenges facing the quality of urban rivers relate to the need for more integrated and radical solutions. The SUDS triangle neatly illustrates scope

BOX 4.4

Emerging pollutants

While it is the case that much progress has been made in cleaning up point-source discharges, new research is showing that there are many chemicals in sewage effluent that have, until recently, not been studied or regulated. Pharmaceuticals are the most prevalent group of these substances and antibiotics, painkillers, cardiovascular drugs and antidepressants have been measured in rivers at concentrations ranging between ng and mg L^{-1} (Kay et al., 2017). This results from their incomplete breakdown in the human body and excretion to the sewer system, as well as disposal of unused drugs. The effects on aquatic ecosystems are uncertain, although there is the potential for chronic impacts and the development of antibiotic resistance (Bundschuh et al., 2009). One of the most striking of these has been the observed feminisation of male fish in the vicinity of sewage discharges in some UK rivers, thought to be due to exposure to endocrine-disrupting chemicals (Tyler et al., 2007) (see also Chapter 8). Regulators have responded to the new information that is available. For instance, three pharmaceuticals (diclofenac, 17alpha-ethinylestradiol and 17beta-estradiol) are being considered for addition to the EU's list of priority substances (EC, 2011). Further emerging contaminants have been identified recently, though, and much research effort is now focusing on microplastics. It is known that sewage works are key sources in river catchments but there are likely to be others, including agricultural and urban runoff (Kay et al., 2018). Ongoing work is looking at the occurrence, fate and effects of microplastics in rivers.

for addressing multiple objectives within the area of source control, but integration must be broader than this. Urban areas cannot be considered in isolation from the wider catchment (see Chapter 13), as too often happens. If programmes of measures to improve water quality are to be cost-effective, we must know the source area loading throughout the catchment, the risk that these spatially defined loadings pose to water quality and ecological objectives, and the cost and efficacy of different interventions, whether technical, regulatory (e.g. enforced rules on urban design, commercial activities), economic (e.g. raising tariffs for storm water disposal) or educational. Integrated management should mean, for

example, incorporating water quality issues into traffic management (as is already done for air quality), working with car manufacturers to promote the use of more benign materials and reduce road-sourced pollution, or working with planners and urban designers so that city compaction to reduce travel and carbon emissions does not constrain delivery of SUDS' benefits and exacerbate flooding. Thus, while SUDS have a role to play in restoring urban river water quality, arguably a greater one than they are currently given, they are only part of the solution, which requires management of multiple sources using multiple tools, both structural and non-structural in nature.

Table 4.8 Typical SUDS devices

Position in control hierarchy	Typical SUDS devices
Source controls Control runoff at or very near the source	• Green roofs (can be extensive or intensive) • Downpipe storage (water butts) • Infiltration devices (soakaways, filter strips, infiltration trenches, pervious pavements)
Site controls Used to manage water in a localised area Often convey water from source to regional controls, or promote local infiltration	• Swales (broad, grassed channels) • Infiltration basins (dry depression storage) • Balancing pond (wet depression storage)
Regional controls Manage water from one or more sites	• Wet pond (primarily for flow attenuation) • Constructed wetland (primarily for water quality)

4.5.3 Acid mine drainage

4.5.3.1 Origins and processes generating pollution from mines

The pollution of surface waters from mines is a problem wherever mineral ores have been exploited. It is a problem throughout the world, including China (He *et al.*, 1997), North America (DeNicola and Stapleton, 2002; Wren and Stephenson, 1991), South America (Van Damme *et al.*, 2008) and Australia (Battaglia *et al.*, 2005). Mines which have been abandoned more commonly pose a greater problem than those that are still in operation. This is because working mines are dewatered by pumps (Younger, 2000). When mines are abandoned, dewatering ceases and a process of water table rebound occurs, which floods the exposed mine workings. When the water table reaches the surface it emerges as springs, **adits** or surface seepage. These polluted waters then enter streams and lakes (**Figure 4.9**).

The activities of mining coal and minerals and mine dewatering expose underground pyrite-bearing (FeS_2) rock strata to oxygen. In the presence of oxygen and water, pyrite becomes oxidised into ferrous sulfate and sulfuric acid. Ferrous sulfate is further oxidised to ferric sulfate; this then reacts with water to produce more acidity and ferric hydroxide (**Box 4.5**). The oxidation of the reduced forms of iron and sulfur is very slow. However, in the presence of acidophilic chemosynthetic bacteria the reactions are accelerated up to a million times. Bacteria which flourish at

Figure 4.9 Acid mine discharge running under an old industrial bridge in Andalusia, Spain.
Source: Shutterstock/Silvia B. Jakiello.

low pH, such as *Thiobaccilus ferroxidans* (Singer and Stumm, 1970) and *Thio bacillus thiooxidans* and *Metallogenium* (Kleinmann *et al.*, 1981), oxidise iron and sulfate, respectively.

4.5.3.2 Environmental consequences of mine pollution

The waters that emerge from mines into receiving surface waters are very variable in chemical composition (Table 4.9) and may depend on the type of mine and the nature of the surrounding geology. They can be acidic and laden with suspended solids and dissolved metals (Gray, 1997). Indeed, they are frequently described as discharges of acid mine drainage. However, although many discharges are acidic, it is not a universal characteristic and in some cases they may be neutral or basic if the surrounding geology is high in carbonates and

able to afford some neutralisation (Kelly, 1988; Schmiermund and Drozd, 1997). The contaminants lead to an increase in the acidity, turbidity, sedimentation and heavy metal concentrations of the receiving water body. They precipitate out as a coloured flocculate, known as ochre, which stains the substrate of receiving waters (e.g. **Figure 4.9**). Ochre can vary in colour from yellow to orange-brown.

It is difficult to predict the duration and chemical composition of mine-water contamination. Younger (1997) separated the impacts into vestigial and juvenile components. Vestigial acidity is that acquired during the water table rebound phase and in theory should decline exponentially with time, with **asymptote** levels (where the curve of the line flattens out and almost reaches zero) being achieved 40 years after the initial discharge. Juvenile acidity arises

---

BOX 4.5

Formation of acid mine drainage

The following chemical reactions summarise the formation of acid mine drainage (Kelly, 1988):

$$FeS_2 + 7/2O_2 + H_2O \rightarrow Fe^{2+} + 2SO_4^{2-} \qquad [1]$$

$$Fe^{2+} + 1/4O_2 + H^+ \rightarrow Fe^{3+} + 1/2\,H_2O \qquad [2]$$

$$Fe^{3+} + 3H_2O \rightarrow Fe(OH) + 3H^+ \qquad [3]$$

$$FeS_2 + 14Fe^{3+} + H_2O \rightarrow 15Fe^{2+} + 2SO_4^{2-} + 16H^+ \qquad [4]$$

Reaction [1] shows the dissociation of iron pyrite (in the presence of oxygen and water). Pyrite breaks down to form sulfuric acid and ferrous iron (Fe^{2+}). When surface waters combine with mine waters the ferrous iron (Fe^{2+}) forms the ferric variety (Fe^{3+}) [2]; a cycle is then established between these two reactions. The Fe^{3+} then further reacts with the available pyrite to form additional acidity, ferrous and sulfate ions [4]. The oxygenation of Fe^{2+} is the rate-determining step because reaction [2] is significantly slower than reaction [1].

Table 4.9 Some chemical data from mine discharges

Chemical parameter	A	B	C	D
pH	2.7	5.3	3.18	7.75
Conductivity (μS cm^{-1})	2590	295	2605	5571
Sulfate*	680	324	1850	3110
Iron*	493	29.04	165	43.7
Aluminium*	59.2	0.25	76	0.03
Zinc*	1.22	0.05	33	0.26
Lead*	—	0.04	—	—
Copper*	0.06	0.04	1.8	—
Cadmium (μg L^{-1})	<0.005	—	49	—
Suspended solids	3158	—	—	—

Notes: * Readings in mg L^{-1}; A = Caulfield mine discharge, Pennsylvania, USA (Letterman and Mitsch, 1978); B = Stoney Heap coal mine water discharge, Durham, UK (Jarvis and Younger, 1997); C = Bally-murtagh mine-water discharge (Gray, 1998); D = Whittle Colliery mine-water discharge (Batty *et al.*, 2005).

from the fluctuation of the water table and may persist for centuries.

The impact of mine drainage contamination on aquatic ecosystems has been studied extensively. Processes of decomposition can be affected by mine drainage. Bacteria, fungi and decomposing invertebrates can be inhibited from colonising leaf litter, consequently reducing the rates of breakdown (Gray and Ward, 1983). Ferreira da Silva *et al.* (2009) reported that diatoms were rare at a location stressed by mine drainage with high levels of cadmium, lead and zinc. However, increased diversity and species richness was observed 6 km downstream of the impact. In Lynx Creek, Arizona, algal communities impacted upon by mine pollution exhibited reduced species richness as compared to locations upstream and those in areas of substantial recovery downstream (Lampkin and Sommerfield, 1982). Koryak and Reilly (1984) observed reduced growth of the aquatic macrophyte *Justica america* as a consequence of coal mine pollution in Ohio. Tremaine and Mills (1991) recorded that

protozoan abundance and grazing rates on bacteria were substantially reduced in a lake affected by acid mine drainage. The impact on benthic invertebrates is widely reported, with reduced abundances, species richness and diversity being common factors. Communities in riffle habitats appear to be affected more than those in pool habitats (Van Damme *et al.*, 2008). Furthermore, species belonging to the orders Ephemeroptera and Plecoptera also appear to be those most affected (Clements, 1994). Letterman and Mitsch (1978) noted that the standing crop of fish communities declined downstream of mine drainage on Ben's Creek, Pennsylvania, from 228.2 kg ha^{-1} to 11.2 kg ha^{-1}. The sculpin, *Cottus bairdi*, showed the greatest decline, from 151.2 kg ha^{-1} to 0.3 kg ha^{-1}.

The human impacts of mine drainage pollution are many and varied. Possible consequences include: reduced water availability for abstraction for public supply from both groundwater and surface water sources; surface waters unsuitable for irrigation and livestock watering; contamination of navigational rivers and canals, for both commercial and recreational use, impacts upon the maintenance of the navigational area and the waterborne craft; ecological damage can reduce the recreational potential for commercial fisheries and angling; and the general recreational and amenity value of watercourses is affected by the visual and aesthetic impact of highly coloured ferruginous mine waters.

4.5.3.3 Mitigation of mine pollution

The reduction of the impact of mine-water pollution falls into three types of system: physical, chemical and biological. Physical systems involve water oxidation through impoundment, and cascades to facilitate the settlement of the contaminants into sludge (Diamond et al., 1993). The costs of these systems can be high, as they require both high capital investment in terms of construction costs and there are further revenue costs from the disposal of the resultant sludge. Chemical systems use the addition of chemicals to actively promote the decontamination of the water. Examples typically include the addition of lime to produce flocculation and settlement of contaminants, followed by pH neutralisation (Chadwick et al.,

1986). Although capital costs are much reduced, they are expensive to run because of the costs of the treatment chemicals and sludge disposal. Biological systems are dominated by the construction of semi-natural or artificial habitats (Figure 4.10) which facilitate bacterial activity to reduce contaminants through reed beds and wetlands (Webb et al., 1998). These systems are relatively cheap to construct and run. Furthermore, the ecological principles behind them also suggest that they may be a sustainable solution to mine drainage pollution (Kalin, 2004). However, such a solution requires the availability of suitable land areas close to the source of pollution and this is not always possible.

In reality, combinations and overlaps of treatment processes are frequently observed. For example, the mine remediation treatment at Bullhouse on the River Don in Yorkshire, UK, involves a physical settlement lagoon and a wetland area (Laine and Dudeney, 2000). Such systems can result in significant reductions in loadings of metals. The Bullhouse scheme illustrates this, with iron being the most effectively removed (Table 4.10). Wiseman et al. (2003) also noted similar reductions in iron loading (82% to 95%) on the River

Figure 4.10 a) Semi-natural wetland and b) artificial reed bed wetlands, River Pelenna, South Wales, UK.

Table 4.10 Changes in typical chemical composition of the Bullhouse mine discharge into the River Don, Yorkshire, UK, (A) before and (B) after remediation

Chemical parameter	A	B
pH	5.9	5.9
Conductivity ($\mu S\ cm^{-1}$)	442	281
Sulfate*	788	475
Iron*	54.6	2.7
Aluminium*	2.31	0.32
Zinc*	91.5	65.3
Lead*	0.147	0.049
Copper*	0.0202	0.0042
Cadmium ($\mu g\ L^{-1}$)	—	0.28
Suspended solids	16	11

Source: Data from Laine and Dudeney, 2000.
Notes: * = Readings in mg L^{-1}.

Pelenna in South Wales. Furthermore, ecological recovery has also been observed in the Pelenna study with the return of sensitive invertebrate species (e.g. the mayfly, *Empherella ignita*), brown trout (*Salmo trutta*) and birds such as dippers (*Cinclus cinclus*).

4.5.4 Atmospheric pollution

4.5.4.1 Source of acid rain

Rainfall is naturally acidic, with a pH of about 5.6, due to its equilibrium with carbon dioxide (CO_2) in the atmosphere:

$$CO_2 + H_2O \leftrightarrow H_2CO_3 \leftrightarrow H^+ + HCO_3^-$$

However, the pH of rain decreases as the amounts of sulfur dioxide (SO_2) and nitrogen oxides (NO_x) in the atmosphere increase. Acid rain is generated from the oxidation of SO_2 and NO_x to sulfuric acid (H_2SO_4) and nitric acid (HNO_3), respectively. Acid rain can be carried great distances in the atmosphere, not just between countries but also from continent to continent. Hence the impact of acid rain on freshwater ecosystems can be far from its source.

Over the last 100 years there have been significant changes to global atmospheric chemistry. The major atmospheric pollutants include acidifiers such as SO_2 and NO_x; toxic substances such as ozone, volatile organic compounds (VOCs) and heavy metals; and fertilising substances such as ammonia (NH_3) and nitrous oxides N_2O. The amount of acidifying pollutants present in the atmosphere has been affected by an increase in the combustion of fossil fuels, changes and intensification in agricultural practices and, more recently, by the introduction of legislation to reduce emissions to the atmosphere such as the 1999 Gothenburg Protocol, which set national mandatory emission reductions for four major air pollutants: SO_2, NO_x, VOCs and NH_3 to be achieved by 2010. In 2012, a revised Gothenburg Protocol was finalised which specifies emission reduction commitments for 2020 in terms of percentage reductions from baseline figures in 2005. It was also extended to cover one additional air pollutant, namely particulate matter ($PM_{2.5}$). By 2020 the EU Member States were expected to jointly cut their emissions of SO_2 by 59%, NO_x by 42%, NH_3 by 6%, VOCs by 28% and particles by 22%. Thus, legislation in Europe and North America has led to a decline in acid rain. In contrast, emissions of acidifying pollutants have increased rapidly in south-east China, north-east India, Thailand, and the Republic of Korea. For example, in China, SO_2 emissions increased rapidly from the 1970s until 2000 (from <4 to ~20 million tonnes) owing to the rapid increase in the consumption of coal and oil; in 2004 coal accounted for 69% of energy production and it contains between 1 and 4% sulfur (Larssen et al., 2006). In 2003, the total emissions of SO_2 to the atmosphere from China were ~22 million tonnes, which was more than the total anthropogenic European emissions of 17 million tonnes in 2002 (excluding shipping) and twice the USA emissions (Larssen et al., 2006). However, since the introduction in China of the National SO_2 Total Emission Control in 2006 and the National NO_x Total Emission Control in 2011, emissions of SO_2 and NO_x have started to decline at the national scale, but emissions of NH_3 in China have continued to increase due to intensive agricultural activity (Yu et al., 2017) and are not currently regulated.

4.5.4.2 Impact of acid rain on surface water

In the 1960s and 1970s, observations of an increase in surface water acidity leading to a decline in the salmon population across areas of southern Scandinavia were widely reported in the media. However, much research was required to establish the cause–effect relationship between acid rain and surface water acidification. Palaeolimnological techniques that reconstructed pH based on diatoms in lake sediment cores allowed the chemical histories of a large number of lakes across Britain, Scandinavia, north-east USA and Canada to be evaluated (Battarbee et al., 1990). This showed that lakes in areas with base-poor geology that had received high rates of acidic deposition had experienced significant acidification.

The effect of surface water acidification on aquatic flora and fauna is pronounced (e.g. Schindler, 1988). The effects on aquatic biota can be direct, such as toxicity associated with changes in water chemistry, or indirect, such as changes in habitat or food availability. Acidification also increases the solubility of heavy metals, such as aluminium, manganese, lead, cadmium and zinc, many of which are toxic to aquatic organisms. For example, acidity and high concentrations of aluminium (Al^{3+}) can lead to deterioration of aquatic life, with losses in the diversity and size of invertebrate and fish populations. At pH less than 5, aluminium is soluble and leached from the soil to surface waters as Al^{3+}. The toxicity of Al^{3+} has been mainly studied with reference to fish; at concentrations <100 μg L^{-1}, it affects the osmoregulation of fish (i.e. their ability to regulate the amount of salt and water within their body) and at concentrations >100 μg L^{-1} a gelatinous precipitate of aluminium hydroxide ($Al(OH)_3$) forms on their gills, leading to suffocation (Jackson and Jackson, 2000). Changes in the population of the dipper, Cinclus cinclus, a riparian bird, in areas that experienced stream-water acidification, such as mid-Wales (Tyler and Ormerod, 1992), are an example of the indirect effect of acidification on aquatic biota.

The susceptibility of a particular surface water body to pollution from acid rain depends on its capacity to act as a buffer and neutralise the effects of acid rain. This capacity is largely determined by the chemical properties of surrounding geology and soil type. Rocks and soils that contain large amounts of base cations (**bases** are substances that can accept hydrogen ions (protons) and are the opposite of acids), such as calcium and magnesium, have a high **buffering capacity**. In addition, surface water bodies in chalk or limestone regions are able to buffer the increase in hydrogen ions (H^+), due to the presence of bicarbonate anions (HCO_3^-):

$$H^+ + HCO_3^- \leftrightarrow H_2CO_3 \leftrightarrow H_2O + CO_2$$

In contrast, water bodies underlain by base-poor **igneous** (e.g. granite) or ancient metamorphic rocks have a lower buffering capacity. Soil development on such lithologies generally results in thin, base-poor soils with a small base cation pool and slow release of base cations from the weathering of the bedrock. Base cation stores in such soils are rapidly depleted in response to incoming acidic rain, resulting in the acidification of soils and surface waters. Hence, areas dominated by igneous and metamorphic rocks and receiving high annual rainfalls that are acidic are particularly vulnerable, such as the Canadian shield, the Appalachians of America and the mountains of Scandinavia and western Britain. This differential ability of surface waters to cope with acid rain has been examined through use of a 'critical load' approach, as discussed in **Box 4.6**.

4.5.4.3 Recovery from surface water acidification

Since the 1980s considerable national and international effort has been made to decrease emissions of acidifying pollutants. This has been challenging, since those countries suffering most from acid rain are not always the main polluters. In 1979, 34 European and North American countries adopted the Convention on Long-Range Transboundary Air Pollution, which bound them to reducing emissions. The convention agreed firm targets in 1983, when 21 European countries agreed to reduce their SO_2 emissions by 30% from 1980 levels by 1993. Twelve countries reached this target by 1988. In the United States, an 'Acid Rain Program' commenced in 1995 that aimed to achieve environmental and public health benefits through reductions in emissions of SO_2 and oxides of nitrogen (NO_x). The convention has been extended by eight protocols, including the 1999 Gothenburg Protocol; all have set specific measures to be taken by countries to cut their emissions of air pollutants. Since 1990 there has been a substantial reduction in SO_2 emissions in Europe (76%), the USA (50%) and UK (92%). As a result, sulfate deposition in the 1990s declined by between 38 and 82% across Europe (Prechtel *et al.*, 2001). In the UK, total sulfur deposition declined by 86% between 1986 and 2006, comprising a 93% reduction in dry deposition and a 57% decrease in wet deposition (Curtis *et al.*, 2014).

The response of surface waters to this decline in sulfur deposition has been monitored via the international UN monitoring programme ICP Waters (International Cooperative Programme on Assessment and Monitoring of Rivers and Lakes). In 12 regions across Europe and North America, a significant decline in sulfate and aluminium concentrations and increase in alkalinity and pH has been observed (Skjelkvåle *et al.*, 2005; Skjelkvåle and De Wit, 2011). However, there appears to be a delay in biological recovery. For example, not all lakes in the Adirondack region of New York State have recovered sufficiently for fish populations to be sustainable despite an increase in water pH and decline in aluminium concentrations (Civerolo and Roy, 2013). Overall, while some improvement in water quality

Evaluating the acidification status of surface waters: the critical load approach

The concept of the critical load approach is based on a 'dose–response' relationship and has been widely applied to addressing pollution problems (Fenn *et al.*, 2011). A critical load can be defined as a 'quantitative estimate of an exposure to one or more pollutants below which significant harmful effects on specified sensitive elements of the environment do not occur according to present knowledge' (Nilsson and Grennfelt, 1988). The critical load of acidity for a freshwater resource is exceeded when the total amount of acid deposition is greater than the ability of the system to neutralise the deposited acidity. Two models have been widely used in critical loads for surface waters: the steady-state water chemistry (SSWC) model (Henriksen *et al.*, 1992) and the first-order acidity balance (FAB) model (Posch *et al.*, 1997). Both models provide an estimate of the level of acid deposition below which the acid neutralising capacity (ANC) will remain above a pre-determined level. ANC, measured in microequivalents per litre (μeq L^{-1}), is a commonly used chemical indicator of sensitivity to acidification, as it characterises the ability of water to neutralise strong acids, including those introduced by atmospheric deposition, and it can be linked to biotic response thresholds such as the health and condition of aquatic biota. Current UK and European critical load applications use the FAB model, and in the UK an ANC limit of 20 μeq L^{-1} is applied to all sites, except those which are believed to be naturally acidic, where a value of ANC = 0 μeq L^{-1} has been used.

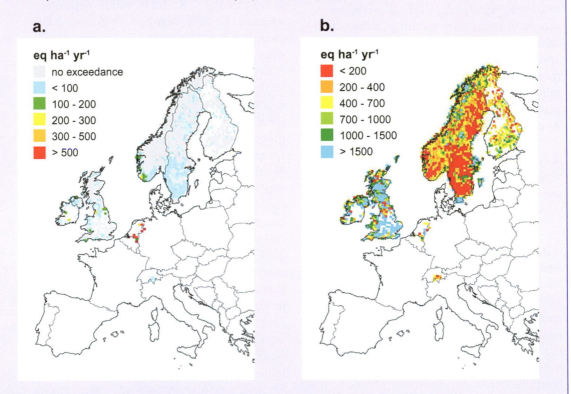

Figure 4.11 Critical load maps for seven European countries based on data available within each grid cell: a) CLmaxS (see box text) and b) average accumulated exceedance of critical loads for 2015.
Source: Norwegian Institute for Water Research (NIVA, 2018).

Continued

The acidity critical loads are only available for seven European countries based on data from national surveys of lakes or streams (Figure 4.11a). The critical load for freshwaters is often expressed in terms of three terms: CLmaxS, CLmaxN and CLminN, where CLmaxS is the critical load of acidity expressed in terms of sulfur (S) only (i.e. where nitrogen (N) deposition is zero), CLmaxN is the critical load of acidity expressed in terms of N only (i.e. where S deposition is zero) and CLminN represents the long-term N removal processes (e.g. N uptake and immobilisation), i.e. the amount of N that can be deposited before it will begin to contribute to surface water acidification (see Henriksen and Posch (2001) for more detail). Because both S and N compounds can contribute to the exceedance of the acidity critical load, these three 'acidity' critical load parameters are required to determine the combinations of deposition that will not exceed the critical load (Figure 4.11b). In general, the lower the values of CLmaxS and CLmaxN, the more sensitive the sites are to acidification from atmospheric deposition. In summary, critical loads have been successfully developed and implemented in Europe and other areas for assessing the impacts of air pollution on essential ecosystem services and for informing public policy for reductions in air pollution emissions.

and, to a lesser extent, biology has occurred in response to large reductions in acid deposition over the last 30 years, surface waters in the most impacted and acid-sensitive regions of northern Europe and North America will remain acidified into the foreseeable future (Curtis *et al.*, 2014). In comparison to Europe and North America, there are few studies that have evaluated the effects of SO_2 and NO_x emission abatement on acid deposition and water acidification in Asia. A recent study by Yu *et al.* (2017) monitored the chemical composition of throughfall, soil and stream water in a forested catchment in south-west China from 2001 to 2013. They observed that, while sulfate deposition decreased significantly over the period, N deposition increased. They also reported a delay in the recovery of soil and surface water probably due to desorption of previously stored sulfate and increase in nitrate leaching from the soil.

The ICP Waters Programme also reported an increase in surface water dissolved organic carbon (DOC) concentrations (Monteith *et al.*, 2007; Skjelkvåle *et al.*, 2005), with concentrations doubling across the UK

(Evans *et al.*, 2006). The primary driver of this trend has been an increase in soil organic matter solubility in response to reductions in acid rain (Monteith *et al.*, 2007). However, it is also clear that DOC concentrations are sensitive to variation and long-term change in soil temperatures, amount and intensity of precipitation, the residence time of lakes and reservoirs, and seasalt deposition events during winter storms (Ritson *et al.*, 2014). As acid deposition in Europe and North America approaches very low levels, it is very likely that these climate related stressors will soon begin to dominate variation and change in DOC in surface waters. The need to determine the extent to which climate change will impact the amount and composition of DOC is becoming increasingly urgent given recent revisions to climate change projections.

While sulfur deposition has declined rapidly over the last 30 years, total deposition of N has shown much smaller changes. For example, N deposition in the UK declined by 13% between 1988 and 2008, despite emission reductions of 50% for NO_x and

18% for NH_3 over the same period. The non-linearities between emissions and deposition of N compounds are poorly understood, but are thought to be related to changes in chemical processing of NO_x in the atmosphere (Curtis *et al.*, 2014). Simulations of N deposition at the global scale indicated that China has become one of the hotspots of N deposition in the world (Dentener *et al.*, 2006) and that rates in the industrialised and agriculturally intensified regions of China are as high as the peak levels of deposition in north-western Europe in the 1980s, before the introduction of mitigation measures (Liu *et al.*, 2017).

Where N deposition exceeds the biological demand of the terrestrial ecosystem (termed N saturation), the excess N is removed in drainage water. This increased N leaching can lead to acidification of surface waters, when the NO_3 is accompanied by Al^{3+} or H^+, and eutrophication, where N is the limiting nutrient. However, identifying the precise impact of increased N leaching on aquatic ecosystems is complicated by the variety of chemical forms in which N may exist in atmospheric deposition, terrestrial biomass and surface waters. In addition, a variety of terrestrial processes separate N deposition from its effects and reduces the ability to attribute observations of aquatic effects to known rates of deposition.

> ## REFLECTIVE QUESTIONS
>
> Why are the culprits for diffuse pollution difficult to determine?
>
> What methods can be adopted to reduce pollution from agriculture, urban areas and mine drainage?
>
> Why does acid rain cause water pollution?

4.6 WATER QUALITY POLICY

Over the last century different strategies have been developed to overcome the conflict between water quality deterioration due to human activities and water use that include a combination of:

1. wastes treated before discharging to a water body;
2. waters treated in order to meet specific water quality requirements (e.g. drinking water);
3. water quality determined and procedures adopted that prevent deterioration (e.g. land management options).

In most countries these strategies have been accompanied by the development of laws and regulation to help to prevent and reduce water pollution. For example, in the United States there are the 1972 Clean Water Act (CWA) and the 1974 Safe Drinking Water Act. Under the CWA, the Environment Protection Agency (EPA) has implemented pollution control programmes such as setting wastewater standards for industry. The CWA made it unlawful to discharge any pollutant from a point source into navigable waters, unless a permit is obtained. The CWA's National Pollutant Discharge Elimination System (NPDES) regulates point sources that discharge pollutants into waters of the United States. NPDES permits are issued to any facility that discharges directly into waters of the US. Regulated entities include industrial and municipal facilities and include the following types of discharges: municipal wastewater overflows, storm water; biosolids and discharges from concentrated animal feeding operations. The EPA has also developed national water quality criteria recommendations for approximately 150 pollutants in surface waters for the protection of aquatic life and human health. To protect human health and aquatic life in these waters, states, territories and authorised tribes establish water quality standards (WQS). WQS consist of three core components: designated uses of a water body, criteria to protect designated uses, and antidegradation requirements to protect existing uses and high-quality/high-value waters. The WQS regulation requires states, territories and authorised tribes to specify goals and expectations for how each water body is used. States, territories and authorised tribes adopt water quality criteria to protect the designated uses of a water body. Water quality criteria can be numeric (e.g. the maximum pollutant concentration levels permitted in a water body) or narrative (e.g. a criterion that describes the desired conditions of a water body being 'free from' certain negative conditions). States, territories and authorised tribes typically adopt both numeric and narrative criteria. One of the principal objectives of the Clean Water Act is to 'maintain the chemical, physical and biological integrity of the Nation's waters'. Antidegradation requirements provide a framework for maintaining and protecting water quality that has already been achieved. Designated uses and water quality criteria are the primary tools states and authorised tribes use to achieve the objectives and goals of the CWA, and antidegradation requirements complement these tools by providing a framework for maintaining existing uses, for protecting waters that are of a higher quality than necessary to support the CWA goals, and for protecting waters identified by states and authorised tribes as Outstanding National Resource Waters.

One of the biggest problems with water pollution is its transboundary nature. Many rivers cross countries, so pollution discharged in one country can cause water quality problems in neighbouring countries. While environmental legislation in one country can make it tougher for people to pollute, to be really effective these laws need to operate across national and international borders (see Chapter 11). The EU Water Framework Directive (WFD), which was introduced in 2000, provides Europe with a framework for integrated water management – ensuring all the issues affecting water bodies are addressed in one plan; it puts ecological health as its indicator of success; it sets targets for the achievement of good health (all water bodies to reach 'good status' by 2021); and it requires stakeholder engagement (providing ownership and sustainability opportunities). The ecological and chemical status of surface waters is assessed according to the following criteria:

1. Biological quality (fish, benthic invertebrates, aquatic flora).
2. Hydromorphological quality such as riverbank structure, river continuity or substrate of the riverbed.
3. Physical-chemical quality such as temperature, oxygenation and nutrient conditions.
4. Chemical quality that refers to environmental quality standards for river basin specific pollutants. These standards specify maximum concentrations for specific water pollutants. If even one such concentration is exceeded, the water body will not be classed as having a 'good ecological status'.

The Directive requires the establishment of river basin districts. These areas have been designated not according to administrative or political boundaries but rather according to the **river basin** (the spatial catchment area of the river) as a natural geographical and hydrological unit, some of which will traverse national boundaries. They are managed according to a river basin management plan, which should provide a clear indication of the way the objectives set for the river basin are to be reached within the required timescale and will be updated every six years. There are four distinct elements to the river basin planning cycle: characterisation and assessment of impacts on river basin districts; environmental monitoring; the setting of environmental objectives; and the design and implementation of the programme of measures needed to achieve them. However, in 2015, despite a 15-year implementation phase, only 53% of European water bodies met good ecological status: a meagre ten point improvement on the 2009 baseline figure of 43%. A number of barriers to achieving good status have been identified (EC, 2012, 2015), including: pollution by new, largely unknown, groups of substances (e.g. pharmaceuticals, endocrine disruptors); hydromorphological and water quantity related pressures, mostly driven by unsustainable practices of agriculture, energy production, transport and industry; the low level of ambition qualified by extensive use of exemptions, by which EU Member States were postponing much-needed management measures and setting less stringent objectives; coordination and governance mechanisms were not clear and RBMPs lacked concrete measures and economic evaluation of such measures; there was a lack of transparency about decision-making, particularly around stakeholder participation; and poor integration with other policies. In 2018, only 38% of surface waters (rivers, lakes and transitional and coastal waters) were in good chemical status (EEA, 2018). This is because if just one of the priority chemicals is present then the water body fails. In most Member States, a few priority substances account for poor chemical status, the most common being mercury. If mercury and other ubiquitous priority substances were omitted, only 3% of surface water bodies would fail to achieve good chemical status (EEA, 2018).

While the WFD has set stringent targets for achieving good ecological status of all surface waters, it is important to remember that surface waters have been used and degraded by humans over decadal and centennial timescales, and thus the 2021 WFD targets need to be considered against these longer timescales of system response. In addition, the impacts of climate change on these targets have not been considered. To date, studies investigating the impact of climate change on the global water cycle have concentrated on water resources and water supply. Hence there is currently much uncertainty about the potential impacts of climate change on surface water quality (Whitehead *et al.*, 2009). However, the following are likely:

1. Changes in the timing, intensity and duration of precipitation will lead to changes in flow regime that will affect water quality and ecology. However, there is a lot of uncertainty about the likely impacts due to changes in regional precipitation.
2. Increases in temperature will cause higher evaporation from open surfaces and soils, and increased transpiration by vegetation potentially reducing water availability.
3. Increase in river and lake water temperatures that will result in changes in water quality (e.g. concentration of dissolved oxygen will deplete faster, and ecology – see Chapter 6).
4. Lower river flows will result in less volume available for dilution of pollution downstream of point discharges.
5. Increased storm events, especially in summer, would lead to more frequent incidences of combined sewer overflows, discharging highly polluted waters into receiving water bodies.
6. More intense rainfall and flooding could result in increased suspended solids,

sediment yields and associated contaminant metal fluxes.

7. There is a need for models to represent the whole ecosystem and the uncertainty in model prediction to be quantified.

Predicting changes in water quality as a result of climate change is difficult because of the inherent complexity and interlinks between the nature and magnitude of the use of freshwater resources by a range of sectors (Cross and Latorre, 2015). In addition, climate impacts on water quality will affect productivity of all sectors, and, without more efficient, innovative and wiser policies and regulations, it is expected that the relative importance of water uses will shift by 2050, increasing competition between uses (OECD, 2012a). Therefore, clear guidance from regulators to reduce the impacts of climate change (such as floods and droughts) that compromise water quality, in turn affecting human and environmental health as well as economic production are urgently needed (Cross and Latorre, 2015). The Great Lakes Water Quality Agreement, between Canada and the USA, is one of the only protocols to clearly address the impacts of climate change on water quality (Annex 9 of the protocol). The protocol establishes an obligation to coordinate efforts between the parties in order to identify, quantify, understand and predict these impacts; the protocol also binds the parties to sharing such information in a timely and comprehensive manner. The system established in the protocol also requires the parties to identify and provide assistance to coastal communities to understand the impacts of climate change on water quality. The protocol is implemented and enforced by an International Joint Commission, established by the same treaty.

Some people believe that global changes in population and economic development will have a much larger impact on water supply and quality than will changes in climate over the next few decades (e.g. Vörösmarty *et al.*, 2000 and see Chapter 2). The greatest increases in exposure to pollutants are expected to occur in low- and lower-middle-income countries, primarily because of higher population and economic growth in these countries, especially those in Africa (UNEP, 2016), and the lack of wastewater management systems (WWAP, 2017). Therefore, providing safe drinking water to urban inhabitants is a major challenge. In cities with large upstream populations, the risk of drinking water containing high concentrations of nutrients and bacteria is high. The lack of sewage systems in villages and towns upstream of cities has serious consequences for drinking water quality downstream. Another way of improving water quality in rivers entering cities is to change land use or management upstream. To obtain a better understanding of how surface water quality will change in the future, an integrated approach is required that brings together climate change, water and land users and socio-economic communities (see also Chapter 13). There is also a need to mobilise financial resources, particularly in the developing world, to support integrated approaches to improving water quality (see Chapter 8).

REFLECTIVE QUESTIONS

How might climate change impact upon water quality?

Why is transnational water quality legislation necessary?

4.7 SUMMARY

Rivers and lakes are vital natural resources; they provide drinking water, habitats for many different types of flora and fauna, and are an important resource for industry and recreation. However, owing to human activities, a significant proportion of them are now polluted or under threat from pollution. Protecting and improving the aquatic environment is an important part of achieving sustainable development and is vital for the long-term health and well-being of humans and wildlife.

The hydrological pathway by which water travels through a watershed has a large control on the solute concentration. For example, water that passes rapidly over or through the watershed has a lower solute content than water that has moved slowly through the watershed. Temperature, redox potential and pH have a strong control on the form and oxidation state of different solutes. Global and regional patterns in surface water chemistry are predominantly controlled by differences in geology and climate, with soil, vegetation and land use being more important controls at the local scale. Surface water chemistry also displays temporal patterns at a range of scales from diurnal to seasonal to long term. These temporal patterns arise because of the processes controlling the supply of solutes available for leaching varying over time, and the hydrological processes that generate runoff within a catchment are dynamic and vary with climate and season.

Surface water chemistry is directly threatened by human activities. River and lake chemistry is affected by widespread land-cover change, agriculture, irrigation, urbanisation, industrialisation and its legacy, engineering schemes like reservoirs, and pollution from atmospheric deposition. Population growth and climate change are all likely to have a big impact on surface water chemistry in the future. National and international water policies and legislation have already played important roles in reducing atmospheric pollution to water bodies. Water quality legislation will continue to have a major role in tackling future point and diffuse pollution.

FURTHER READING

Ahuja, S. (ed.) 2013. *Monitoring water quality.* Elsevier; Amsterdam.

Fifteen chapters from different authors providing good coverage of water quality issues, monitoring and analysis techniques and water pollution mitigation measures.

Boyd, C.E. 2015. *Water quality: an introduction.* 2nd edition. Springer; New York.

This text covers a wide range of water quality variables and also has a chapter about water quality regulation.

Campbell, N., D'Arcy, B., Frost, A., Novotny, V. and Sansom, A. 2005. *Diffuse pollution: theory, control measures, practical experiences.* IWA Publishing; London.

This is an introductory text covering the nature, causes and significance of urban and rural diffuse pollution. Best management practices to tackle the problems are examined, as are the ways in which the adoption of such practices may be brought about. It makes use of case studies, from an EU perspective, to examine the strengths and weaknesses of various approaches, including SUDS.

Kelly, M. 1988. *Mining and the freshwater environment.* Elsevier Applied Science; London.

Although this book concentrates to a significant extent on the impacts of metal mining, it also provides good coverage of the general principles and processes that lead to acid mine drainage and the extent of their impacts on the freshwater environment.

OECD. 2012b. *Water quality and agriculture: meeting the policy challenge. OECD studies on water.* OECD Publishing; Paris.

The book examines the need to link farm management and practices with policies and economic assessments of water quality impacts from agriculture. Available for free download here: www1.oecd.org/publications/water-quality-and-agriculture-9789264168060-en.htm.

Classic papers

Burt, T.P., Arkell, B.P., Trudgill, S.T. and Walling, D.E. 1988. Stream nitrate concentrations in a small stream in Southwest England over a 15 year period (1970–1985). *Hydrological Processes* 2: 155–163.

An early paper showing the value of long-term water quality data for detecting change and determining the drivers of change.

Gibbs, R. 1970. Mechanisms controlling world water chemistry. *Science* 170: 1088–1090.

A classic global assessment showing the geological and climatological drivers of spatial variability in water quality.

Hasler, A.D. 1947. Eutrophication of lakes by domestic drainage. *Ecology* 28: 383–395.

A paper which showed how agricultural and urban pollution could lead to eutrophication, and hence lake ecosystems required good land management practice.

O'Connor, D.J. 1960. Oxygen balance of an estuary. *Journal of the Sanitary Engineering Division*, ASCE 86: Proceedings Paper 2472.

The foundation of modern water-quality modelling for surface waters.

PROJECT IDEAS

- Measure the solutes and sediment concentrations of two tributaries just upstream of a confluence and also measure the same things immediately downstream of the confluence. Have there been chemical changes at the confluence or has the system just mixed?

- In an agricultural catchment, sample several tributaries and the main river channel at different points for key nutrient concentrations (e.g. nitrate and phosphate). Can you relate differences to land use?

- Test some water quality variables immediately upstream and downstream of a BMP/SUDS device and evaluate its performance.

REFERENCES

Ahuja, S. (ed.) 2013. *Monitoring water quality*. Elsevier; Amsterdam.

Armitage, P.D., Moss, D., Wright, J.F. and Furse, M.T. 1983. The performance of a new biological water quality score system based on macroinvertebrates over a wide range of unpolluted running-water sites. *Water Research* 17: 333–347.

Battaglia, M., Hose, G.C., Turak, E. and Warden, B. 2005. Depauperate macroinvertebrates in a mine affected stream: clean water may be the key to recovery. *Environmental Pollution* 138: 132–141.

Battarbee, R.W., Mason, J., Renberg, I. and Talling, J.F. 1990. *Palaeolimnology and lake acidification*. ENSIS Publishing; London.

Batty, L.C., Atkin, L. and Manning, D.A.C. 2005. Assessment of the ecological potential of mine-water treatment wetlands using a baseline survey of macroinvertebrate communities. *Environmental Pollution* 138: 412–419.

Boyd, C.E. 2015. *Water quality: an introduction.* 2nd edition. Springer; New York.

Bundschuh, M., Hahn, T., Gessner, M.O. and Schulz, R. 2009. Antibiotics as a chemical stressor affecting an aquatic decomposer-detrivore system. *Environmental Toxicology and Chemistry* 28: 197–203.

Burns, D.A., McHale, M.R., Driscoll, C.T. and Roy, K.M. 2006. Response of surface water chemistry to reduced levels of acid precipitation: comparison of trends in two regions of New York, USA. *Hydrological Processes* 20: 1611–1627. doi: 10.1002/hyp.5961.

Burt, T.P., Arkell, B.P., Trudgill, S.T. and Walling, D.E. 1988. Stream nitrate concentrations in a small stream in Southwest England over a 15 year period (1970–1985). *Hydrological Processes* 2: 155–163.

Burt, T.P., Howden, N.J.K., Worrall, F., Whelan, M.J. and Bieroza, M. 2011. Nitrate in United Kingdom rivers: policy and its outcomes since 1970. *Environmental, Science and Technology* 45: 175–181.

Campbell, N., D'Arcy, B., Frost, A., Novotny, V. and Sansom, A. 2005. *Diffuse pollution: theory, control measures, practical experiences.* IWA Publishing; London.

Chadwick, J.W., Canton, S.P. and Dent, R.L. 1986. Recovery of benthic invertebrate communities in Silver Bow Creek, Montana, following improved metal mine wastewater treatment. *Water, Air and Soil Pollution* 28: 427–438.

Chapman P.J. and Edwards, A.C. 1999. The impact of atmospheric N deposition on the behaviour of N in surface water. *In:* Langan, S.J. (ed.) *Impacts of nitrogen deposition on natural and semi-natural ecosystems.* Kluwer Academic Publishers; Dordrecht, 153–212.

Chapman, P.J., Reynolds, B. and Wheater, H.S. 1993. Hydrochemical changes along storm-flow pathways in a small moorland headwater catchment in mid-Wales, UK. *Journal of Hydrology* 151: 241–265.

Civerolo, K.L. and Roy, K.M. 2013. On the road to recovery: acid rain and the Adirondacks. *New York State Conservationist* 68: April 2013, 17–19.

Clements, W.H. 1994. Benthic invertebrate community responses to heavy metals in the upper Arkansas River basin, Colorado. *Journal of the North American Benthological Society* 13: 30–44.

Colwill, D.M., Peters, C.J. and Perry, R. 1984. Water quality of motorway runoff. *Supplementary Report 823.* Transport Road Research Laboratory; Crowthorne.

Cross, K. and Latorre, C. 2015. Which water for which use? Exploring water quality instruments in the context of a changing climate. *Aquatic Procedia* 5: 104–110.

Curtis, C.J., Battarbee, R.W., Monteith, D.T. and Shilland, E.M. 2014. The future of upland water ecosystems of the UK in the 21st century: a synthesis. *Ecological Indicators* 37: 412–430.

DeNicola, D.M. and Stapleton, M.G. 2002. Impact of acid mine drainage on benthic communities in streams: the relative roles of substratum vs. aqueous effects. *Environmental Pollution* 119: 303–315.

Dentener, F., Drevet, J., Lamarque, J.F., Bey, I., Eickhaut, B., Fiore, A.M. *et al.* 2006. Nitrogen and sulfur deposition on regional and global scales: a multimodel evaluation. *Global Biogeochemical Cycles* 20: GB4003.

Diamond, J.M., Bower, W. and Gruber, D. 1993. Use of man-made impoundment in mitigating acid mine drainage in the North Branch Potomac River. *Environmental Management* 17: 225–238.

Dillon, P.J. and Rigler, F.H. 1974. The phosphorous–chlorophyll relationship in lakes. *Limnology and Oceanography* 19: 767–773.

Dobson, M. and Frid, C. 2009. *Ecology of aquatic systems.* 2nd edition. Oxford University Press; Oxford.

Driscoll, C.T., Schaefer, D.A., Molot, L.A. and Dillion, P.J. 1989. Summary of North America data. *In:* Malanchuk, J.L. and

Nilsson, J. (eds) *The role of nitrogen in the acidification of soil and surface waters.* Nordic Council of Ministers; Sweden.

Du Preez, L.H., Jansen van Rensburg, P.J., Jooste, A.M., Carr, J.A., Giesy, J.P., Gross, T. S., Kendall, R.J., Smith, E.E., Van Der Kraak, G. and Solomon, K.R. 2005. Seasonal exposures to triazine and other pesticides in surface waters in the western Highyield corn-production region in South Africa. *Environmental Pollution* 135: 131–141.

EC. 1991. *Council Directive 91/676/EEC concerning the protection of waters against pollution by nitrates from agricultural sources.* OJ L375. European Commission; Brussels.

EC. 2011. *Chemicals and the Water Framework Directive: draft environmental quality standards, diclofenac.* European Commission; Brussels.

EC. 2012. Report from the Commission to the European Parliament and the Council on the Implementation of the Water Framework Directive (2000/60/EC) river basin management plans (COM(2012)670 final). European Commission; Brussels.

EC. 2013. Report from the Commission to the Council and the European Parliament on the Implementation of Council Directive 91/676/ EEC concerning the protection of waters against pollution caused by nitrates from agricultural sources based on Member State reports for the period 2008–2011. European Commission; Brussels.

EC. 2015. Communication from the Commission to the European Parliament and the Council on the Water Framework Directive and the Floods Directive: Actions towards the 'good status' of EU water and to reduce flood risks (COM(2015) 120 final). European Commission; Brussels.

Edwards, A.C. and Withers, P.J.A. 2008. Transport and delivery of suspended solids, nitrogen and phosphorus from various sources to freshwaters in the UK. *Journal of Hydrology* 350: 144–153.

EEA (European Environment Agency). 2018. *European waters – assessment of status and pressures 2018.* EEA; Copenhagen.

Ellis, J.B. and Mitchell, G. 2006. Urban diffuse pollution: key management issues for the Water Framework Directive. *Journal of the Chartered Institution of Water and Environmental Management* 20: 19–26.

Evans, C.D., Chapman, P.J., Clark, J.M., Monteith, D.T. and Cresser, M.S. 2006. Alternative explanations for rising dissolved organic carbon export from organic soils. *Global Change Biology* 12: 2044–2053.

Fenn, M.E., Lambert, K.F., Blett, T.F., Burns, D. A., Pardo, L.H., Lovett, G.M., Haeuber, R. A., Evers, D.C., Driscoll, C.T. and Jeffries, D.S. 2011. Setting limits: using air pollution thresholds to protect and restore US ecosystems. Issues in Ecology, Report Number 14. J.S. Baron (ed.) Ecological Society of America. Washington, D.C. 21 p.

Ferreira da Silva, E., Almeida, S.F., Nunes, M.L., Luís, A.T., Borg, F., Hedlund, M., de Sá, C.M., Patinha, C. and Teixeira, P. 2009. Heavy metal pollution downstream the abandoned Coval da Mo nine (Portugal) and associated effects on epilithic diatom communities. *Science of the Total Environment* 407: 5620–5636.

Garrod, G.D., Garratt, J.A., Kennedy, A. and Willis, K.G. 2007. A mixed methodology framework for the assessment of the voluntary initiative. *Pest Management Science* 63: 157–170.

Gibbs, R. 1970. Mechanisms controlling world water chemistry. *Science* 170: 1088–1090.

Gray, L.J. and Ward, J.V. 1983. Leaf litter breakdown in streams receiving treated and untreated metal mine drainage. *Environment International* 9: 135–138.

Gray, N.F. 1997. Environmental impact and remediation of acid mine drainage: a management problem. *Environmental Geology* 30: 62–71.

Gray, N.F. 1998. Acid mine drainage composition and the implications for its impact on lotic systems. *Water Research* 32: 2122–2134.

Hasler, A.D. 1947. Eutrophication of lakes by domestic drainage. *Ecology* 28: 383–395.

Hatch, D., Goulding, K. and Murphy, D. 2002. Nitrogen. *In:* Haygarth, P.M. and Jarvis, S.C.

(eds) *Agriculture, hydrology and water quality*. CABI Publishing; Wallingford, 8–27.

He, M., Wang, Z. and Tang, H. 1997. Spatial and temporal patterns of acidity and heavy metals in predicting the potential for the ecological impact on the Le An river polluted by acid mine drainage. *The Science of the Total Environment* 206: 67–77.

Henriksen, A. and Posch, M. 2001. Steady-state models for calculating critical loads of acidity for surface waters. *Water, Air and Soil Pollution* 1: 375–398.

Henriksen, A., Kamari, J., Posch, M. and Wilander, A. 1992. Critical loads of acidity: Nordic surface waters. *Ambio* 26: 304–311.

Hinton, M.J., Schiff, S.L. and English, M.C. 1997. The significance of storms for the concentration and export of dissolved organic carbon from two Precambrian Shield catchments. *Biogeochemistry* 36: 67–88.

Hooda, P.S., Edwards, A.C., Anderson, H.A. and Miller, A. 2000. A review of water quality concerns in livestock farming areas. *Science of the Total Environment* 250: 143–167.

IPCC. 2007. *Climate change 2007: impacts, adaptation and vulnerability. Contribution of Working Group II to the fourth assessment report of the Intergovernmental Panel on Climate Change* [Parry, M.L. *et al.* (eds)]. Cambridge University Press; Cambridge.

Irvine, K., Moss, M. and Balls, H. 1989. The loss of submerged plants with eutrophication II. Relationships between fish and zooplankton in a set of experimental ponds, and conclusions. *Freshwater Biology* 22: 89–107.

Jackson, A.R.W. and Jackson, J.M. 2000. *Environmental science: the natural environment and human impact*. Pearson Education; Harlow.

Jarvie, H.P., Withers, P.J.A., Hodgkinson, R., Bates, A., Neal, M., Wickham, H.D., Harman, S.A. and Armstrong, L. 2007. Influence of rural land use on streamwater nutrients and their ecological significance. *Journal of Hydrology* 350: 166–186.

Jarvis, A.P. and Younger, P.L. 1997. Dominating chemical factors in mine water induced impoverishment of the invertebrate fauna of two streams in the Durham coalfield, UK. *Chemistry and Ecology* 13: 249–270.

Jeffries, D.S., Clair, T., Couture, S., Dillon, P., Dupont, J., Keller, W., McNicol, D., Turner, R., Vet, R. and Weever, R. 2003. Assessing recovery of lakes in southeastern Canada from acidic deposition. *Ambio* 32: 176–182.

Kalin, M. 2004. Passive mine water treatment: the correct approach? *Ecological Engineering* 22: 299–304.

Kasprzyk-Hordern, B., Dinsdale, R.M. and Guwy, A.J. 2008. The occurrence of pharmaceuticals, personal care products, endocrine disruptors and illicit drugs in surface water in South Wales, UK. *Water Research* 42: 3498–3518.

Kay, P., Blackwell, P.A. and Boxall, A.B.A. 2004. Fate of veterinary antibiotics in a macroporous tile drained clay soil. *Environmental Toxicology and Chemistry* 23: 1136–1144.

Kay, P., Edwards, A.C. and Foulger, M. 2009. A review of the efficacy of contemporary agricultural stewardship measures for addressing water pollution problems of key concern to the UK water industry. *Agricultural Systems* 99: 67–75.

Kay, P., Grayson, R., Phillips, M., Stanley, K., Dodsworth, A., Hanson, A., Walker, A., Foulger, M., McDonnell, I. and Taylor, S. 2012. The effectiveness of agricultural stewardship for improving water quality at the catchment scale: experiences from an NVZ and ECSFDI watershed. *Journal of Hydrology* 422–423: 10–16.

Kay, P., Hughes, S.R., Ault, J.R., Ashcroft, A.E. and Brown, L.E. 2017. Widespread, routine occurrence of pharmaceuticals in sewage effluent, combined sewer overflows and receiving waters. *Environmental Pollution* 220: 1447–1455.

Kay, P., Hiscoe, R., Moberley, I., Bajic, L. and McKenna, N. 2018. Wastewater treatment plants as a source of microplastics in river catchments. *Environmental Science and Pollution Research* 25: 20264–20267.

Kelly, M. 1988. *Mining and the freshwater environment*. Elsevier Applied Science; London.

Kladivko, E.J., Vanscoyok, G.E., Monke, E.J., Oates, K.M. and Pask, W. 1991. Pesticide and nutrient movement into subsurface tile drains on a silt loam soil in Indiana. *Journal of Environmental Quality* 20: 264–270.

Kleinmann, R.L.P., Crerar, D.A. and Pacelli, R.R. 1981. Biogeochemistry of acid mine drainage and a method to control acid formation. *Mining Engineering* 33: 300–305.

Koryak, M. and Reilly, R.J. 1984. Vascular riffle flora of Appalachian streams: the ecology and effects of acid mine drainage on *Justicia Americana* (L.) Vahl. *Proceedings of the Pennsylvania Academy of Science* 58: 55–60.

Laine, D.M. and Dudeney, A.W.L. 2000. Bullhouse minewater project. *Transactions of the Institution of Mining and Metallurgy Section A* 109: A224–227.

Lampkin, A.J. III and Sommerfield, M.R. 1982. Algal distribution in a small, intermittent stream receiving acid mine drainage. *Journal of Phycology* 18: 196–199.

Larssen, T., Lydersen, E., Tang, D., He, Y., Gao, J., Liu, H., Duan, L., Seip, H.M., Vogt, R.D., Mulder, J. and Shao, M. 2006. Acid rain in China. *Environmental Science and Technology* 40: 418–425.

Letterman, R.D. and Mitsch, W.J. 1978. Impact of mine drainage on a mountain stream in Pennsylvania. *Environmental Pollution* 17: 53–73.

Likens, G.E., Bormann, F.H., Pierce, R.S., Eaton, J.S. and Johnson, N.M. 1977. *Biogeochemistry of a forested ecosystem*. Springer; New York.

Littlewood, I.G. 1992. *Estimating contaminant loads in rivers: a review*. Report No. 117, Institute of Hydrology; Wallingford.

McDonald, R., Douglas, I., Revenga, C., Hale, R., Grimm, N., Gronwall, J. and Feketa, B. 2011. Global urban growth and the geography of water availability, quality and delivery. *Ambio* 40: 537–446.

Liu, X., Xu, W., Duan, L., Du, E., Pan, Y., Lu, X., Zhang, L., Wu, Z., Wang, X., Zhang, Y. and Shen, J. 2017. Atmospheric nitrogen emission, deposition, and air quality impacts in China: an overview. *Current Pollution Reports* 3: 65–77.

Meybeck, M. 1981. Pathways of major elements from land to oceans through rivers. *In: River inputs to ocean systems*, UNEP/UNESCO Report. 18–30.

Meybeck, M., Kuusisto, A., Makela, A. and Malkki, E. 1996. Water quality. *In:* Bartram, J. and Balance, R. (eds) *Water quality monitoring: a practical guide to the design and implications of freshwater quality studies and monitoring programmes*. Chapman and Hall; London.

Mitchell, G. 2001. *The quality of urban stormwater in Britain and Europe: database and recommended values for strategic planning models*. Technical Report, School of Geography; University of Leeds.

Mitchell, M.J., Driscoll, C.T., Kahl, J.S., Likens, G.E., Murdoch, P.S. and Pardo, L.H. 1996. Climatic control of nitrate loss from forested watersheds in the Northeast United States. *Environmental Science and Technology* 30: 2609–2612.

Monteith, D.T., Reynolds, B. and Evans, C.D. 2000. Are temporal variations in the nitrate content of UK upland freshwaters linked to the North Atlantic Oscillation? *Hydrological Processes* 14: 1745–1749.

Monteith, D.T., Stoddard, J.L., Evans, C.D., de Wit, H.A., Forsius, M., Hogasen, T., Wilander, A., Skjelkvåle, B.L., Jeffries, D.S., Vuorenmaa, J., Keller, B., Kopácek, J. and Vesely, J. 2007. Dissolved organic carbon trends resulting from changes in atmospheric deposition chemistry. *Nature* 450: 537–539.

Morgan, J.J. and Stumm, W. 1965. The role of multivalent metal oxides in limnological transformations, as exemplified by iron and manganese. *Advances in Water Pollution Research, Proceedings of the Second International Conference on Water Pollution Research* 1: 103–118.

Moss, B. 2010. *Ecology of freshwaters: a view for the twenty-first century*. Wiley-Blackwell; Chichester.

Mullaney, J.R., Lorenz, D.L. and Arntson, A.D. 2009. Chloride in groundwater and surface water in areas underlain by the glacial aquifer system, Northern United States. U.S. Geological Survey Scientific Investigations Report 2009–5086.

Neal, C., Jarvie, H.P., Neal, M., Hill, L. and Wickham, H. 2006. Nitrate concentrations in river waters of the upper Thames and its tributaries. *The Science of the Total Environment* 365: 15–32.

Nilsson, J. and Grennfelt, P. (eds) 1988. *Critical loads for sulphur and nitrogen*. UNECE/ Nordic Council workshop report, Skokloster, Sweden. Nordic Council of Ministers; Copenhagen.

NIVA. 2018. *ICP Waters Report 135/2018 Regional assessment of the current extent of acidification of surface waters in Europe and North America*. Norsk Institutt for Vann forskning; Oslo.

Novotny, V. 2002. *Water quality: diffuse pollution and watershed management*. 2nd edition. John Wiley and Sons; New York.

O'Connor, D.J. 1960. Oxygen balance of an estuary. *Journal of the Sanitary Engineering Division*, ASCE 86: Proceedings Paper 2472.

OECD (Organisation for Economic Co-operation and Development). 2012a. *Environmental outlook to 2050: the consequences of inaction*. OECD Publishing; Paris. doi.org/10.1787/9789264122246-en.

OECD. 2012b. *Water quality and agriculture: meeting the policy challenge*. OECD studies on water. OECD Publishing; Paris.

OECD. 2017. *Diffuse pollution, degraded waters: emerging policy solutions*. OECD Publishing; Paris [online, accessed April 2019]. doi:10.1787/9789264269064-en.

Øygarden, L., Kvaerner, J. and Jenssen, P.D. 1997. Soil erosion via preferential flow to drainage systems in clay soils. *Geoderma* 76: 65–86.

Posch, M., Hettelingh, J.P., de Smet, P.A.M. and Downing, R.J. 1997. *Calculation and mapping of critical thresholds in Europe*. RIVM Report No. 259101007. National Institute of Public Health and the Environment; Bilthoven.

Prechtel, A., Alewell, C., Armbruster, M., Bittersohl, J., Cullen, J.M., Evans, C.D., Helliwell, R., Kopácek, J., Marchetto, A., Matzner, E. and Meesenburg, H. 2001. Response of sulphur dynamics in European catchments to decreasing sulphate deposition. *Hydrology and Earth System Sciences* 5: 311–326.

Pretty, J.N., Brett, C., Gee, D., Hine, R.E., Mason, C.F., Morison, J.I.L., Raven, H., Rayment, M.D. and van der Bijl, G. 2000. An assessment of the total external costs of UK agriculture. *Agricultural Systems* 65: 113–136.

Reynolds, B., Emmett, B.A. and Woods, C. 1992. Variations in streamwater nitrate concentrations and nitrogen budgets over 10 years in a headwater catchment in mid-Wales. *Journal of Hydrology* 136: 155–175.

Ritson, J.P., Graham, N.J.D., Templeton, M.R., Clark, J.M., Gough, R. and Freeman, C. 2014. The impact of climate change on the treatability of dissolved organic matter (DOM) in upland water supplies: a UK perspective. *Science of the Total Environment* 473–474: 714–730.

Rule, K.L., Comber, S., Ross, D., Thornton, A., Makropoulos, C.K. and Rautiu, R. 2006. Sources of priority substances entering an urban wastewater catchment – trace organic chemicals. *Chemosphere* 63: 581–591.

Schindler, D.W. 1988. Effects of acid rain on freshwater ecosystems. *Science* 239: 149–157.

Schmiermund, R.L. and Drozd, M.A. 1997. Acid mine drainage and other mining-influenced waters (MIW). *In*: Marcus, J.J. (ed.) *Mining environmental handbook*. Imperial College Press; London, 599–617.

SEPA. 1996. *State of the environment report*. Scottish Environmental Protection Agency; Stirling.

Singer, P.C. and Stumm, W. 1970. Acidic mine drainage: the rate-determining step. *Science* 167: 1121–1123.

Skjelkvåle, B.L. *et al.* 2005. Regional scale evidence for improvements in surface water chemistry 1990–2001. *Environmental Pollution* 137: 165–176.

Skjelkvåle, B.L. and De Wit, H.A. 2011. *Trends in precipitation chemistry, surface water chemistry and aquatic biota in acidified areas in Europe and North America from 1990 to 2008* (NIVA-rapport OR-6218). Norsk Institutt for Vann forskning (NIVA); Oslo.

SNIFFER. 2006. Retrofitting sustainable urban water solutions. Report UE3(05)UW5, Scottish and Northern Ireland Forum For Environmental Research [online]. Available from: www.sniffer.org.uk.

Sprague, L.A., Hirsch, R.M. and Aulenbach, B.T. 2011. Nitrate in the Mississippi River and its tributaries, 1980–2008; are we making progress? *Environmental Science and Technology* 45: 7209–7216.

Stoddard, J.L., Jeffries, D.S., Lükewille, A., Clair, T.A., Dillon, P.J., Driscoll, C.T., Forsius, M., Johannessen, M., Kahl, J.S., Kellogg, J.H., Kemp, A., Mannio, J., Monteith, D., Murdoch, P.S., Patrick, S., Rebsdorf, A., Skjelkvåle, B.L., Stainton, M.P., Traaen, T., van Dam, H., Webster, K.E., Wieting, J. and Wilander, A. 1999. Regional trends in aquatic recovery from acidification in North America and Europe. *Nature* 401: 575–578.

Stutter, M.I., Chardon, W.J. and Kronvang, B. 2012a. Riparian buffer strips as a multifunctional management tool in agricultural landscapes: introduction. *Journal of Environmental Quality* 41: 297–303.

Stutter, M.I., Dunn, S.M. and Lumsdon, D.G. 2012b. Dissolved organic carbon dynamics in a UK podzolic moorland catchment: linking storm hydrochemistry, flow path analysis and sorption experiments. *Biogeosciences* 9: 2159–2175.

Tremaine, S.C. and Mills, A.L. 1991. Impact of water column acidification on protozoan bacterivory at the lake sediment–water interface. *Applied and Environmental Microbiology* 57: 775–784.

Tyler, C.R., Lange, A., Paull, G.C., Katsu, Y. and Iguchi, T. 2007. The roach (Rutilus rutilus) as a sentinel for assessing endocrine disruption. *Environmental Sciences* 14: 235–253.

Tyler, S.J. and Ormerod, S.J. 1992. A review of the likely causal pathways relating the reduced density of breeding dippers *Cinclus cinclus* to the acidification of upland streams. *Environmental Pollution* 78: 49–56.

UNEP (United Nations Environment Programme). 2016. *A snapshot of the world's water quality: towards a global assessment.* UNEP; Nairobi. uneplive.unep.org/media/docs/assessments/unep_wwqa_report_web.pdf.

UNPD. 2007. *World urbanization prospects: the 2007 revision.* United Nations Population Division; New York.

US Senate. 2002. Federal Water Pollution Control Act, as amended through P.L. 107–303, 27 November, 2002.

Van Damme, P.A., Hamel, C., Ayala, A. and Bervoets, L. 2008. Macroinvertebrate community response to acid mine drainage in the rivers of the High Andes (Bolivia). *Environmental Pollution* 156: 1061–1068.

Vörösmarty, C.J., Green, P., Salisbury, J. and Lammers, R.B. 2000. Global water resources: vulnerability from climate change and population growth. *Science* 289: 284.

Walling, D.E. and Webb, B.W. 1985. Estimating the discharge of contaminants to coastal waters by rivers: some cautionary comments. *Marine Pollution Bulletin* 16: 488–492.

Walling, D.E. and Webb, B.W. 1986. Solutes in river systems. *In:* Trudgill, S.T. (ed.) *Solute processes.* Wiley and Sons Ltd; Chichester, 251–327.

Webb, B.W., Phillips, J.M., Walling, D.E., Littlewood, I.G., Watts, C.D. and Leeks, G.J.L. 1997. Load estimation methodologies for British rivers and their relevance to the LOIS RACS(R) programme. *The Science of the Total Environment* 194: 379–389.

Webb, J.S., McGuiness, S. and Lappin-Scott, H.M. 1998. Metal removal by sulphate-reducing bacteria from natural and constructed wetlands. *Journal of Applied Microbiology* 84: 240–248.

Wen, Y., Schoups, G. and van de Giesen, N. 2017. Organic pollution of rivers: combined threats of urbanization, livestock farming and global climate change. *Scientific Reports* 7: 43289; doi: 10.1038/srep43289.

Wetzel, R.G. 2001. *Limnology lake and river ecosystems.* Academic Press; New York.

White, I. and Alacon, A. 2009. Planning policy, sustainable drainage and surface water management: a case study of Greater Manchester, UK. *Built Environment* 35: 516–530.

Whitehead, P.G., Wilby, R.L., Battarbee, R.W., Kernan, M. and Wade, A.J. 2009. A review of the potential impact of climate change on surface water quality. *Hydrological Sciences Journal* 54: 101–123.

Wiseman, I.M., Edwards, P.J. and Rutt, G.P. 2003. Recovery of an aquatic ecosystem following treatment of abandoned mine drainage with constructed wetlands. *Land Contamination and Reclamation* 11: 221–229.

Worrall, F., Spencer, E. and Burt, T.P. 2009. The effectiveness of nitrate vulnerable zones for limiting surface water nitrate concentrations. *Journal of Hydrology* 370: 21–28.

Wren, C.D. and Stephenson, G.L. 1991. The effect of acidification on the accumulation and toxicity of metals to freshwater invertebrates. *Environmental Pollution* 71: 205–241.

WWAP. 2017. *The United Nations World Water Development Report 2017. Wastewater: the untapped resource.* UNESCO; Paris [online, accessed April 2019]. Available from: www.unesco.org/new/en/natural-sciences/environ ment/water/wwap/wwdr/2017-wastewater-the-untapped-resource/

Younger, P.L. 1997. The longevity of minewater pollution: a basis for decision-making. *The Science of the Total Environment* 194/195: 457–466.

Younger, P.L. 2000. Holistic remedial strategies for short- and long-term water pollution from abandoned mines. *Transactions of the Institution of Mining and Metallurgy, Section A. Mining Industry* 109: 210–218.

Yu, Q., Zhang, T., Ma, X., Kang, R., Mulder, J., Larssen, T. and Duan, L. 2017. Monitoring effect of SO_2 emission abatement on recovery of acidified soil and streamwater in southwest China. *Environmental Science & Technology* 51: 9498–9506.

CHAPTER FIVE

Groundwater

*L. Jared West and
Sumit Sinha*

5

LEARNING OUTCOMES

After reading this chapter you should be able to:

- describe the role of groundwater in the hydrological cycle;
- understand factors influencing recharge and water balance, and the potential impacts of groundwater abstraction and injection;
- appreciate which characteristics of rocks and soils determine whether they are groundwater aquifers;
- understand causes of groundwater movement and the factors that affect flow rates below the ground surface, including the important concepts of hydraulic head, hydraulic gradient and Darcy's Law;
- appreciate that groundwater flow can be represented using hydrogeological maps and numerical simulations;
- explain the basics of groundwater abstraction/injection techniques and their associated problems, and how groundwater can be used for heating and cooling;
- describe the factors that influence the natural chemical constituents of potable groundwater;
- describe key groundwater contaminants, the major threats to groundwater quality such as seawater intrusion and the main factors influencing groundwater vulnerability to pollution.

5.1 INTRODUCTION

Around 22% of the world's population rely solely on groundwater. An estimated 10% of global groundwater consumption is unsustainable in that water is being extracted faster than it is being replenished, and 10% of global food supply is based on such unsustainable groundwater exploitation (USAID, 2011). Given growing demand from population growth and development, coupled with reductions in sustainable yields due to climatic change, the twenty-first century is likely to see increased conflicts and other undesirable socio-economic consequences arising from insufficient and dwindling groundwater resources.

This chapter provides an introduction to groundwater resources and the physical and chemical processes which influence their availability. Beginning with the nature and occurrence of groundwater, and concepts of recharge, discharge and water balance, the idea of groundwater management is introduced. Then, the physical processes of groundwater flow are dealt with in more detail, in order to explain the distribution of extractable groundwater resources (aquifers). Flow processes that occur in aquifers, how they are represented using maps or numerical models, and methods for groundwater abstraction or reinjection are explained (including the physical construction of wells and managed aquifer recharge approaches). The use of groundwater resources for ground-source heating and cooling, increasingly prevalent in the twenty-first century, is described. The second part of the chapter deals with the chemical processes in groundwater, including natural constituents and how they are derived, and groundwater pollutants. The concepts of groundwater vulnerability to pollution and well protection are described and the factors which influence these are discussed.

5.2 GROUNDWATER RESOURCES

5.2.1 Recharge and discharge to groundwater

Groundwater constitutes an important component of the hydrological cycle, shown schematically in Figure 5.1, and is principally derived from precipitation, surface runoff and standing water bodies such as lakes and reservoirs. The addition of water to groundwater reserves is known as recharge, and humans also make a contribution to this

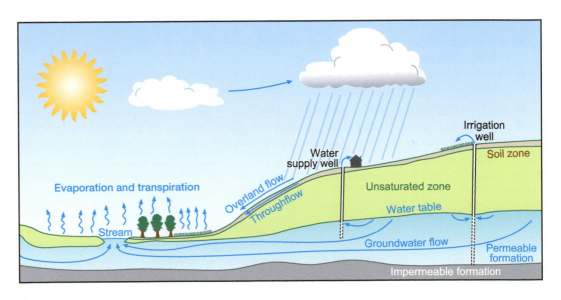

Figure 5.1 Role of groundwater in the hydrological cycle.

via surplus **irrigation**, pipe, sewer and canal leakage, and deliberate aquifer augmentation schemes (artificial recharge; see Section 5.5). In the upper layers of soil and rock there may be pores and voids which are not filled with water, but with air. This is known as the unsaturated or **vadose zone**. In this zone, water tends to flow downward under the force of gravity (although it may also be diverted laterally: West and Truss, 2006). In the saturated (or **phreatic**) zone beneath this, all pores and voids are water filled (except for occasional small pockets of trapped gas) and groundwater tends to flow laterally, typically following the topographic gradient. Groundwater may eventually intercept the ground surface, forming springs or seeps, or enter stream beds where it contributes to stream **baseflow**. Groundwater can also

emerge as springs on the seabed. Artificial abstraction, for irrigation or domestic and industrial supply, also constitutes an important means of groundwater discharge.

5.2.2 Rainfall recharge

The absorption of surface water into the ground is called **infiltration** (see **Figure 5.2**). Not all of the water that is present on the surface infiltrates: some runs off as **overland flow**, and some evaporates (see Chapter 3). Of the water that does infiltrate, some is drawn back to the surface by evaporation and some is drawn back to the surface by plants. Excess water from this process is then transpired by the plant leaves. A mature oak tree can draw water from as much as 10 m deep into the

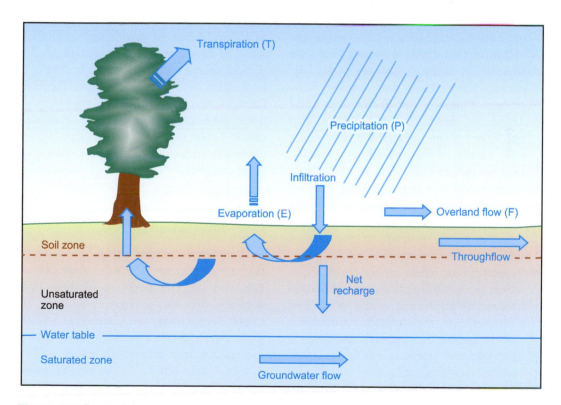

Figure 5.2 Factors influencing rainfall recharge to groundwater.

ground. The maximum rate at which infiltration can occur for a particular medium is known as its **infiltration capacity**. If the rate of precipitation exceeds infiltration capacity, then surplus water may lead to overland flow. There are a number of factors which determine how much water contributes to overland flow compared to that which infiltrates or is otherwise removed (e.g. by evaporation and transpiration):

1 The properties of the ground surface, for example, its porosity (void space) and **permeability** (ability to allow water to flow through or permeate, see Section 5.4).
2 Rainfall intensity and duration. The more intense the rainfall is, the more likely it is to exceed the infiltration capacity of the ground. This is why tropical regions, which commonly experience high-intensity rainfall, have a high percentage of **infiltration-excess overland flow** (see Chapter 3) compared to temperate climates, which have less intense rainstorms. Duration of precipitation and degree of ground saturation are also important; if the upper soil layer reaches its maximum water-holding capacity, then no further infiltration will be able to occur. Surplus water will then run off from the surface as **saturation-excess overland flow**.
3 Gradient of the ground. Steeper ground will encourage greater overland flow.
4 Vegetation (or other) surface cover. Vegetation in particular influences infiltration because a) plant roots create small flow pathways into the ground, encouraging and increasing infiltration compared to non-vegetated surfaces, and, b) conversely, less rainfall may reach the ground since it is intercepted by plant leaves.

Some of this intercepted rainfall may reach the ground indirectly by leaf-drip, while some of it may be used directly in **transpiration** and never reach the ground.

Water which infiltrates into the soil may move laterally as **throughflow** (also known as **interflow**) within the soil. See Chapter 3 for a more detailed treatment of such runoff production processes. Throughflow in the soil zone (see **Figure 5.2**) occurs where the soil is relatively permeable, as compared with the underlying bedrock. Throughflow does not recharge deeper groundwater but moves laterally to enter drainage channels. Water that passes through the soil zone and down to the phreatic zones contributes to groundwater recharge. In temperate climates this occurs mainly during the winter; in tropical climates, during the wet season. In drier seasons in both cases, the soil zone is below its water-holding capacity (**field capacity**) and tends to absorb any infiltration, which is later evaporated or transpired. Prediction of net recharge to groundwater is a complex field in its own right; for more information on how such prediction is done, see Scanlon *et al.* (2002). For an early classic paper on how evapotranspiration is estimated from meteorological data, see Penman (1948).

5.2.3 Groundwater – stream interactions

In general, rivers can either lose water to the surrounding areas or gain water (**Figure 5.3a** and **b**). The direction of these flows depends on the difference between the stream level (**stage**; see Chapter 3) and the groundwater level. During a flood period, increased stream stage may induce infiltration of stream water into the aquifer; subsequent declines in stream stage cause a reverse motion of infiltrated water. The

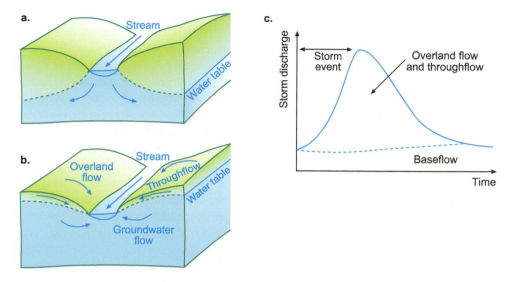

Figure 5.3 Interactions between streams and groundwater: a) losing stream; b) gaining stream; c) the stream hydrograph response showing the baseflow component due to groundwater discharges.

interaction between groundwater and stream water will be dependent on the geology and the topographic characteristics of the area. Streams on permeable catchments in temperate countries often show net gain of water from groundwater, and hence are called 'gaining streams'. Ephemeral streams (dry for some parts of the year) in arid climates more often lose water to groundwater and may be the main source of groundwater recharge in some areas. An increasing body of literature exists on methods of measuring exchange between streams and aquifers (e.g. Jones and Mulholland, 2003). One approach is hydrograph separation, which can be a useful tool to determine groundwater contributions at different flow stages during a flood event (**Figure 5.3c** and see also Chapter 3).

5.2.4 Groundwater management

The simplest approach to evaluation of sustainable groundwater resources is to determine inputs and outputs to groundwater in a defined area known as a groundwater catchment and by doing so to conduct a water balance evaluation. Groundwater catchments represent defined areas from which the in- and outflows of water can be specified, i.e. they are conceptually similar to the idea of a surface water catchment or drainage basin. Hence, they vary from just a portion of a geological formation, where this can be isolated by the presence of groundwater divides, to several geological units, where these are hydraulically interconnected. A groundwater catchment may be confined to a portion of a stream drainage catchment, such as only where permeable rocks are present. However, in extensive areas of highly permeable rocks such as limestones, surface and groundwater catchments may have completely different configurations so that groundwater divides bear little relation to the topographic divides that control surface drainage. Definition of catchments is now regarded as an essential first step in the holistic management of water resources.

Evaluation of the water balance allows management of groundwater resources via prediction of the impact of a change, such as increased abstraction, or climatic change, on other abstractions and the aquatic environment (e.g. river baseflows). For example, any increase in abstraction must ultimately reduce one of the other outputs from groundwater (e.g. river baseflow), or increase one of its inputs (e.g. by pulling water in from an adjacent groundwater body or, worse, a saltwater-saturated formation). In order to evaluate the water balance, the change in volume of groundwater stored over a given time period (typically a year) is found by:

$$\text{Change in storage} = (P - E - T - F - I) \\ + G - L + A,$$

$$[5.1]$$

where P is precipitation, E is evaporation, T is transpiration, F is overland flow, I is throughflow (also known as interflow) (see **Figure 5.2**), G is gains from surface water, L is losses to surface water, and A is artificial abstraction (where negative) or injection (where positive), with each item having units of volume per unit of time. If the change in storage is negative, a reduction in groundwater storage will take place. However, in non-arid climates, other terms will adjust to make the change in storage zero again. For example, the baseflow component of a gaining stream may be reduced. What needs to be decided in groundwater management is whether the environmental consequences of such changes are acceptable. In order to implement groundwater management, it is usual for regulatory bodies to license groundwater abstractions. It is also increasingly common to consider groundwater and surface water resources in a holistic way,

called integrated catchment management. This reflects the interaction of streams and groundwater outlined above. In the European Union such an approach is now mandatory as a result of the Water Framework Directive (European Commission, 2000) and the Groundwater Directive (European Commission, 2006). Although groundwater is a natural resource this does not imply that all groundwater on Earth has the potential to satisfy human needs or desires. There are mainly three factors that inhibit us from considering all of the groundwater as an exploitable resource. First, some of the groundwater may be discarded because it might not achieve an acceptable quality even after conventional treatments. Second, the abstraction of groundwater in some places might be technically and financially unviable. Finally, abstraction of groundwater almost always has an environmental consequence, which might be detrimental for the health of the ecosystem. Furthermore, the distribution of groundwater reserves worldwide is quite heterogeneous (**see Box 5.1**). A different approach is required for 'fossil' aquifer systems, where water is effectively being mined as a non-renewable resource, such as the Nubian Sandstone Aquifer in North Africa. In such cases, sustainable usage is not possible, so demand needs to be managed in order to provide supply for a finite period.

REFLECTIVE QUESTIONS

What are the key ways groundwater is recharged?

What are the main factors that determine the availability of groundwater resources in a particular area?

BOX 5.1

World aquifers

There is an immense diversity in aquifer sizes all over the world. Aquifer extent ranges from a few km to more than 1 million km^2, while aquifer thickness can vary from a few metres to several kilometres. Furthermore, the structure of an aquifer system may be simple (monolayer) or more complex (multilayer configuration). An aquifer may be composed of different aquifers and aquitards, between which both the reservoir capacity and conductive properties that allow groundwater to flow from recharge areas to one or more discharge zones might vary. It is difficult to present a summary of the world's hydrogeology because of the immense heterogeneity in the available hydrogeological information between regions. However, it is universally acknowledged that groundwater is the largest accessible and often still untapped freshwater reservoir on the Earth. The volume of worldwide groundwater reserves is assessed to be 10.5 million km^3. It is widely thought that regional water shortages and crises can only be met by enhanced understanding and sustainable use of our groundwater reserves. Yet, the information on this critical and crucial resource is poor in many places. There has been an international standardisation effort, initiated by the International Association of Hydrogeologists (IAH) and UNSECO, that aims to promote a uniform mapping methodology (Struckmeier and Margat, 1995). In order to provide data and information about the major groundwater resources of the world and devise a sound strategy for its management, the World-wide Hydrogeological Mapping and Assessment Programme (WHYMAP) was launched in 2000. The programme compiles data on groundwater from diverse sources and is instrumental in communicating groundwater-related issues to water experts, decision makers and the general public (e.g. see Figure 5.4).

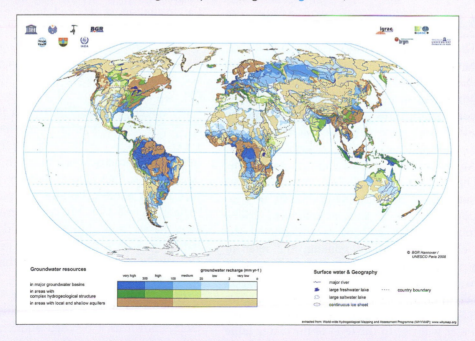

Figure 5.4 Groundwater resources of the world. Please see box text for an explanation of the legend.
Source: Reproduced with kind permission from WHYMAP: www.whymap.org.

The two main factors affecting the hydrogeological setting of a continent or a region are its geological conditions and climate. There are large differences in structure, hydraulic behaviour and water resources development potential of different aquifers worldwide. In order to better understand the geography of groundwater in the world, it is useful to look at each major aquifer type independently from a global perspective.

The groundwater resources map of the world (**Figure 5.4**) combines the related groundwater data published so far. The map depicts the spatial heterogeneity of the groundwater reserves worldwide. On the map blue is used for large and rather uniform groundwater basins in large sedimentary basins that might be conducive for groundwater exploitation. Green symbolises areas of complex geological structure. These areas might have productive aquifers (including **karst** aquifers; see Section 5.3.4) found adjacent to non-aquiferous strata. The combined use of remote sensing with detailed ground survey in conjunction with spring and stream flow analysis might help in identifying zones of high-yielding aquifers. Brown outlines areas with local and shallow aquifers in which crystalline bedrock is exposed. In these areas groundwater is confined to alteration zones and overlying shallow layers of weathered bedrock.

The mean recharge intensities are superimposed and are depicted by difference in colour tone varying from pale (very low recharge, i.e. less than 2 mm/year) to dark (very high recharge, i.e. more than 300 mm/year). The intervals in between are low (2–20 mm/year), medium (20–100 mm/year) and high rates of mean recharge (100–300 mm/year).

5.3 AQUIFERS

5.3.1 Aquifers and confining layers

The term 'aquifer' refers to a geological stratum which contains water and which allows the movement of water within it. It is porous, in that it has space represented by pores or fractures which can be occupied by water, and permeable (see Section 5.4), in that the water can move through this space. An example would be sand or sandstone. Other types of geological formation that make good aquifers are discussed in Sections 5.3.2 to 5.3.4. If a well were drilled into such strata and the water extracted, the well would keep on refilling. We refer to strata with relatively low permeability, such as silt or siltstone, as confining layers (**Figure 5.5**). In well A, shown in **Figure 5.5**, drilled in the permeable formation, the water will rise to the level of the **water table**

(uppermost part of the saturated zone) and the formation is referred to as an unconfined, or phreatic, aquifer. In well B, drilled through the low-permeability confining layer into the aquifer beneath, the water will rise to a greater elevation than the top of the aquifer, because this layer contains water under pressure. The aquifer is referred to as a confined aquifer at this point. The water in well B rises towards a **piezometric** (or potentiometric) surface; this is an imaginary surface determined by the pressure in the aquifer. Because the piezometric surface at well C is above the ground surface, water from this well would overflow naturally. Such a well is known as an **artesian well** and the aquifer in this area is called an artesian aquifer.

Low-permeability strata above the water table may produce a locally elevated water table, known as a perched aquifer. Perched

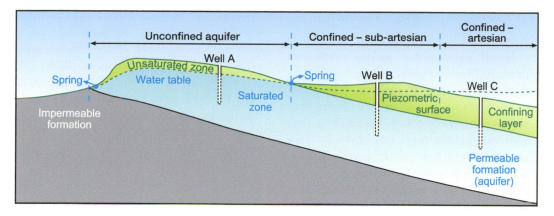

Figure 5.5 Confined and unconfined aquifers.

aquifers may be of limited lateral extent, and sometimes may exist only in response to particularly high infiltration (after a prolonged and heavy rainstorm, for instance). They may also be of considerable lateral extent and thickness, and be occasionally mistaken for permanent aquifers. Because of their variability, perched aquifers do not provide reliable long-term water supply sources.

5.3.2 Igneous and metamorphic rock aquifers

Most of the best aquifers around the world are found in sediments and sedimentary rocks. Igneous rocks and metamorphic rocks are commonly poor aquifer rocks. This is largely because of the mode of formation of these broad rock groups. Igneous rocks form from the solidification of magma, a process which usually creates a texture of interlocking crystals. This produces very little primary porosity (void space produced at the time of rock genesis), except for occasional vesicles in vesicular rocks. However, some volcanic rocks such as basalt have a well-developed fracture network formed during cooling, and this provides good

permeability. For example, the Columbia Plateau Basalts in the north-west USA are one of the highest-yielding aquifers in the world (Foxworthy, 1983). Other extrusive igneous rocks, such as pumice, may have very high porosities, but, owing to poor connectivity between void spaces, permeability is low. Metamorphic rocks form from the alteration of existing rocks by the application of heat and/or pressure. This tends to reduce their porosity and permeability below that of the parent rock.

Secondary porosity may later develop in some igneous and metamorphic rocks, due to alteration and decomposition of minerals during weathering. Tectonic activity and stress-release processes may also lead to the generation of fractures which can act as conduits for groundwater. This is the means by which some of these rocks can be developed, usually locally and on a small scale, as fractured aquifers. Assessment of the yield of such aquifers, however, is costly because changes in rock properties due to weathering and other secondary processes are extremely variable. However, these aquifers are vitally important in many developing countries, where they are a useful local supply of water (Wright and Burgess, 1992).

5.3.3 Unconsolidated aquifers

On a global scale, as much as 90% of developed aquifers are in unconsolidated materials (Todd, 1980) such as sediments in river valleys and floodplains, buried valleys and inter-montane valleys. These consist predominantly of sands and gravels (clays can have a high porosity but, since they have very fine pores, permeability is low and they make poor aquifers). The high permeability commonly found adjacent to watercourses generally gives very high aquifer yields, and this is often augmented by inflow from streams. Floodplain deposits also provide high yields, but they are more variable and can be thin. The High Plains Aquifer in the USA is a very important water source and consists of fluvial sands and gravels stretching from South Dakota to Texas (Miller and Appel, 1997). In western USA, inter-montane valleys contain aquifers. They comprise deep valley deposits derived from erosion of adjacent rock walls and can be very productive. Groundwater in such aquifers is replenished from streams seeping into **alluvial fans** at the mouth of the valley.

5.3.4 Sedimentary rock and karst aquifers

Sedimentary rocks form from the deposition of particles or sediments derived from the weathering and erosion of other rocks. Sediments initially become consolidated and cemented, which usually reduces porosity, though later **diagenesis**, fracturing and weathering may increase the porosity and the permeability. Most sandstones and **conglomerates** will yield some water. Those that make the best aquifers are usually poorly cemented, such as the Permo-Triassic sandstones in Northern Europe (**Figure 5.6a**; Allen *et al.*, 1997) and the Nubian Sandstone Aquifer

System in North Africa. Limestones, which are composed of carbonate rather than silicate minerals, have special properties as aquifers because of their high solubility in rainwater. Limestone areas may contain large solution cavities, and dissolution along joints can lead to large fissure-flow permeability. Such aquifers are sometimes referred to as karst aquifers. In such aquifers it is difficult to predict whether a borehole sunk at random will intercept the water-bearing fissures and cavities, so these types of aquifer are often exploited by harnessing spring flows. However, because karst cavities in limestone are relatively large, the rock tends to drain quickly after rainfall, and so they may represent an unreliable water source. Chalk is a type of limestone with

Figure 5.6 Sedimentary rock aquifers in the United Kingdom: a) Permo-Triassic sandstone; b) Cretaceous Chalk.

relatively closely spaced fractures (**Figure 5.6b**) and makes a very good aquifer, although it can be more susceptible to drought than sandstone aquifers, for the reasons just given. The extent of karstic development in chalk is variable and remains under investigation.

REFLECTIVE QUESTIONS

What types of aquifer are there and which rocks or sediments make the best aquifers?

Under what circumstances might water pour out of a newly drilled well, while the surrounding ground surface layer is unsaturated?

5.4 GROUNDWATER FLOW PRINCIPLES

5.4.1 Fluid potential and hydraulic head

M. King Hubbert of the US Geological Survey investigated the flow of water in porous media and defined, theoretically, the potential of water to flow (see classic paper: Hubbert, 1940). He argued that the potential for water flow was a measurable, physical quantity and that it should also be some quantity whose properties were such that water would flow from regions where it was high to regions where it was low. This quantity was referred to as fluid potential (by analogy with electrical potential, or voltage, which causes electricity to flow). The fluid potential of groundwater is seen by the elevation to which water will rise in a well, which is called **hydraulic head** (units of length, L). Given constant fluid density, water will

generally flow from regions of high hydraulic head to regions of low hydraulic head. The absolute value of the hydraulic head is always arbitrary because it depends on the datum from which well-water level is measured (ground surface or mean sea level, for example). However, this does not matter, because it is the difference in hydraulic head, rather than its absolute value, which causes water to flow.

5.4.2 Darcy's Law

In 1856, Henri Darcy published the results of some experimental work in which he was able to calculate the amount of flow through sand (Darcy, 1856). The purpose of the work was to determine the size of sand filter beds for public water supply. The measured rule that he discovered turned out to be very widely applicable for quantifying flow of fluids in both porous and fractured media (Holden, 2005). Darcy's experiment is illustrated in **Figure 5.7**; Darcy's Law can be written as:

$$q = \frac{Q}{A} = -K\frac{\Delta h}{\Delta l}, \qquad [5.2]$$

where Q is total discharge [units of length^3 and time^{-1}, L^3T^{-1} e.g. m^3 day^{-1}], q is **specific discharge** [LT^{-1}] (volume of flow rate divided by cross-sectional area), A is cross-sectional area [L^2], Δh is the difference in hydraulic head between the measurement points [L], Δl is the distance between measurement points [L] and K [LT^{-1}] is a constant of proportionality representing permeability, called **hydraulic conductivity**. The negative sign in Equation [5.2] indicates that water flows from high to low hydraulic head. The value of K depends on the properties of the porous medium and fluid properties (e.g. viscosity and density). However, in hydrogeology the fluid concerned is

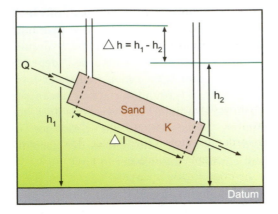

Figure 5.7 Column experiment that can be set up in the laboratory illustrating Darcy's Law for flow in sand.

usually water at environmental temperatures, so hydraulic conductivity can be considered to be a property of the rock or soil. Although the original Darcy experiment was on sand, fractures also show a linear relationship between hydraulic gradient and flow. For typical values of hydraulic conductivity for various geological media, see Domenico and Schwartz (1998). Another useful parameter in hydrogeology is the transmissivity, T, of an aquifer system (units L^2T^{-1}); this is the product of the hydraulic conductivity K and the saturated thickness of an aquifer b. It represents the overall capacity of the aquifer system to transmit water. Typical values for productive aquifers range from a few tens to greater than 10,000 m^2 per day.

5.4.3 Hydraulic gradient and specific discharge

In order for water to flow in an aquifer, it has to be in response to a gradient in hydraulic head. Horizontal hydraulic gradient in an unconfined aquifer is given by the slope of the water table (Figure 5.8a). In

a confined aquifer it is similarly given by the slope of the piezometric surface. In well A in Figure 5.8a, the water table is at height h_A above a datum, and, in well B, at a horizontal distance Δl from well A; the water table is at height h_B above the same datum. The hydraulic gradient is given by:

$$\frac{h_A - h_b}{\Delta l} = \frac{\Delta h}{\Delta l} = i \qquad [5.3]$$

Where the water table is flat, there is no horizontal hydraulic gradient and hence no horizontal flow. However, hydraulic gradients also occur in the vertical direction. In areas where surface and rainwater is recharging groundwater (Figure 5.8b), vertical hydraulic gradients are downward. In areas where groundwater discharges (typically areas of relatively low topography, Figure 5.8c), vertical hydraulic gradients are upward. Very strong upward hydraulic gradients (e.g. exceeding values of one) can produce dangerous quicksand conditions in unconsolidated sediments, because the force exerted by the moving water on the sediment grains overcomes the intergranular contact forces. In order to estimate the volumetric flow rate per unit area in an aquifer, sometimes called the specific discharge, it is simply necessary to multiply the hydraulic gradient with the hydraulic conductivity. A wide range of hydraulic conductivities are found in formations used as aquifers (10^{-10} m s^{-1} to 1 m s^{-1}: Singhal and Gupta, 1999). With a typical hydraulic gradient for a lowland aquifer of around 10^{-3} (i.e. 1 m hydraulic head change per km) these yield specific discharges of 10^{-13} to 10^{-3} m s^{-1}. These are smaller than typical stream-specific discharge values, but, as the cross-sectional areas of aquifers are often much larger than those of streams, there are broadly similar amounts of water flow within them. This can certainly be

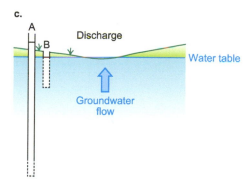

Figure 5.8 Hydraulic gradients in unconfined aquifers: a) horizontal gradient causing lateral flow; b) downward gradient in recharge zone; c) upward gradient in discharge zone (dashed well casing indicates where wells are open to surrounding aquifer).

expected in permeable catchments in temperate climates, where much river baseflow originates from groundwater.

5.4.4 Hydrogeological maps

Hydrogeological maps can be constructed from seasonally averaged groundwater heads, in order to show groundwater flow patterns and available resources. An example of a hydrogeological map in **Figure 5.9** shows a section of the River Wylye, which flows over the outcrop of the unconfined Cretaceous Chalk aquifer in southern England. In this case, the hydraulic head contours represent the seasonally averaged water-table elevation. As flow generally occurs in the direction of the hydraulic gradient, we can infer that in this example flow is exiting the aquifer to the River Wylye, because the contours indicate that flow is towards the river. Hence, the River Wylye is an example of a 'gaining stream' (see Section 5.2.3), as groundwater flows from the aquifer into the stream. In permeable catchments like this one, the groundwater component forms an important contribution to flow and ensures that river flow continues throughout the year. Unfortunately, it also means that river flows are highly sensitive to groundwater elevations. If groundwater resources are depleted by over-abstraction, there can be a highly detrimental effect on aquatic ecosystems by reducing minimum river flows.

5.4.5 Groundwater flow models

In recent decades, the advent of digital computing resources has led to static representations of hydrogeological maps being largely replaced by more sophisticated representations of groundwater flow models (e.g. see Anderson and Woessner, 1992). Such models can be used to make predictions

Figure 5.9 Hydrogeological map showing the River Wylye flowing over an unconfined Cretaceous Chalk aquifer in southern England. Contours are water-table elevations in metres above mean sea level.

about the effects of changing the stresses (e.g. recharge rates, pumping rates, river stage) on the system. Changes in water levels and, hence, changes in the amount of water stored in aquifers are simulated numerically. The explicit modelling of changes in hydraulic head with time allows seasonal fluctuations in water levels, for example, to be simulated, or they can be used to provide information on the timescale of impacts due to climate or land-use change. However, such models require much more data in order to be calibrated properly (e.g. monthly water-level measurements and recharge estimates). An example of a groundwater modelling output showing the effects of adjusting groundwater abstraction rates on river flows, for the Permo-Triassic sandstone aquifer of the Worfe catchment in the West Midlands, UK, is shown in **Figure 5.10** (derived by using the US Geological Survey's MODFLOW modelling code, which simulates both flows in the aquifer and the river: Rolf Farrell, UK Environment Agency, pers. comm. 2011).

REFLECTIVE QUESTIONS

What is Darcy's Law, and why is it of importance to groundwater science?

What are hydrogeological maps and numerical simulations (models) used for?

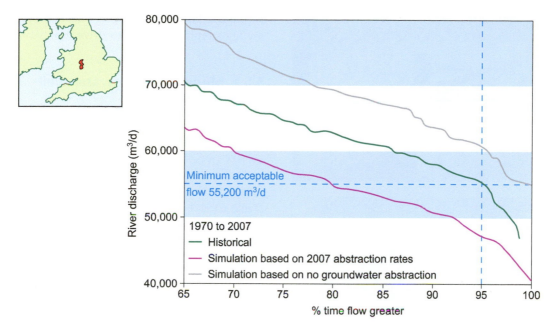

Figure 5.10 Impact of groundwater abstraction on flow at a gauging station in the River Worfe, West Midlands, UK, based on groundwater flow modelling of the underlying Permo-Triassic sandstone aquifer.
Source: Data supplied by the UK Environment Agency.
Note: The difference between 'historical' and 'simulation using 2007 abstraction rates' arises because actual abstraction rates rose over the period 1970 to 2007.

5.5 ARTIFICIAL ABSTRACTION AND RECHARGE

5.5.1 Low-technology groundwater abstraction

Groundwater has been exploited for supply and irrigation for many thousands of years, via hand-dug wells, and via sloping tunnels or infiltration galleries, which have various local names but are called 'qanats' in Iran and 'falaj' in Oman; these are used for both domestic water supply and irrigation. In developing countries, hand-dug wells are still perhaps the most important sources of relatively clean drinking water (**Figure 5.11**). The lining of these wells is commonly formed from concrete rings or masonry to support unconsolidated material, with appropriate pathways for water to pass through the lining below the water

table. The wells may be unlined at depth in consolidated rock aquifers. They tend to have relatively large diameters (1 to 1.5 m) to accommodate the well diggers, and also to allow the well to act as a reservoir in low-yielding aquifers so that peak demand (i.e. cooking and washing times) can be accommodated (Misstear *et al.*, 2017). The annulus (ring) behind the well lining should be filled with cement grout to prevent surface water contamination and the well head should also be designed to divert spilled water away from it. The method used to lift the water needs to be of appropriate technology to allow maintenance, while protecting the well water from contamination. Hand pumps may be difficult to repair in some rural economies, and bucket and winding arrangements may be more suitable where these can be designed to

Figure 5.11 Hand-dug wells: a) a shallow, manually drilled borehole, equipped with an Afridev hand pump, at Chiaquelane, near Chokwe, Mozambique; b) a timber-lined dug well in the village of Malii Anzas, southern Siberia, Russia.
Source: Photos by and © David Banks (currently employed at Holymoor Consultancy Ltd., Chesterfield, UK).

prevent contaminated privately owned containers from coming into direct contact with the well water.

An alternative approach to low-technology groundwater water supply for small communities is to develop gravity-driven pipework from natural springs which are topographically above the village concerned. Here, a major design consideration is ensuring that the route of the pipework across the landscape produces an appropriate topographic gradient to drive flow.

5.5.2 Modern groundwater well construction

Modern groundwater abstraction wells are commonly boreholes drilled using either percussive or rotary drilling methods, or methods which combine both actions. The 'intake section' of the well may be unlined in stable rock, but will be supported with permeable casing called well screen, surrounded by permeable pea gravel, in less stable formations (**Figure 5.12**). The pump (or pump intake) will normally be located above the intake section of the well, in a section lined with impermeable casing, in order to reduce the chances of any aquifer solids entering the pump. However, it is important that the pump is sufficiently far below the resting water level to avoid running dry when the water level in the well reduces due to seasonal fluctuations or abstraction. As with hand-dug wells, the well head needs to be protected from ingress of contaminants; this is achieved by using concrete headworks which direct surface water away from the well head, and by filling the annulus around the impermeable, upper section of the well casing with cement grout. The diameter and depth of the abstraction well will depend on the geological setting, such as the depth to the water table and most-permeable sections of the aquifer, but also on the required discharge rate (i.e. maximum pumping rate). More extensive site investigation will be appropriate for design of a deep, drilled well than for a shallow, hand-dug well, in order to ensure that the economic investment of well construction is worthwhile. The maximum discharge rate will determine the size of pump required, and hence the minimum

Figure 5.12 Modern groundwater abstraction wells: a) well with self-supporting open section in a consolidated confined aquifer; b) well with screen and gravel pack in an unconsolidated unconfined aquifer.

casing diameter for the section of the well containing the pump. However, the diameter of the intake section of the well may need to be larger than this, in order to ensure that the well is hydraulically efficient, enabling the water to enter the well from the surrounding formation at the required rate. Low well efficiency will mean that the water level in the well falls excessively during pumping, despite there being sufficient water in the surrounding formation.

5.5.3 Well testing

There are various reasons for testing groundwater abstraction wells, but the main reason from a water supply perspective is to ensure that the well performs as expected. In other words, can the well yield the required flow rate without the water level in the well falling excessively (this water-table fall is referred to as 'drawdown')? There are two components to total well drawdown – they are drawdown

in the aquifer near the well, and the additional drawdown between the aquifer and the water level in the well (**Figure 5.13a**); the latter is called 'well loss'. The efficiency of the well is defined as the ratio of the aquifer drawdown to total well drawdown; efficiencies of >80% are considered desirable. Generally, larger diameter wells are more efficient, but efficiency also reduces (i.e. well loss increases) as the pumping rate increases. Hence, there will be some upper limit to the pumping rate above which well drawdown will become large and well efficiency will reduce sharply (**Figure 5.13b**). In order to measure this limit, and define the yield characteristics of the well, it is usual to perform a 'step pumping test', both immediately after well construction and periodically to check that well performance has not deteriorated (Clark, 1977). This involves pumping the abstraction well until the drawdown has stabilised at a range of (usually increasing) pumping rates. The initial portion of the drawdown versus pumping rate plot,

Figure 5.13 Well loss and well efficiency: a) cross-section illustrating well loss; b) example plot showing variation of well drawdown with pumping rate. Well losses become large, and efficiency reduces sharply as pumping rates become high.

where well losses are small, is sometimes assumed to be linear in order to define the **specific capacity** of the well, which is the yield per unit of drawdown (units L^2T^{-1}, e.g. $m^3\ s^{-1}$ per m of drawdown). As the well losses are relatively small for low pumping rates, the specific capacity under such conditions depends primarily on the properties of the aquifer such as its hydraulic conductivity (see Section 5.4.2). To an approximation, if well losses are small, the specific capacity of the well can be estimated as 1.22 Kb where K is hydraulic conductivity and b is the saturated thickness of aquifer penetrated by the open section of the well (Misstear *et al.*, 2017). Note that Kb is the effective aquifer transmissivity intersected by the well (see Section 5.4.2).

In addition to the response of the abstraction well itself, it is also necessary (especially for larger-scale abstractions) to predict the impact of groundwater abstraction on the surrounding water table and any other wells or water bodies (such as rivers) in the vicinity. The response of the aquifer surrounding an abstraction well to pumping is complex, as the drawdown will depend on not only the properties of the aquifer and the pumping rate, but also the duration of pumping,

any surrounding hydrological features such as rivers, lakes and other abstraction wells, and recharge from rainfall. Understanding and predicting these responses requires longer-term pumping tests (typically at least a week in duration), where the water level is monitored not only in the pumped well itself but also in surrounding 'observation' wells. The theoretical responses of an idealised confined aquifer to pumping an abstraction well were described in a seminal paper by Charles Theis of the US Geological Survey (see classic papers: Theis, 1935) and still forms the basis of predicting effects of abstraction and injection of groundwater via wells today.

5.5.4 Problems with groundwater wells

Several physical, chemical and microbiological processes can interfere with the performance of groundwater wells and cause a reduction in efficiency and water quality over time. Physical processes include the clogging of the aquifer, gravel pack, well screen or the well itself, due to the ingress of fine particles which have been removed from their original positions in the aquifer by high groundwater velocities induced by pumping. In addition to reducing the efficiency of the well, such fine particles can also cause abrasion of the pump. Chemical fouling can result from the precipitation of minerals caused by a change in the chemical conditions between the aquifer and water in the well, for example, caused by degassing of carbon dioxide from the water in the well, which is at a lower pressure than in the aquifer, or mixing of shallower, oxygenated groundwater with deeper, anoxic groundwaters within the well. Chemical processes can also cause corrosion of metallic well casing, leading to leaking casing joints, which may allow ingress of shallower, contaminated water. Microbiological processes include the formation of **biofilms** (**Figure 5.14**) and the precipitation of biominerals such as ferric hydroxide, which can lead to biofouling of gravel packs, well screens and pumps. The above factors all relate to the performance of the well and the pump system, and hence can be addressed by

replacing parts, cleaning operations (for example, using chlorine to treat biofouling) or, in the worst instance, decommissioning and the construction of a new well. More serious are problems related to the aquifer itself. Wells can go dry as a result of regional groundwater-level falls, which may be caused by over-abstraction, or reductions in aquifer recharge related to climate change. Water quality can decline regionally as a result of excessive use of agrochemicals, salt-water intrusion and ingress of surface waters. The latter effects are themselves often related to over-abstraction. For example, arsenic contamination of groundwaters in Bangladesh and West Bengal (see Chapter 7) is thought to be driven by dissolution of arsenic-bearing iron oxy(hydr)oxide minerals, caused by ingress of surface water from rice paddy

Figure 5.14 Build-up of iron bacteria on the intake of a submersible pump at Great Heck Borehole, Triassic Sherwood sandstone, East Yorkshire, England. The rising main is 15 cm in diameter. The biofilm has accumulated in the form of large compact mass of gel.
Source: Photo courtesy of Gerd Cachandt, Arup.

Figure 5.15 Artificial recharge in the Llobregat Delta Aquifer, Barcelona, Spain: map showing location of river, recharge ponds, and injection wells.

fields, which itself results from groundwater abstraction (Nickson *et al.*, 1998). In order to identify and deal with problems associated with groundwater abstractions, Misstear *et al.* (2017) recommended that abstraction wells should be monitored for water level (either manually or automatically using pressure sensors and data recorders), discharge rate (i.e. with a flow meter), water chemistry and microbiology, via a sampling tap or line in the well head. Periodic CCTV inspection of the borehole, and step pumping tests, are also recommended.

5.5.5 Artificial recharge and conjunctive use of groundwater and surface water

Artificial or enhanced recharge represents an attempt to increase the amount of rainfall recharge or surface water recharge to groundwater, for example by modifications to topography to encourage infiltration rather than runoff, or by direct injection via pumping wells (Box 5.2). Several reasons exist for artificially recharging groundwater, including replenishment of supplies on a long-term

---

BOX 5.2

Managed aquifer recharge (MAR)

There are two main ways for enhancing the potential of exploitable groundwater resources. Both involve transforming surface water into groundwater, but they differ based on whether there is direct recharge into the system or whether there is an indirect recharge process (*induced recharge*). Direct methods might include creating permeable surface infiltration areas to enhance the natural percolation of water into the subsurface, or injecting water into wells to replenish deeper aquifers. Indirect methods of artificial recharge often involve pumping to lower groundwater levels near lakes and streams to induce percolation from surface water bodies into groundwater. In recent years, the term artificial recharge has been replaced by a more comprehensive term, managed aquifer recharge (MAR). There is a broader and more comprehensive understanding associated with MAR and it tends to include induced recharge and sometimes also the recovery of stored water. MAR can be significantly helpful in restoring the balance of an excessively exploited aquifer, sometimes creating conditions of surplus, which can then allow for greater groundwater abstraction than warranted by the natural recharge. However, the extent and efficacy of artificial recharge are highly influenced by the aquifer characteristics. Unconfined aquifers with relatively deep water levels offer the largest storage capacity. MAR projects are operational in many countries in Europe, in the United States, India, Australia, the Maghreb, the Middle East and Africa. Many of the MAR projects are located near a well field, with the primary aim of increasing the production capacity of the well. However, many other MAR projects are primarily intended to increase the resources of an entire regional or local aquifer system. Examples of the latter are recharge facilities in the Santa Clara Valley in California, the large recharge dams in Oman and numerous MAR sites scattered all over rural India. A classification of various MAR techniques is shown in Figure 5.16. A global inventory of successful MAR sites is accessible through the International Groundwater Resources Assessment Centre, a UNESCO Global Groundwater Centre supported by the government of the Netherlands: www.un-igrac.org/special-project/mar-portal.

Figure 5.16 Overview of MAR techniques and subtypes.
Source: Based on classifications outlined by IGRAC (2007).

The different artificial recharge schemes shown in **Figure 5.16** can be broadly subdivided into *extensive* and *intensive* techniques. The so-called extensive techniques (such as ponding and infiltration basins; see **Figure 5.17**) are cheap and better adapted to arid conditions, but require more space. On the other hand, the intensive techniques (shallow wells, shafts, pits or deep well injection) are more expensive, less durable (due to clogging) and more localised. The artificial recharge scheme adopted for a given site will depend on the source of water to be used (flow regime, physical quality), the type of aquifer, available means and the desired result.

Figure 5.17 Recharge pond in Bhabar zone, Sunwal area of Western Nepal.
Source: Photo courtesy of Moti L. Rijal.

basis, or shorter-term storage of water underground (known as managed underground storage, MUS, or aquifer storage and recovery, ASR). For water storage, regions can be used where the ground contains non-potable water, as well as natural groundwater aquifers, in order to store water in times of excess (for example, high river flows) for later extraction and usage. Use of groundwater as temporary storage facilitates conjunctive use of groundwater with surface water resources, which means the coordinated operation of both resources to meet demand. For example, surface water supplies can be used to meet supply requirements in most years, with some water being transferred to groundwater storage via artificial recharge. In years of low precipitation where surface water supplies are inadequate, stored groundwater can be used to supplement supply (Todd, 2005). However, artificial recharge is also used in other contexts, for example to prevent or limit the spread of saline intrusion (see Section 5.7.3). For example, freshwater from the Llobregat River near Barcelona, Spain, is artificially recharged to a deltaic aquifer, to limit the spread of saline water from the Mediterranean Sea, which has previously been inducted into the aquifer as a result of over-abstraction for municipal water supply. Two approaches are used: water is encouraged to infiltrate the Llobregat Delta aquifer during periods of high flow via recharge ponds, and it is also injected directly into the deltaic aquifer via wells, to form a freshwater 'barrier' (Figure 5.15).

Various issues arise that are relevant to the design and management of artificial recharge schemes. First, an appropriate source of recharge water is required: this can be rainfall, storm runoff, river water or treated wastewater. The variety of methods for enhancing recharge or injecting water into shallow or deeper aquifers, some of which have been in use since the early nineteenth century, are described by Todd (1980). One common issue is clogging of the recharge basins or wells with fine particles or biofilms, which reduces their efficiency over time; this can be dealt with by removing the finer particles by settling prior to injection, periodic drying of the recharge pond and scarification of its base; injection wells can be cleared by backflushing and/or via chemical treatments. Another issue concerns the quality of recharge water; while some degradable pollutants (e.g. coliform bacteria) are effectively removed via filtration in granular aquifers, this process is less efficient in fracture flow systems and ineffective in karstic systems. There is increasing concern about the impact on groundwater quality of recalcitrant and emerging pollutants such as pharmaceuticals, most notably endocrine system disruptors (e.g. oestrogens) (see Chapters 4 and 6). These are not removed by conventional wastewater treatments and may survive for long timespans in the environment (National Research Council, 1999); their threat to groundwater may be made much worse by artificial recharge, especially of treated wastewater. Issues specific to the relatively new technology of ASR include mixing and diffusion effects with *in situ* waters, which may be non-potable, and water–rock interactions (e.g. release of arsenic and fluoride from aquifer minerals); both of these effects tend to degrade water quality and reduce the volume of usable water below that injected. The ratio of usable water to that injected in ASR is called the recovery efficiency, and most working ASR schemes have shown increased efficiency with each cycle of water injection and abstraction. Matrix-flow systems such as poorly cemented sandstones have generally been found to give higher recovery efficiencies than fracture flow aquifers. For further information concerning the ASR/ MUS techniques, see Pyne (2005) and National Research Council (2008).

5.5.6 Use of groundwater for thermal control of buildings

Groundwater has historically been utilised for its thermal properties, for heating purposes from hot springs and deep wells (e.g. see Brassington 2007 for details of UK geothermal springs at Harrogate); and forms a significant part of the societal need for heating in some countries. A more recent development has been the exploitation of groundwater from aquifers where the groundwater temperature is not itself sufficient to provide heating, but the relatively high specific heat capacity of water allows it to act as a source or sink for heat using heat pumping technology, which can either be used to heat or cool buildings, or potentially both, at different times of the year. The principles of the application of heat pumping technology to aquifers are described by Banks (2009). The technology is now considered to be well established and many major public buildings in some countries have their temperature controlled in this way, including the Tate Modern in London (Birks *et al.*, 2015). Systems include 'closed-loop' designs where heat is simply exchanged with groundwater without injection or abstraction of fluid, and 'open-loop' designs where groundwater is abstracted and reinjected (Box 5.3). Both designs require hydrogeological input as groundwater flow rates and hence heat flow rates in the subsurface influence their effectiveness. For example, it is getting increasingly difficult to assess efficiency and longevity of open-loop systems in London, owing to the large number of systems that inject heat into confined chalk, which are interfering with one another, while introducing heat pollution into the aquifer (Fry, 2009).

REFLECTIVE QUESTIONS

What are the key considerations when constructing a well for water supply?
    How can humans help to recharge aquifers that are being depleted?
    How can groundwater be used as a source of heating and cooling?

5.6 GROUNDWATER CHEMISTRY

5.6.1 Groundwater quality

Groundwater used for public supply is often extremely clean (high quality) compared with surface water, and therefore represents a valuable resource. Most of this 'clean' groundwater is 'meteoric' water (supplied from the atmosphere) from relatively shallow depths of up to a few hundred metres. Rocks at greater depth usually contain saline brines, which are unsuitable for drinking. The reason for this is that, if water remains in contact with rocks for very long time periods, the minerals in the rock dissolve and increase the salinity. Unfortunately, over-abstraction of potable groundwater can cause upwelling and mixing of these brines with fresher groundwaters (e.g. see Bottrell *et al.*, 2006), as well as influx of seawater, known as marine incursion. Therefore, it is very important to be able to characterise groundwater quality and ensure that groundwater abstraction and land-use management prevent practices which lead to deterioration, for example, over-abstraction or inappropriate application of agrochemicals such as nitrate.

BOX 5.3

Groundwater for heating and cooling

The use of groundwater for implementing building thermal control is a low-carbon alternative to traditional heating and chilled water technologies and has significant potential to enhance renewable energy sources for heating. However, lack of precedence and associated operational uncertainties are perceived as two major prohibiting factors in uptake of this technology. A ground-source heat pump (GSHP) is simply a heat pump that is connected to the ground or groundwater. It should be noted that a GSHP can be used for either cooling or heating, as per the requirement, with minor modifications. The exploitation of ground source for space heating can be done via either a closed-loop system (**Figure 5.18**) or an open-loop system (**Figure 5.19**). In a closed-loop system the carrier fluid (usually a solution of antifreeze) is passed through a circulatory tube, laid in a horizontal trench or vertical borehole, that conducts heat from the ground source. In this instance, it is only heat that is exchanged, rather than groundwater, to achieve either heating or cooling. The performance of such closed-loop systems is, however, affected by groundwater regime. Where installed in the saturated zone (often the case of vertical borehole systems) groundwater flow removes heat from the closed-loop area, hence influencing efficiency. The use of groundwater, abstracted via open-loop systems, is increasingly being considered as a means to provide heating and cooling for buildings, particularly in major metropolitan areas. In open-loop systems the subsurface is accessed by abstracting water from one or more wells, passing it through a heat pump and then either injecting it back into the aquifer or discharging it to a surface water body.

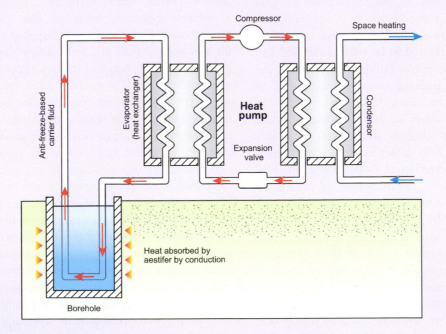

Figure 5.18 Schematic diagram of a heat pump, connected to an indirect circulation, closed-loop, borehole heat exchanger.
Source: Adapted and modified after Banks (2009).

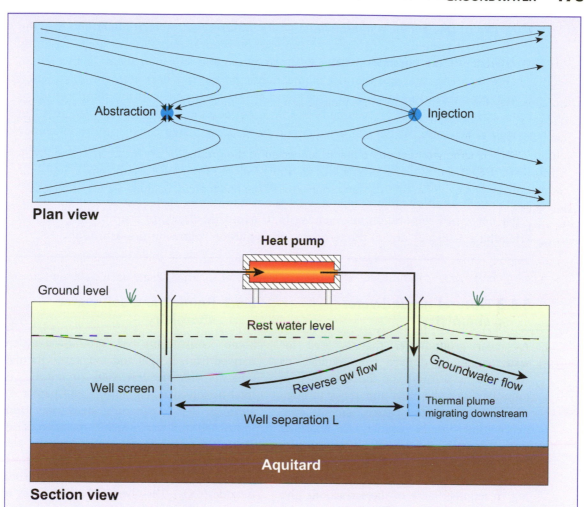

Plan view

Section view

Figure 5.19 A well doublet system for cooling where a heat pump rejects heat into a groundwater flux, abstraction (Ab) and reinjection (In) are large enough and separation distance between wells (L) is small enough for thermal feedback to occur.
Source: Adapted and modified after Banks (2009).

The renewable energy targets for new developments published in the UK Energy Strategy of 2004 led to significant increase in applications for installation of open-loop GSHP (Fry, 2009). In groundwater-based open-loop heat-pump systems (**Figure 5.19**) the hydrogeological aspect is significantly more important. A typical well doublet scheme (**Figure 5.19**) is comprised of groundwater abstracted from one well and the chilled (or heated) water being reinjected into an adjacent well. The reinjection is primarily done on the insistence of the regulatory authority to conserve groundwater resources. If the spacing between the abstraction and injection well is sufficient, the thermal plume originating at the site of reinjection will migrate down the natural hydraulic gradient and it is the job of hydrogeologists to compute the spatiotemporal evolution of such a plume. However, if the separation between abstraction and injection wells is too small, there is a risk that the thermal plume might be sucked back into the abstraction well, as shown in **Figure 5.19**. This thermal feedback can lead the temperature of the abstracted water to become hotter (in a net cooling scheme) or cooler (in a net heating scheme), which can eventually lead to progressive decline in the efficiency of the whole system. Therefore, prior to installation of such a system a risk assessment of the likelihood of significant thermal feedback is warranted, which can be tackled by numerical heat or flow models. However, for simple systems robust analytical solutions have been known since the 1960s that help in calculating a) the necessary separation between wells to avoid feedback, b) the time for thermal feedback to occur and c) the temperature evolution following breakthrough.

5.6.2 Natural groundwater chemistry

As most groundwater that is used for water supply is derived from rainfall, its chemistry is a result of the chemistry of rainwater and the soils and rocks through which the water has passed, and the residence time in the aquifer. As residence time increases, groundwater tends to pick up more and more dissolved solids as a result of mineral dissolution (or weathering) reactions. Some of these reactions give waters their distinctive tastes, characterised, for example, by the recent trend to market 'natural' mineral waters as healthy. However, many mineral dissolution reactions result in a deterioration of water quality. Similarly, meteoric derived waters can mix with old, connate water (deposited within sedimentary rock, and saline), or magmatic waters produced by volcanic activity, or natural brines present in deeper crustal rocks. It is useful to be able to identify the sources for a particular water composition, as this tells us something about where the groundwater originates, and to distinguish anthropogenic pollutants from natural groundwater constituents.

5.6.3 Rainfall and soil water chemistry

Some constituents of groundwater are derived from rainfall and soil water. Rainfall typically contains less than 20 mg L^{-1} of total dissolved solids (TDS) (see Chapter 4), although it is less in rural areas away from industrial centres, which are sources of industrial atmospheric emissions. It also contains dissolved gases from the atmosphere, notably oxygen (O_2) and carbon dioxide (CO_2). Dissolved CO_2 dissociates to form bicarbonate **ions** (HCO_3^-) and hydrogen ions, i.e. acidity (H^+):

$$H_2O + CO_2(g) = H_2CO_3(aq)$$
$$= H^+(aq) + HCO_3^-(aq) \quad [5.4]$$

which reduces the pH below neutral (pH 7). The equilibrium pH of rainwater in contact with atmospheric CO_2 is pH 5.7 (Freeze and Cherry, 1979). Dissolved oxygen makes the rainwater an oxidising solution as well as being acidic. When a well-developed soil horizon is present, bacteria in the soil produce CO_2 and increase the amount that is dissolved in the water, reducing the pH further. Where the atmosphere contains oxides of sulfur as a result of industrial emissions or volcanic activity, these also dissolve in rainwater, producing more H^+ ions, which can cause the pH to fall to between 3 and 4; this is known as acid rain (see Chapter 4). As a result of its dissolved constituents, water that recharges aquifers from the soil zone is both acidic and oxidising and it is these elements which drive many of the mineral dissolution reactions described below.

5.6.4 Mineral dissolution (weathering) reactions

The composition of groundwater derived from rainfall varies according to the lithology of the aquifer and the residence time, due to mineral dissolution reactions (also known as mineral weathering reactions). Such reactions result in a characteristic suite of dissolved ions and lead to different TDS for groundwaters, according to the geology. Both TDS and ion concentrations are given on branded mineral waters and it is an interesting exercise to try to relate these to the geology of the area from which the water originates. Another property of groundwaters that results from geology is their hardness, which reflects the content of **cations** Ca^{2+} and Mg^{2+} that can precipitate and cause mineral scaling, for example as a result of heating or reaction with soaps. Scaling causes problems with pipe blockages as well as producing a characteristic 'water mark' or 'scum line' in baths and sinks when soaps are used. Hard waters can be recognised in domestic use because they require far more soap in order to produce suds (e.g. the instructions on washing powder indicate that larger amounts are required for hard water). Hardness is usually expressed as mg L^{-1} equivalent dissolved

$CaCO_3$, with waters containing <60 mg L^{-1} classified as soft, 60–150 mg L^{-1} as medium and >150 mg L^{-1} as hard; hard waters often have to be treated to reduce their hardness before they can be used in industrial processes, although they are quite potable and may represent a useful source of dietary calcium and magnesium. The key mineral dissolution reactions that control the major components of most natural groundwaters are described below.

5.6.4.1 Carbonate dissolution

$$CaCO_3(s) \ (calcite) + H^+(aq) \\ = Ca^{2+}(aq) + HCO_3^-(aq) \quad [5.5]$$

Where there are carbonate minerals such as calcite ($CaCO_3$) or dolomite ($MgCa(CO_3)_2$) within the aquifer (even a small amount) they tend to dominate the groundwater chemistry. This is because these weathering reactions are relatively fast. Resulting groundwaters are called 'calcium–magnesium–bicarbonate' waters because the major ions are calcium and/or magnesium and bicarbonate. Usually, the final bicarbonate concentration in such waters is between 170 and 260 mg L^{-1}, giving TDS between ~300 and 700 mg L^{-1}; the higher concentrations result where biological production of carbon dioxide allows more H^+ to be produced via reaction 5.4 and then consumed by reaction 5.5 (Freeze and Cherry, 1979). As the dominant cations are calcium and magnesium, which precipitate with bicarbonate on heating, these waters are classified as hard. This type of hardness can be reduced by boiling the water, and hence is known as temporary hardness. Groundwaters dominated by carbonate dissolution are common in the shallow subsurface and in groundwaters with short residence time, because of the ubiquitous occurrence of carbonate minerals in soils.

5.6.4.2 Gypsum and halite dissolution

Where rock salts such as gypsum ($CaSO_4.2H_2O$) and halite (NaCl) are present, these will dominate the groundwater chemistry, e.g. gypsum dissolution:

$$CaSO_4.2H_2O(s) \text{ (gypsum)}$$
$$= Ca^{2+}(aq) + SO_4{}^{2-}(aq) + 2H_2O$$
$$[5.6]$$

As such salts are very soluble, this can lead to very high concentrations of ions in groundwater; TDS of up to 5000 mg L^{-1} or more, which makes these 'calcium–magnesium–sulfate' waters unpotable. However, highly soluble salts will not be present in active flowing zones in groundwater because they will have already been dissolved out. It is more common to encounter the effects of such salts where over-abstraction causes water from less permeable units to enter aquifers. Given their high content of cations such as calcium, these waters are classified as hard in that they precipitate with soaps. However, unlike 'temporary hardness', where calcium or magnesium ions are balanced by bicarbonate ions, here they are balanced by sulfate and chloride ions, so this hardness cannot be removed by boiling the water and is known as permanent hardness. Despite the name, permanent hardness can be removed by passing the water through ion exchange resins known as water softeners.

5.6.4.3 Silicate weathering

In rocks comprised of silicate minerals (e.g. sedimentary rocks such as quartz sandstones, and most igneous rocks), silicate mineral weathering reactions become important controls on groundwater chemistry (Garrels and Mackenzie, 1971):

Aluminosilicates:

$$2NaAlSi_3O_8(s) \text{ (albite, Na − feldspar)}$$
$$+ 2H^+(aq) = Al_2Si_2O_5(OH)_4(s)$$
$$\text{(kaolin clay)} + 2Na^+(aq) + 2H_4SiO_4(aq)$$
$$2NaAlSi_3O_8(s) \text{ (albite, Na − feldspar)}$$
$$+ 2H^+(aq) = Al_2Si_2O_5(OH)_4(s)$$
$$\text{(kaolin clay)} + 2Na^+(aq) + 2H_4SiO_4(aq)$$
$$[5.7]$$

Non-aluminous silicates:

$$Mg_2SiO_4(s) \text{ (olivine)} + 4H^+(aq)$$
$$= 2Mg^{2+}(aq) + H_4SiO_4(aq) \quad [5.8]$$

Ions derived from silicates will be present in the resulting groundwater (e.g. depending on the type of silicate, Na^+, K^+, Mg^{2+} or Ca^{2+}), but often in relatively low concentrations because these reactions are relatively slow. In the absence of other reactions, the major anion present is bicarbonate HCO_3^-, derived from rainfall and soil zone processes. Where silicate weathering reactions are limited, giving overall low TDS of <100 mg L^{-1}, the resulting types of water can be marketed as 'pure' and represent a high proportion of branded mineral waters; these waters are very soft. However, in many silicate rocks such as volcanic rocks and sandstones the groundwater may have longer residence times, producing intermediate TDS values and hardness; many also contain sulfide or sulfate minerals which can produce much higher TDS values (see reaction 5.10 below).

5.6.4.4 Cation exchange

$$ClayNa(s)+Ca^{2+}(aq)$$
$$= Clay_2Ca(s)+ 2Na^+(aq)$$
$$[5.9]$$

Cation exchange reactions are relatively important in sedimentary rock aquifers such as sandstones, which contain some clay minerals. As these rocks were deposited in the ocean basins, such clays have their exchangeable surface sites saturated with sodium (Na^+) and potassium (K^+) ions because these predominate in seawater. As meteoric water ingresses, cations like Ca^{2+} and Mg^{2+}, derived from, for example, carbonate weathering in the soil zone, are progressively exchanged onto clay minerals, and Na^+ and K^+ are released in their place as the groundwater moves deeper into the aquifer. This process is called water softening because the 'sodium/potassium waters' produced are relatively soft (i.e. these cations do not react with soaps). A good example of this sequence of reactions is described by Price (1993) for the Cretaceous Chalk aquifer of Northern Europe, and results in the groundwater chemistry illustrated in Figure 5.20. As cation exchange reactions simply replace one ion with another, they do not themselves increase the TDS load of groundwaters, which will therefore depend on other mineral dissolution reactions.

5.6.4.5 Oxidation reactions

As infiltrating water contains dissolved oxygen, some minerals such as sulfides of iron (Fe) or manganese (Mn) in the soil or deeper aquifer can be oxidised by bacteria, to produce sulfate and acidity:

$$FeS_2(s) \text{ (pyrite)} + 3.5O_2(aq) + H_2O$$
$$= Fe^{2+}(aq) + 2SO_4^{2-}(aq) + 2H^+(aq)$$

[5.10]

Iron (II) produced is oxidised to iron (III), which, if oxygen is still present, is then precipitated as a hydroxide:

$$Fe^{2+}(aq) + 0.25O_2(aq) + 2.5H_2O$$
$$= Fe(OH)_3(s) + 2H^+(aq)$$

[5.11]

Where carbonate minerals are present, the acidity produced is neutralised by carbonate dissolution, leading to groundwater dominated by calcium, sulfate and bicarbonate ions; this may remain potable but will contain both temporary and permanent hardness. Where no carbonate is present the groundwater may become acidic, with the pH dropping as low as 4. This can produce unpotable waters because of the tendency to dissolve trace elements such as arsenic (Nordstrom, 2002). Where the oxygen in the water is all used up, such that the groundwater becomes anoxic, iron will remain in solution as Fe^{2+}, but this will then precipitate as iron (III) hydroxide where such waters come into contact with oxygen. An extreme case of this occurs where acid mine drainage emerges (see Figure 4.9 in Chapter 4). The resulting iron (III) hydroxide precipitate can smother benthic organisms, destroying

Figure 5.20 Schematic diagram of dominant chemistry in the Cretaceous Chalk aquifer of southern England. Calcium and bicarbonate ions in the unconfined aquifer arise from carbonate dissolution. In the confined aquifer calcium ions are exchanged for sodium and potassium ions present on clay minerals. Deeper within the confined aquifer connate waters derived from seawater are present.

stream ecosystems. Where put into supply, Fe and Mn precipitation can stain porcelain, so waters have to be oxified by treatment. A good example illustrating the effects of oxygen consumption on groundwater chemistry of a carboniferous sandstone aquifer is given by Banks (1997). Sulfide weathering reactions typically result in groundwaters with TDS of 400 to 1000 mg L^{-1}, but can ultimately produce TDS values of several thousand mg L^{-1} where rocks are exposed to excess oxygen via mining processes and produce acid mine drainage (see Chapter 4, Section 4.5.3 for a more detailed treatment of acid mine drainage).

REFLECTIVE QUESTIONS

Why does the chemistry of groundwater naturally vary from one place to another?

What are hard water and soft water?

5.7 GROUNDWATER POLLUTION

5.7.1 Groundwater pollutant properties

There are a very wide range of groundwater pollutants, including those that can harm humans, animals or plants and those that have detrimental effects on natural ecosystems where water emerges into rivers. Pollutants enter groundwater as a result of human activities such as mining, industry, waste and sewage disposal, agriculture, or where excessive water is pumped out for supply purposes, causing influx of 'naturally' contaminated water (e.g. salt water from the

sea – called marine intrusion or incursion – or from deeper, saline groundwaters).

Two important characteristics influence the extent of detrimental effects of groundwater contaminants. These are the extent of interaction and attachment to the solid matrix (known as sorption) and the rate of transformation of contaminants into less harmful substances, for example by biological processes. The extent of sorption depends on the mineralogy of the aquifer and on the type of contaminant. Good sorbents include clay minerals and iron minerals such as those responsible for the red coloration of many sandstones, as well as the natural organic matter that is present within soils and rocks. Generally, positively charged ionic contaminants, for example toxic metal ions such as cadmium, are strongly adsorbed by the negatively charged surfaces of clay minerals and iron oxides; neutral contaminants such as organic chemicals may be adsorbed by natural organic matter. Negatively charged anionic contaminants such as nitrate, chromate and arsenate ions are typically not adsorbed and migrate freely in groundwater. Strongly adsorbed contaminants are often effectively immobilised because the vast majority of the contaminant is attached to the solid phase. However, remobilisation of such adsorbed contaminants can occur as a result of chemical changes such as changes in the concentration of dissolved oxygen in the groundwater, for example arsenic sources from the natural minerals within aquifers in Bangladesh (Nickson *et al.*, 1998) and radioactive technetium released from nuclear industry-contaminated sites in the UK (Burke *et al.*, 2005).

Transformation of organic contaminants into less harmful substances is usually via biological processes, although radioactive decay is also important for radioisotopes. Substances that decay rapidly are less of a problem than those that persist for long periods or indefinitely because contaminants may be unable to reach groundwater wells before decay takes place. For example, many human **pathogens** cannot survive for very long in groundwater and, provided wells are sited far enough away from sources such as manure spreading and latrines, passage through the subsurface acts as an effective water treatment. Unfortunately, some types of contaminants, such as the anti-knock additive in unleaded petrol, methyl-tertiary butyl ether (MTBE), are not transformed quickly enough and so can reach supply wells. Biological decay processes are often dependent on local chemical conditions, such as the presence or absence of oxygen or nutrients (Spence *et al.*, 2001), which can result in highly variable or unpredictable decay rates.

5.7.2 Contaminant source characteristics

It is useful to classify sources of groundwater contamination according to their spatial characteristics. Spatially, the size of a source of contamination varies hugely, from a few metres for a leaking septic tank, for example, to several hundred km^2 for a large area of agricultural chemical application to farmland, but generally we can speak of point sources (relatively small compared to the thickness of the aquifer) and diffuse sources, which are large compared with this thickness. Point sources give rise to 'plumes' of contamination which migrate away and spread out, whereas diffuse sources give rise to very large areas of groundwater contamination (**Figure 5.21**). Generally, point sources may lead to the shutdown of individual pumping wells, but diffuse sources can cause whole aquifers to be damaged.

Figure 5.21 Contaminant source characteristics: a) landfill site representing point source; b) agrochemical application representing diffuse source. The numbers on contour lines are contaminant concentrations in mg L^{-1}.

5.7.3 Types of groundwater contaminant

The following broad groups of groundwater contaminants can be recognised.

5.7.3.1 Radionuclides

Elements such as uranium, plutonium, and some isotopes of strontium, caesium and technetium are subject to radioactive decay. They are produced mainly by the nuclear industry in mining and processing uranium ore, fuel reprocessing, and waste and weapons disposal. As the radioactive decay process produces ionising radiation it is especially damaging where the elements get inside living cells by consuming contaminated groundwater. Some radioisotopes, such as those of strontium (i.e. $^{90}Sr^{2+}$), are positively charged and are thus subject to sorption reactions. However, many others, such as uranium, form negative ions (e.g. uranyl UO_4^-), which are highly soluble and can thus migrate for long distances in groundwater.

5.7.3.2 Trace metals

The trace metals group of contaminants includes 'metalloid' elements such as arsenic (As), as well as metals such as lead (Pb) and copper (Cu) which are toxic in relatively low concentrations in groundwater. The major sources are mining, industrial wastewater streams, runoff from industrial solid wastes (e.g. slagheaps) and agriculture. However, some of the most widespread groundwater contamination from these types of contaminants is natural, but made worse by groundwater abstraction, such as arsenical contamination of groundwaters in Bangladesh and West Bengal (Nickson *et al.*, 1998) (see Chapter 7). The main problem with trace metals such as Pb is their ability to bioaccumulate within organisms.

5.7.3.3 Nutrients

This nutrient group of contaminants includes ions and compounds containing nitrogen (N) and phosphorus (P), such as nitrate (NO_3^-) and ammonium (NH_4^+). The main sources of these contaminants are agriculture (artificial fertiliser application and cultivation of virgin soils) and organic wastes such as sewage (leaking sewers and sewage disposal) and landfill sites. The main health effect of nitrate in drinking water is methemoglobinemia (Schwartz and Zhang, 2003) – 'blue baby syndrome' (see Chapter 8). Nitrate is currently threatening many large aquifers in developed countries as a result of the extensive increases in mineral fertiliser applications to arable land in the latter part of the last century. The World Health Organization recommends a maximum concentration in drinking water of 50 mg L^{-1} (see Table 7.2 in Chapter 7). Many wells have had to be shut down for exceeding this concentration. Denitrification of abstracted groundwater is possible, but it is expensive. As an alternative, many large water suppliers currently pursue a policy of 'blending' abstracted waters from various sources in order to reduce the concentration below the limit. However, the sustainability of this policy is unclear, given the continuing rising trends seen for nitrate in some aquifers.

5.7.3.4 Major ions (salts) and saltwater incursion

This group includes ions such as Ca^{2+}, Mg^{2+} and Na^+, SO_4^{2-} and Cl^-. Extremely high concentrations of these make the water unfit for human consumption, and for many industrial uses. The health effects of consuming such water are usually not so serious or long term. However, very high concentrations of these substances give the water a bad taste, and ultimately make it unpotable. The main sources of major ion contaminants to groundwater are industrial or domestic waste (in landfills), natural saline waters such as seawater being drawn into the aquifer by over-abstraction and mixing with deep groundwater (brines). Where saline waters are drawn into groundwater aquifers from the sea this is known as marine incursion (see Ergil, 2000, and the example of the Llobregat Delta in Barcelona illustrated in Figure 5.15). Many coastal aquifers across the world have been seriously damaged by this process (e.g. Ergil, 2000) and it represents one of the biggest threats to groundwater quality worldwide. In theory, saline water induced from the sea sits beneath freshwater in the aquifer because it is denser (Figure 5.22). Where this scenario is realised, abstraction of freshwater can cause up-coning of the underlying saline 'wedge'. However, in reality aquifer heterogeneity and dynamic flow processes cause variation from this simple

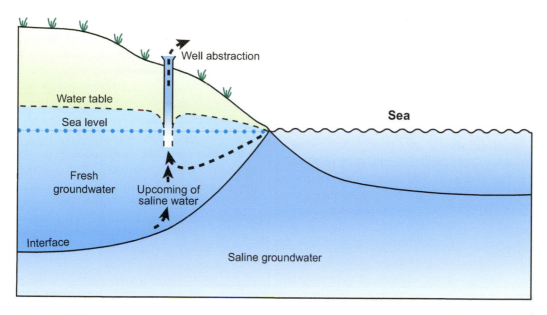

Figure 5.22 Schematic of seawater intrusion in a homogeneous unconfined aquifer. The depth of the interface between fresh and saline water is theoretically 40 times the height of the freshwater above sea level.

scenario; typically, a wide zone of dispersion or mixing between freshwater and seawater is present. Salinisation can also occur as a result of evaporative recycling of irrigation water, where water is continuously extracted, used for irrigation and recharged to groundwater (Milnes and Renard, 2004). Salt concentrations increase progressively because each irrigation results in a proportion of the water being evaporated or transpired. Salinisation problems are among the greatest threats to groundwater worldwide, and their effects are likely to have massive socio-economic consequences as pressures on groundwater increase as a result of climate changes over the coming century.

5.7.3.5 Organic compounds

Organic compounds (those containing carbon, hydrogen and sometimes oxygen, sulfur and nitrogen) represent a complex group of groundwater contaminants. There

are a huge range of sources of organic compounds: they include fossil biomass such as petroleum and coal, as well as biological decay of modern biomass (e.g. in landfills). Two main subgroups can be recognised according to solubility: compounds that are relatively insoluble in water, which thus form a separate liquid phase within the subsurface, called non-aqueous phase liquids (NAPLs), and organic compounds that are soluble in water. The behaviour of NAPLs depends on whether their specific gravity is greater or less than that of water. Those that are less dense – light non-aqueous phase liquids (LNAPLs) – float on the water table; those that are more dense than water – dense non-aqueous phase liquids (DNAPLs) – sink through the groundwater to the base of the aquifer (**Figure 5.23a** and **b**). As even NAPLs are slightly soluble, they progressively dissolve (**Figure 5.23c**), but this process can result in the contamination

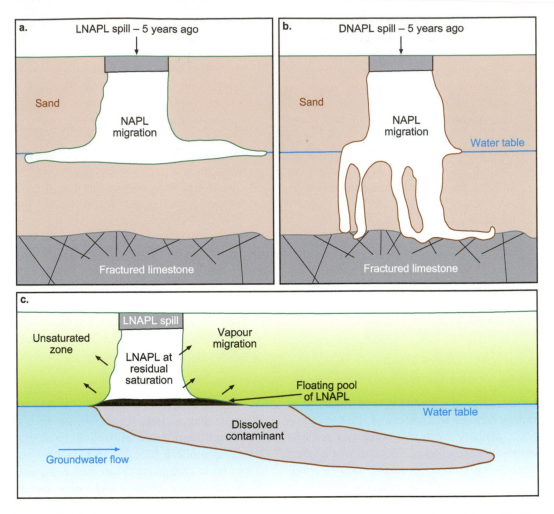

Figure 5.23 Behaviour of non-aqueous phase liquids (NAPLs) in groundwater aquifers: a) dense NAPLs; b) light NAPLs; c) dissolution of LNAPL to form a plume of contaminated groundwater.

of huge volumes of groundwater and take many years.

Organic contaminants can also be subdivided, depending on their chemical structure, into:

- Aliphatic hydrocarbons – these are chains of carbon atoms with hydrogen attached. They tend to have low solubility and are less dense than water (i.e. they form LNAPLs). The smallest molecules (e.g.

methane CH_4 and ethane C_2H_6) are gases at 20°C, the middle-sized molecules such as octane (in petroleum) are volatile liquids and the heavier molecules are tarry non-volatile liquids. The main sources of these compounds are leaks and spills at oil refineries, petrol stations and airfields and from pipelines.

- Aromatic hydrocarbons – these compounds contain at least one benzene ring (a ring structure of six carbon atoms with specific properties). They are more soluble in water than aliphatic compounds and are extremely toxic. The most common groundwater contaminant compounds in the group are benzene, toluene, ethylbenzene and xylene – these are called the BTEX compounds. These are slower to degrade in groundwater than aliphatic compounds, and so are often the most significant contaminants. Another set of aromatic compounds that are often found in groundwater are the poly-aromatic hydrocarbons (PAHs), which are hydrocarbons containing several benzene rings, often derived from coal processing. These compounds are very toxic to humans and aquatic ecosystems.

- Halogenated hydrocarbons – where some of the hydrogen atoms have been replaced by chlorine, bromine or fluorine atoms, are denser than water and often form DNAPLs, although they may be soluble. Halogenated aliphatic compounds such as trichloroethane (TCE) are used as solvents for dry cleaning and as cleaning agents in metal manufacturing. TCE is quite soluble in water and is the biggest cause of groundwater abstraction well shutdown in developed countries such as the USA. Other important groundwater contaminants include halogenated aromatic compounds (e.g. chlorobenzene), which is extensively used for agricultural purposes, and polychlorinated biphenyls (PCBs), which are chlorinated hydrocarbons containing two benzene rings and are extremely toxic. These have often leaked into groundwater from large-scale electrical equipment present at old industrial sites (they have now been phased out of use, due to their high toxicity).
- Biological degradation products – compounds produced by the degradation of biological materials, including organic acids, characterised by the COOH (carboxylate) group; alcohols, characterised by the OH (hydroxyl) group; and esters (which contain the C–O–C linkage), as well as more complex molecules such as sugars and proteins. These compounds are quite water soluble; one environmental effect is their **biochemical oxygen demand** (BOD) (see Chapter 4), which means that microorganisms use them as a food source and use up the oxygen present in groundwaters. This produces a chemically reducing environment, which can cause inorganic contaminants to dissolve or change into more toxic forms (e.g. sulfates to sulfides). The sources of biological decay products entering groundwater include landfill sites and sewage disposal lagoons, slaughterhouses and tanneries.
- Environmentally persistent pharmaceutical pollutants – these are relatively new chemical compounds, which include pharmaceutical and personal care products, such as antibiotics, painkillers and antidepressant medicines, and which are derived from wastewater. They are increasingly present in surface and groundwaters because they are not removed by current wastewater treatment (see Chapter 4). Many are endocrine disruptors (synthetic hormones), which can have a relatively large impact on microorganisms, wildlife and humans, even in low concentrations $\ll 1$ mg L^{-1}. Our inability to remove these chemicals from wastewater streams and their unpredictable environmental effects are likely to be a key challenge in future water management. Section 7.4.3 in Chapter 7 provides more details about these new pollutants and potential health impacts.

The extent to which organic compounds are adsorbed or undergo biological decay once they are dissolved in the groundwater is very variable and this group of compounds have presented great difficulties where spills from industrially contaminated sites have degraded large volumes of groundwater and impacted upon surface water or groundwater abstraction wells up to several kilometres distant. A good example where this occurred is the Hyde Park Landfill Site in Niagara, New York State, where chemical wastes containing chlorinated aromatic hydrocarbons were dumped in an unlined landfill site in 1954. These DNAPLs and LNAPLs migrated downward into a glacial sand aquifer underlain by a limestone aquifer. As well as creating problems for housing subsequently built near the site, these contaminants also migrated several hundred metres in groundwater to reach the Niagara River (Domenico and Schwartz, 1998).

5.7.3.6 *Pathogenic organisms*

Viruses, bacteria and parasites can be transported in groundwater and there are instances of groundwater contamination producing most of the major diseases that are carried in water (typhoid, cholera, polio, hepatitis) (see Chapter 7). The sources are usually human or animal sewage or landfills. Fortunately, pathogenic microbes cannot survive for long in the subsurface, so contamination is usually local – and incidents are usually due to poor well siting or well-head protection. An incident of pathogenic organism contamination of a municipal groundwater supply which led to 2300 cases of illness and seven deaths at Walkerton, Ontario, is reported by Worthington *et al.* (2002).

5.7.4 Groundwater vulnerability to pollution

Groundwater vulnerability to pollution is the likelihood of contaminants reaching groundwater after introduction at the ground surface. Vulnerability to contamination from the surface is an increasingly important issue as more stringent regulations on water quality and the need to protect groundwater resources clash with the economic incentives to use plant-protection products (fertilisers, pesticides, herbicides) in the agricultural industry. Vulnerability is controlled by complex interactions between anthropogenic activities, natural stresses (rainfall, runoff) and intrinsic characteristics of the aquifer system. The intrinsic groundwater vulnerability is a function of the following parameters: the nature and thickness of any overlying confining layer, depth to groundwater and characteristics of the aquifer materials. Generally, fractured aquifers are relatively vulnerable, as compared with systems where flow is mainly via the matrix pores. Approaches to groundwater vulnerability assessment include semi-quantitative empirical (index-based) and statistical methods on large scales, usually used to produce maps of intrinsic groundwater vulnerability, and numerical groundwater flow and solute transport modelling methods to predict vulnerability of specific wells (Robins, 1998; Pavlis *et al.*, 2010). This allows well-head or source protection zones to be defined around groundwater wells (Frind *et al.*, 2006); potentially polluting land uses such as petrol stations and landfills are then restricted within these zones. Figure 5.24 illustrates how source protection zones can be defined, based on the ease with which contaminants can reach the supply well.

> ### REFLECTIVE QUESTIONS
>
> What are the main threats to groundwater quality and how do they arise?
>
> How is groundwater pollution vulnerability defined, and how can it be assessed?

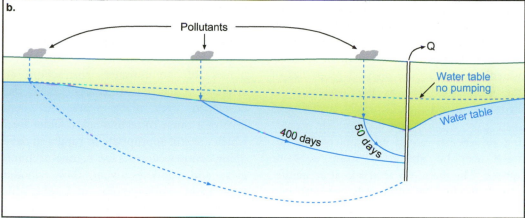

Figure 5.24 Well-head or source protection zones as defined by the Environment Agency, England: a) plan view; b) cross-sectional view. Zone I: 50-day travel time to well; Zone II: 400-day travel time to well; Zone III: well catchment area. Zones are used for land-use planning such as waste disposal site licensing.

5.8 SUMMARY

Groundwater plays an important role in the hydrological cycle, and has provided sources of water for supply and irrigation for several millennia. Most potable groundwater is sourced from meteoric water (rainfall) and thus its sustainable use ultimately depends on ensuring that groundwater losses do not exceed net recharge from rainfall and surface water. Rainfall recharge to groundwater is highly seasonal, as it depends strongly on evapotranspiration; little rainfall recharge occurs during drier seasons in temperate climates. In arid climates, recharge from temporary surface water bodies during the wet season is the main source. As well as acting as a direct supply, groundwater also acts as a reservoir that stores water during the wetter seasons and releases it to surface watercourses throughout the year, and in this way river baseflows are maintained.

Hence, over-abstraction of groundwater or reductions in recharge due to climatic change can damage surface water resources and ecosystems, as well as reducing or damaging the groundwater resource. Unfortunately, many of the world's groundwater aquifers are currently being over-abstracted, and combined with reductions in recharge due to climatic change in some areas, this is likely to lead to severe future ecological, sociological and humanitarian consequences. Modern approaches to groundwater management, aimed at ameliorating such threats, include managed aquifer recharge (MAR) and conjunctive use of groundwater and surface water resources.

The geographical distribution of groundwater depends on the physical properties of the ground as well as the availability of hydrological recharge. Generally, hydrologists distinguish aquifers and non-aquifers according to whether useful amounts of groundwater can be abstracted from wells or springs. The main characteristic of an aquifer is that it must be sufficiently permeable in order to allow groundwater to be extracted. Generally, sediments of relatively large grain size (sands and gravels) are suitable but silts and clays are not. With consolidated rocks, fracture state and the resulting fracture connectivity are relatively more important than grain size, and many rock aquifers are dominated by flow through fracture networks. In so-called karst aquifers, dissolution of the rock by groundwater flow results in flow via conduit or cave systems; these typically develop in limestones. Aquifers are also categorised as confined or unconfined systems, depending on whether the permeable layer is overlain by a less permeable, confining layer. Aquifers are typically recharged by rainfall in high topographic areas, and discharge via springs or directly into rivers, lakes or seabeds in lower areas. Hydrologists use the concept of hydraulic head, which can be quantified by water level in wells or boreholes, in order to predict the direction and quantity of groundwater flow. The relationship between hydraulic gradient (which is expressed as the slope of the water table in unconfined aquifer systems) and flow is linear in most subsurface systems (excluding karst aquifers); this relationship is known as Darcy's Law.

Hydraulic heads can be represented as contours on a hydrogeological map, and flows and heads can be simulated using numerical simulations (called groundwater models). These models are useful management tools to predict the impacts of, for example, groundwater abstraction or injection from boreholes on river flows. Techniques of groundwater abstraction and artificial reinjection (recharge) have been developed over many years and hence involve a range of technologies. Older forms which are still appropriate in rural and developing country contexts use manual technologies and gravity-driven flow; modern approaches more typically use wells which are made from drilled boreholes with electrical pumps. The efficiency of a well depends on its geometry (diameter and screen length) as well as the pumping rates; excessive pumping rates for a given size lead to poor efficiency. Various processes lead to the degradation of groundwater wells and pumps including loss of fine particles from the aquifer, corrosion and biofouling.

As well as uses for irrigation and drinking water, groundwater has been increasingly exploited for its thermal properties. Geothermal springs have long been exploited for hot water and form a significant part of the societal need for heating in some countries. More recently, shallower groundwaters have been used in ground-source heating and cooling schemes, which utilise heat-pump technology. Understanding aquifer hydrogeology is particularly important in the design of ground-source heating and cooling systems because it affects their efficiency.

Groundwater derived from atmospheric input at the surface is often of relatively high quality compared to surface water, as a result of the natural filtration capacity of the subsurface. However, it is important to recognise that such groundwater contains a range of natural substances from mineral dissolution. Which mineral dissolution reactions are operative depends on the mineralogy of the aquifer and the residence time of the water. While many reactions give bottled mineral waters their distinctive taste, others can make the water unpotable. Similarly, mixing with deeper waters which are not derived from rainfall (e.g. original waters present when the rock was formed) and induction of seawater into groundwater aquifers can also reduce quality and render groundwater unpotable. Seawater intrusion, in particular, represents a major threat to groundwater quality worldwide.

Furthermore, a range of natural and artificial substances are introduced into groundwater as a result of anthropogenic activities such as agriculture and industry, and increasingly via artificial recharge schemes. These include heavy metals and radioisotopes, nutrients, major ions (salts), pathogens and organic compounds (including pharmaceutical products), some of which do not degrade and interfere with the operation of biological systems. Both intrinsic groundwater vulnerability to pollution and the vulnerability of individual sources (e.g. wells) depend on the properties of the aquifer and the overlying soil, with fracture and conduit flow systems being relatively vulnerable as compared with matrix-flow systems, because of relatively rapid groundwater movement and poor filtration. A variety of approaches have been developed for characterising the vulnerability to pollution of groundwater aquifers and of specific well abstractions.

FURTHER READING

Freeze, R.A. and Cherry, J.A. 1979. *Groundwater*. Prentice-Hall; Englewood Cliffs, NJ.

   Comprehensive textbook on physical and chemical hydrogeology; notable for natural groundwater chemistry coverage, which is not always included in more recent textbooks.

Hiscock, K. 2014. *Hydrogeology: principles and practice*. 2nd edition. Wiley-Blackwell; Chichester and New York.

   Textbook which includes discussion of climate change issues.

Misstear, B., Banks, D. and Clark, L. 2017. *Water wells and boreholes*. 2nd edition. Wiley-Blackwell; Chichester and New York.

   Covers all aspects of water well drilling and well design; lots of good photographs.

Price, M. 2016. *Introducing groundwater*. 2nd edition. Routledge; Abingdon.

   This book provides a good basic introduction to physical hydrogeology for undergraduate students.

Schwartz, F.W. and Zhang, H. 2003. *Fundamentals of ground water*. John Wiley and Sons; Chichester and New York.

   This book provides a balanced coverage of both physical and chemical hydrogeology from a United States perspective; suitable for undergraduate- and master's-level courses.

Younger, P.L. 2007. *Groundwater in the environment*. Blackwell; Oxford.

   Thorough coverage of groundwater hydrology fundamentals, catchment hydrology, interaction with surface water, and groundwater management.

Classic papers

Hubbert, M.K. 1940. The theory of ground-water motion. *Journal of Geology* 48: 785–944.

This seminal paper describes the application of Bernoulli's concept of 'fluid potential' to groundwater.

Penman, H.L. 1948. Natural evaporation from open water, bare soil and grass. *Proceedings of the Royal Society of London, A* 194: S120–1145.

Highly influential paper on estimation of evaporation from meteorological data; gave rise to methods for assessing sustainable groundwater resources.

Rushton, K.R., Connorton, B.J. and Tomlinson, L. M. 1989. Estimation of the groundwater resources of the Berkshire Downs supported by mathematical modelling. *Quarterly Journal of Engineering Geology, London* 22: 329–341.

An example of how field data informed modelling to examine the groundwater resource and how it might respond to different management interventions.

Taylor R.G. *et al.* 2012. Ground water and climate change. *Nature Geoscience* 3: 322–329.

A review of research on climate change impacts on groundwater including an examination of the future use of groundwater resources in climate adaptation strategies.

PROJECT IDEAS

- Measure water level in a well or borehole through the year (you can use a dip-meter if available, but a weight attached to a measuring tape is a good substitute). Plot the water level against the date through the year and see if you can identify the main recharge period (i.e. when the water level is increasing). If more than one well/borehole/pond is available, compare the water-level responses to see how they vary for wells in different geological deposits (identify these using a geological map of the area if available) and with the topography (normally, areas that are higher with respect to valley base level will show larger variations).

- Find a local aquifer using a geological/hydrogeological map. Identify, for example, from map information, its relationship to nearby surface water features (springs, wetlands, streams, lakes, the sea); i.e. is the aquifer likely to interact with these features? Produce a list of all possible environmental impacts of taking water from this aquifer via pumped wells. What strategies might be adopted to mitigate any adverse impacts of groundwater abstraction?

- Examine the chemical composition listed on the side of a bottle of mineral water. By identification of the major ions present, their relative concentrations and the total dissolved solids, try to determine which minerals may have dissolved and the lithology of the aquifer. If possible, check whether you are correct by examining the location of the spring that supplied the mineral water on a geological map.

REFERENCES

Allen, D.L., Brewerton, L.J., Coleby, L.M., Gibbes, B.R., Lewis, M.A., MacDonald, A.M., Wagstaff, S.J. and Williams, A.T. 1997. *The physical properties of major aquifers in England and Wales*. British Geological Survey Technical Report WD/97/34.

Allshorn, S.J.L. 2008. *Flow and solute transport in the unsaturated zone of the chalk in East Yorkshire*. PhD thesis; University of Leeds.

Anderson, M.P. and Woessner, W.W. 1992. *Applied groundwater modelling*. Academic Press; New York.

Banks, D. 1997. Hydrogeochemistry of Millstone Grit and Coal Measures groundwater in South Yorkshire. *Quarterly Journal of Engineering Geology* 30: 237–256.

Banks, D. 2009. An introduction to 'thermogeology' and the exploitation of ground source heat. *Quarterly Journal of Engineering Geology and Hydrogeology* 42: 283–293.

Birks, D., Whittall, S., Savill, I., Younger, P.L. and Parkin, G. 2015. Groundwater cooling of a large building using a shallow alluvial aquifer in central London. *Quarterly Journal of Engineering Geology and Hydrogeology* 46: 189–193.

Bottrell, S.H., West, L.J. and Yoshida, K. 2006. Combined isotopic and modelling approach to determine the source of saline groundwaters in the Selby Triassic sandstone aquifer, UK. *In:* Barker, R.D. and Tellam, J.H. (eds) *Fluid flow and solute movement in sandstones: the Onshore UK Permo-Triassic Red Bed Sequence*. Geological Society; London; Special Publications 263: 325–338.

Brassington, F.C. 2007. A proposed conceptual model for the genesis of the Derbyshire thermal springs. *Quarterly Journal of Engineering Geology and Hydrogeology* 40: 35–46.

Burke, I.T., Boothman, C., Lloyd, J.R., Mortimer, R.J.G., Livens, F.R. and Morris, K. 2005. Effects of progressive anoxia on the solubility of technetium in sediments. *Environmental Science and Technology* 39: 4109–4116.

Clark, L. 1977. The analysis and planning of step drawdown tests. *Quarterly Journal of Engineering Geology* 10: 125–143.

Darcy, H. 1856. *Les fontaines publiques de la ville de Dijon*. V. Dalmont; Paris.

Domenico, P. and Schwartz, F.W. 1998. *Physical and chemical hydrogeology*. John Wiley and Sons; Chichester and New York.

Ergil, M.E. 2000. The salinisation problem of the Guzelyurt aquifer, Cyprus. *Water Research* 34: 1201–1214.

European Commission. 2000. *Water framework directive*. Directive 2000/60/EC. OJ L 327, 22.12.2000: 1–73.

European Commission. 2006. *Groundwater directive*. Directive 2006/118/EC. OJ L 372, 27.12.2006: 19–31.

Foxworthy, B.L. 1983. Pacific North-West region. *In:* Todd, D.K. (ed.) *Ground water resources of the United States*. Premier Press; Berkeley, CA, 590–629.

Freeze, R.A. and Cherry, J.A. 1979. *Groundwater*. Prentice-Hall; Hemel Hempstead.

Frind, E.O., Molson, J.W. and Rudolph, D.L. 2006. Well vulnerability: a quantitative approach for source water protection. *Ground Water* 44: 732–742.

Fry, V.A. 2009. Lessons from London: regulation of open loop ground source heat pumps in central London. *Quarterly Journal of Engineering Geology and Hydrogeology* 42: 325–334.

Garrels, R.M. and Mackenzie, F.T. 1971. *Evolution of sedimentary rocks*. Norton; London.

Hiscock, K. 2014. *Hydrogeology: principles and practice*. 2nd edition. Wiley-Blackwell; Chichester and New York.

Holden, J. 2005. Darcy's law. *In:* Lehr, J.H. and Keeley, J. (eds) *Water encyclopedia, Volume 5*. John Wiley and Sons; New York, 63–64.

Hubbert, M.K. 1940. The theory of ground-water motion. *Journal of Geology* 48: 785–944.

IGRAC. 2007. *Global inventory of artificial recharge* [online, accessed April 2019]. Available from: www.un-igrac.org/downloads.

Jones, J.B. and Mulholland, P.J. (eds) 2003. *Streams and groundwater*. Academic Press; New York.

Miller, J.A. and Appel, C.L. 1997. *Ground water atlas of the United States, Segment 3, Kansas,*

Missouri and Nebraska, *Hydrologic Investigations Atlas, 730-D*. U.S. Geological Survey; Reston, VA.

Milnes, E. and Renard, P. 2004. The problem of salt recycling and seawater intrusion in coastal irrigated plains: an example from the Kiti aquifer (Southern Cyprus). *Journal of Hydrology* 288: 327–343.

Misstear, B., Banks, D. and Clark, L. 2017. *Water wells and boreholes*. 2nd edition. Wiley-Blackwell; Chichester and New York.

National Research Council. 1999. *Hormonally active agents in the environment*. National Academy Press; Washington, DC.

National Research Council. 2008. *Prospects for managed underground storage of water*. National Academy Press; Washington, DC.

Nickson, R., McArthur, J., Burgess, W., Ahmed, K.M., Ravenscroft, P. and Rahmann, M. 1998. Arsenic poisoning of Bangladesh groundwater. *Nature* 395: 338.

Nordstrom, D.K. 2002. Public health – worldwide occurrences of arsenic in ground water. *Science* 296: 2143–2145.

Patterson, L.J., Sturchio, N.C., Kennedy, B.M., van Soest, M.C., Sultan, M.T., Lu, Z-T., Lehmann, B., Purtschert, R., El Alfy, Z., El Kaliouby, B., Dawood, Y. and Abdallah, A. 2005. Cosmogenic, radiogenic and stable isotope constraints on groundwater residence time in the Nubian Aquifer, Western Desert of Egypt. *Geochemistry Geophysics Geosystems* 6: 1–19.

Pavlis, M., Cummins, E. and McDonnell, K. 2010. Groundwater vulnerability of plant protection products: a review. *Human and Ecological Risk Assessment* 16: 621–650.

Penman, H.L. 1948. Natural evaporation from open water, bare soil and grass. *Proceedings of the Royal Society of London, A* 194: S120–1145.

Price, M. 1993. The chalk as an aquifer. *In*: Downing, R.A., Price, M. and Jones, G.P. (eds) *Hydrogeology of the chalk of NW Europe*. Oxford University Press; Oxford, 35–58.

Price, M. 2002. *Introducing groundwater*. 2nd edition. Nelson Thornes; Cheltenham.

Pyne, D. 2005. *Aquifer storage recovery: a guide to groundwater recharge through wells*. 2nd edition. ASR Press; Gainesville, FL.

Robins, N.S. 1998. Recharge: the key to groundwater pollution and aquifer vulnerability. *In*: Robins, N.S. (ed.) *Groundwater pollution, aquifer recharge and vulnerability*. Geological Society; London, Special Publications 130: 71–76.

Rushton, K.R., Connorton, B.J. and Tomlinson, L.M. 1989. Estimation of the groundwater resources of the Berkshire Downs supported by mathematical modelling. *Quarterly Journal of Engineering Geology, London*, 22: 329–341.

Scanlon, B.R., Healy, R.W. and Cook, P.G. 2002. Theme issue: groundwater recharge. *Hydrogeology Journal* 10.

Schwartz, F.W. and Zhang, H. 2003. *Fundamentals of ground water*. John Wiley and Sons; Chichester and New York.

Singhal, B.B.S. and Gupta, R.P. (eds) 1999. *Applied hydrogeology of fractured rocks*. Kluwer Academic Publishers; Dordrecht, 151–168.

Spence, M.J., Bottrell, S.H., Higgo, J.J.W., Harrison, I. and Fallick, A.E. 2001. Denitrification and phenol degradation in a contaminated aquifer. *Journal of Contaminant Hydrology* 53: 305–318.

Struckmeier, W.F. and Margat, J. 1995. Hydrogeological maps: a guide and standard legend. *IAH International Contributions to Hydrogeology 17*. Heise; Hannover.

Taylor, R.G. *et al.* 2012. Ground water and climate change. *Nature Geoscience* 3: 322–329.

Theis, C.V. 1935. The relation between the lowering of the piezometric surface and rate and duration of discharge of a well using groundwater storage. *Transactions of the American Geophysical Union* 2: 519–524.

Todd, D.K. 1980. *Groundwater hydrology*. 2nd edition. John Wiley & Sons; New York.

Todd, D.K.T. 2005. *Groundwater hydrology*. 3rd edition. John Wiley and Sons; Chichester and New York.

USAID. 2011. *USAID website* [online, accessed August 2011]. Available from: www.usaid.gov.

West, L.J. and Truss, S.W. 2006. Borehole time domain reflectometry in layered sandstone: impact of measurement technique on vadose zone process identification. *Journal of Hydrology* 319: 143–162.

Worthington, S.R.H., Smart, C.C. and Ruland, W. W. 2002. Assessment of groundwater velocities to the municipal wells at Walkerton. *In:* Stolle, D., Piggott, A.R. and Crowder, J.J. (eds) *Proceedings of the 55th Canadian Geotechnical and 3rd Joint IAH-CNC and CGS Groundwater Specialty Conferences*, Niagara Falls, Ontario, October 20–23, 2002.

Wright, E.P. and Burgess, W.G. (eds) 1992. *The hydrogeology of the crystalline basement aquifers in Africa*. Geological Society Special Publication No. 66. Geological Society; London.

CHAPTER SIX

Aquatic ecosystems

6

Lee E. Brown, Colin S. Pitts,
Alison M. Dunn and
Christopher Hassall

LEARNING OUTCOMES

After reading this chapter you should be able to:

■ understand the main groups of aquatic organisms and their interactions;
■ evaluate some of the differences between river and lake ecosystems;
■ outline the key aspects of aquatic ecosystem functional processes;
■ recognise some major ways in which humans have altered aquatic ecosystems.

6.1 INTRODUCTION

Rivers and lakes, the main focus of this chapter, constitute less than 0.01% of the world's water resources and cover only ~0.8% of the Earth's surface. Wetlands (areas where the water table reaches the surface and persists long enough to support aquatic plants) can also be considered freshwater ecosystems (see Dobson and Frid, 2009) but these are not considered in this chapter. The study of any 'ecosystem' unavoidably requires some knowledge of both the living organisms and their effective environment. Where necessary, this chapter makes reference to the aquatic environment to place understanding of the biota into relevant context, but the reader is directed to Chapters 2 to 5 for relevant information on hydrology and water quality.

The terms 'stream' and 'river' are often used interchangeably when referring to flowing waters, because in reality there is no obvious distinction, although the latter term is typically used when describing a 'larger' running watercourse. Some ecologists use the term lotic to describe running water systems of any size but the term 'river' is used for consistency in this chapter. The world's longest river ecosystem is thought by many to be the 6695-kilometre-long River Nile, which flows through north-east Africa and reaches the Mediterranean Sea in Egypt. However, some Brazilian scientists proposed that the Amazon is longer, at 6800 km (Roach, 2007). The disagreement stems in part from imprecise definitions of a river's ultimate source and the point at which the estuarine zone becomes 'sea'. Nevertheless, the Amazon River is the world's largest in terms of volume, with an annual average discharge of >175,000 m^3 s^{-1}, and it is likely to have the highest biodiversity.

Rivers are extremely diverse in their geomorphological form (see Chapter 3) and physicochemical characteristics (see Chapter 4), which both influence the remarkable diversity of organisms that we find in flowing water ecosystems. Describing the characteristics of an individual river ecosystem can be difficult, owing to this diversity of characteristics. A widely used framework to aid understanding of nested river ecosystem geomorphological units is that of hierarchical organisation (Frissell *et al.*, 1986). Spatial units include the whole catchment, river segments, river reaches, mesohabitats and patches (or microhabitats) (Figure 6.1). Streams and rivers are recognisable by their major unidirectional (upstream to downstream) movement of water, sediment and other material. This 'continuum' has formed the basis of many conceptual ideas about how river ecosystems are structured and function, most notably the river continuum concept, which suggested that changes in the river environment from source to sea were accompanied by a distinct zonation of stream invertebrates, fish and material processing (see Vannote *et al.*, 1980). In general, the headwaters of river systems are typically shallow and steep, with fast velocities and coarse bed sediments (Figure 6.2). With distance, downstream width and discharge increase but gradient and sediment sizes

decrease. The changing river environment can lead to notable zonation of aquatic biodiversity and ecosystem functioning along a river's course.

Lakes and ponds move much more slowly than rivers and are considered as still or standing bodies of water. They can be described as being lentic, from the Latin word meaning 'sluggish'. Lakes are much larger than ponds and, while there is not an internationally recognised standard, they are usually more than 2 ha in area and so are influenced strongly by wind turbulence. Lake Superior in the Great Lakes region of North America is the largest freshwater lake ecosystem, with a surface area of 82,414 km^2 and a maximum depth of 406 m. The largest freshwater lake by volume is Lake Baikal, southern Siberia, which is 31,500 km^2 in surface area and up to 1637 m deep. Lake Baikal is over 25 million years old and hosts over 2700 species of plant and animal, many of which are endemic (found nowhere else on the planet). Natural lakes can be created by tectonic processes, such as the fault line that is responsible for the creation of Lake Baikal. They may also be the result of the influence of repeated glaciations on the landscape, such as the Great Lakes of North America. Alternatively, they may arise from fluvial processes, in the case of oxbow lakes (see Chapter 3). Lakes of human origin are created by impounding a river behind a dam wall, creating an artificial water body in the river valley.

While lakes may appear to be more homogeneous than rivers, they can nevertheless possess a high diversity of habitat. The littoral zone of a lake is the area in the upper part of the water body that is marginal to the lake shore and receives abundant sunlight to the bed (Figure 6.3). This area also provides habitat for a wide variety of invertebrates and fish species. The pelagic photic zone is the area in the upper part of the water body, which

Stream system 10³m

Segment system 10²m

Reach system 10¹m

Debris dam

Boulder cascade

Pool/riffle system 10⁰m

Pool

Riffle

Leaf and stick detritus in margin

Sand-silt over cobbles

Tranverse bar over cobbles

Fine gravel patch

Moss on boulder

Microhabitat system 10⁻¹m

receives sunlight but is open water and away from the shoreline. This is the area in which most primary production takes place. The **profundal zone** is in the lower and deep open water away from the shoreline. The **benthic zone** is the area at the bottom of the lake, at the transition between the water body and sediment and substrate (**Figure 6.4**). Ponds, being smaller, are occasionally overlooked in terms of their ecological importance. Nevertheless, both permanent and temporal ponds play an important role in sustaining the life cycles of a number of plants, insects and amphibians. In the case of temporary ponds there are many species that have developed adaptations and life history strategies to cope with changes in hydrological conditions. These include invertebrates that burrow into deeper, moist sediment and insects and amphibians which use water for part of the life cycle, before metamorphosis into their mature forms.

This chapter considers river and lake ecosystems in terms of their major biotic groups, including primary producers, consumers and parasites. The way in which these groups interact is considered with an introduction to food web structure. The chapter subsequently considers the role that aquatic organisms play in ecosystem processes, including organic-matter production, **respiration** and nutrient cycling, and introduces the concept of ecosystem services. Finally, the focus of the chapter turns to systems that have been altered significantly through the actions of humans, with a consideration of the effects of flow regulation, land-use alteration, point-source pollution,

Figure 6.1 Hierarchical structure of river ecosystems.

Source: Reprinted by permission from Springer Science +Business Media: *Environmental Management*, 10: 199–214, A hierarchical framework for stream habitat classification: viewing streams in a watershed context, Frissell, C.R., Liss, W.J., Warren, C.E. and Hurley, M.D., Figure 2. © 1986.

Figure 6.2 Visible changes occur to the river environment from headwaters to lowland: a) an open-canopy UK moorland stream; b) a meltwater-fed river in the French Pyrenees; c) a densely forested stream in south-east Alaska; d) a forested river in north-west North America; e) a mid-reach on the River Aire, UK; f) the River Danube in the lowlands of Austria; and g) the braided Matukituki River in New Zealand.

Figure 6.3 Diversity of lake habitats in the Windermere catchment, north-west England: a) the stony littoral zone of Rydal Water; b) a fringing littoral reed bed in Windermere; c) littoral reed swamp in Rydal Water; and d) the expansive pelagic zone of Windermere.
Source: Photos a)–c) courtesy of Gary Rushworth.

eutrophication, climate change, invasive species and aquaculture. Boxed features throughout the chapter provide more detailed aquatic ecosystem methodological and management insights, including ways in which freshwater biota can be used to assess the condition of freshwater bodies.

6.2 LIFE IN AQUATIC ECOSYSTEMS

While freshwaters cover only a small area of the Earth's surface, they contain a disproportionately high diversity of plants and animals, with >10% of known species (Mittermeier *et al.*, 2010; Dijkstra *et al.*, 2014). There are >10,000 known fish species and some 90,000 species of invertebrates (with insects, crustaceans, molluscs and mites being particularly common; Dudgeon *et al.*, 2006). Other notable groups include amphibians, mammals, birds, macrophytes (plants) and algae (e.g. diatoms, phytoplankton), and there are numerous bacteria, fungi, protozoa and rotifer species inhabiting freshwaters. Despite this knowledge, understanding of freshwater diversity remains incomplete and new species are identified every year. Yet there is also a

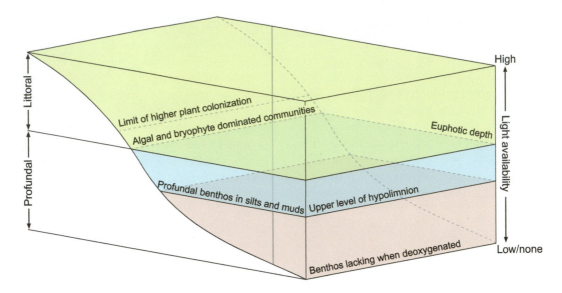

Figure 6.4 A general scheme of littoral and profundal habitats in a freshwater lake.
Source: After Moss (1998). Reproduced with kind permission by John Wiley & Sons; Moss, B. 1998. *Ecology of freshwaters: man and medium, past to future.* © Blackwell Publishing; Oxford.

freshwater biodiversity crisis, with many species at critical risk of extinction due to unsustainable exploitation of freshwaters by humans (Darwall *et al.*, 2018).

Organisms inhabiting freshwaters can be grouped according to their trophic role within aquatic **food webs** (see Section 6.2.3). For example, **producers** (or autotrophs) are the plants and algae that synthesise biomass from inorganic compounds and light. Organisms that obtain organic matter as **consumers** are collectively called heterotrophs, and this group includes species that feed on producers (i.e. herbivores) or other consumers (i.e. carnivores or predators). A third group of consumers consumes dead organic matter and these are collectively known as **detritivores**. The diets of carnivores and detritivores can be heavily subsidised with organic materials from adjacent terrestrial ecosystems (e.g. terrestrial insects or leaf litter from trees, respectively). A fourth

group are the **parasites**, and these exist on or in living organisms, benefiting at their host's expense. These groups are discussed in further detail below. These classifications are a convenient starting point, but it should be borne in mind that often many aquatic organisms have omnivorous diets, feeding on a variety of resources. Resources originating within the aquatic ecosystem and directly providing energy to the aquatic food web are termed **autochthonous**, while those originating from adjacent terrestrial ecosystems are termed **allochthonous**.

A variety of terms are used to explain where aquatic organisms spend the majority of their existence. Benthic organisms (or the **benthos**) live on, in or near the bed sediments of rivers or lakes. **Nektonic** organisms (or the **nekton**) is the term for organisms that can *actively* move around within the water column, whereas **planktonic** organisms live suspended in the water column and *passively* float or drift around in the water column. The **neuston** is a term reserved for those organisms that exist predominantly on, or just beneath, the surface of water bodies. Many terrestrial plants and animals are intricately linked with freshwater habitats, in particular existing in, on or around the bank habitats (the **riparian** zone; Naiman *et al.*, 2005), but in this chapter we focus predominantly on those organisms that spend the majority of their life cycle in aquatic habitats.

6.2.1 Primary producers

Primary producers are organisms that convert inorganic compounds into biomass through **photosynthesis** (see Chapter 1, Section 1.3 for details of the photosynthesis process). In rivers, there are two major groups of producers: macrophytes, which are large aquatic plants, and much smaller organisms, which can be collectively considered algae. Aquatic macrophytes predominantly consist of flowering plants, mosses and liverworts (Allan and Castillo, 2007), and, while as a group they are most successful in lakes and slower-flowing parts of rivers, some mosses and liverworts are often found in swift-flowing rivers. They have four different growth habits (**Figure 6.5**): a) emergent plants with roots; b) floating plants with roots; c) floating plants without roots; and d) submerged plants with roots (Murphy, 1998). Angiosperms (flowering plants) are typical of slower, more nutrient-rich waters, while plants with emergent vegetation or floating leaves are found only in rivers if they are deep and slow-flowing. Macrophytes play important roles in influencing aquatic habitats by slowing the velocity of water flow and encouraging the deposition of fine sediments. Macrophytes are found where light is able to reach the bed sediments, and so they are found in most areas of rivers except for the larger lowland stretches and

estuaries that carry large quantities of suspended sediment (highly turbid rivers). In lakes, aquatic macrophytes are a characteristic feature of the littoral zone, and emergent macrophytes with root systems often emerge and grow upward above the water surface. Examples include the common reed *Phragmites* and bulrush *Typha*. Submersed and floating-leaved macrophytes, such as waterlilies *Nymphaea* and pondweed *Potamogeton* also inhabit these locations. Some lakes are shallow across their entire area and so macrophytes grow extensively. Slower-flowing stretches of river may also be colonised by macrophytes. Macrophytes serve as food sources and habitat for some animals, as well as providing substrate for some algae.

Most aquatic algae in rivers grow attached to, or on, inorganic substrate and are termed epilithic (Wehr and Sheath, 2003). Others can grow on fine sediments (epipelic), sand (episammic) or wood (epixylic). A major group of the aquatic algae is the Bacillariophyta (diatoms); these are microscopic organisms that construct a shell (frustule) from silica. Owing to their diversity and varied responses to environmental conditions, they are widely used in water quality assessments. Other groups include green algae (Chlorophytes), yellow-brown algae (Chryophytes), red algae (Rhodophyta) and blue-green algae (Cyanobacteria). Algal distribution in rivers is strongly influenced by light availability as well as water chemistry, water temperature, current velocity and the abundance of aquatic herbivores. Some algae are free-living in the water column in slow-flowing parts of rivers, and these are collectively known as phytoplankton. However, this group of algae is more common in lakes than in rivers.

The open water habitat of lakes contains a complex assortment of phytoplankton species, but many of these are invisible to the naked eye except for where they form immense colonies or blooms. Phytoplankton

Figure 6.5 Examples of aquatic producers: a) emergent reeds (*Phragmites*) and floating rooted waterlilies; b) close-up view of the rooted waterlily with floating leaves and flowers; c) floating duckweed; d) filamentous algae; and e) rock with a thin covering of diatoms and green algae, visible below the water line.

range in size from 1 μm (millionth of a metre) (picoplankton), <5 μm (ultraplankton), 5–60 μm (nanoplankton) and >60 μm (net plankton). This group includes a hugely diverse array of bacteria, green, yellow-green and golden-yellow algae, diatoms, cryptophytes, euglenoids and dino-flagellates (Moss, 1998). The growth of phytoplankton is strongly influenced by light and dissolved nutrients such as nitrogen and phosphorous. Periphytic algae such as *Cladophora* can be found attached to the substrate in the shallower littoral areas. Floating macrophytes, such as duckweed *Lemna*, can be found in the deeper profundal waters. However, free-floating algae, known as phytoplankton, carry out most of the primary production in this part of a lake. These include diatoms such as *Tabellaria* and *Asterionella*, desmids such as *Staurastrum*, dinoflagellates such as *Peridinium*, and blue-green algae such as *Microcystis* and *Anabaena*. These are grazed upon by small, free-floating animals known as zooplankton, a group which includes a variety of cladocerans, copepods and rotifers.

6.2.2 Consumers

Consumers are organisms that are heterotrophic, obtaining their energy from the consumption and processing of organic matters originating from either primary producers or other heterotrophs. This broad group has a diverse array of feeding modes, including herbivory (feeding on plant matter), frugivory (fruits), detritivory (non-living matter), carnivory (animal matter) and parasitism. The latter consumption strategy occurs at the expense of the 'prey' or host, but, unlike in the other feeding modes, the prey often remains alive for a prolonged period. Aquatic consumers play key roles in processing organic matter in aquatic ecosystems and in linking the energy and biomass created by primary producers, or input from adjacent terrestrial systems, to consumers higher up the food web (either in aquatic systems or even back to those in the terrestrial ecosystem; see Section 6.2.3).

Herbivorous aquatic organisms include zooplankton which feed on phytoplankton in lakes and slow-flowing rivers, macroinvertebrates that graze on diatoms and algal growth, and fish such as the grass carp which feed on larger macrophytes and algae. Aquatic organisms that consume fruits can subsequently act as seed dispersers, including, for example, many fish species in the Amazon River, some of which have diets composed of up to 98% fruit (Reys *et al.*, 2009). Predatory (carnivorous) aquatic organisms include fish such as the pike and perch which feed on smaller fish, many fish and amphibians that feed on aquatic and terrestrial invertebrates, and invertebrate larvae such as some stoneflies and dragonflies that prey on smaller aquatic and terrestrial animals.

Detritivory is a major feeding strategy adopted by many aquatic organisms, with feeding possible on a variety of non-living organic-matter sources, including terrestrial plant litter, woody debris, pollen, excreted waste and dead animals. In particular, detritivores play major roles in processing organic matter that falls into water bodies from riparian (bankside) trees. In streams where plant leaves form a large component of the terrestrial plant input, leaves usually aggregate around obstructions, forming leaf 'packs'. Leaves are broken down in various ways that have been studied in detail (Gessner *et al.*, 1999; Benfield, 2006). For example, deciduous leaf litter that falls in autumn is largely nutritionally poor because the tree will have reabsorbed most of the soluble nutrients. However, the remaining leaf biomass still contains cellulose and lignin, which is partly consumed by fungi and bacteria. This consumption can serve to *condition* the litter, making coarse pieces of leaf litter more palatable to aquatic invertebrate consumers. Organisms that feed on this material have mouthpart adaptations that allow them to consume large pieces of plant material and small pieces of wood

Table 6.1 Functional feeding groups (FFGs) of invertebrates, their feeding mechanisms, food sources and typical size range of particles ingested

FFG	Feeding mechanism	Main food source	Particle size range (mm)
Collector-filterer	Suspension feeders – filtering particles from the water column	FPOM, detritus, algae, bacteria, fungi	0.01–1.0
Collector-gatherer	Deposit feeders – collecting deposited particles of FPOM or sediment	FPOM, detritus, algae, bacteria, fungi	0.05–1.0
Grazer	Graze surfaces of rocks, plants and wood	Producers such as attached algae, and associated detritus, bacteria and fungi	0.01–1.0
Predator	Capture prey and either engulf whole or pierce it and ingest body fluids	Living animals	>0.5
Shredder	Chewing	CPOM or live plant tissue	>1.0
Parasite	Microparasites absorb nutrients, often living within host cells. Macroparasites and endo-parasites attach to and absorb nutrients from host gut or tissues. Ectoparasites feed on host blood or tissues (e.g. gills)	Producers and consumers	N/A

Source: After Cummins *et al.* (2005).
Note: FPOM and CPOM = fine/coarse particulate organic matter, respectively.

(>1 mm) known as coarse particulate organic matter (CPOM). CPOM is consumed predominantly by a group of organisms known as shredders, such as freshwater shrimp (*Gammarus*), louse (*Asellus*) and some stoneflies (e.g. Leuctridae; Table 6.1).

The processes of shredding and/or physical breakdown due to the action of moving water and sediment break the coarse material into fine particulate organic matter (FPOM; <1 mm). FPOM is easily maintained in suspension in rivers or lakes, and can be collected passively from the water column by organisms with filtering appendages, such as the invertebrate larvae of blackflies (Simuliidae) or those that collect food using nets (e.g. hydropsychid caddis flies). Other filterers, such as freshwater mussels (e.g. the zebra mussel *Dreissena polymorpha*), actively pump water and associated suspended particles through their

mouth parts and in doing so remove the organic particles. Fine particulates can also be deposited on the bed of lakes and streams, where flows are slower, and a third group of detritivorous consumers (collector-gatherers) feed on these deposits. Organic matter also forms part of a river's or lake's dissolved load, and the vast microbial component of aquatic food webs is fuelled largely by dissolved organic carbon (DOC) originating from soils, water and sediments (Meyer, 1994). Carbon and other nutrients in river water can be absorbed and assimilated by plants and microbes or adsorbed to sediments. When these nutrients are eventually released back to the water, some may be immediately recycled into the biota, or they remain as dissolved nutrients for later uptake by biota or abiotic mineralisation.

Parasitism is a major consumer strategy, although it is often neglected by community ecologists, despite an estimated 40% of all species being parasitic (Dobson *et al.*, 2008). Parasites exploit hosts at all trophic levels, and can affect host population dynamics as well as behavioural interactions with other species. In a study of estuarine salt marsh communities in North America, Kuris *et al.* (2008) reported 199 free-living animal species, 15 free-living vascular plants and 138 parasite species. Parasite biomass made up ~1% of total biomass, which does not seem large. However, the biomass of parasitic trematode worms was, in fact, comparable to that of the top predators in the estuaries. In addition to exploiting the host, parasites provide food for species at higher trophic levels, either when they are eaten along with the host or as free-living life stages. For example, in the horn snail *Cerithidea californica*, the dominant invertebrate grazer in the estuaries, trematode parasites averaged 22% of the body mass of their host. Hence, when the host is consumed the parasites also provide a large resource for predators (Hechinger *et al.*, 2009).

Parasites can change the functional role of their hosts, and the strength of interactions of the host with other species in an ecosystem, either by reducing host population density or by changing the behaviour of the host (Hatcher and Dunn, 2011). These indirect effects of parasites can cascade throughout the trophic levels, leading to changes in both productivity and community structure. For example, detritus processing by the isopod crustacean *Caecidotea communis* was reduced by >40% as a result of infection by the acanthocephalan parasite *Acanthocephalus tahlequalhensis* (Hernandez and Sukhdeo, 2008a). In streams in Michigan, USA, parasite outbreaks caused crashes in populations of caddis flies (*Glossoma nigrior*), dominant grazers of periphyton. The resulting reduction in grazing led to increased periphyton biomass, as well as increased densities of the other organisms which compete with the caddis flies for food. This in turn led to increased densities of predatory caddis flies and stone flies (Kohler, 2008). Parasites can influence the flow of energy between terrestrial and aquatic ecosystems (Section 6.2.3, food webs). Terrestrial crickets infected by nematomorph worms show behavioural manipulation, leaping suicidally into nearby water bodies, where the parasitic worm emerges and enters the free-living stage of its life cycle. These crickets provided an additional food supply for fish which, as a result, ate fewer benthic invertebrates. The resulting release of invertebrates from fish predation led indirectly to higher leaf shredding and decreased algal biomass (Sato *et al.*, 2012). Conversely, die back of riparian alder trees (*Alnus sp*) infected by the phytophthora fungus (*Phytophthora alni*) is associated with a reduction in leaf litter reaching the aquatic ecosystem as well as a reduction in riverbank stability (Bjelke *et al.*, 2016).

Parasites can have indirect effects on predator–prey interactions because many use both members of the relationship as hosts, and they can manipulate anti-predator behaviour of their intermediate host to enhance trophic transmission to the next host. For example, *Gammarus pulex* (amphipod crustaceans) that are infected by the acanthocephalan parasite *Echinorynchus truttae* show increased activity and move towards light (uninfected amphipods seek out shade to avoid predation); these changes in behaviour increase the risk of predation by trout, and hence increase the rate of parasite transmission (MacNeil *et al.*, 2003). The strength of predator–prey linkages can also be affected by parasites. *G. pulex* that are infected by *E. truttae* (**Figure 6.6**) were found to eat 30% more isopods than did uninfected hosts, probably as a result of the metabolic demands of the parasite and of

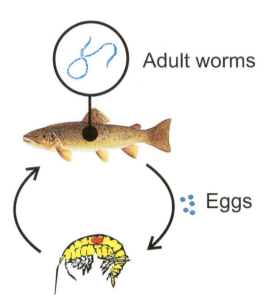

Figure 6.6 The life cycle of the trophically transmitted parasitic worm *Echinorhyncus truttae*. The adult worms live in the definitive fish hosts, which egest *E. truttae* eggs in the faeces. The intermediate amphipod host becomes infected when it eats the eggs, which subsequently develop into a bright orange cystacanth. The parasite makes the amphipod host more active and it seeks light. As a result of these changes, the host is at risk of predation by fish, and the parasite has an increased chance of trophic transmission back to a definitive fish host.

their increased activity (Dick *et al.*, 2010). Conversely, white-clawed crayfish (*Austropotomobius pallipes*) that harboured the microsporidian parasite *Thelohania contejeani* have been seen to eat less prey than their uninfected competitors, probably as a result of muscle damage by the parasite (Hatcher and Dunn, 2011).

6.2.3 Food webs

Food chains are representations of interactions between consumers and resources, linking autotrophic producers to top predators via a series of intermediate consumers (Figure 6.7a).

In reality, food chains do not exist in isolation, because consumers feed on multiple resources and thus form food webs, which are complex networks of species connected to one another by their feeding interactions (Figure 6.7b). The study of food webs is important because the complex interactions mean it is difficult to predict how ecosystems respond to environmental change by simply extrapolating from studies of individual populations or groups of species (Woodward *et al.*, 2010). Many of the best-described food webs originate from running waters, and structural characteristics have been catalogued from a variety of habitats, including upland and lowland rivers, forested and open-canopy river basins, and tropical, temperate and Arctic environments (Brown *et al.*, 2011). Numerous food webs have also been documented from lakes and river estuaries (Table 6.2), and comparative analysis of food webs available from different aquatic habitats shows comparable food web size (number of species, number of links) and complexity (connectance), but generally lower mean food chain length in rivers. Some key influences on aquatic food webs include flow regimes, river catchment size and disturbance regime, as well as ecosystem size, thermal regime, pH, invasion by new species and surrounding land use (e.g. Thompson and Townsend, 2005).

Connectance food webs are those which illustrate individual feeding links between species (e.g. Figures 6.7b and Figures 6.8a), and they are constructed by sampling numerous individuals of constituent species, then examining their diet through microscopic analysis of gut contents to observe dietary links. Species can then be assigned trophic positions as basal species (those with consumers but no prey – i.e. primary producers), intermediate species (those with both consumers and prey) and top species (those with prey but no consumers – i.e. top predators). Food webs from rivers and lakes typically show trophic pyramids, whereby the number of species and their biomass

Table 6.2 Food web summary statistics for rivers and lakes

Study locations (and no. of published food webs from that location)	No. of taxa	No. of links between species	Connectance (links per (species)^2 to give proportion of all possible links observed)	Mean food chain length
Rivers				
Alaska, USA (n = 4)	38–45	150–193	0.10–0.11	1.65–3.10
Appalachian Mountains, USA (n = 4)	35–41	125–200	0.10–0.15	2.45–2.97
Artificial streams, Dorset, UK (n = 4)	61–71	320–492	0.09–0.11	2.03–2.17
Broadstone, UK (n = 11)	24–128	90–721	0.04–0.21	4.88–5.38
Burdekin, Australia (n = 12)	13–17	18–43	0.23–0.39	—
Duffin Creek, Canada (n = 7)	31–39	101–146	0.09–0.11	—
Hengill, Iceland (n = 7)	46–70	360–860	0.13–0.22	—
Muskingham Brook, USA (n = 8)	33–48	97–171	0.07–0.09	2.86
Odenwinkelkees, Austrian Alps (n = 4)	13–23	16–85	0.05–0.19	2.00–2.36
Kennet, UK (n = 4)	60–73	635–1060	0.17–0.19	—
Southeast USA (n = 4)	58–105	126–343	0.03–0.04	1.52–3.13
South Island, New Zealand (n = 20)	49–113	110–950	0.03–0.08	1.56–4.42
UK streams (n = 20)	19–87	56–1653	0.12–0.29	—
Lakes and ponds				
Lake Tahoe, USA	172	3885	0.13	—
Little Rock Lake, USA	92	1008	0.12	>10.0
Lochnagar, UK (n = 18)	45–81	270–831	0.14–0.38	—
Mirror Lake, USA	586	14 709	0.15	—
North American lakes (n = 50)	10–74	17–571	0.06–0.17	4.04
Sierra de Gredos, Spain (n = 3)	64–96	344–534	0.06–0.08	6.22
Tuesday Lake, USA	56	269	0.09	4.2–4.6
Skipwith Pond, UK	35	276	0.23	—
Lake Tatvan	50	432	0.17	—

Source: Data from Amundsen *et al.*, 2009, Blanchett *et al.*, 2014; Brown *et al.*, 2011; Clitheroe *et al.*, 2013; Grey, 2016; Pitchler, 2012; Sánchez-Hernández *et al.*, 2015.

decreases with food web height, due mainly to inefficiencies in energy transfer between trophic levels (i.e. not all energy synthesised or consumed by prey is available to their consumers, due to respiration – energy used for oxygen uptake and carbon release – and growth costs).

One major drawback with connectance food webs is that they assume that all feeding links are equal, because they do not provide any information on relative strength. Consequently, ever more detailed food webs are being developed with quantified links (**Figure 6.8b**) in attempts to develop more detailed knowledge of how aquatic food webs vary in space and over time. Furthermore, there are developing ideas about how aquatic food webs interact with other types of ecological networks such as host–parasite webs (Hernandez and

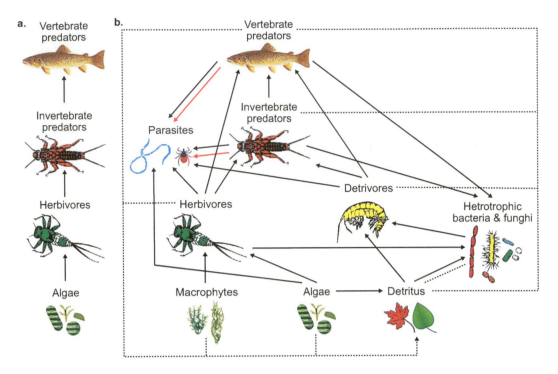

Figure 6.7 Schematic outline of a) a simple aquatic food chain, and b) a simple aquatic food web. Solid lines represent feeding links and broken lines represent decomposition pathways. Red arrows denote energy flows from parasites to (non-host) predator when consumed along with the host.

Sukhdeo, 2008a; Kuang and Zhang 2011; Figure 6.9), and the implications of these cross-overs for food web dynamics. Many food web studies of aquatic (and indeed terrestrial) ecosystems have tended to omit parasites, perhaps because of their relatively small biomass. However, parasitism is the most common consumer strategy (Kuris *et al.*, 2008), and parasites also influence the trophic interactions of their hosts (Section 6.2.2). Inclusion of parasites in a stream food web (**Figure 6.9**) increased species richness (28 free-living and 10 parasitic species were recorded), linkage density and the number of possible and observed links (Hernandez and Sukhdeo, 2008b). Similarly, including parasites in a lake food web

increased species richness (37 free-living and 13 parasitic species), food chain length and connectance (Amundsen *et al.*, 2009).

Aquatic food webs and those of the adjacent terrestrial ecosystem (or the riparian zone) are often strongly linked through the consumption of allochthonous leaf litter by shredders. Additionally, the diet of aquatic predators, such as fish and some invertebrates, can also be supported by terrestrial invertebrates which fall into the water. In this sense it can be said that the terrestrial ecosystem feeds, or subsidises, the aquatic ecosystem (Nakano *et al.*, 1999), and some studies have shown that the diet of river-dwelling fish can be made up of >50% terrestrial invertebrates (Baxter *et al.*, 2005). Research has also indicated reciprocal subsidies

a.

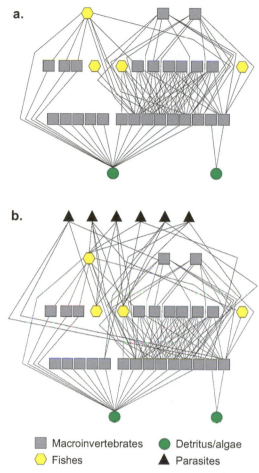

b.

	<0.1%
	0.1–0.9%
	1–10%

Figure 6.8 Connectance food web from Broadstone Stream, UK. Panel (b) shows the Broadstone Stream web with feeding links quantified based on numbers of prey eaten per capita 24 hr^{-1} (as a percentage of numbers m^{-2}). Filled circles represent species or groups of species and lines represent feeding links. Food webs are arrayed hierarchically, based on species' trophic levels.

Source: After Woodward *et al.*, 2005. Reprinted from Woodward, G., Speirs, D.C. and Hildrew, A.G. Quantification and resolution of a complex, size-structured food web. *Advances in Ecological Research* 36: 85–135, © 2005, with permission from Elsevier.

■ Macroinvertebrates   ● Detritus/algae
⬡ Fishes   ▲ Parasites

Figure 6.9 Food web structure of Muskingham Brook, New Jersey, USA, during autumn 2003. Panel (a) shows the food web excluding parasites, and (b) shows the same food web with parasitic links included. Relative positions are not arrayed according to trophic level.

Source: International Journal for Parasitology 38, Hernandez, A.D. and Sukhdeo, M.V.K. Parasite effects on isopod feeding rates can alter the host's functional role in a natural stream ecosystem, pp 683–690, © Elsevier, 2008.

from aquatic to terrestrial ecosystems (Nakano and Murakami, 2001). Aquatic insects develop in the stream as larvae (nymphs), but towards the end of their life cycle they emerge as adult flies with a terrestrial aerial stage. The emergence of adult aquatic insects is typically concentrated into a small part of the year in temperate regions (peaking in early summer), whereas in tropical areas it may occur year

round, providing food for terrestrial organisms such as spiders, bats, birds, amphibians and beetles. Some birds and amphibians also prey on aquatic invertebrate larvae. It has also been shown that fish can exert an influence on the pollination success of terrestrial plants through their consumption of aquatic dragonfly larvae densities. Thus, fewer adult dragonflies emerge into the terrestrial environment to consume terrestrial insect pollinators than is the case in fishless ponds (Knight *et al.*, 2005).

REFLECTIVE QUESTIONS

What are functional feeding groups? How might functional feeding groups differ in a river before and after deforestation of the surrounding riparian zone?

6.3 AQUATIC ECOSYSTEM FUNCTIONAL PROCESSES

6.3.1 Primary production

Primary production represents an important source of autochthonous energy to aquatic ecosystems. Primary productivity can be defined as the rate of formation of organic matter from inorganic carbon by photosynthesising organisms (Bott, 2006) and can be measured in a variety of ways (Box 6.1). Aquatic primary production is often thought of as being less important than allochthonous inputs of terrestrial plant litter, but this is not the case for open-canopy (non-forested) sites, some stretches of forested rivers where light penetration is relatively high, or large lakes (Murphy, 1998). Producer turnover (i.e. replacement of dead, consumed or eroded plants or algae) can be rapid compared to detritus stores, even though standing stocks of the latter may be much higher (Hershey and Lamberti, 1998). In rivers, primary production often shows a general increase with distance downstream into the middle reaches, particularly where the

headwaters are forest covered and turbidity is relatively low. However, there can be high variation owing to the character of the river at different locations and over time (e.g. Vannote *et al.*, 1980; Murphy, 1998). The edges and bottoms of lakes are considered to be relatively productive and diverse where light penetration is high and there are abundant nutrients in sediments and water (Moss, 2010).

The rate of aquatic primary production is influenced by a multitude of factors, many of which are common to running and standing waters. Regional and catchment-scale features such as climate, topography, geology and land use affect key proximate influences such as light availability, water chemistry/nutrient concentrations and temperature (Allan and Castillo, 2007). Light is vital for photosynthesis, and seasonal changes in primary production can be attributed to the annual solar cycle, which also drives thermal dynamics. The amount of light reaching the water surface is strongly affected by shading either from overhanging vegetation or from topography, while light reaching the bottom of rivers and lakes is a function of water depth and turbidity (a measurement of the amount of suspended particles in the water) (Wetzel, 2001). There can, however, be light-saturation effects, and thus other variables may serve to limit production (Murphy, 1998). Phosphorous (as dissolved inorganic P) and nitrogen (N mainly as nitrate or ammonia) are often key limiting nutrients and are usually implicated in eutrophication, where excess algal growth causes severe water quality problems (see Section 6.5). Silica is also important because it is required for the construction and maintenance of diatom frustules, and in some lakes spring diatom blooms can deplete silica concentrations significantly, such that other forms of algae start to become more abundant (Wetzel, 2001). Other important influences on the rate of primary production can include water velocity, substrate type and quality and grazing

BOX 6.1

Measuring aquatic primary productivity

Primary production can be represented as:

$$GPP = NPP + R$$

where: *GPP* = gross primary production, *NPP* = net primary production, and *R* = respiration. NPP represents the energy or biomass available to herbivorous consumers after the producer's respiration 'costs' are taken into account. Net ecosystem production (NEP) can also be determined by calculating the difference between NPP and the respiration of all heterotrophs in an aquatic system. Measurement of primary production can involve quantifying the accrual of periphyton on introduced substrate (e.g. unglazed tiles or microscope slides), determining chlorophyll a (the primary pigment for photosynthesis) concentrations, measuring changes in dissolved O_2 or CO_2 concentrations *in situ* or in enclosed chambers, or adding the isotope ^{14}C and tracking its uptake into plants and algae (Howarth and Michaels, 2000; Bott, 2006; Figure 6.10). The biomass of primary producers is sometimes used as a measure of production but this can be misleading because, while some plants and algae may maintain only low standing stocks (e.g. perhaps due to high grazing pressures), they might also have a high turnover (e.g. constant and rapid replacement of lost biomass) and thus high production (Murphy, 1998). Standing stock measurements are also problematic because there is no way to differentiate the amount of production that has been respired or consumed.

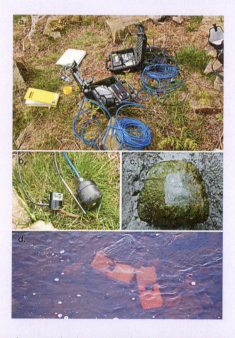

Figure 6.10 Examples of the various methods used by freshwater ecologists to study primary production: a) dual-station oxygen measurements using battery-powered dataloggers connected to dissolved-oxygen sensors; b) from left to right, a photosynthetically active radiation sensor to measure incoming light, a thermistor used to measure stream temperature and a dissolved-oxygen sensor; c) a 3 × 3 cm patch of algae has been sampled from a rock for later analysis of algal community composition and extraction of chlorophyll a; d) unglazed 10 × 10 cm^2 quarry tiles attached to bricks to allow algal accrual over time.

pressures from herbivores (Murphy, 1998; Allan and Castillo, 2007). In a study of 30 streams from 30–50°N, Lamberti and Steinman (1997) showed that gross primary production (GPP) ranged from 3.5 to 5400 g C m^{-2} y^{-1}, averaging 560 g C m^{-2} y^{-1}. Estimated average GPP for lakes worldwide is 260 g C m^{-2} y^{-1}, with 40 g C m^{-2} y^{-1} of this attributed to excess nutrient input and eutrophication (Lewis, 2011).

6.3.2 Secondary production

Secondary production involves the formation of heterotrophic biomass (microbial and animal) for a given area and over a given time, and it represents the processes of recruitment and growth, respiration, emigration and mortality (Benke, 1984; see Box 6.2). While the approach has been commonly used to calculate production of individual populations of heterotrophs (Huryn and Wallace, 2000), it is, nevertheless, possible to calculate secondary production for entire trophic levels or assemblages, and to use these estimates to quantify energy flow through whole food webs (e.g. Hall *et al.*, 2000). The latter method is particularly important because it reveals that the 'weighting' of individual links in aquatic food webs is not equal, with most being weak but some being extremely strong. These differences in link weighting can have implications for the stability of ecosystems under disturbance conditions (Montoya *et al.*, 2009).

Reported values of secondary production for aquatic ecosystems are common for invertebrates, with values from across the world ranging between approximately 1 to 1000 g m^{-2} y^{-1} (Huryn and Wallace, 2000). Meiofauna (invertebrates <250 μm when fully grown) can be abundant in streams, and their intermediate role between microbial/detrital resources and larger organisms means that they can contribute significantly

to secondary production (e.g. Tod and Schmid-Araya, 2009). In small, cold, stony open-canopy moorland streams, such as those in the North Pennines of the UK, fish production is low, ranging from just 1.02–3.50 g m^{-2} y^{-1} for trout and 0.48–7.43 g m^{-2} y^{-1} for bullhead (Crisp *et al.*, 1975). In contrast, in river systems where there are abundant prey production and favourable habitat conditions, trout production can reach >10 g m^{-2} y^{-1} (Huryn, 1996). Many estimates of trout production seemingly exceed estimates of prey production, a discrepancy which has been termed the Allen Paradox (Hynes, 1970). This situation arises because fish diet can be subsidised heavily from terrestrial (riparian) prey inputs, invertebrate production in the hyporheic zone (the sediments beneath the stream bed) or by cannibalism. All of these prey sources can be difficult to quantify.

6.3.3 Community respiration

Respiration is the biological process in which organic matter is utilised by organisms, typically in the presence of dissolved oxygen, to provide energy for metabolism, growth and reproduction. Both autotrophs (algae, macrophytes) and heterotrophs (animals, bacteria, fungi) respire organic matter, and together these account for community respiration. Quantifying community respiration (CR) is particularly important in studies that are also measuring primary production (P), because it allows the operator to examine ratios of P:R (or, more precisely, GPP:CR). Where GPP > CR there is a net addition of energy to an ecosystem and it can be considered net autotrophic. In contrast, the system can be said to be net heterotrophic where GPP < CR (Bott, 2006). Respiration is fuelled by primary production that occurs within the river (autochthonous

BOX 6.2

Measuring aquatic secondary production

Similar to producers, where only a fraction of GPP is converted to biomass, some of the material consumed by heterotrophs is not converted to biomass because of inefficiencies in assimilation and the use of energy and matter for respiration (Benke, 1984). Secondary production can thus be considered as:

$$I = A + F$$

where: I = ingestion, A = assimilation, and F = faeces.

Biomass that is assimilated (A) is then processed by the individual:

$$A = P + R + E$$

where: P = production of biomass, R = respiration, and E = energy lost during excretion.

At the individual level, P can be considered to be growth, while at the population or community level it represents the collective growth of all individuals. Measuring growth involves quantitative sampling of species cohorts over time (usually a minimum of 12 months) and calculating production from changes in species density and biomass. The mean biomass over the entire year (B) can be calculated for each species, and used to calculate P/B ratios to illustrate species turnover (Benke, 1984). P is determined by availability of food, but also how efficiently that food is converted to tissue, which is determined by assimilation efficiency (A/I), net production efficiency (P/A) and gross production efficiency (P/I). Indicative values of assimilation efficiency for aquatic invertebrates are usually in the order of 10% for vascular plant detritus, 30% for diatoms and algae, 50% for fungi, >70% for animals and 50% for microbes, with net production efficiencies of ~40% (Benke and Wallace, 1997; Benke, 2010).

production), or by allocthonous resources (e.g. leaf litter, dissolved organic carbon from soils) that enter the river from adjacent terrestrial ecosystems.

Community respiration of aquatic systems can be determined from measurements of dissolved oxygen over daily time periods. Other methods for monitoring the consumption of oxygen include enclosing surface stones in respiration chambers, enclosing bed sediments in respiration tubes followed by incubation, or collecting water samples and incubating for fixed time periods (e.g. Naegeli and Uehlinger, 1997). Smaller streams are dominated by benthic metabolism, whereas planktonic metabolism predominates in larger rivers and lakes (Thamdrup and Canfield, 2000).

6.3.4 Nutrient cycling

Advances in our understanding of in-stream nutrient supply, transformation and retention have developed around concepts of nutrient *spiralling*, in the sense that nutrient *cycles* in streams are inevitably coupled with the downstream movement of water, and are therefore better conceptualised as spirals (Newbold *et al.*, 1982). The approach, which is founded on experimental additions of nutrients to streams (**Figure 6.11**), followed by sampling at fixed locations downstream of the addition, allows estimates of: the distance required for nutrients to complete cycling through dissolved to particulate to consumer phases and then return to the water column (so called spiralling lengths); and the **bioavailability** (and/or demand) for nutrients within the ecosystem (uptake velocities). Ensign and Doyle (2006) reviewed >400 studies reporting spiralling lengths and uptake velocities for streams and the vast majority of these had been undertaken in small first- and second-order headwater streams (see glossary for explanation of **stream order**). More remains to be done to understand nutrient spiralling in larger rivers further down the river network. Nutrient limitation can also be assessed using diffusing substrates. In essence, nutrient mixtures (N, P, N+P) are dissolved in agar and then placed into small containers (**Figure 6.12**). The lid of the container is replaced with a porous substrate (usually a clay tile or wood, which enables comparative assessments of epilithic and epixylic biofilm growth, respectively) through which the nutrients can diffuse into the water column before being utilised by producers and heterotrophic microorganisms.

Figure 6.11 A battery-powered peristaltic pump being used to deliver small quantities of dissolved NO_3^-, NH_4^+ and organic carbon from a tank into an Arctic tundra stream as part of an experiment to measure nutrient spiralling lengths and uptake velocity.

REFLECTIVE QUESTION

How could changes in water quality due to runoff from an agricultural catchment impact upon the processes of primary production and community respiration?

Figure 6.12 Nutrient-diffusing pots with wooden (brown) and clay (orange) substrate.

6.4 AQUATIC ECOSYSTEM SERVICES

Ecosystem services are the benefits that the human population is able to gain from a functioning ecosystem. The Millennium Ecosystem Assessment provides a framework which divides the various services into *provisioning*, *regulating*, *cultural* and *supporting* services (Millennium Assessment, 2005). The main provisioning services associated with freshwater ecosystems include food (fish and other edible biota production), fuel (wood and peat), biochemical (material from biota), genetic (resistant genes and medicine) and biodiversity (diversity of species and gene pools). Important regulating services include flow regulation (wetland storage and groundwater recharge, floodplain water storage), sediment transport (form and function of river channel and floodplain) and waste assimilation (pollutants and excess nutrient cycling). The main cultural services associated with aquatic ecosystems include recreation associated with angling, and amenity values (education, research). Supporting services include primary production, secondary production and nutrient cycling. Many freshwaters have been, and continue to be, managed poorly by humans (see Section 6.5), and in many locations this has led to degradation or loss of some ecosystem services at the expense of others (e.g. Maltby *et al.*, 2011). Balancing the competing demands for ecosystem services from multiple stakeholders is emerging as a major political, economic and scientific challenge, not least because freshwater services are estimated to be worth >$4 trillion annually (Darwall *et al.*, 2018). Other examples of ecosystem services provided by freshwaters and of ecosystem service approaches are provided in Chapter 10.

> ### REFLECTIVE QUESTION
>
> Thinking of a local river or lake – how do you benefit personally from that water body and, more widely, how does society benefit from the ecosystem services it provides?

6.5 HUMAN MODIFICATION AND MANAGEMENT OF AQUATIC ECOSYSTEMS

6.5.1 Overview

Humans have severely altered aquatic ecosystems in a variety of ways. These include hydrological (e.g. water abstraction, dams), geomorphological (e.g. channelisation, culverts, dams, gravel extraction), water quality (e.g. acidification, organic pollution, nutrients, thermal pollution) and biological modifications (e.g. introduced species). Vörösmarty *et al.* (2010) conducted a global analysis of threats to aquatic biodiversity and concluded that 65% of freshwater habitats have moderate to high threats. Freshwater biodiversity is disproportionately at risk from habitat modification, because while freshwaters cover only 0.8% of the Earth's surface, they are home to an estimated >10% of known species (Mittermeier *et al.*, 2010; Dijkstra *et al.*, 2014). Additionally, aquatic ecosystems are rarely affected by individual modifications. In reality, multiple stressors can combine to exacerbate the threat to aquatic biodiversity (e.g. Ormerod *et al.*, 2010). For example, an urban stream may have been channelised to reduce flood risk, it can have a modified flow regime owing to runoff from impermeable surfaces and it may receive pollution from point sources associated with manufacturing plants or sewage treatment works. This section provides an overview of some of the most common human modifications of rivers and lakes and the biological consequences of such actions.

6.5.2 River flow regulation

Many river ecologists consider the natural river regime (see Chapter 3) as a 'master' environmental variable, owing to its influence on a suite of physical, chemical and biological processes. Key aspects of the natural flow regime are the *magnitude* of flow extremes and the *rate of change* between different flow states, plus the *frequency*, *duration* and *timing* of a given flow magnitude (see Poff *et al.*, 1997). High flows of a natural flow regime (e.g. due to spring snowmelt) can be important for cleaning river gravels prior to fish spawning, whereas lower flows are often necessary for juvenile fish to be reared (Acreman *et al.*, 2009). River systems and their constituent species have adapted to the natural flow regime over long timescales, and so alteration of flows by river regulation (e.g. dam construction, abstractions or diversions) can severely alter habitats by changing the thermal or sediment transport regime, impeding the migration of organisms and leading to large-scale changes in the structure and functioning of aquatic ecosystems (Lytle and Poff, 2004). Dams constructed for water storage constitute a major impact on river ecosystems globally (Figure 6.13), with an

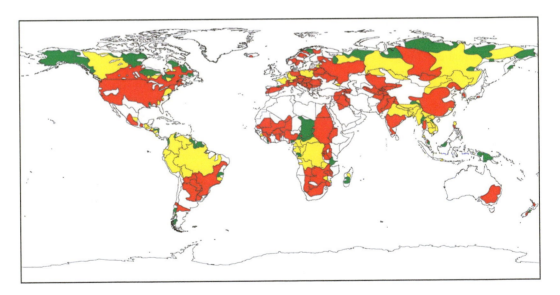

Figure 6.13 River channel fragmentation and flow regulation effects of dams on 292 of the world's major river systems. Green = unimpacted, yellow = moderately impacted and red = strongly impacted catchments. Uncoloured areas are not covered by large river systems but may also be strongly affected by river regulation.

Source: Modified with permission from American Association for the Advancement of Science, Nilsson, C., Reidy, C.A., Dynesius, M. and Revenga, C. Fragmentation and flow regulation of the world's large river systems. *Science* 308: 405–408, 2005. Permission conveyed through Copyright Clearance Center, Inc.

estimated 15% of river runoff held in large reservoirs (>15 m height; Nilsson *et al.*, 2005). In the USA, there are an estimated 2.5 million structures controlling river flow, and only 2% of rivers are considered to be unimpacted (Lytle and Poff, 2004). A further 10% of global runoff is extracted through abstractions (Vörösmarty and Sahagian, 2000). These human interventions in the water cycle are thought to have contributed to major negative impacts on aquatic biodiversity and some ecosystem services (Vörösmarty *et al.*, 2010) and they may even be responsible for alterations to marine ecosystems due to nutrients such as C, N, P and Si being retained in reservoirs rather than transported downstream to the ocean (Maavara *et al.*, 2017).

A detailed study by Poff and Zimmerman (2010) evaluated the links between river flow regulation and ecological response by considering 165 studies conducted worldwide across a range of different rivers. Negative responses of river ecosystems due to flow alterations were reported in 152 (92% of) studies. Dams intercept large quantities of sediment, and sections of river downstream of impoundments can be characterised by coarsening of bed sediments, due to the removal of fines, or alternatively the loss of peak flow events may see a progressive infilling of the interstices with finer materials (Stanford and Ward, 2001; Poff *et al.*, 1997). Similar changes can occur when water is abstracted from rivers, and the effects can be particularly impactful to fish, whose eggs develop in stream bed sediments, or invertebrate larvae, which live, migrate and feed in and on bed sediments. Fish populations in particular can be seriously impacted upon by alterations to flow, with low flows and flow regulation structures impeding migration, altered thermal regimes seriously affecting physiology, and loss of prey as a result of changed aquatic

and riparian food webs. Lowered flows can also reduce dissolved oxygen concentrations, due to reduced turbulence/re-aeration, and because of a lowered heat capacity (warmer waters can hold less dissolved oxygen). Changes to the flow regime can lead to major shifts in invertebrate and algal community composition, alterations to the cycling of nutrients and a depression of primary and secondary production (Ledger *et al.*, 2011). In contrast, there can be some 'positive' changes to ecosystems, including increases in non-native species cover, encroachment of plants onto river sediments that are no longer inundated (Poff and Zimmerman, 2010; see **Figure 6.14**, for example), increased abundance of small, short-lived invertebrates and enhanced growth of algae and bryophytes on stabilised riverbed sediments (Armitage, 2006).

Recognition of the negative effects of river regulation on aquatic and riparian ecosystems has resulted in the development of policy to manage flows and attempt to restore degraded ecosystems and maintain those that are least affected (Petts, 1996; Arthington *et al.*, 2006). The approach involves regulation of water abstraction to ensure that there is enough water to maintain aquatic ecosystems. In some heavily regulated systems, water is released from dams to create artificial floods in an attempt to restore some aspects of the natural flow regime. For example, in Switzerland, periodic water releases on the Spöl River are made during the course of the year to raise the continuous baseline flow from ~1.6 m^3 s^{-1} to >40 m^3 s^{-1} (**Figure 6.15**). Associated river ecosystem studies have shown that such interventions can cause shifts in the aquatic ecosystem, including the elimination of mosses from previously stabilised river sediments and the development of invertebrate communities that are more resilient to flood disturbances

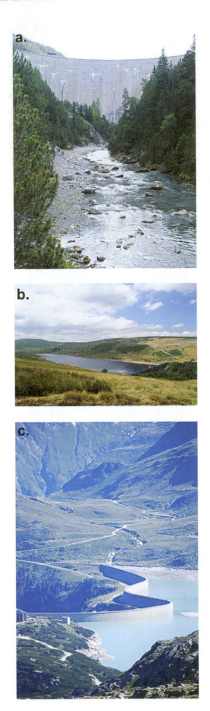

Figure 6.14 Examples of regulated rivers and reservoir infrastructure: a) Regulated river channel below Punt da Gall, Switzerland, b) Scar House reservoir, England, and c) Tauernmoossee, Austria.

Figure 6.15 The Spöl River, Switzerland, a) before, b) during and c) after an artificial flood generated by ~30 m³ s⁻¹ water release from the Ova Spin reservoir, September 2018.

(Robinson *et al.*, 2018). Other approaches being adopted in some developed regions of the world include the removal of smaller dams (e.g. Bednarek, 2001; Babbitt, 2002) so as to restore natural flows and allow free passage of migratory fish to upstream reaches. Short-term effects of dam removal include increases in fine sediment loads downstream, although some of these sediments may have associated pollutants, necessitating sediment remediation. Despite the known negative effects of regulation on river ecosystems, new large dams continue to be proposed and constructed each year in developing countries, including in Brazil on the Amazon, where aquatic ecosystems are particularly rich in biodiversity.

6.5.3 Land-use alteration: urbanisation

River catchments worldwide have been altered significantly as a consequence of land-use change (e.g. Harding *et al.*, 1998; Paul and Meyer, 2001; Allan, 2004), with major effects on water quality (see Chapter 4) and aquatic ecosystems. Some major examples of land-use change which impact upon aquatic ecosystems include afforestation and deforestation, the turning over of land to agriculture, and the development of towns and cities and their associated infrastructure. The *urban stream syndrome* is a term used to describe the ecological impact that an urbanised environment has upon the watercourses that receive their runoff (Meyer *et al.*, 2005). The symptoms include: a) hydrological change, with reduced baseflow and enhanced peak flows; b) modified thermal regimes linked to (a), in addition to changes in shading/exposure, point-source discharges and runoff from paved areas; c) altered water chemistry with enhanced nutrients and contaminants; d) transformed channel morphology (e.g. culverts, straightening of channels, reinforcement of banks) with reduced natural habitat (**Figure 6.16**); e) altered ecosystem functioning; and f) a reduction of biodiversity.

Changes to ecosystem functional processes following urbanisation can include increased gross primary productivity and community respiration when compared to forested streams. Increased algal biomass has been recorded in urban streams with a

Figure 6.16 Channelised urban streams showing: a) stone reinforcements of the bed and banks; b) sheet piling reinforcement of banks; and c) stone-filled gabion basket reinforcement of the left-hand bank and a brick wall forming the right-hand bank.

prevalence of filamentous algae as compared to forested streams, which have a high proportion of diatoms (Taylor *et al.*, 2004). Heavily urbanised locations have P:R ratios of much less than 1, suggesting that heterotrophic metabolism is dominant (Meyer *et al.*, 2005). Nutrient uptake is also reduced in urbanised streams. Studies on aquatic macrophytes in urban watercourses are rare but they do appear to be affected by the loss of habitat, increased turbidity, chemical contamination, increased velocities and coarser sediment profile that arise as a consequence of urbanisation.

Patterns of the impact of urbanisation on macroinvertebrate assemblages have shown consistent patterns worldwide. Robson *et al.* (2006) measured the degree of urbanisation of a catchment as the area covered by impermeable surface upstream of a series of monitoring sites. Chemical contamination increased with levels of impermeable surface cover, and that contamination was greatest during rainfall events. A reduction in macroinvertebrate diversity was also observed moving upstream to downstream. There was a decline in the numbers of taxa intolerant to disturbance, including Gammaridae (shrimp) and families from the orders of Ephemeroptera (mayflies), Plecoptera (stoneflies) and Trichoptera (caddis flies). In the degraded section of the river the macroinvertebrate assemblages were dominated by Asellidae (louse), Chironomidae (non-biting midges) and oligochaete worms. Walsh *et al.* (2005) demonstrated that even small levels of impermeable cover (<10% of a catchment area) can have overwhelming impacts on macroinvertebrates in streams. Fish communities are also impacted (Morgan and Cushman, 2005). Contaminants in both water and sediment can be directly toxic to fish. Channelisation and enhanced water velocity mean that there is a reduction in slow-flow zones and off-channel habitat for refuge (particularly important for juvenile fish) and spawning. Predator fish such as pike are compromised by the loss of tree and woody debris cover for concealment, and through indirect trophic links such as the decline in numbers of prey caused by chemical contamination.

6.5.4 Point-source pollution

Humans have a long history of discharging waste products from industrial processes and sewage treatment works through pipes or culverts into rivers, and such inputs are collectively termed 'point-source pollution' (see Chapter 4). A major source of point-source pollution is a sewage treatment works. Waste from these sources can be rich in organic effluent, which leads to severe increases in **biochemical oxygen demand** (see Chapters 4 and 8), and thus dissolved oxygen concentration decreases in river reaches downstream of the outlet pipe, due to enhanced respiration rates (Mason, 2002). The severe lack of oxygen that can ensue following organic enrichment typically leads to losses of most invertebrates and fish, with the exception of a few tolerant groups such as worms and some midge larvae. The effects of aquatic point-source pollution have even been shown to propagate into riparian (terrestrial/riverbank) ecosystems (Paetzold *et al.*, 2011). However, in the last two decades there have been major improvements in river water quality in many parts of the developed world, linked to increased investment in enhanced sewage treatment technologies and recovery of aquatic ecosystems.

Problems still remain in rivers that receive sewage works effluent because treatment processes focus predominantly on reducing organic pollution by facilitating enhanced respiration in the treatment plant (see Chapter 8). These treatment processes are largely unable to remove the active compounds contained in human pharmaceuticals, recreational drugs, and personal care products such as toothpaste, shower gels and shampoo. Research in the last couple of decades has, for example, found that male roach (*Rutilus rutilus*) have developed female characteristics in many UK rivers, due to steroidal oestrogens and chemicals that mimic oestrogens reaching the river in sewage effluent (Tyler and Jobling, 2008). Other studies show reduced swimming speeds in amphipods (De Castro-Catala *et al.*, 2017) and reduced reaction speeds of fish when avoiding predators in water polluted by antidepressant drugs (Painter *et al.*, 2009), changes to the breakdown of plant litter by microbes and fungi in the presence of antibiotics (Bundschuh *et al.*, 2009) and some alterations to animal physiology and immunology, rendering them more susceptible to parasites (Morley, 2009). The full ecological effects of the many hundreds of synthetic chemicals released into rivers remain to be studied in detail. Chapter 7 provides more detail on the human health effects of these modern chemicals.

6.5.5 Eutrophication

Eutrophication, the process by which waters become enriched by nutrients, in particular nitrogen and phosphorus (see Chapter 4), leads to major changes to aquatic ecosystem structure and function. As nutrient loadings increase, the nature of the phytoplankton communities begins to change from low-productivity types of diatoms and green algae to highly competitive, high-productivity types such as Cyanobacteria (also known as blue-green algae; Table 6.3). Blue-green algae have mechanisms that are able to provide them with a competitive advantage over other forms of phytoplankton. They are able to maintain optimum buoyancy for photosynthesis using internal gas vesicles which expand and contract to alter their density and position in the water column. They are able to store phosphorus in their cells as polyphosphates, and then utilise them when phosphate levels in the water become very low. In addition, in circumstances when phosphorus levels are very high and the Redfield Ratio is less than 16:1, then nitrogen availability becomes the limiting factor. Many types of blue-green algae are able to take advantage of these conditions and utilise nitrogen by fixation of N_2. Eutrophic conditions are characterised by frequent and intense algal blooms, high productivity, high turbidity and low visibility, and are dominated by species of blue-green algae.

Changes to the phytoplankton community also have effects on the composition of the zooplankton species which consume them. The mechanisms involved could include one or a combination of the following: the ability of the zooplankton species to graze on the algae; nutritional deficiencies of the phytoplankton; production of toxins by phytoplankton; and external factors, including

Table 6.3 Examples of algal types associated with trophic status

Productivity	Main algal groups	Examples of typical algal genera
Oligotrophic	Diatoms, desmids	*Tabellaria, Cyclotella, Staurastrum*
Mesotrophic	Dinoflagellates, diatoms	*Ceratium, Peridinium, Asterionella, Cyclotella*
Eutrophic	Cyanobacteria, diatoms	*Microcystis, Anabaena, Asterionella, Fragilaria*

other anthropogenic influences. The invertebrate community alters from one of high diversity and low biomass to one of low diversity and high biomass. In oligotrophic conditions, invertebrate species inhabit the macrophytes in lake shores, which provide refuge from predating fish. However, in eutrophic conditions, these are lost and the community becomes dominated by decomposers such as tubificids and chironomids in the benthic sediment.

Changes in fish communities have also been noted with the onset of eutrophication, due to a number of factors. The ability to prey upon different forms of zooplankton varies according to the foraging behaviour and feeding adaptations of each individual fish species. Consequently, changes in zooplankton species as prey will cause changes in fish species as predators. Furthermore, increased turbidity and loss of visibility, in combination with changes in the invertebrate community, mean that there is a shift in predation from species using visual means to seek prey to those which forage among the benthos using olfactory means. The pattern of change among fish is from coregonids and salmonids in oligotrophic conditions to cyprinids such as bream in eutrophic conditions.

A series of wider issues are problematic in eutrophic aquatic systems. The high biomass of phytoplankton may become a problem during hours of darkness, when respiration is far in excess of photosynthesis. Severe oxygen depletion has resulted in fish kills around the world. Eutrophication gives rise to high production in the epilimnion in lakes with thermal stratification. This exacerbates the depletion of oxygen in the hyperlimnion, when the dead phytoplankton sink into the hypolimnion to decompose. Algal blooms have also been associated with the production of harmful toxins because the blue-green algae such as *Mycrocystis* and *Anabaena* produce hepatotoxins and neurotoxins. Algal blooms can reduce the effectiveness of water treatment processes for supply by causing coagulation, sedimentation and filter blocking. They can also reduce the amenity and recreational value of freshwaters (**Figure 6.17**).

6.5.6 Climate change (environmental warming and altered water availability)

The effects of climate change on freshwaters can be thought of as a complex amalgam of stressors, including acidification resulting from elevated atmospheric sulfate and nitrate concentrations, elevated dissolved CO_2 and the more obvious changes in water availability (see Chapter 2) and water temperature (Woodward *et al.*, 2010). Freshwater ecosystems are likely to be highly vulnerable to climate change because they are relatively isolated and physically fragmented within a largely terrestrial landscape, and dispersal of organisms to other systems may be difficult, owing to a lack of hydrological connectivity and/or the poor dispersal abilities of many aquatic organisms. Freshwaters are also already heavily

Figure 6.17 Dense algal growth which has turned lake edge waves green.
Source: Photo courtesy of Shutterstock/smspsy.

exploited by humans for the provision of 'goods and services' and they may be less resistant/resilient to change than might be the case if they were entirely pristine. A comprehensive discussion of how climate change can be expected to alter aquatic ecosystems could take up an entire book; therefore, this section refers only to selected studies to illustrate some possible responses. Those wishing to explore the effects of climate change on aquatic ecosystems in more detail can find good starting points in comprehensive review papers such as Heino *et al.* (2008), Palmer *et al.* (2009), Whitehead *et al.* (2009) and Woodward *et al.* (2010).

A major driver of aquatic ecosystem response to climate change will be alterations to river and lake thermal regimes. Evidence that aquatic ecosystems are warming is growing, with studies to date indicating significant warming of streams in the USA (e.g. Kaushal *et al.*, 2010) and central Europe (Hari *et al.*, 2006; Webb and Nobilis, 2007) and of various lakes around the world (Schneider and Hook, 2010). These thermal changes are expected to have strong effects on aquatic plants and animals because the metabolism and foraging behaviour of individual organisms are strongly influenced by temperature, and in freshwaters many organisms are ectothermic (i.e. unable to regulate their body temperature and therefore reliant on environmental temperature; see Woodward *et al.*, 2010). For example, fish are particularly sensitive to changes in environmental temperature and some studies have forecast that up to 75% of local fish biodiversity could be lost by 2070, due to combined changes in flow and thermal regimes linked to climate change (Xenopoulos *et al.*, 2005). O'Reilly *et al.* (2003) observed 0.1°C warming per decade since 1913 in the upper water column of Lake Tanganyika, East Africa. This warming was associated with decreases in the primary productivity of the lake, and consequent declines in catches of fish. The latter organisms are a vital source of protein to humans living in countries surrounding the lake, and so further warming and ecosystem change could have major consequences for food supply in this region.

Recent studies in experimental ponds have also illustrated how climate change might shift the metabolic balance of some freshwaters from net autotrophic (i.e. carbon sinks) to net heterotrophic (i.e. carbon sources) because, under warming scenarios, ecosystem respiration increases at a faster rate than primary production (Yvon-Durocher *et al.*, 2010). In contrast, a study of >200 rivers globally suggested that photosynthesis and respiration respond to warming in a similar manner (Demars *et al.*, 2016), while warming experiments in stream mesocosms (small experimental channels) have suggested there could be significant unexpected increases in primary production due to changes in nutrient use efficiency as well as changes in algal species composition (Hood *et al.*, 2018). In rivers, warming may lead to changes in the processing of accumulated terrestrial plant litter. This is because, as temperature increases, microbial breakdown of leaf litter increases but consumption by detritivores decreases (Boyero *et al.*, 2011). Results from these and other studies (e.g. Box 6.3) suggest that we still have a lot to learn about how aquatic ecosystems respond to changes in their thermal regime.

In north Wales, UK, some of the longest continuous records of stream habitat data and aquatic invertebrates have been collected since 1981. By 2005, streams from open moorland had seen average temperature increases of 1.7°C, while forest streams increased by up to 1.4°C (Durance and Ormerod, 2007). The effects of this warming appear to be strongest in streams with circumneutral pH, with

BOX 6.3

Using geothermal heating to study the effects of climate change on river ecosystems

Laboratory experiments can be criticised for being unrealistic, but observing changes in whole rivers or lakes due to climate warming requires observations to be made over many years. Geothermally influenced river systems provide natural experimental settings where natural temperature gradients that may have been in existence for long periods of time allow us to examine the effects of warming on whole streams in their natural environment (O'Gorman *et al.*, 2015).

A series of geothermally influenced rivers in the Hengill catchment, south-west Iceland, have been studied by teams of international researchers over the last 15 years to increase understanding of how temperature changes influence aquatic biodiversity and functional processes. Friberg *et al.* (2009) found similar numbers of macroinvertebrate species in rivers ranging from 6–23°C but there were clear differences in the composition of the assemblages. Filter feeder density increased with temperature but scrapers remained unchanged, perhaps linked to more rapid breakdown of leaf litter into fine particulates in the warmer streams. Further studies by Woodward *et al.* (2010) across an extended temperature range (5–45°C) showed changes in the size of top predators, longer food chains but also decreases in the density of some species at the highest temperatures. Despite low nutrient concentrations, whole river metabolism increased strongly with temperature due to faster nutrient spiralling rates (Demars *et al.*, 2011). These natural laboratories could offer unprecedented insights into the effects of temperature on aquatic ecosystems, allowing us to make predictions about the likely impacts of future climate change around the world.

reductions in the abundance of invertebrates and changes in the composition of communities observed. The authors hypothesised that for each future temperature increase of 1°C there could be a reduction of spring invertebrate abundance by 21%. Similar long-term studies of invertebrates in rivers of south-east Alaska have documented significant changes with widespread glacial retreat and subsequent stream ecosystem development since the late 1970s (Milner *et al.*, 2008, 2011). More broadly, glacier retreat is occurring across much of the world and is strongly linked to a warming climate. Further retreat is expected to lead to major shifts in the water sourcing of Arctic and alpine rivers, with glacier- and snowmelt reductions and changes in river hydrological, geomorphological and

water quality dynamics. These changes could have significant, widespread consequences for plants and animals of river ecosystems, leading to changes in species diversity (potentially including some species extinctions) and abundance and the structure of food webs (Milner *et al.*, 2017).

6.5.7 Invasive species

Invasive alien species are those that have been introduced into a new environment with serious negative consequences. Aquatic ecosystems are particularly at risk of invasions by plants, invertebrates and vertebrates. Invasions are a major economic cost, they threaten ecosystem services and they rank second only to habitat destruction as drivers of biodiversity loss. Species that are deliberately introduced to a new habitat for aquaculture (see Section 6.5.8) often escape and become invaders in the new habitat. Furthermore, accidental transportation and introduction as a result of trade and shipping, particularly as a result of ballast water exchange, is a major source of aquatic species introductions.

Invasive species can change the physical properties of the habitat (ecosystem 'engineering'; Jones *et al.*, 1994), as well as competing with, becoming prey for, preying on or parasitising native species. They can also drive changes in the productivity and biodiversity of aquatic ecosystems. Successive invasions have transformed freshwater communities; for example, the Rhine running through Central and Western Europe has suffered continual introductions of new species as a result of shipping traffic as well as the building of canals which have increased connections with other rivers in Europe. As a result, by 2006, invasive species were estimated to make up 90% of macroinvertebrate numbers (van Riel *et al.*, 2006).

Invasive species come from all trophic groups and can affect native ecosystems through trophic interactions and through ecosystem engineering, with these effects potentially cascading through the ecosystem (Figure 6.18). For example, water hyacinth (*Eichhornia crassipes*) has been described as the world's worst aquatic weed. It is an ecosystem engineer that affects the habitat structure of lakes and rivers, with knock-on effects on other species and great economic costs (Lowe *et al.*, 2000). It originated in South America and has been introduced to new habitats as an ornamental plant. Once introduced to tropical or subtropical habitats, it can spread rapidly and populations can double in size in only 12 days. The water hyacinth impedes water flow, navigation and fishing and reduces biodiversity, as it deprives native plants of space, sunlight and

oxygen. For example, the spread of this weed in Lake Victoria in Kenya has led to closures of a hydroelectric plant as well as creating habitats for mosquitoes and snails, vectors for the human diseases malaria and schistosomiasis. Invasive animals can also act as ecosystem engineers. The Chinese mitten crab (*Eriocheir sinensis*) damages banks (including flood defences) by burrowing into sediments, creating a network of tunnels that can eventually cause bank collapse. Banks along the River Thames in south-east UK have receded up to 6 m as a result of mitten crab burrowing. The native range of this crab covers the Far East, but it was reported in Germany in 1912 and has spread through northern Europe (predominantly in ballast water) over the last 100 years and across North America (Lowe *et al.*, 2000). This invader exerts a stronger predatory impact on native invertebrate prey (Rosewarne *et al.*, 2016) and also shreds more detritus than does the native decapod crayfish *Austropotamobious pallipes* (Doherty-Bone *et al.*, 2018).

For other invaders, their trophic interactions with other species may affect ecosystem structure and functioning. The amphipod *Dikerogammarus villosus* (also known as 'the killer shrimp') invaded Western Europe via the Danube–Main–Rhine canal, which opened in 1992. The rate of downstream movement has been estimated at 124 km per year, as a result of drift and transport on boats (van Riel *et al.*, 2006). It was first recorded in the UK in 2010. In some habitats, there is evidence that the beds of the invasive zebra mussel, *Dreissena polymorpha*, facilitate invasion by the killer shrimp by providing habitat and food; such mutual facilitation by invaders is termed 'invasional meltdown'. The main impact of *D. villosus* stems from its predatory behaviour, which affects the abundance of a range of invertebrates and also has cascading effects through the ecosystem. Stable isotope

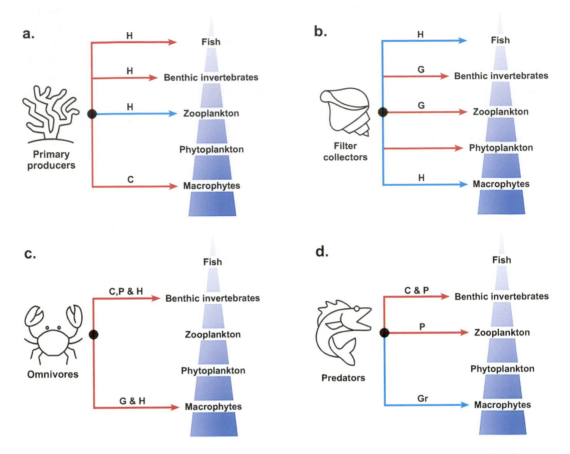

Figure 6.18 A framework summarising the positive (red) and negative (blue) links between invasive species at different trophic positions. Impacts result from direct ecological (C = competition, P = predation, G = grazing, Gr = grazer release) and indirect physicochemical impacts (H = habitat alteration) of invasive species.

Source: After Gallardo, B., Clavero, M., Sánchez, M.I., Vilà, M. 2016. Global ecological impacts of invasive species in aquatic ecosystems. *Global Change Biology* 22: 151–163. © John Wiley and Sons.

analysis revealed the killer shrimp to be on the same trophic level as predatory fish (van Riel *et al.*, 2006). It has been shown to eat more prey in the laboratory than does the native amphipod *G. pulex*, and in the field it may displace native amphipods both by outcompeting them and by preying on them (Dodd *et al.*, 2014). Native amphipods are weaker predators, but process more detritus by shredding than does the killer shrimp. Hence, displacement of native amphipods

by this invader could have knock-on effects on ecosystem functioning within invaded ecosystems (Kenna *et al.*, 2017).

Parasites (or a loss of parasites) can also affect invasions (Hatcher *et al.*, 2019). Animal and plant population growth is often regulated by parasites which cause increased mortality. However, invading species may lose their parasites either because, by chance, only uninfected hosts are introduced to the new habitat or because parasitised (sick) individuals are more likely to die during translocation to, and during establishment in, the new range. This process is called 'enemy release' and may be one factor that explains the success of an invader and its ability to outcompete native species. For example, in Northern Ireland, UK, the native freshwater amphipod crustacean competes with three invasive amphipods: a European amphipod, *Gammarus pulex*, which is gradually replacing the native species, and two other species that arrived from North America via ballast water and the transport of ornamental pond plants. In Northern Ireland the native species harbours five different species of parasite which reduce its growth and survival, while the invaders are affected by only three (Dunn and Dick, 1998). Invasive species can also facilitate the spread of parasites. For example, the invasive signal crayfish has spread through large areas of Europe, driving native crayfish extinct as a result of competition and disease transmission. Signal crayfish carry the fungus *Aphanomyces astaci* but are not killed by it; they act as reservoirs of infection, transmitting the parasite to endangered native crayfish (Roy *et al.*, 2016). Similarly, the amphibian fungal disease Chytridiomycosis is spread by the invasive American bullfrog, driving a global decline in the diversity and abundance of frogs and toads (Roy *et al.*, 2016).

Once an invasive species becomes established, control can be extremely costly and eradication rarely successful. Legislation (e.g. EU invasive alien species legislation EU1143, the IMO Ballast Water Management Convention) highlights *prevention* of the introduction and spread of invasive species. Human activity is the main driver of invasive species spread. Propagules (small fragments of invasive plants, plant seeds or animals that can go on to form a new population) can be accidentally moved to new habitats on equipment or clothing used in trade, transport, recreation and education. Biosecurity refers to practices that reduce the risk that our activities spread invasive propagules (Dunn and Hatcher, 2015). The biosecurity campaigns 'Check Clean Dry' in the UK and 'Check Drain Dry' in the US encourage good biosecurity practice among water users. Activities undertaken by students (from a field course in the local environment to eco-tourism and recreation) may also pose a risk of acting as accidental vectors of invasive species. It is important that students embed good biosecurity practices into their field work and carry this knowledge with them into their careers (see Further reading).

6.5.8 Aquaculture

Aquaculture is typically thought of as the production of fish under captive and controlled conditions. However, it also includes the production of aquatic plants, and animals such as crustaceans and molluscs, under similar circumstances. Over 80% of global aquaculture is in Asia, with >50% occurring in China. The scale and intensity of aquaculture-based primary and secondary production may vary enormously, from low-intensity extensive ponds to highly controlled production on an industrial scale. In some situations the temperature, light and feeding regimes are completely optimised to ensure efficient production. Production through aquaculture methods has increased more than sixfold since 1990 (Table 6.4). In comparative terms, the increase in levels of production by capture methods are much lower. This could be the result of overfishing, leading to the depletion of some wild stocks and the recognition that production under more controlled conditions is more reliable in terms of satisfying demand.

Table 6.4 Comparison of global aquaculture and capture production (million tonnes) between 1990 and 2016 (excluding aquatic plants and mammals)

Production method	1990	2016
Freshwater aquaculture	7.6	51.4
Marine aquaculture	5.5	28.7
Total aquaculture	13.1	80.1
Freshwater capture	5.3	11.6
Marine capture	80.1	79.3
Total capture	85.4	90.9

Source: United Nations Food and Agriculture Organization.

The environmental impacts of aquaculture can be complex and far-reaching. Aquaculture requires large volumes of water, and abstraction of water from the environment can lead to significant changes in hydrological regimes and so disturb ecosystems. The wastewater from aquaculture operations will have high levels of ammonia and is also contaminated with excess feed and faecal material. This can lead to high levels of suspended sediments, elevation of biochemical oxygen demand and reduced levels of oxygen concentration. Camargo (1992) observed the impacts of a trout farm's discharge on the macroinvertebrate community of the stream receiving the wastewater. In the area immediately upstream and downstream it was found that species richness was reduced from 38 to 19 and the BMWP score (a measure of water quality; Box 6.4) was reduced from 168 to 19. The dense numbers of fish in same-age monocultures are also vulnerable to disease and parasite infestation. This may be exacerbated where populations have been manipulated genetically in order to improve productivity, but at the expense of the ability to resist diseases.

It is common for the feed used in aquaculture to be produced artificially, with high levels of nitrogen and phosphorus to improve the potential productivity. Enrichment of freshwaters with these nutrients in areas around rearing pens in lakes or in rivers receiving farm effluent can lead to issues of eutrophication if they are not controlled. The resultant algal blooms may produce toxins and, as they die and decompose, they can cause oxygen depletion. In addition, antibiotics and other therapeutic components are frequently added to artificial feed pellets. The toxic impact of such chemicals released from uneaten pellets as they decompose is still to be properly assessed and understood in freshwater environments.

A much more significant issue that arises from aquaculture is the ecological and genetic impact on the environment of fast-growing alien species and genotypes that escape from captivity (Naylor *et al.*, 2001). The minimum impact that escaped organisms may have is to disturb the balance of the natural ecosystem through interactions with the communities that exist in their vicinity. This can be propagated on a longer term, perhaps even permanently, if the escapees are able to establish a sustainable breeding population. This has been a common problem associated with species of *Tilapia* which have originated in Africa but are now widespread through the

Biomonitoring and the European Water Framework Directive

The Water Framework Directive (WFD) is a piece of legislation developed by the European Union which uses ecosystem-based objectives to manage water resources (Kallis and Butler, 2001). It is intended to provide a common basis for the protection and sustainable use of water, in terms of quantity and quality, across the member states of the EU. The approach aims to rank the ecological status of rivers and lakes onto an ordinal scale of high, good, moderate, poor and bad. These status classifications are determined on the basis of observed versus expected values of biological groups, including diatoms, macrophytes, macroinvertebrates and fish. The classification system also considers hydrogeomorphological and water chemistry parameters. The lowest classification across the biological, hydrogeomorphological and water chemistry parameters determines the 'ecological' status of a water body. Expected values are determined from databases of high-quality reference sites. To ensure comparability between test and reference sites, a typology based on catchment geology, catchment size and altitude is used (Logan and Furse, 2002).

There are a number of different methods that have been developed to assess ecological status using various organisms and across a variety of aquatic habitats. One system that was developed and used widely prior to the introduction of the WFD focuses on macroinvertebrate benthic fauna and utilises the Biological Monitoring Working Party (BMWP) scoring system (Armitage *et al.*, 1983) to assess water quality. The BMWP has subsequently informed approaches such as the River Invertebrate Prediction and Classification System (RIVPACS; Wright *et al.*, 2000), which can be used to derive the type-specific 'expected' values that feed into the determination of ecological status classification. Similar approaches are used in Australia (Australian River Assessment System – AUSRIVAS). In the USA, individual states have different monitoring approaches, but some have adopted RIVPACS-type approaches.

In the UK, the BMWP system has recently been updated by Whalley–Hawkes–Paisley–Trigg (WHPT; Paisley *et al.*, 2013). The WHPT score for a location can be determined by taking a standard 3-minute kick sample from the riverbed and identifying all the families of macroinvertebrates that are present. The families of macroinvertebrates are then given scores in relation to their tolerance to pollution (e.g. a higher score is given to organisms that are least tolerant to pollution, predominantly stoneflies, mayflies and caddis flies). The score also takes into account the abundance of each family. A total WHPT score is then derived for the location from the sum of the scores of each family in the sample. An unpolluted location may have a WHPT score in excess of 100. Additional measures that can be calculated from the sample are the number of scoring taxa in the sample (N taxa) and the average score per taxon (ASPT). The WHPT-ASPT is calculated as the total WHPT/N taxa. The ASPT has long been used in biomonitoring applications as it is considered to be more robust to variations in sample size, sampling effort and seasonal fluctuations (Armitage *et al.*, 1983).

One of the main aims of the WFD is to prevent the deterioration of water bodies that are currently at good or better status, and to improve those that fall below this level. Member states were expected to consider ways to improve failing water bodies up to a minimum of good status by 2015, in the first cycle. For the second cycle (2016–2021), evaluations are ongoing to inform river basin management planning. Detailed statistics are available via the Water Information System for Europe

(WISE; www.eea.europa.eu/data-and-maps/dashboards/wise-wfd). For example, in November 2018 these showed that only 27% of UK rivers were classed as having 'good' or 'high' status. Romania reported the highest percentage (63%) of rivers in 'good' or 'high' status, while the Netherlands reported only 0.3% in these categories. The WFD does allow Member States to exempt some water bodies from the classification. Examples include surface waters where necessary improvements cannot reasonably be achieved over a few years, those where improvements in status are unfeasible or unreasonably expensive, if unforeseen or exceptional circumstances occur, or where new modifications are necessary to a water body (e.g. flood control).

continents around the tropics (Pullen et al., 1997; De Silva et al., 2004). Where aquaculture is used to increase the production of native species, then when individuals escape from the aquaculture system into the wild there can be large problems. The genetics of a cultured population of a species often exhibit a high level of inbreeding and abnormally high growth rates, and this can be at the expense of other important genetic traits. Consequently, these individuals compete aggressively with indigenous populations for food and habitat. Moreover, cross-breeding between captive and wild populations can lead to decreased fitness of the indigenous population because of the impact of the alien genotypes on the local gene pool (Ferguson et al., 2007).

The typical aquaculture techniques described above usually involve isolated areas of freshwater (e.g. ponds, lakes) or coastal ecosystems (using nets or cages) to culture organisms. Hydroponics, on the other hand, seeks to grow plants in an environment that lacks soil by suspending the roots in a solution that provides essential nutrients and minerals (**Figure 6.19a**). Thus, the traditional definitions of what constitutes a terrestrial versus an aquatic ecosystem become debatable. Hydroponics is often carried out indoors with artificial lighting or in greenhouses, leading to increased energy costs. However, the approach offers a considerable opportunity for high yield crop production where energy is plentiful and water

is scarce, such as might be the case in arid environments. By way of example, lettuce yields were 11 times greater in hydroponic compared to conventional growing, and this required 8% of the water but 8200% more energy (Lages Barbosa et al., 2015). A major limiting factor is that the majority of crop plants that might be grown have varieties that require specific conditions within which to maximise yield (Jensen, 1999). Hydroponics as a field is still at a relatively early stage of research, although considerable innovations are occurring at small scales (Chenin and Omaye, 2016).

A recent growth area in the field of aquaculture has integrated cultured organisms typical of traditional aquaculture with hydroponic systems to create small-scale ecosystems that are, to a degree, self-sustaining (e.g. **Figure 6.19b**). This practice of 'aquaponics' (a combination of 'aquaculture' and 'hydroponics') often involves the addition of nitrifying bacteria that convert toxic excretions, such as ammonia, into compounds that can be absorbed as nutrients by plants, such as nitrates. Nutrients are therefore recycled into plant biomass, while water is cleaned rather than disposed of, thereby reducing water use. While the practice has evolved over 2000 years, beginning with the culturing of fish in rice paddy fields in China, the scale of aquaponics facilities has remained relatively small and the methods have been heterogeneous (Love et al., 2014). Recently, innovations in the practice of aquaponics have led to a

Figure 6.19 Hydroponic and aquaponic systems: a) Lettuce crops growing in nutrient-rich water with no soil substrate (hydroponics); b) a schematic of an aquaponics system where fish and plant cultivation are connected. Fish waste provides nutrients for the crop plants. Both fish and crops are harvested from the system.
Source: a) Southern SARE; b) Shutterstock/Scharfsinn.

standardisation of the definition to include only those practices where 50% or more of the nutrients provided to the plants are derived from animal waste, and where the following components are used as a minimum: tanks for the rearing of animals (usually fish, but molluscs and crustaceans are also used), a mechanical filter for large pieces of detritus, a biological filter for chemical waste, plant cultivation to extract nutrients, and a sump (Palm *et al.*, 2018). Aquaponics come with several important challenges. The first is that there is still a lack of clear best practice for maximising yield from both plant and aquatic components of the system. Recent research has demonstrated a potential trade-off between species (Knaus and Palm, 2017). Pest control is also complicated by the coupled aquatic-hydroponic system, where chemical pesticides cannot be used on the plant component of the system owing to pollution of the rearing tank water. There are also issues with the welfare of animals (particularly fish) that are deliberately exposed to high levels of pollutants in the holding tank (Yavuzcan Yildiz *et al.*, 2017). However, as with hydroponics, there is considerable interest in the potential of aquaponics for high-intensity urban food production through the creation of green walls and towers (Palm *et al.*, 2018).

REFLECTIVE QUESTIONS

In what ways is it likely that humans have had an impact on the aquatic ecosystem of your nearest lake or river?

What do you think the key impacts of warming and more extreme conditions (more droughts and more intense rainfall events) would be on the aquatic ecosystem in the river closest to your home?

Why is it important to adopt biosecurity measures for aquatic ecosystems?

6.6 SUMMARY

This chapter has provided an introduction to some of the diverse habitats and biotic groups that can be found in rivers and lakes, and the differences between these two types of freshwater system. The trophic roles of various organisms have been considered, and the chapter has explored how freshwater ecosystem structure can be considered holistically through the study of food webs. The main functional processes that are studied in freshwater systems, namely primary production, secondary production, community respiration and nutrient cycling, have been explored. Some of the key methods used by aquatic scientists to study these processes have been outlined, and a consideration of the ecosystem services that these processes provide to humans has been introduced. The later sections of the chapter have provided overviews of some of the main ways in which human activities have led to the modification of aquatic ecosystems. It is important to remember that rivers and lakes are particularly diverse and well-studied ecosystems. The ideas discussed herein for the most part provide only introductions to general frameworks or ideas, rather than in-depth blueprints for how individual river or lake systems can be expected to work. The reader is encouraged to consult the recommended further reading and key research papers/references to build up a more detailed understanding and knowledge of freshwater ecosystems.

FURTHER READING

Allan, D.J. and Castillo, M.M. 2007. *Stream ecology: structure and function of running waters.* Springer; Dordrecht.

Provides a detailed review of most topics relevant to the study of stream ecology, including coverage of stream hydrology, water chemistry and human modifications of river systems.

Closs, G., Downes, B. and Boulton, A. 2004. *Freshwater ecology: a scientific introduction.* Blackwell Publishing; Oxford.

An introductory text to both rivers and lakes, aimed at those who have not previously studied freshwater ecology, including discussion of scientific methodology, key concepts and case studies. Each chapter is based around a key research question in aquatic ecology.

Hauer, F.R. and Lamberti, G. 2006. *Methods in stream ecology.* Elsevier; New York.

This book provides an overview of the scientific rationale and applications of most common methods used by stream ecologists, as well as more technical information on how to process data.

Mason, C.F. 2002. *Biology of freshwater pollution.* Pearson; London.

Discusses the multiple alterations of freshwater ecosystems such as through pollution, habitat modification and the introduction of invasive species.

Moss, B. 2010. *Ecology of freshwaters: a view for the twenty-first century.* Wiley-Blackwell; Chichester.

Mainly discusses lakes and wetlands but has some chapters on rivers. The book provides a detailed overview and is aimed at students who wish to gain an integrated view of freshwaters.

Dunn, A. and Shannon, C. 2019. *Better biosecurity.* https://openeducation.blackboard.com/mooc-catalog/courseDetails/view?course_id=_ 1189_1

A short (1–2 hour) e-learning module which provides information and guidance on invasive species and on biosecurity to slow their spread.

Classic papers

Resh, V.H., Brown, A.V., Covich, A.P., Gurtz, M.E., Li, H.W., Minshall, G.W., Reice, S.R., Sheldon, A.L. and Wallace, J.B. 1988. The role of disturbance in stream ecology. *Journal of the North American Benthological Society* 7: 433–455.

An examination of how disturbance impacts on aquatic ecosystems.

Scheffer, M., Hosper, S.H., Meijer, M.L., Moss, B. and Jeppesen, E. 1993. Alternative equilibria in shallow lakes. *Trends in Ecology and Evolution* 8: 275–279.

Very important theory about non-linearity in lake systems which makes restoration of damaged lakes more difficult.

Schindler, D.W. 1974. Eutrophication and recovery in experimental lakes: implications for lake management. *Science* 184: 897–899.

An early but important paper on eutrophication and how to get lakes out of a eutrophic state.

Vannote, R.L., Minshall, G.W., Cummins, K.W., Sedell, J.R. and Cushing, C.E. 1980. The river continuum concept. *Canadian Journal of Fisheries and Aquatic Sciences* 37: 130–137.

Very important paper of relevance to the introductory material within this chapter.

PROJECT IDEAS

- Using the kick net sampling method, collect and identify samples of macroinvertebrates at several locations along your local river. At each site, measure variables such as dissolved oxygen concentration, pH, water temperature and electrical conductivity using handheld probes, and collect a sample of water for analysis of nutrients and metals in the laboratory. Hydrological and geomorphological measurements (e.g. width, depth, discharge) can also be measured. When analysing your datasets you could consider how the macroinvertebrates vary along the river, and how they are related to changes in hydrology, geomorphology and/or water quality.

- Contact your local environmental monitoring agency and request hydrological, water quality and/or ecological datasets for your local river/lake if they are available. Using such datasets you may be able to analyse whether there have been any notable changes over time in relation to changes in catchment land use or year-to-year weather conditions. Alternatively, you could compare your local river/lake to others in your region/country to assess what are the reasons for spatial differences.

- Collect freshly fallen leaves from one or more species of tree growing adjacent to your local river and then dry them out for a few weeks. You can then construct litter bags using known weights of leaf litter (a few grams per bag will suffice). Litter bags can be constructed from fine mesh to exclude invertebrates and gain a measure of fungal/microbial decomposition processes, and/or a coarse mesh which allows invertebrates to colonise, giving a measure of total breakdown. Litter bags can then be anchored to the bed of your local river for several weeks, during which time they will begin to decompose. You could leave all the bags in place for a fixed period of time, or remove some at various intervals to assess how litter decomposes over time. After retrieval of the litter bags from the river, return to the laboratory and carefully rinse off any sediment and attached invertebrates. Dry the remaining leaf litter, reweigh the material and subtract the value from the litter weight prior to placing them in the river, to assess how much decomposition has taken place.

REFERENCES

Acreman, M.C., Aldrick, J., Binnie, C. *et al.* 2009. Environmental flows from dams: the Water Framework Directive. *Engineering Sustainability* 162: 13–22.

Allan, J.D. 2004. Landscapes and riverscapes: the influence of land use on stream ecosystems. *Annual Review of Ecology, Evolution and Systematics* 35: 257–284.

Allan, J.D. and Castillo, M.M. 2007. *Stream ecology: structure and function of running waters.* Springer; Dordrecht.

Amundsen, P.A., Lafferty, K.D., Knudsen, R. *et al.* 2009. Food web topology and parasites in the pelagic zone of a subarctic lake. *Journal of Animal Ecology* 78: 563–572.

Armitage, P.D. 2006. Long-term faunal changes in a regulated and an unregulated stream – Cow Green thirty years on. *River Research and Applications* 22: 947–966.

Armitage, P.D., Moss, D., Wright, J.F. and Furse, M.T. 1983. The performance of a new biological water quality score system based on macroinvertebrates over a wide range of unpolluted running-water sites. *Water Research* 17: 333–347.

Arthington, A.H., Bunn, S.E., Poff, N.L. *et al.* 2006. The challenge of providing environmental flow rules to sustain river ecosystems. *Ecological Applications* 16: 1311–1318.

Babbitt, B. 2002. What goes up, may come down. *BioScience* 52: 656–658.

Barbosa, G., Gadelha, F., Kublik, N. *et al.* 2015. Comparison of land, water, and energy requirements of lettuce grown using hydroponic vs. conventional agricultural methods. *International Journal of Environmental Research and Public Health* 12: 6879.

Baxter, C.V., Fausch, K.D. and Saunders, W.C. 2005. Tangled webs: reciprocal flows of invertebrate prey link streams and riparian zones. *Freshwater Biology* 50: 201–220.

Bednarek, A. 2001. Undamming rivers: a review of the ecological effects of dam removal. *Environmental Management* 27: 803–814.

Benfield, E.F. 2006. Leaf breakdown in stream ecosystems. *In*: Hauer, F.R. and Lamberti, G.A. (eds) *Methods in stream ecology.* Elsevier; New York, 579–590.

Benke, A.C. 1984. Secondary production of aquatic insects. *In*: Resh, V.H. and Rosenberg, D.M. (eds) *Ecology of aquatic insects.* Praeger; New York, 289–322.

Benke, A.C. 2010. Secondary production. *Nature Education Knowledge* 3: 23.

Benke, A.C. and Wallace, J.B. 1997. Trophic basis of production among riverine caddisflies: implications for food web analysis. *Ecology* 78: 1132–1145.

Bjelke, U., Boberg, J., Oliva, J. *et al.* 2016. Dieback of riparian alder caused by the *Phytophthora alni* complex: projected consequences for stream ecosystems. *Freshwater Biology* 61: 656–579.

Blanchette, M.L., Davis, A.M., Jardine, T.D. and Pearson, R.G. 2016. Omnivory and opportunism characterize food webs in a large dry-tropics river system. *Freshwater Science* 33: 142–158.

Bott, T.L. 2006. Primary productivity and community respiration. *In*: Hauer, F.R. and Lamberti, G.A. (eds) *Methods in stream ecology.* Elsevier; New York, 263–290.

Boyero, L., Pearson, R.G., Gessner, M.O. *et al.* 2011. A global experiment suggests climate warming will not accelerate litter decomposition in streams but might reduce carbon sequestration. *Ecology Letters* 14: 289–294.

Brown, L.E., Edwards, F.K., Milner, A.M. *et al.* 2011. Food web complexity and allometric-scaling relationships in stream mesocosms: implications for experimentation. *Journal of Animal Ecology* 80: 884–895.

Bundschuh, M., Hahn, T., Gessner, M.O. *et al.* 2009. Antibiotics as a chemical stressor affecting an aquatic decomposer-detritivore system. *Environmental Toxicology and Chemistry* 28: 197–203.

Camargo, J.A. 1992. Temporal and spatial variations in dominance, diversity and biotic indices along a limestone stream receiving a trout farm effluent. *Water, Air and Soil Pollution* 63: 343–359.

Clitherow, L.R., Carrivick, J.L. and Brown, L.E. 2016. Food web structure in a harsh glacier-fed river. *PLoS ONE* 8: e60899.

Closs, G., Downes, B. and Boulton, A. 2004. *Freshwater ecology: a scientific introduction*. Blackwell Publishing; Oxford.

Crisp, D.T., Mann, R.H.K. and McCormack, J.C. 1975. The populations of fish in the River Tees system on the Moor House National Nature Reserve, Westmorland. *Journal of Fish Biology* 7: 573–593.

Hood, J.M., Benstead, J.P., Cross, W.F. et al. 2018. Increased resource use efficiency amplifies positive response of aquatic primary production to experimental warming. *Global Change Biology* 24: 1069–1084.

Cummins, K.W., Merritt, R.W. and Andrade, P.C.N. 2005. The use of invertebrate functional groups to characterise ecosystem attributes in selected streams and rivers in south Brazil. *Studies on Neotropical Fauna and Environment* 40: 69–89.

Darwall, W., Bremerich, V., de Wever, A. et al. 2018. The *Alliance for Freshwater Life*: A global call to unite efforts for freshwater biodiversity science and conservation. *Aquatic Conservation* 28: 1015–1022.

De Castro-Catala, N., Munoz, I., Riera, J.L. and Ford, A.T. 2017. Evidence of low dose effects of the antidepressant fluoxetine and the fungicide prochloraz on the behavior of the keystone freshwater invertebrate *Gammarus pulex*. *Environmental Pollution* 231: 406–414.

Demars, B.O.L., Gíslason, G.M., Ólafsson, J.S. et al. 2016. Impact of warming on CO_2 emissions from streams countered by aquatic photosynthesis. *Nature Geoscience* 9: 758–761.

Demars, B.O.L., Manson, J.R., Ólafsson, J.S. et al. 2016. Temperature and the metabolic balance of streams. *Freshwater Biology* 56: 1106–1121.

De Silva, S.R., Subasinghe, R., Bartley, D. and Lowther, A. 2004. *Tilapias as alien aquatics in Asia and the Pacific: a review*. FAO; Rome.

Dick, J.T.A., Armstrong, M., Clarke, H.C. et al. 2010. Parasitism may enhance rather than reduce the predatory impact of an invader. *Biology Letters* 6: 636–638.

Dijkstra, K.-D.B., Monaghan, M.T. and Pauls, S.U. 2014. Freshwater biodiversity and aquatic insect diversification. *Annual Review of Entomology* 59: 143–163.

Dillon, P.J. and Rigler, F.H. 1974. The phosphorous–chlorophyll relationship in lakes. *Limnology and Oceanography* 19: 767–773.

Dobson, A., Lafferty, K.D., Kuris, A.M. et al. 2008. Homage to Linnaeus: how many parasites? How many hosts? *Proceedings of the National Academy of Sciences of the United States of America* 105: 11482–11489.

Dobson, M. and Frid, C. 2009. *Ecology of aquatic systems*, 2nd edition. Oxford University Press; Oxford.

Dodd, J.A., Dick, J.T.A., Alexander, M.E., MacNeil, C., Dunn, A.M. and Aldridge, D.C. 2014. Predicting the ecological impacts of a new freshwater invader: functional responses and prey selectivity of the 'killer shrimp', *Dikerogammarus villosus*, compared to the native *Gammarus pulex*. *Freshwater Biology* 59: 337–352.

Doherty-Bone, T.M., Dunn, A.M., Liddell, C. and Brown L.E. 2018. Transformation of detritus by a European native and two invasive alien freshwater decapods. *Biological Invasions* 20: 1799–1808.

Dudgeon, D., Arthington, A.H., Gessner, M.O. et al. 2006. Freshwater biodiversity: importance, threats, status and conservation challenges. *Biological Reviews* 81: 163–182.

Dunn, A.M., and Hatcher, M.J. 2015. Parasites and biological invasions: parallels, interactions, and control. *Trends in Parasitology* 31: 189–199.

Dunn, A.M. and Dick, J.T.A. 1998. Parasitism and epibiosis in native and non-native gammarids in freshwater in Ireland. *Ecography* 21: 593–598.

Dunn, A. and Shannon, C. 2019. Better biosecurity. https://openeducation.blackboard.com/mooc-catalog/courseDetails/view?course_id=1189_1.

Durance, I. and Ormerod, S.J. 2007. Effects of climatic variation on upland stream invertebrates over a 25 year period. *Global Change Biology* 13: 942–957.

Ensign, S.H. and Doyle, M.W. 2006. Nutrient spiralling in streams and river networks. *Journal of Geophysical Research* 111: G04009.

EU Invasive Alien Species Regulation 1143/2014 [online, accessed April 2019]. http://ec.europa.eu/environment/nature/invasivealien/index_en.htm.

Ferguson, A., Fleming, I., Hindar, K. *et al.* (eds) 2007. *The Atlantic salmon: genetics, conservation and management*. Blackwell Publications Ltd; Oxford, 357–398.

Friberg, N., Dybkjaer, J.B., Ólafsson, J.S. *et al.* 2009. Relationships between structure and function in streams contrasting in temperature. *Freshwater Biology* 54: 2051–2068.

Frissell, C.A., Liss, W.J., Warren, C.E. and Hurley, M.D. 1986. A hierarchical framework for stream habitat classification: viewing streams in a watershed context. *Environmental Management* 10: 199–214.

Gallardo, B., Clavero, M., Sánchez, M.I. *et al.* 2016. Global ecological impacts of invasive species in aquatic ecosystems. *Global Change Biology* 22: 151–163.

Gessner, M.O., Chauvet, E. and Dobson, M. 1999. A perspective on leaf litter breakdown in streams. *Oikos* 85: 377–384.

Grey, C. 2016. *Towards a networks based approach to biomonitoring*. PhD thesis, Queen Mary University of London; London.

Hall, R.O., Wallace, J.B. and Eggert, S.L. 2000. Organic matter flow in streams with reduced detrital resource base. *Ecology* 81: 3445–3463.

Harding, J.S., Benfield, E.F., Bolstad, P.V. *et al.* 1998. Stream biodiversity: the ghost of land use past. *Proceedings of the National Academy of Sciences* 95: 14843–14847.

Hari, R.E., Livingstone, D.M., Siber, R. *et al.* 2006. Consequences of climatic change for water temperature and brown trout populations in alpine rivers and streams. *Global Change Biology* 12: 10–26.

Hatcher, M.J. and Dunn, A.M. 2011. *Parasites in ecological communities; from interactions to ecosystems*. Cambridge University Press; Cambridge.

Hatcher, M.J., Dick, J.T.A., Bojko, J. *et al.* 2019. Infection and invasion: study cases from aquatic communities. *In*: Wilson, K., Fenton, A. and Tompkins, D. (eds) *Wildlife disease ecology: linking theory to data and application* (Ecological Reviews). Cambridge University Press; Cambridge.

Hauer, F.R. and Lamberti, G. 2006. *Methods in stream ecology*. Elsevier; New York.

Hechinger, R.F., Lafferty, K.D., Mancini III, F.T., Warner, R.R. and Kuris, A.M. 2009. How large is the hand in the puppet? Ecological and evolutionary factors affecting body mass of 15 trematode parasitic castrators in their snail host. *Evolutionary Ecology* 23: 651–667.

Heino, J., Virkkala, R. and Toivonen, H. 2008. Climate change and freshwater biodiversity: detected patterns, future trends and adaptations in northern regions. *Biological Reviews* 84: 39–54.

Hernandez, A.D. and Sukhdeo, M.V.K. 2008a. Parasite effects on isopod feeding rates can alter the host's functional role in a natural stream ecosystem. *International Journal for Parasitology* 38: 683–690.

Hernandez, A.D. and Sukhdeo, M.V.K. 2008b. Parasites alter the topology of a stream food web across seasons. *Oecologia* 156: 613–624.

Hershey, A.E. and Lamberti, G.A. 1998. Stream macroinvertebrate communities. *In*: Naiman, R.J. and Bilby, R.E. (eds) *River ecology and management: lessons from the Pacific coastal ecoregion*. Springer-Verlag; New York, 169–199.

Howarth, R.W. and Michaels, A.F. 2000. The measurement of primary production in aquatic ecosystems. *In*: Sala, O.E., Jackson, R.B., Mooney, H.A. and Howarth, R.W. (eds) *Methods in ecosystem science*. Springer; New York, 72–85.

Huryn, A.D. 1996. An appraisal of the Allen Paradox in a New Zealand trout stream. *Limnology and Oceanography* 41: 243–252.

Huryn, A.D. and Wallace, J.D. 2000. Life history and production of stream invertebrates. *Annual Review of Entomology* 45: 83–110.

Hynes, H.B.N. 1970. *The ecology of running waters*. University of Toronto; Toronto.

IMO. Ballast Water Management Convention [online, accessed April 2019]. www.imo.org/About/Conventions/ListOfConventions/Pages/International-Convention-for-the-Control-and-Management-of-Ships'-Ballast-Water-and-Sediments-(BWM).aspx.

Irvine, K., Moss, M. and Balls, H. 1989. The loss of submerged plants with eutrophication II. Relationships between fish and zooplankton in a set of experimental ponds, and conclusions. *Freshwater Biology* 22: 89–107.

Jensen, M.H. 1999. Hydroponics worldwide. *Acta Horticulturae* 481: 719–729.

Jones, C.G., Lawton, J.H. and Shachak, M. 1994. Organisms as ecosystem engineers. *Oikos* 69: 373–386.

Kallis, G. and Butler, D. 2001. The EU Water Framework Directive: measures and implications. *Water Policy* 3: 125–142.

Kaushal, S.S., Likens, G.E., Jaworski, N.A. *et al.* 2010. Rising stream and river temperatures in the United States. *Frontiers in Ecology and the Environment* 8: 461–466.

Kenna, D., Fincham, W.N., Dunn, A.M. *et al.* 2016. Antagonistic effects of biological invasion and environmental warming on detritus processing in freshwater ecosystems. *Oecologia* 183: 875–886.

Knaus, U. and Palm, H.W. 2017. Effects of fish biology on ebb and flow aquaponical cultured herbs in Northern Germany (Mecklenburg Western Pomerania). *Aquaculture* 466: 51–63.

Knight, K.M., McCoy, M.W., Chase, J.M. *et al.* 2005. Trophic cascades across ecosystems. *Nature* 437: 880–883.

Kohler, S.L. 2008. The ecology of host–parasite interactions in aquatic insects. *In*: Lancaster, J., Briers, R. and Macadam, C. (eds) *Aquatic insects: challenges to populations.* CAB International; Wallingford.

Kuang, W. and Zhang, W. 2011. Some effects of parasitism on food web structure: a topological analysis. *Network Biology* 1: 171–185.

Kuris, A.M., Hechinger, R.F., Shaw, J.C. *et al.* 2008. Ecosystem energetic implications of parasite and free-living biomass in three estuaries. *Nature* 454: 515–518.

Lafferty, K.D., Dobson, A.P. and Kuris, A.M. 2006. Parasites dominate food web links. *Proceedings of the National Academy of Sciences of the United States of America* 103: 11211–11216.

Lamberti, G.A. and Steinman, A.D. 1997. Comparison of primary production in stream ecosystems. *Journal of the North American Benthological Society* 16: 95–104.

Ledger, M.E., Edwards, F.K., Brown, L.E. *et al.* 2011. Impact of simulated drought on ecosystem biomass production: an experimental test in stream mesocosms. *Global Change Biology* 17: 2288–2297.

Lewis, W.M. 2011. Global primary production of lakes: 19th Baldi Memorial Lecture. *Inland Waters* 1: 1–28.

Logan, P. and Furse, M. 2002. Preparing for the European Water Framework Directive: making the links between habitat and aquatic biota. *Aquatic Conservation: Marine and Freshwater Ecosystems* 12: 425–437.

Love, D.C., Fry, J.P., Genello, L. *et al.* 2014. An International Survey of Aquaponics Practitioners. *PLoS ONE* 9: pe102662.

Lowe, S.J., Browne, M. and Boudjelas, S. 2000. *100 of the world's worst invasive alien species* IUCN/SSC Invasive Species Specialist Group (ISSG). Auckland; New Zealand.

Lytle, D.A. and Poff, N. 2004. Adaptation to natural flow regimes. *TRENDS in Ecology and Evolution* 19: 94–100.

Maavara, T., Lauerwald, R., Regneir, P. and Van Cappellen, P. (2017) Global perturbation of organic carbon cycling by river damming. *Nature Communications* 8: 15347.

MacNeil, C., Fielding, N.J., Hume, K.D. *et al.* 2003. Parasite altered micro-distribution of Gammarus pulex (Crustacea: Amphipoda). *International Journal for Parasitology* 33: 57–64.

Maltby, E., Omerod, S., Acreman, M. *et al.* 2011. Freshwaters – open waters, wetlands and floodplains. *In*: *UK national ecosystem assessment.* UNEP-WCMC; Cambridge.

Mason, C.F. 2002. *Biology of freshwater pollution.* Pearson; London.

Meyer, J.L. 1994. The microbial loop in flowing waters. *Microbial Ecology* 28: 195–199.

Meyer, J.L., Paul, M.J. and Taulbee, W.K. 2005. Stream ecosystem function in urbanizing landscapes. *Journal of the North American Benthological Society* 24: 602–612.

Millennium Ecosystem Assessment. 2005. *Ecosystems and human well-being: current state and trends: findings of the Condition and Trends Working Group.* Hassan, R., Scholes, R. and Ash, N. (eds). Island Press; New York.

Milner, A.M., Robertson, A., Monaghan, K. *et al.* 2008. Colonization and development of a stream community over 28 years; Wolf Point Creek in Glacier Bay, Alaska. *Frontiers in Ecology and the Environment* 6: 413–419.

Milner, A.M., Robertson, A.L., Brown, L.E. *et al.* 2011. Evolution of a stream ecosystem in recently deglaciated terrain. *Ecology* 92: 1924–1935.

Milner, A.M., Khamis, K., Battin, T.J. *et al.* 2017. Glacier shrinkage driving global changes in downstream systems. *Proceedings of the National Academy of Sciences of the United States of America* 114: 9770–9778.

Mittermeier, R.A., Farrell, T.A., Harrison, I. J., Upgren, A.J. and Brooks, T.M. 2010. *Fresh water: the essence of life* (Vol. 18). Conservation International, US; Arlington, VA.

Montoya, J.M., Woodward, G., Emmerson, M.C. and Solé, R.C. 2009. Press perturbations and indirect effects in real food webs. *Ecology* 90: 2426–2433.

Morgan, R.P. and Cushman, S.F. 2005. Urbanisation effects on fish assemblages in Maryland, USA. *Journal of the North American Benthological Society* 24: 643–655.

Morley, N.J. 2009. Environmental risk and toxicology of human and veterinary waste pharmaceutical exposure to wild aquatic host–parasite relationships. *Environmental Toxicology and Pharmacology* 27: 161–175.

Moss, B. 1998. *Ecology of freshwaters: man and medium, past to future.* Blackwell Publishing; Oxford.

Moss, B. 2010. *Ecology of freshwaters: a view for the twenty-first century.* Wiley-Blackwell; Chichester.

Murphy, M.L. 1998. Primary production. *In:* Naiman, R.J. and Bilby, R.E. (eds) *River ecology and management: lessons from the Pacific coastal ecoregion.* Springer-Verlag; New York, 144–168.

Naegeli, M.W. and Uehlinger, U. 1997. Contribution of the hyporheic zone to ecosystem metabolism in a prealpine gravel-bed river.

Journal of the North American Benthological Society 16: 794–804.

Naiman, R., Décamps, H. and McClain, M. 2005. *Riparia.* Elsevier Academic Press; New York.

Nakano, S. and Murakami, M. 2001. Reciprocal subsidies: dynamic interdependence between terrestrial and aquatic food webs. *Proceedings of the National Academy of Sciences of the United States of America* 98: 166–170.

Nakano, S., Miyasaka, H. and Kuhara, N. 1999. Terrestrial–aquatic linkages: riparian arthropod inputs alter trophic cascades in a stream food web. *Ecology* 80: 2435–2441.

Naylor, R.L., Williams, S.L. and Strong, D.R. 2001. Aquaculture – a gateway for exotic species. *Science* 294: 1655–1656.

Newbold, J.D., O'Neill, R.V., Elwood, J.W. and Van Winkle, W. 1982. Nutrient spiralling in streams: implications for nutrient limitation and invertebrate activity. *American Naturalist* 120: 628–652.

Nilsson, C., Reidy, C.A., Dynesius, M. and Revenga, C. 2005. Fragmentation and flow regulation of the world's large river systems. *Science* 308: 405–408.

O'Gorman, E.J., Benstead, J.P., Cross, W.F. *et al.* 2015. Climate change and geothermal ecosystems: natural laboratories, sentinel systems, and future refugia. *Global Change Biology* 20: 3291–3299.

O'Reilly, C.M., Alin, S.R., Plisnier, P.D. *et al.* 2003. Climate change decreases aquatic ecosystem productivity of Lake Tanganyika, Africa. *Nature* 424: 766–768.

Ormerod, S., Dobson, M., Hildrew, A.G. and Townsend, C.R. 2010. Multiple stressors in freshwater ecosystems. *Freshwater Biology* 55: 1–4.

Paetzold, A., Smith, M., Warren, P.H. and Maltby, L. 2011. Environmental impact propagated by cross-system subsidy: chronic stream pollution controls riparian spider populations. *Ecology* 92: 1711–1716.

Painter, M.M., Buerkley, M.A., Julius, M.L. *et al.* 2009. Antidepressants at environmentally relevant concentrations affect predator avoidance of larval fathead minnows (*Pimephales promelas*). *Environmental Toxicology and Chemistry* 28: 2677–2684.

Paisley, M.F., Trigg, D.J. and Walley, W.J. 2013. Revision of the Biological Monitoring Working Party (BMWP) score system: derivation of present-only and abundance-related scores from field data. *River Research and Applications* 30: 887–904.

Palm, H.W., Knaus, U., Appelbaum, S. *et al.* 2018. Towards commercial aquaponics: a review of systems, designs, scales and nomenclature. *Aquaculture International* 26: 813–842.

Palmer, M.A., Lettenmaier, D.P., Poff, N.L. *et al.* 2009. Climate change and river ecosystems: protection and adaptation options. *Environmental Management* 44: 1053–1068.

Paul, M.J. and Meyer, J.L. 2001. Streams in the urban landscape. *Annual Review of Ecology and Systems* 32: 333–365.

Petts, G.E. 1996. Water allocation to protect river ecosystems. *Regulated Rivers – Research and Management* 12: 353–367.

Pitchler, D.E. 2012. *Stream food webs in a changing climate: the impacts of warming on Icelandic freshwaters.* PhD thesis. Queen Mary University of London; London.

Poff, N.L. and Zimmerman, J.K.H. 2010. Ecological responses to altered flow regimes: a literature review to inform the science and management of environmental flows. *Freshwater Biology* 55: 194–205.

Poff, N.L., Allan J.D., Bain M.B. *et al.* 1997. The natural flow regime: a paradigm for river conservation and restoration. *BioScience* 47: 769–784.

Pullen, R.S.V., Palomares, M.L., Casal, C.V. *et al.* 1997. Environmental impacts of tilapias. *In:* Fitzsimmons, K. (ed.) *Tilapia aquaculture volume 2.* Northeast Regional Agricultural Engineering Service (NRAES) Cooperative Extension; New York, 554–570.

Rawcliffe, R., Sayer, C.D., Woodward, G. *et al.* 2010. Back to the future: using palaeolimnology to infer long-term changes in shallow lake food webs. *Freshwater Biology* 55: 600–613.

Resh, V.H., Brown, A.V., Covich, A.P. *et al.* 1988. The role of disturbance in stream ecology. *Journal of the North American Benthological Society* 7: 433–455.

Reys, P., Sabino, J. and Galetti, M. 2009. Frugivory by the fish *Brycon hilarii* (Characidae) in western Brazil. *Acta Oecologica* 35: 136–141.

Roach, J. 2007. Amazon longer than Nile River, scientists say. *National Geographic News.* June 18. http://news.nationalgeographic.com/news/2007/06/070619-amazon-river.html.

Robinson, C.T., Siebers, A.R. and Ortlepp, J. 2018. Long-term ecological responses of the River Spöl to experimental floods. *Freshwater Science* 37: 433–437.

Robson, M., Spence, K. and Beech, L. 2006. Stream quality in a small urbanised catchment. *Science of the Total Environment* 357: 194–207.

Rosewarne, P.J., Mortimer, R.J., Newton, R.J. *et al.* 2016. Feeding behaviour, predatory functional responses and trophic interactions of the invasive Chinese mitten crab (*Eriocheir sinensis*) and signal crayfish (*Pacifastacus leniusculus*). *Freshwater Biology* 61: 426–443.

Roy, H.E., Hesketh, H., Purse, B.V. *et al.* 2017. Alien pathogens on the Horizon: opportunities for predicting their threat to wildlife. *Conservation Letters* 10: 477–484.

Sánchez-Hernández, J., Cobo, F., Amundsen, P.-A. 2015. Food web topology in high mountain lakes. *PLoS ONE* 10: e0143016.

Sato, T., Egusa, T., Fukushima, K. *et al.* 2012. Nematomorph parasites indirectly alter the food web and ecosystem function of streams through behavioural manipulation of their cricket hosts. *Ecology Letters* 15: 786–793.

Scheffer, M., Hosper, S.H., Meijer, M.L. *et al.* 1993. Alternative equilibria in shallow lakes. *Trends in Ecology and Evolution* 8: 275–279.

Schindler, D.W. 1974. Eutrophication and recovery in experimental lakes: implications for lake management. *Science* 184: 897–899.

Schneider, P. and Hook, S.J. 2010. Space observations of inland water bodies show rapid surface warming since 1985. *Geophysical Research Letters* 37: L22405.

SEPA. 2009. *The river basin management plan for the Scotland river basin district 2009–2015.* Scottish Government; Edinburgh.

Stanford, J.A. and Ward, J.V. 2001. Revisiting the serial discontinuity concept. *Regulated Rivers: Research and Management* 17: 303–310.

Taylor, S.L., Roberts, S.C., Walsh, C.J. and Hatt, B.E. 2004. Catchment urbanisation and increased benthic algal biomass in streams: linking mechanisms to management. *Freshwater Biology* 49: 835–851.

Thamdrup, B. and Canfield, D.E. 2000. Benthic respiration in aquatic sediment. *In*: Sala, O.E., Jackson, R.B., Mooney, H.A. and Howarth, R.W. (eds) *Methods in ecosystem science*. Springer; New York, 86–103.

Thompson, R. and Townsend, C.R. 2005. Energy availability, spatial heterogeneity and ecosystem size predict food-web structure in streams. *Oikos* 108: 137–148.

Tod, S.P. and Schmid-Araya, J.M. 2009. Meiofauna versus macrofauna: secondary production of invertebrates in a lowland chalk stream. *Limnology and Oceanography* 54: 450–456.

Trefts, C. and Omaye, S.T. 2016. Hydroponics: potential for augmenting sustainable food production in non-arable regions. *Nutrition & Food Science* 46: 672–684.

Tyler, C.R. and Jobling, S. 2008. Roach, sex, and gender-bending chemicals: the feminization of wild fish in English rivers. *BioScience* 58: 1051–1059.

van Riel, M.C., van der Velde, G., Rajagopal, S., Marguillier, S., Dehairs, F. and de Vaate, A.B. 2006. Trophic relationships in the Rhine food web during invasion and after establishment of the Ponto-Caspian invader *Dikerogammarus villosus*. *Hydrobiologia* 565: 39–58.

Vannote, R.L., Minshall, G.W., Cummins, K.W., Sedell, J.R. and Cushing, C.E. 1980. The river continuum concept. *Canadian Journal of Fisheries and Aquatic Sciences* 37: 130–137.

Vörösmarty, C.J. and Sahagian, D. 2000. Anthropogenic disturbance of the terrestrial water cycle. *BioScience* 50: 753–765.

Vörösmarty, C.J., McIntyre, P.B., Gessner, M.O. et al. 2010. Global threats to human water security and river biodiversity. *Nature* 467: 555–561.

Walsh, C.J., Fletcher, T.D. and Ladson, A.R. 2005. Stream restoration in urban catchments through redesigning stormwater systems: looking to the catchment to save the stream. *Journal of the North American Benthological Society* 24: 690–705.

Webb, B.W. and Nobilis, F. 2007. Long-term changes in river temperature and the influence of climatic and hydrological factors. *Hydrological Sciences Journal* 52: 74–85.

Wehr, J.D. and Sheath, R.D. 2003. *Freshwater algae of North America. Ecology and classification*. Academic Press; San Diego, CA.

Wetzel, R.G. 2001. *Limnology. Lake and river ecosystems, 3rd edition*. Academic Press; San Diego, CA.

Whitehead, P.G., Wilby, R.L., Battarbee, R.W., Kernan, M. and Wade, A.J. 2009. A review of the potential impacts of climate change on surface water quality. *Hydrological Sciences Journal* 54: 101–123.

Woodward, G., Dybkjaer, J.B., Ólafsson, J.S., Gíslason, G.M., Hannesdóttir, E.R. and Friberg, N. 2010. Sentinel systems on the razor's edge: effects of warming on Arctic geothermal stream ecosystems. *Global Change Biology* 16: 1979–1991.

Woodward, G., Perkins, D. and Brown, L.E. 2010. Climate change and freshwater ecosystems: impacts across multiple levels of organisation. *Philosophical Transactions of the Royal Society B* 365: 2093–2106.

Wright, J.F., Sutcliffe, D.W. and Furse, M.T. 2000. *Assessing the biological quality of fresh waters: RIVPACS and other techniques*. Freshwater Biological Association; Ambleside.

Xenopoulos, M.A., Lodge, D.M., Alcamo, J., Märker, M., Schulze, K. and van Vuuren, D.P. 2005. Scenarios of freshwater fish extinctions from climate change and water withdrawal. *Global Change Biology* 11: 1557–1564.

Yavuzcan Yildiz, H., Robaina, L., Pirhonen, J. et al. 2017. Fish welfare in aquaponic systems: its relation to water quality with an emphasis on feed and faeces—a review. *Water* 9: 13.

Yvon-Durocher, G., Woodward, G., Jones, J.I., Trimmer, M. and Montoya, J.M. 2010. Warming alters the metabolic balance of ecosystems. *Philosophical Transactions of the Royal Society B* 365: 2117–2126.

CHAPTER SEVEN

Water and health

Rebecca J. Slack

LEARNING OUTCOMES

After reading this chapter you should be able to:

- understand the important links between water and human health;
- describe key water-related health issues and distinguish between biological, chemical and other health effects;
- describe the historical context of improved public health through enhanced sanitation and water supply;
- recognise the different types and scales of health issues related to water around the world today and the complex interactions with climate science, human geography and agriculture among many other areas;
- appreciate the role of research and development in maintaining the health of growing global, particularly urban, populations.

7.1 INTRODUCTION

All life on Earth requires water to survive. However, water can also be harmful to life. Chemical and physical hazards can threaten the health of individuals, populations and even whole ecosystems through changes in water quality (the characteristics of freshwater that make it suitable to sustain life) and water quantity (the availability of freshwater resources). Physical hazards such as drought and flooding, along with chemical hazards such as saline intrusion or pollution events, impact on the health of many organisms including humans. It is also true that many pathogenic organisms use water as a pathway to infect host organisms; for human hosts, water-related diseases that result from such infection make significant contributions to morbidity and mortality.

Ensuring a reliable and safe supply of water and removal and treatment of wastewater via adequate sanitation is essential for all human populations. If the quantity or quality of water is insufficient, impacts on health can be considerable. This chapter considers the dual issues of water quantity and water quality in relation to human health. The key causes of disease and ill-health associated with water will be discussed and, where appropriate, measures used to improve the water–health interface will be examined.

7.2 THE HISTORY OF WATER AND HEALTH

The association between water and health has a long history. Human habitations have always been associated with freshwater, whether springs or wells, streams, rivers, or lakes. However, as populations increased rapidly in the nineteenth century, water and sanitation conditions in most cities became a major public health concern. There were frequent cases of contamination of the water supply by sewage, particularly as watercourses were used for both abstraction and wastewater disposal. While there were exceptions, notably the cities of Japan, whose economic, cultural and religious attitudes resulted in more sanitised environments, most cities suffered from high death rates, particularly of young children. In the UK, the work of social reformers such as Edwin Chadwick (1800–1890) led to the Public Health Act 1848, which sought to improve sanitation and, hence, the health of the poor. This was some of the earliest legislation that dealt with public health and the environment, but the scientific understanding of the causes of disease was still very unclear.

There were many theories about the spread of infectious disease, with one of the most popular being the **miasmatic theory of disease**, prevalent at the time of the Public Health Act 1848. This theory linked foul smells and stagnant air with the cause of disease. As miasmas were supposedly a product of environmental factors such as vapours from cesspits and open sewers, it was believed that diseases could not be passed from person to person, and could be prevented by purifying the air. While this did lead to some improvements in sanitary conditions and an attempt to clean up many towns and cities (as dirty water frequently smells unpleasant), there were still significant incidences of disease. The **germ theory of disease** challenged the miasmatic theory but started to gain momentum only in the 1850s, when the work of Dr John Snow linked cholera with contaminated water supplies (see **Box 7.1**; Smith, 2002). However, it was not until the 1860s and 1870s that the work of Louis Pasteur, Robert Koch and Joseph Lister confirmed that microorganisms were responsible for many diseases and the miasmatic theory was finally laid to rest. Since then, public health has improved considerably in the developed world. In particular, the discovery of penicillin in 1928 initiated the identification and application of antibiotics to the treatment of all bacterially associated water-related disease.

The development of water and wastewater treatment and distribution systems resulted in the most significant impact on public health and quality of life. The first recognised drinking water treatment system was the slow sand filter devised by James Simpson (1799–1869) and still used today (see Chapter 8, **Figure 8.3a**). Wastewater treatments had been used sporadically from the early 1800s, when sewage sludge was cited to be an effective agricultural fertiliser. Subsequent sewage collection and treatment developments, including the London sewer network designed by Joseph Bazalgette, were a result of the recognition that faecal contamination of the water supply led to foul smells and disease (Halliday, 2001). By the late nineteenth and early twentieth centuries, most towns and cities in the developed world had some form of sewer system and safe public water supply. Further development of drinking water treatment systems included

BOX 7.1

John Snow and Cholera

Cholera spread from the Indian subcontinent in the early nineteenth century, rapidly becoming responsible for the deaths of millions worldwide. In 1849, Dr John Snow published a pamphlet called *On the Mode of Communication of Cholera*, which stressed water as the medium for the spread of cholera. It was not met with much enthusiasm – the miasmatic theory of disease was still prevalent in most scientific circles. It was not until an 1854 outbreak of cholera in the Soho area of London that Snow could put his theory to the test. Snow mapped the locations of the homes of those who had died in the outbreak and determined that many cases were centred on a water pump in Broad Street (Figure 7.1). Despite some anomalies, including no cases reported in a nearby

Figure 7.1 John Snow mapped the area surrounding the Broad Street pump, representing each cholera death by a bar. The greatest number of bars can be seen clustering around the infected pump.

Continued

workhouse, workers at the local brewery escaping infection, and two cases in distant parts of London, Snow successfully petitioned the authorities to remove the handle of the pump and hence disable it from use. The number of cases of cholera, and subsequent deaths, rapidly fell, confirming the link to the pump. It later became apparent that the workhouse and brewery had private wells and the two deaths at a distance from the pump were a result of the householders preferring the taste of water from the Broad Street pump over their local water supplies. Further examination of the pump revealed that a few days before the first cases of disease, the water was reported to be cloudy and foul smelling. Water was supplied to the pump by the Southwark and Vauxhall Water Company, who extracted sewage-contaminated water from the River Thames. Subsequent legislation required that all water companies in the London region take water from the river upstream of the city.

Snow's work on the 1854 cholera outbreak not only demonstrated how cholera was spread and how it could be prevented, it also provided support for the germ theory of disease and founded the science of epidemiology, which is the cornerstone of public health research. It also demonstrated the importance of water quality for human health. More information about John Snow's work is provided in the biographies by Vinten-Johansen and colleagues (2003) and Hempel (2007).

rapid gravity filters (see Chapter 8, Figure 8.3b), used extensively in the USA from the 1890s; disinfection of water supplies from the 1900s; and, for wastewater, biological wastewater treatment from the 1880s and activated sludge in the 1920s (see Chapter 8).

REFLECTIVE QUESTIONS

What were the key discoveries in understanding the links between water and health?

How might your understanding of what caused disease have changed if you had lived between the years 1825 and 1875?

What do you think was the single, most important discovery or change linked to contaminated water that led to improvements in public health?

7.3 INFECTIOUS DISEASES

There are four categories of infectious diseases associated with water. These are categorised by method of distribution rather than by microorganism type:

- waterborne diseases – infections spread by water;
- water-washed diseases – infections spread by lack of water (for personal hygiene) and insanitary conditions that may not directly involve water as a means of infection transfer;
- water-based diseases – infections transmitted by aquatic invertebrates;
- vector-based water-related diseases – infections spread by insects that need water to complete their life cycle.

All categories described in Table 7.1 have the potential to cause illness and death anywhere around the world. However, the

Table 7.1 Water-related diseases: pathogens, treatment and prevention strategies

Infection type/sub-category	Disease	Pathogen	Treatment regime	Prevention
Water-borne				
Bacterial infections	*E. coli* infection	*Escherichia coli*	Antibiotics	Good personal hygiene Improved water quality
	Cholera	*Vibrio cholerae* (serogroups O1 and O139)	Oral rehydration therapy Antibiotics	Improved water quality Improved sanitation Vaccination
	Vibrio illness	*Vibrio vulnificus*, *Vibrio alginolyticus*, *Vibrio parahaemolyticus*	Antibiotics	Improved water quality Improved sanitation Good personal hygiene
	Typhoid fever	*Salmonella enteritica serovar* Typhi (called *S. typhi*)	Antibiotics	Improved water quality Improved sanitation Good personal hygiene Vaccination
	Paratyphoid	*Salmonella paratyphi A, S. paratyphi B* (or *S. schottmuelleri*) *S. paratyphi C (S. hirschfeldii)*	Antibiotics	Improved water quality Improved sanitation Good personal hygiene Vaccination
	Salmonellosis	*Salmonella* spp	Antibiotics	Good personal hygiene
	Leptospirosis	*Leptospira* spp	Antibiotics	Avoiding contact with water in likely areas of infection; covering cuts; preventing splashing to mucous membranes
	Bacillary dysentery	*Shigella dysenteriae*	Antibiotics Oral rehydration therapy	Good personal hygiene
	Botulism	*Clostridium botulinum*	Antibiotics	Good personal hygiene
	Campylobacteriosis	*Campylobacter jejuni*	Antibiotics	Good personal hygiene
	M. marinum infection	*Mycobacterium marinum*	Antibiotics	Good personal hygiene
	Tularaemia	*Francisella tularensis*	Antibiotics – streptomycin	Good personal hygiene
	Diarrhoeal disease	*Shigella flexneri, Shigella sonnei*	Antibiotics	Good personal hygiene
Protozoan infections	Giardiasis	*Giardia lamblia*	Self-limiting Antibiotics	Improved water quality

(Continued)

Table 7.1 (Cont.)

Infection type/subcategory	Disease	Pathogen	Treatment regime	Prevention
	Amoebiasis (amoebic dysentery)	*Entamoeba histolytica*	Amoebicides (e.g. diloxanide)	Better hygiene practices Improved water and sanitation Good personal hygiene
	Balantidiasis	*Balantidium coli*	Antibiotics (e.g. tetracycline)	Improved water and sanitation Good personal hygiene
	Cryptosporidiosis	*Cryptosporidium parvum*	Supportive (e.g. oral rehydration therapy, antibiotics, etc.)	Effective water treatment/quality Avoiding contact with water in likely areas of infection
	Cyclosporiasis	*Cyclospora cayetanensis*	Cotrimoxazole	Effective water treatment
Fungal infection	Microsporidiosis	Microsporidia: *Enterocytozoon bieneusi, Encephalitozoon cuniculi Encephalitozoon intestinalis*	Albendazole and supportive care	Good personal hygiene Avoiding contact with water in likely areas of infection
Viral infections	Adenovirus infection	Adenovirus	Supportive	Chlorination of water Avoiding contact with infected water
	Gastroenteritis	Astrovirus Calicivirus Enteric adenovirus Parvovirus Rotavirus Norovirus	Self-limiting: supportive care Oral rehydration therapy, antibiotics	Vaccination (e.g. rotavirus) Good personal hygiene Improved water quality
	SARS	Coronavirus	Supportive	Improved sanitation Good personal hygiene
	Hepatitis	Hepatitis A	Supportive: diet, hydration, therapeutic drugs	Vaccination Good personal hygiene Improved sanitation
	Poliomyelitis	Poliovirus	Supportive	Vaccination
	Polyomavirus infection	JC virus, BK virus, SV40	Supportive only	Good personal hygiene and sanitation
Parasitic infections	Ascariasis	*Ascaris lumbricoides*	Ascaricides	Good personal hygiene Improved sanitation

(Continued)

Table 7.1 (Cont.)

Infection type/sub-category	Disease	Pathogen	Treatment regime	Prevention
	Taeniasis	*Taenia solium* *T. saginata*	Praziquantel and/or antihelminthics	Good personal hygiene Improved sanitation
	Hymenolepiasis	*Hymenolepis nana* *H. diminuta*	Praziquantel and/or antihelminthics	Good personal hygiene Improved sanitation
	Echinococcosis	*Echinococcus granulosus* *E. multilocularis* *E. vogeli*	Surgery with albenda-zole and/or mebendazole	Good personal hygiene Improved sanitation
	Coenurosis	*Taenia multiceps* *T. serialis* *T. brauni* *T. glomerata*	Praziquantel and/or antihelminthics	Good personal hygiene Improved sanitation
	Enterobiasis	*Enterobius vermicularis*	Albendazole and/or mebendazole	Good personal hygiene
Other waterborne infections	Otitis externa	Various bacteria (e.g. *Pseudomonas aeruginosa*)	Antibiotics	Preventing water entering the ear canal
	Pseudomonas dermatitis/folliculitis	*Pseudomonas aeruginosa*	Self-limiting Antibiotics	Chlorination of recreational facilities Good personal hygiene
	Legionellosis	*Legionella* ssp. (e.g. *L. pneumophilia*)	Antibiotics	Good general hygiene
	Pontiac fever	*Legionella* ssp. (e.g. *L. pneumophilia*)	Antibiotics	Good general hygiene
Water-washed				
Skin and eye infections	Nosocomia infections	*Staphylococcus aureus* Methicillin-resistant *Staphylococcus aureus* (MRSA) *Pseudomonas aeruginosa* *Acinetobacter baumannii* *Stenotrophomonas maltophila* *Clostridium difficile*	Isolation and care	Good personal hygiene
	Conjunctivitis	Not necessarily caused by pathogen (see microsporidiosis)	Water/eye drops Self-limiting	Water/eye drops Good personal hygiene

(Continued)

Table 7.1 (Cont.)

Infection type/sub-category	Disease	Pathogen	Treatment regime	Prevention
	Leprosy	*Mycobacterium leprae, M. lepromatosis*	Antibiotics: rifampicin, clofazimine, dapsone	Good personal hygiene (although not very contagious)
	Scabies	*Sarcoptes scabiei*	Insecticide cream Spermethrin	Good personal hygiene and general cleaning
	Tinea	Tinea species: *T. pedis, T. cruris, T. corporis, T. capitis*	Antifungal cream: miconazole, clotrimazole	Good personal hygiene
	Trachoma	*Chlamydia trachomatis*	Antibiotics: azithromycin, erythromycin, doxycycline/tetracycline	Good personal hygiene Improved water supply and sanitation
	Yaws (Frambesia tropica)	*Treponema pallidum*	Antibiotics	Good personal hygiene
Biting invertebrate	Typhus	*Rickettsia* species	Antibiotics: azithromycin, doxycycline/tetracycline	Good personal hygiene
	Relapsing fever (typhinia)	*Borrelia recurrentis*	Antibiotics: tetracycline	Good personal hygiene
	Spotted fevers	*Rickettsia* species *Orientia* species	Antibiotics: azithromycin, doxycycline/tetracycline	Good personal hygiene
	Trench fever	*Bartonella quintana*	Antibiotics: doxycycline/tetracycline	Good personal hygiene
	Tularaemia	*Francisella tularensis*	Antibiotics: streptomycin	Good personal hygiene
	Bubonic plague	*Yersinia pestis*	Antibiotics: streptomycin	Good personal hygiene and general cleaning
Water-based				
Skin penetration	Schistosomiasis	*Schistosoma spp.: S. mansoni, S. intercalatum, S. haematobium, S. japonicum, S. mekongi*	Praziquantel and/or antihelminthics	Molluscicides Prevent contact with infected water Improved water quality and sanitation
		Dracunculus medinensis		Improved water supply (filtration/boiling)

(Continued)

Table 7.1 (Cont.)

Infection type/sub-category	Disease	Pathogen	Treatment regime	Prevention
Intestinal penetration	Dracunculiasis (Guinea worm disease)		Surgical or traditional-winding removal of worm	
	Clonorchiasis	*Clonorchiasis sinensis*	Chloroquine (partially effective)	Avoid uncooked fish
	Diphyllobothriasis	*Diphyllobothrium latum*	Diatrizoic acid or praziquantel	Avoid uncooked fish
	Fasciolopsiasis	*Fasciolopsis buski*	Praziquantel and/or antihelminthics	Improved water supply and cooking raw vegetation
	Paragonimiasis	*Paragonimus westermani*	Praziquantel and/or antihelminthics	Avoid undercooked crustaceans
Vector-based				
Breeding in water	Malaria	*Plasmodium falciparum* *P. vivax* *P. ovale* *P. malariae* *P. knowlesi*	Mefloquine, doxycycline and the atovaquone/proguanil	Vector-control – insecticides, mosquito nets, covering or managing potential mosquito breeding pools
	Dengue fever	Flavivirus – Dengue fever virus (spread by *Aëdes aegypti, A. albopictus, A. polynesiensis, A. scutellaris*)	Oral/intravenous rehydration therapies	Vector-control – insecticides, mosquito nets, covering or managing potential mosquito breeding pools
	Yellow fever	Flavivirus – Yellow fever virus (spread by *Aëdes aegypti, A. albopictus, A. africanus, Haemagogus* spp. *Sabethes* spp.)	Anti-viral drugs	Vaccination; Vector-control – insecticides, mosquito nets, covering or managing potential mosquito breeding pools
	Zika virus	Flavivirus spread by *Aëdes aegypti*	No treatment but usually only mild symptoms	Vector-control – insecticides, mosquito nets, covering or managing potential mosquito breeding pools
	Chikungunya	Alphavirus – spread by *Aëdes* spp.	Chloroquine	Vector-control – insecticides, mosquito nets, covering or managing potential mosquito breeding pools

(Continued)

Table 7.1 (Cont.)

Infection type/sub-category	Disease	Pathogen	Treatment regime	Prevention
	Viral encephalitis/ viral haemorrhagic fever (other)	Arboviruses: Asfivirus, Phlebovirus, Ortho-bunyavirus, Nairovirus, Flavivirus, Coltivirus, Orbivirus, Alphavirus, Uukuvirus	Antivirals	Vector-control
	Lymphatic filariasis	Wuchereria bancrofti Brugia malayi B. timori	Albendazole/ivermectin Diethylcarbamazine/ albendazole	Vector-control – insecticides, mosquito nets, covering or managing potential mosquito breeding pools
	Subcutaneous filariasis	Loa loa Mansonella streptocerca Onchocerca volvulus Dracunculus medinensis	Albendazole/ivermectin Diethylcarbamazine/ albendazole	Vector-control – insecticides, mosquito nets, covering or managing potential mosquito breeding pools
	Serous cavity filariasis	Mansonella perstans M. ozzardi	Albendazole/ivermectin Diethylcarbamazine/ albendazole	Vector-control – insecticides, mosquito nets, covering or managing potential mosquito breeding pools
	Onchocerciasis (river blindness)	Onchocerca volvulus – Wolbachia pipientis	Ivermectin Doxycycline	Vector-control – insecticides, mosquito nets, covering or managing potential mosquito breeding pools
Bite near water	African trypano-somiasis (sleeping sickness)	Trypanosoma brucei	Pentamidine Suramin Eflornithine Melarsoprol Nifurtimox	Vector-control Medical intervention

resilience of the water supply and sewerage network of developed countries ensures that the impact of all water-related diseases is kept to a minimum through the supply of high-quality water in sufficient quantities for high standards in hygiene and sanitation. Most developed countries are found in temperate regions where water-based diseases such as schistosomiasis and vector-borne diseases such as trypanosomiasis are not reported. Developing countries in tropical regions with poor water supply and sanitation are at the greatest risk of water-related infectious diseases, as discussed in subsequent sections.

Water-related infections are the main causes of death from pathogenic infections worldwide (malaria, schistosomiasis, amoebic dysentery). Although there is an emphasis on human health and infectious disease, most pathogens also affect a wide number of mammals, including cattle and sheep (typhoid and cholera are, however, human specific). Disease outbreaks can therefore have important consequences for the livelihoods of farming communities, particularly in non-industrialised countries. Livestock can also be a source of contamination, particularly in rural areas where surface water runoff to drinking water sources can be contaminated by faecal matter (**Figure 7.2**).

7.3.1 Waterborne disease

For centuries, outbreaks of typhoid and cholera were responsible for the deaths of hundreds of thousands of people. Improvements in water quality and the provision of separate supply and sanitation networks have eradicated incidences of these diseases in many countries. The last major outbreak of cholera in the United States was in 1910–1911, but outbreaks still occur today, usually following natural disasters (cholera in Haiti in late 2010) or after conflict/civil unrest (the Democratic

Republic of Congo cholera outbreak in 2011) (WHO 2011a). Small, localised outbreaks are frequent where appropriate sanitation facilities are lacking. Outbreaks of these two waterborne diseases can result in high mortality rates, but treatments are available and epidemiological approaches can usually identify the source of the outbreak to prevent reoccurrence.

There are 3 to 5 million cases of cholera per year worldwide, resulting in about 120,000 deaths. Spread of the disease is a result of faecal contamination of drinking water; the bacteria that cause cholera are relatively fragile and a large infective dose is required. The classic symptom of cholera is acute diarrhoea, which, if left untreated, will result in death from dehydration. The large volumes of diarrhoea produced also increase the likelihood of the spread of infection if appropriate disposal facilities are not available (e.g. in refugee camps) (WHO, 2011b). The disease is **endemic** to the Indian subcontinent, Central and South America and sub-Saharan Africa. Infection can be reduced by implementing relatively simple measures to prevent faecal contamination of water supplies, such as bunds to prevent surface water drainage into wells, separating sanitation and water facilities, provision of hand pumps, etc. Treatment of the disease with oral rehydration therapies is very effective and relatively cheap. Antibiotics can be used to lessen the severity of the disease, but the spread of antibiotic-resistant mutations is raising concern. In areas where there is the greatest risk of infection, vaccination is possible, such as in Dhaka, Bangladesh, or in some places already well established, such as in Zimbabwe (WHO, 2010a).

Typhoid fever resulted in many deaths in the nineteenth century in the United States and Europe but was largely eradicated through the provision of water and sanitation

Figure 7.2 Faecal contamination of water supplies is one of the commonest routes for the spread of water-related disease; farmed animals are a common source of such contamination in rural locations. *Source:* Photo courtesy of Shutterstock/baxys.

services. With an estimated 16 to 33 million cases resulting in over 200,000 deaths per year, typhoid and paratyphoid (the 'enteric fevers') remain a major public health concern around the world today, and are one of the main causes of death in young children (Crump *et al.*, 2004). Typhoid and paratyphoid are caused by faecal or urinal contamination of drinking water, but only a small dose is required to cause infection. If the patient recovers, bacteria will still be present in faeces and urine and can cause infection of others via poor hygiene, contaminated food or contaminated water. Healthy people can act as typhoid carriers (bacteria remains in the gall bladder in 2–5% of patients), as shown by Mary Mallon ('Typhoid Mary') in New York in the early twentieth century. Antibiotics are the main form of treatment,

drastically cutting the number of people who die from the disease. Other preventive measures are associated with provision of clean water (typhoid bacteria are killed by chlorination and removed by most filtration methods), adequate sanitation and hygiene education (e.g. hand-washing).

Leptospirosis can infect a range of mammals and is spread by water contaminated by infected animal urine (particularly rodents). Although rare, the disease occurs in both temperate and tropical climates (Levett, 2001). The bacteria can be ingested or can penetrate the skin or mucous membranes, with infection starting with influenza-like symptoms and progressing to jaundice and fever leading to liver failure ('Weil's disease'). The number of cases correlates with warm and wet conditions leading to epidemics most recently in

Nepal (2010), Ireland (2010) and the Philippines (2009). Leptospirosis may have been responsible for almost stopping the British colonisation of North America (Marr and Cathey, 2010). Prevention of leptospirosis is aided by provision of clean water, general good hygiene and awareness of risks when engaging in recreational activities around natural water systems (e.g. water sports).

There are many other waterborne diseases associated with human/animal faecal contamination of drinking or recreational waters, or simply poor personal hygiene (Pruss, 1998; Black *et al.*, 1989). Cryptosporidiosis, usually an acute self-limiting infection causing diarrhoea and abdominal pains, is estimated to be responsible for over 50% of all waterborne disease. Unlike many other waterborne infections, it is not uncommon in developed countries and there are numerous reports of cases associated with public swimming pools and water supply sources, despite water treatment processes (Fayer *et al.*, 2000). Certain strains of *Escherichia coli*, one of the coliform group of bacteria used to assess the quality of food and water and indicative of faecal contamination, can result in serious illness and death in humans, as demonstrated by an outbreak in Germany in May 2011. The protozoan disease giardiasis causes diarrhoea but is rarely fatal, despite infecting millions of people worldwide: in the United States alone, up to 2 million people are estimated to be infected at any one time (Regli *et al.*, 1991). Amoebiasis or amoebic dysentery, however, is a more common cause of death, with approximately 50 million people infected worldwide and 70,000 deaths per year. Bacillary dysentery (shigellosis) and related bacterial infections resulting in severe diarrhoeal infections cause an estimated 160 million cases and 1 million deaths per year (Kotloff *et al.*, 1999). Gastroenteritis is used to describe a number of bacterial infections associated with contaminated water, including *Shigella, Campylobacter jejuni, Salmonella, Staphyl ococcus, Clostridium* and *Vibrio* species. Viral gastroenteritis may also be waterborne, including the highly infectious norovirus, polio, hepatitis A and hepatitis E. Many of these microorganisms are also found in shellfish grown in waters contaminated with sewage and this is a frequent cause of 'food poisoning' in developed countries.

Ascariasis is caused by the ingestion of eggs of the roundworm *Ascaris* species. It is estimated that 25% of the world's population may be infected by *Ascaris* ssp., particularly in Latin America and Africa, and some cases can result in pneumonia and respiratory failure. Enterobiasis, another roundworm disease and one of the most common childhood parasitic diseases around the world (known as pinworm or threadworm), can be spread by contaminated water but is more likely to result from poor personal hygiene or contaminated food. Other diseases resulting from the ingestion of parasite eggs include taeniasis and, less commonly, hymenolepiasis, echinococcosis and coenurosis, all of which result from tapeworm infections. While water is one mode of infection, it is usually undercooked food or poor personal hygiene which lead to infection (Bath *et al.*, 2010).

There is a sub-set of waterborne diseases that are not associated with faecal contamination of water. These can rarely be controlled by improving personal hygiene but can be significantly reduced by the cleaning of water pipes and other infrastructure. These diseases are spread by the inhalation of infected water droplets rather than from drinking or other contact with contaminated water (CDC, 2018). The pathogens causing Legionnaires' disease and Pontiac fever commonly occur in natural water sources but rarely at concentrations causing disease. However, in warm human-made systems such as industrial cooling systems, hot tubs, windscreen washer systems and ornamental pools, particularly if water is stagnant or the system is sealed, bacteria numbers can grow to infectious levels. Regular cleaning and disinfection of pipes and pools help to reduce the population of pathogens.

Hot tub rash, *Pseudomonas* dermatitis or folliculitis, is a specific disease which occurs when skin is in contact with contaminated water for long periods of time. The bacteria are commonly found in the environment, particularly in water and other moist environments. This tends to be a self-limiting skin disease and can be prevented by regular disinfection of hot tubs or swimming pools, combined with showering to minimise infection. Showering before and after using hot tubs or swimming pools can help to reduce cross-contamination and infection rates.

7.3.2 Water-washed disease

Many waterborne diseases may also be considered to be water-washed diseases, as they are also spread by unwashed hands during food preparation and human-to-human contact (Mackintosh and Hoffman 1984). Provision of adequate supplies of water to maintain good personal hygiene, such as hand-washing before leaving the bathroom or washing of infected clothing, is essential to prevent infection. This is particularly important for many of the diarrhoea-causing bacteria, viruses and protozoa as well as tapeworm infections. Provision of adequate, clean water for washing and cleaning can also help to reduce the spread of infections more generally. Transmission of airborne infections such as influenza can be reduced by frequent hand-washing, as can the spread of hospital-acquired (or **nosocomial**) infections such as *Clostridium difficile* and methicillin-resistant *Staphylococcus aureus* (MRSA).

Other water-washed diseases are infections of the body surface, the skin and eyes. These can be fungal infections such as tinea, conditions caused by bacteria, including yaws and leprosy, or parasitic invertebrates such as the skin-burrowing mite causing scabies; all result in skin sepsis or ulcers. The eye disorder conjunctivitis is a particular problem in arid countries, where the eyes can become dry and so unable to 'self-cleanse'. However, trachoma is a much more debilitating disease, leading to blindness in infected individuals, with 41 million cases worldwide and 8 million suffering severe vision impairment, particularly in sub-Saharan Africa (Resnikoff *et al.*, 2004). Infection rates for all diseases can be reduced through provision of improved access to water and sanitation facilities and good personal hygiene practices such as regular hand and face washing.

Biting arachnids and insects, principally mites, ticks, lice and fleas, are responsible for a number of vector-related infections that can largely be prevented through good personal hygiene, which is dependent on access to suitable water supplies. Epidemics of typhus and other louse- or flea-borne diseases have been reported at intervals around the world, especially at times of war or civil unrest or in slum communities when access to water for washing is restricted. Tularaemia is largely spread via tick bites but can also be spread by flies and, in about 5 to 10% of US cases, by contaminated water sources (Ellis *et al.*, 2002). Symptoms of tularaemia are similar to bubonic plague, a disease spread by rat fleas and largely thought to be responsible for killing 30–60% of the European population in the fourteenth century (in an epidemic called the Black Death) and the European Great Plagues of the seventeenth century. Although now rare, isolated outbreaks of bubonic plague are still reported.

7.3.3 Water-based disease

Water-based diseases are all worm infections (flukes or trematodes), which can be spread by aquatic invertebrates, particularly water snails, with humans or other mammals acting as the definitive host. Such diseases are rarely life-threatening but can cause discomfort and have a negative impact on livelihood.

Schistosomiasis (or bilharziasis) ranks just behind malaria in public health importance, infecting 200 million in many tropical and subtropical countries (Gryseels *et al.*, 2006). Although the mortality rate associated with the disease is low, it does result in chronic illness, which can affect the livelihoods of infected individuals. The most common route of infection involves the larval parasite, released from water snails, penetrating the skin during contact with infected waters (for fishing, rice planting, recreation). The subsequent development of the worm may be in the liver, blood vessels, bladder or kidneys, from where eggs will be released to infect faeces or urine (**Figure 7.3**). Programmes of eradication, focusing on eliminating water snail populations and mass treatment of infected populations, have been initiated in a number of countries where the disease is endemic. Limiting human contact with untreated water through better water quality and sanitation facilities is a further approach. The building of dams and **irrigation** schemes in endemic countries has also resulted in new habitats for water snails, and hence levels of the disease close to such developments have increased, an issue that has led to the design of new irrigation systems and approaches to reduce or prevent water contact.

Other trematode infections, including fasciolopsiasis, clonorchiasis, the fish tapeworm infection diphyllobothriasis, paragonimiasis and anisakis, result from larval ingestion in drinking water, raw or undercooked fish or crustaceans, or raw aquatic plants.

Dracunculiasis is spread through drinking water contaminated with infectious larvae of the nematode guinea worm, which mature in the human body. The adult female migrates to the skin, usually of the foot, to release larvae into any water that the person may enter, to repeat the life cycle. Prevention of infection through improved water quality by filtering or boiling water is very effective in African countries, where dracunculiasis is endemic, and the Indian subcontinent, as is protecting well supplies from surface water runoff (Esrey *et al.*, 1991).

7.3.4 Water-vector disease

Disease vectors, rather than the pathogens, may require water to complete their life cycle. One example of this is the mosquito, which acts as a disease vector for malaria (**Box 7.2**). Malaria is the world's biggest public health concern, with over 250 million cases per year, resulting in almost 1 million deaths (2.23% of deaths worldwide) (WHO, 2010b). While malaria is endemic in most tropical or subtropical countries (**Figure 7.4**), the majority of deaths are among young children living in sub-Saharan Africa. While treatments of malaria are available, preventive strategies focusing on the control of the mosquito vector, such as use of mosquito nets and spraying of insecticides, have been the target of recent attempts to reduce infection rates, but research is ongoing in a number of areas including engineering of genes in mosquitoes that induce infertility in female offspring in later generations (Kyrou *et al.*, 2018). Water development projects can increase as well as decrease breeding habitats for disease-causing vectors. Human-made lakes and irrigation measures such as for rice cultivation may result in malaria epidemics; new dams in the Tennessee Valley in the 1930s led to an increase in malaria infections until control measures such as altering the water level were instigated (Hall *et al.*, 1946).

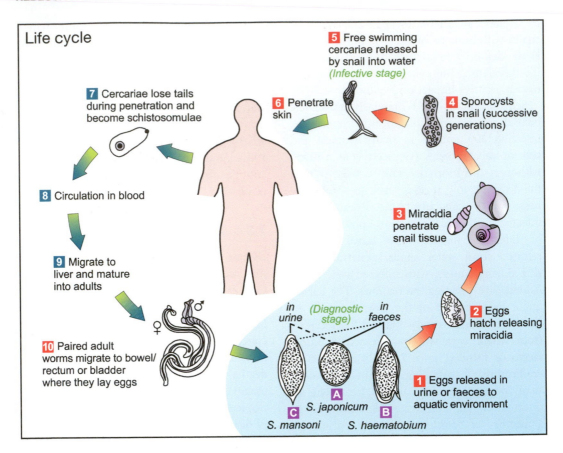

Figure 7.3 The life cycle of the *Schistosoma* spp. trematode worm, demonstrating infection of human hosts by the infective cercariae and subsequent release of eggs from human hosts to infect intermediate water snail hosts.
Source: Adaptation of an original figure by CDC/Alexander J. da Silva/Melanie Moser.

Mosquitoes are responsible for the spread of other diseases, such as arboviruses and filariasis. Mosquito control measures are often the only effective disease management measures for these diseases. Culicine mosquitoes are the vector responsible for the spread of many arboviruses, such as dengue virus and yellow fever. There are four types of dengue virus spread by mosquitoes of the *Aëdes* genus, principally *Aëdes aegypti*. About 5% of cases, usually a result of infection by more than one type of the virus (multiple infection),

result in severe illness involving fever, joint and muscle pain (hence 'breakbone fever'), and bleeding, which may lead to death from shock and haemorrhage. Of the 50 to 100 million infected across the tropics (Asia, Africa and Latin America), fewer than 25,000 deaths from dengue fever are reported (Gubler, 1998). However, dengue fever has increased 30-fold over the last 50 years, due to deforestation (*A. aegypti* is an African forest mosquito), urbanisation and international travel. For yellow fever, a vaccine does exist, although

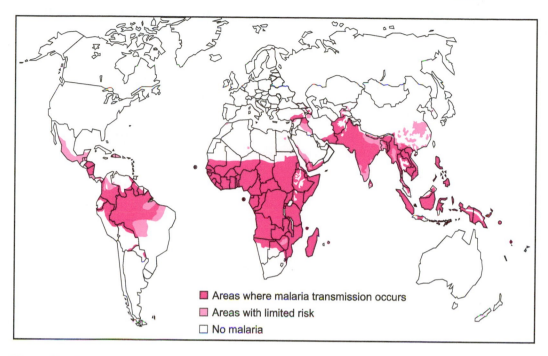

Figure 7.4 Map showing countries affected by malaria.

vaccination programmes do not exist in all the countries where the disease is endemic in Africa and South America. A haemorrhagic fever that causes jaundice in severe cases and can lead to death, yellow fever is also spread by the *A. aegypti* mosquito. It was a major cause of death in the nineteenth century, with epidemics across the United States and Europe; it is rarely reported in these countries today. It affects an estimated 200,000 people per year, with 30,000 deaths, but an effective vaccine does exist (Robertson *et al.*, 1996). More recently, Zika virus has been identified as another viral disease spread by the day-flying *A. aegypti* mosquito. First seen in Uganda in 1952, it was originally considered a mild disease causing a rash, fever and joint pain. However, an outbreak in Brazil in 2015 resulted in Guillain-Barré syndrome, an auto-immune neuralgic response, in some infected individuals, and an increase in pregnancy

complications and congenital abnormalities including microcephaly (WHO, 2018). Other mosquito-vector diseases include lymphatic filariasis (LF), a chronic nematode disease which results in swelling in different parts of the body. Treatments exist and preventive strategies can be implemented, but it can take many years to diagnose, as clinical symptoms develop slowly. Other filariasis diseases include dracunculiasis and onchocerciasis.

Incidence of the nematode-bacteria dis-ease onchocerciasis (river blindness), once a major cause of blindness worldwide, has been massively reduced in the last 20 years, due to international eradication programmes. Transmitted via the *Simulium* genus of black-flies, which breed in fast-flowing rivers in sub-Saharan Africa, Yemen and Latin Amer-ica, the nematodes can persist in the skin for many years, causing itching, swelling

Mosquitoes and malaria

The mosquito is the vector (an agent, usually an arthropod, which carries and transmits disease but remains unaffected itself) for malaria-causing *Plasmodium* species. There are almost 1 million deaths from malaria per year, with 85 to 90% of these occurring in sub-Saharan Africa. Many large organisations around the world are looking at methods to reduce the number of people who die or whose lives are blighted by this disease. The larval and pupal stages of the mosquito's life cycle are dependent on freshwater (Figure 7.5). By reducing the amount of standing water in malaria-affected regions, covering water-storage ponds to restrict access by mosquitoes, or use of biological controls in surface water, the number of mosquitoes can be reduced and rates of malaria infection lowered. Preventive medications are available but the high cost and long-term side effects of these drugs prohibit their widespread use for all but the short-term traveller.

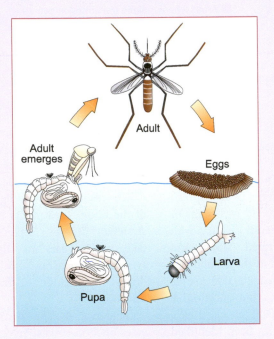

Figure 7.5 The life cycle of the mosquito is dependent on the presence of a water body.

Malaria is spread by the *Anopheles* species of mosquito (Figure 7.6). While there are in excess of 400 species of Anopheles, only 30–40 species are common vectors for the malaria plasmodium. Of these, *Anopheles gambiae* is the disease vector for the most dangerous malarial plasmodium *Plasmodium falciparum*. Only the adult female mosquitoes feed on blood. Adult female mosquitoes become infected with *Plasmodium* sporozoites after feeding on the blood of an infected human. The sporozoites are carried in the salivary gland of the mosquito for infection of subsequent humans. Of the over 200 species of *Plasmodium*, only about 11 infect humans. Of these, *P. falciparum* causes the severest type of malaria and is responsible for 90% of deaths from malaria; *Plasmodium vivax, Plasmodium ovale, Plasmodium malariae* and *Plasmodium knowlesi* result in a milder version of the disease.

Figure 7.6 Adult female mosquitoes become infected after feeding on human blood.
Source: Photo courtesy of Shutterstock/Kletr.

The mosquito is still found in areas where malaria has been eradicated (e.g. Europe, United States), and hence its reintroduction is possible. Climate change may make northern latitudes more susceptible to malarial infection.

and tissue damage, particularly in the lower limbs and eyes. When the worms die, they release a symbiotic bacterium that increases the inflammatory response, which, in chronic cases, can result in blindness. While the disease is not lethal, it can greatly alter the quality of life for the estimated 18 million people infected, and especially the 300,000 who are permanently blinded (Burnham 1998). Effective treatments for the disease are widely available, but blackfly control measures, particularly the use of insecticides, can greatly reduce transmission, as seen in the Onchocerciasis Control Programme in West Africa. For countries where vector-control is not considered feasible or cost-effective, it has been suggested that water-engineering projects such as slowing flow may help to reduce fly numbers by changing the habitat of blackfly breeding sites. Such measures may be expensive and damaging to the wider environment and may also increase breeding sites for other disease vectors preferring slow or still water, such as mosquitoes.

African trypanosomiasis (sleeping sickness) and American trypanosomiasis (Chagas disease) are protozoan diseases spread by the tsetse fly and assassin bug, respectively. Neither disease vector is dependent on water for its life cycle but the tsetse flies associated with the spread of both chronic and acute sleeping sickness are the riverine tsetse flies of the subgenus *Palpalis*. Insecticide use and land management regimes have resulted in temporary reduction of tsetse fly populations (Kagbadouno *et al.*, 2011), but medical

260 REBECCA J. SLACK

and veterinary programmes may circumvent the eradication of the fly.

7.4 CHEMICAL CONTAMINANTS

Water rarely contains just hydrogen and oxygen. Components that reflect the surrounding geology of a body of water (particularly if a groundwater resource); the interaction with aquatic organisms, sediment and atmospheric inputs; runoff from land; and anthropogenic inputs (Figure 7.7) can all determine the chemical load of water (see Chapter 4). Box 7.3 provides guideline values for the many different compounds found in drinking water. Most of these substances are present at low levels and hence will have little impact on human health but there are incidences where levels exceed water quality guidelines and hence pose a threat to health. This section

Figure 7.7 The River Yumana has been declared 'dead' due to pollution, yet humans still use the untreated water from the river. Disposal of solid waste to watercourses, as shown here, will alter chemical water quality, while human sewage effluent will alter both the biological and chemical content of water.
Source: Shutterstock/pzAxe.

considers the main issues concerning the chemical content of water and its impact on human health.

7.4.1 Chemicals derived from the environment

Arsenic is found in many freshwater resources but high levels are recorded in groundwater-dependent areas with strata rich in arsenic rock deposits (Smedley and Kinniburgh 2002; Kavcar *et al.*, 2009; see Chapter 5). High levels of inorganic arsenic have been reported in drinking water supplies in the Americas (Argentina, Chile, Mexico, the United States) and Asia (India, China, Thailand, Pakistan), but it is in Bangladesh that the problem has reached epidemic proportions. In the 1970s, large numbers of deep tube-wells were dug to tap groundwater reserves relatively free of pathogens found in existing water sources and responsible for one of the world's highest child mortality rates from waterborne disease. However, from the early 1990s, it

BOX 7.3

Maintaining water quality

While we may initially be concerned with the biological quality of drinking water, particularly the microorganism load contained in water resources, chemical compounds found in water are also important considerations for human health and well-being. A number of the chemicals found in water may be derived from natural sources such as arsenic and fluoride. However, many chemicals occur in water as a result of human activities, particularly organic compounds such as hydrocarbons and pesticides. In all developed countries and many developing countries, water quality legislation or guidelines have established limits for these chemicals. The limits and guidelines are continuously reviewed and updated to ensure the safety of drinking water. The World Health Organization has established the guideline values shown in Table 7.2 for chemicals that are of health significance in drinking water (WHO, 2017). National legislation in developed and many developing countries has set legally enforced limits for a number of chemicals based on available toxicological data, based on the World Health Organization guideline values (Table 7.3). In the United States, National Primary Drinking Water Regulations (NPDWRs or primary standards) are legally enforceable standards that apply to public water systems. NPDWRs protect public health by limiting the levels of contaminants in drinking water, including organic chemicals, inorganic chemicals including heavy metals, radionuclides, disinfectants and disinfection by-products, as well as microorganisms. The legally enforced limits are called maximum contaminant levels (MCLs) and represent the highest level of a contaminant allowed in drinking water. Maximum contaminant levels goals (MCLGs) represent the level below which there is no appreciable risk to health. Including a margin of safety and forming the basis of MCLs, MCLGs are not enforceable but do represent a preferred contaminant limit for drinking water. The Drinking Water Directive applies across all European Union (EU) Member States, ensuring harmonisation of high-quality water across the continent. Each EU Member State has implemented these regulations in national legislation. It is worth noting that a number of these regulatory instruments utilise the 'precautionary principle', which urges the adoption of cautious approaches when a risk is suspected but scientific consensus is still lacking (Principle 15 of the Rio Declaration on Environment and Development).

was apparent that a large number of these wells, an estimated 27% of all wells, were heavily contaminated with arsenic (at levels above 0.05 mg L^{-1} or 50 parts per billion (ppb)). Around contaminated tube-wells in Bangladesh and other countries, symptoms of arsenicosis were common: skin lesions (which, if infected, can become gangrenous), skin hardening, dark spots on limbs, swollen limbs and loss of feeling in hands and feet (Rasheed *et al.*, 2017). Children may show impaired cognitive development, and long-term exposure increases the risk of cancers among the estimated 35 to 77 million people exposed to high arsenic levels. Various programmes have been initiated to prevent further exposure to arsenic-contaminated water in Bangladesh, including development of alternative water supply options and arsenic removal systems. In the United States, arsenic levels rarely exceed the 50 ppb level but some areas do exceed the 'maximum contaminant level' permitted by the Environmental

Table 7.2 World Health Organization guideline concentrations for chemicals found in drinking water

Chemical	Guideline value (mg L^{-1})	Chemical	Guideline value (mg L^{-1})
Acrylamide	0.0005	Dimethoate	0.006
Alachlor	0.02	1,4-Dioxane	0.05
Aldicarb	0.01	Edetic acid	0.6
Aldrin and dieldrin	0.00003	Endrin	0.0006
Antimony	0.02	Epichlorohydrin	0.0004
Arsenic	0.01	Ethylbenzene	0.3
Atrazine and its chlorostriazine metabolites	0.1	Fenoprop	0.009
		Fluoride	1.5
Barium	0.7	Hexachlorobutadiene	0.0006
Benzene	0.01	Hydroxyatrazine	0.2
Benzo[a]pyrene	0.0007	Isoproturon	0.009
Boron	2.4	Lead	0.01
Bromate	0.01	Lindane	0.002
Bromodichloromethane	0.06	MCPA	0.002
Bromoform	0.1	Mecoprop	0.01
Cadmium	0.003	Mercury	0.006
Carbofuran	0.007	Methoxychlor	0.02
Carbon tetrachloride	0.004	Metolachlor	0.01
Chlorate	0.7	Microcystin-LR	0.001
Chlordane	0.0002	Molinate	0.006
Chlorine	5	Monochloramine	3
Chlorite	0.7	Monochloroacetate	0.02
Chloroform	0.3	Nickel	0.07
Chlorotoluron	0.03	Nitrate	50
Chlorpyrifos	0.03	Nitrilotriacetic acid	0.2
Chromium	0.05	Nitrite	3
Copper	2	N-Nitrosodimethylamine	0.0001
Cyanazine	0.0006	Pendimethalin	0.02
2,4-Db	0.03	Pentachlorophenol	0.009
2,4-DB	0.09	Selenium	0.04
DDTd and metabolites	0.001	Simazine	0.002
Dibromoacetonitrile	0.07	Sodium dichloroisocyanurate	50
Dibromochloromethane	0.1		
1,2-Dibromo-3-chloropropane	0.001	Sodium (as cyanuric acid)	40

(*Continued*)

Table 7.2 (Cont.)

Chemical	Guideline value (mg L^{-1})	Chemical	Guideline value (mg L^{-1})
1,2-Dibromoethane	0.0004	Styrene	0.02
Dichloroacetate	0.05	2,4,5-T	0.009
Dichloroacetonitrile	0.02	Terbuthylazine	0.007
1,2-Dichlorobenzene	1	Toluene	0.7
1,4-Dichlorobenzene	0.3	Tetrachloroethene	0.04
1,2-Dichloroethane	0.03	Trichloroacetate	0.2
1,2-Dichloroethene	0.05	Trichloroethene	0.02
Dichloromethane	0.02	2,4,6-Trichlorophenol	0.2
1,2-Dichloropropane	0.04	Trifluralin	0.02
1,3-Dichloropropene	0.02	Trihalomethanes	1 (max)
Dichlorprop	0.1	Uranium	0.3
Di(2-ethylhexyl)phthalate	0.008	Vinyl chloride	0.0003
		Xylenes	0.5

Protection Agency (Box 7.3), which was lowered in 2006 from 50 ppb to the World Health Organization's guideline value of 10 ppb (0.01 mg L^{-1}), leading to the development of a range of water treatment options, from point of use to large-scale treatment.

Fluoride occurs naturally in some water resources, often at levels exceeding recommended health levels of 0.7–1.2 parts per million (ppm) (0.7–1.2 mg L^{-1}); the World Health Organization's guideline value is 1.5 mg L^{-1} (WHO, 2017). While surface waters generally contain less than 0.5 ppm (0.5 mg L^{-1}), some groundwater reserves in Argentina, western states of the United States, northern Africa, eastern Africa, the Indian subcontinent and parts of China have fluoride levels exceeding 1.5 mg L^{-1}. High levels of fluoride in water can result in dental fluorosis, an unaesthetic brown mottling stain on the tooth surface, but a greater resistance to dental decay. As a consequence of lower levels of cavities, fluoride has been added at levels of between 0.5 and 1.0 ppm to the water supply of a number of regions/countries, including the United States, Brazil and Australia, to reduce the public health problem of tooth decay. However, fluoridation of water supplies has been met with mixed support – often welcomed by public health authorities but viewed with suspicion by the public. In a number of European countries and Japan, water fluoridation has been discontinued. Although the negative health impacts from water fluoridation are not well reported, high-level fluoride exposure can lead to skeletal fluorosis, a reduction in bone strength caused by calcium displacement, which is seen in a number of countries with naturally occurring high levels of fluoride.

Table 7.3 Limits and guidelines set for drinking water for selected pesticides and other organic substances

Contaminants	mg L^{-1}			
	MCLG	MCL	WHO	EU
Pesticides				
Atrazine	0.003	0.003	0.002	0.0001
2,4-D	0.07	0.07	0.03	0.0001
Dichlorprop			0.1	0.0001
Glyphosate	0.7	0.7		0.0001
Malathion				0.0001
MCPA			0.002	0.0001
Mecoprop			0.01	0.0001
Simazine	0.004	0.004	0.002	0.0001
Cumulative pesticides				0.0005
Other organic pollutants				
Benzene	0	0.005	0.01	0.001
Ethylbenzene	0.7	0.7	0.3	—
Toluene	1	1	0.7	—
Xylene(s)	10	10	0.5	—
1,1,1-trichloroethane	0.2	0.2	—	—
1,2-dichloroethane	0	0.005	0.03	0.003
Dichloroethylenes	0.1	0.1	0.05	—
Trichloroethylene	0	0.005	0.07	0.01
Tetrachloroethylene	0	0.005	0.04	0.01
Chlorobenzene	0.1	0.1	—	—
Dichloromethane	0	0.005	0.02	—
Pentachlorophenol	0	0.001	0.009	—
DEHP	0	0.006	0.008	—
Vinyl chloride	0	0.002	0.0003	0.0005
Polycyclic aromatic hydrocarbons	0	0.0002		0.0001
Polychlorinated biphenyls	0	0.0005	—	—

Notes: MCLG = maximum contaminant level goal (US-EPA); MCL = maximum contaminant level (US drinking water); WHO = World Health Organization guidelines; EU = European Union drinking water limits (applied to all Member States).

Radon is a naturally occurring radioactive gas formed during the decay of small amounts of uranium found in rock and soils, particularly granite and granite-derived alluvial soils. After tobacco smoking, radon is cited as the second most common cause of lung cancer. While levels in the environment are easily vented, levels in enclosed spaces such as homes or mines can build up, due to lack of ventilation. For example, areas of the UK, particularly Cornwall and Derbyshire, and the USA, particularly Iowa and the Appalachian Mountains, demonstrate elevated levels of radon gas in homes well in excess of the World Health Organization action or reference level of 100 Bq m^{-3}, but generally within the World Health Organization upper limit of 300 Bq m^{-3} (WHO, 2017). Drinking water sourced from groundwater in radon-affected areas may contain elevated levels of radon. While risks from ingestion are low, radon can be volatilised during showering and other activities, and may contribute 1 to 5% of indoor airborne radon. In the USA, a maximum contaminant level of 300 pCi L^{-1} (approximately 11,000 Bq m^{-3}) is used to restrict radon levels in drinking water. Water can be treated before entering homes if levels are particularly high.

Large-scale groundwater extraction in coastal areas can lead to saline intrusion and threaten water supplies. When freshwater is rapidly drawn from groundwater reserves along the coast, the water table is drawn down, causing seawater from the coastal fringe to be drawn into the aquifer (see Chapter 5). Saline intrusion renders the water supply unfit for consumption and for use in irrigation. Problems occur around the world, with areas of North Africa, Mediterranean Europe, the Middle East, the United States, China and Mexico all reporting saline contamination of groundwater reserves. Careful management of groundwater supplies and saltwater monitoring programmes can ensure that the risks of saline contamination are minimised. Coastal flooding may also temporarily increase the salinity of surface water sources, an issue that is mitigated by coastal sea defences but is likely to be a growing issue with climate change.

Cyanobacteria, often referred to as blue-green algae, occur in both freshwater and marine habitats (see Section 4.2.7 of Chapter 4). When conditions are favourable – usually a combination of nutrient availability, sunlight and warm temperature – cyanobacteria can reproduce rapidly to form a bloom (see Figure 6.17 in Chapter 6). Between 30 and 50% of blooms are relatively harmless to human health. However, as certain cyanobacteria species produce toxins (the most common is called microcystin), the bloom may pose a health threat to humans and other organisms that drink or swim in water contaminated by cyanobacteria. As water sources used for drinking water abstraction in developed countries are generally managed to control for blooms, risks to humans are low, unless they are exposed through recreational activities such as wild swimming and water sports. Even exposure via these means is unlikely to result in death, although exposure to high levels of cyanobacterial hepatotoxins may result in liver damage, particularly in children, while neurotoxins may affect the nervous system. Generally, symptoms will start with skin and/or eye irritation, followed by nausea, abdominal pain, diarrhoea and fever. A number of deaths have been reported in South America and China, associated with cyanobacteria blooms in reservoirs and dams supplying drinking water (Chorus and Bartram, 1999; WHO/UNICEF, 2003). However, the effect on animals is greater, as they may not have access to managed drinking water resources, and there have been many reports of fatal poisonings (due to liver or nervous system damage) in wild and domestic animals that drink from lakes and reservoirs affected by blooms.

7.4.2 Classical anthropogenic chemicals

Aquatic contaminants derived from human activities, particularly activities associated with agriculture, have long been a problem. While high nitrate levels in groundwater and surface water may result from natural processes, most nitrate contamination is a result of human activities, particularly the over-application of inorganic fertilisers (see Chapter 4). Nitrate is also a degradation product of other nitrogen compounds derived from animal manure or sewage sludge, runoff from animal husbandry, or wastewater treatment or septic tank release. High levels of nitrate in drinking water can lead to the development of methaemoglobin-aemia, or 'blue-baby syndrome', in infants under six months; after ingestion, nitrate is converted to nitrite, which reacts with haemo-globin to form methaemoglobin, resulting in significantly reduced oxygen-carrying capacity of the blood, and hence discoloration of the skin (Gupta *et al.*, 2000). Elevated nitrate in the environment contributes to **eutrophica-tion**, and hence affects the health of the aquatic ecosystem (see Section 6.5.5 in Chapter 6). Strict water quality guidelines for nitrate are in place in most developed countries as a safeguard against methaemoglobin-aemia, and these are usually based on the World Health Organization guideline value of 50 ppm (50 mg L^{-1}) of nitrate in drinking water (WHO, 2017). While water and waste-water treatment can remove nitrate or reduce levels, measures to prevent nitrate entering water, such as catchment-sensitive farming in nitrate vulnerable zones (a legal requirement in the UK) and other land and water manage-ment practices can reduce the need for and therefore the cost of treatment. Other nutri-ents, particularly phosphates, benefit from land management strategies to reduce input to water resources, reducing levels of eutrophication and the need for treatment.

Pesticides are very commonly used in agriculture and horticulture to reduce yield loss due to pests and diseases. In 1962, Rachel Carson's book *Silent Spring* first voiced the concern that pesticide use was having a wider environmental impact than was initially intended (Carson, 1962). Since then, most industrialised countries have instigated the monitoring of natural waters and drinking waters for pesticides, intro-duced legislation to control the types of pesti-cides that can be used, and set limits on the amount of pesticide residues permitted in drinking water. In the EU, individual pesti-cides must not be present in water above 0.0001 mg L^{-1} (equivalent to 0.1 µg L^{-1}; limits are lower, at 0.03 µg L^{-1}, for aldrin, dieldrin and heptachlor), with a cumulative pesticide limit of 0.0005 mg L^{-1}; limits set elsewhere are based on the individual toxicity of each pesticide chemical (Table 7.3). Despite this, it has been reported that almost 90% of US sur-face water bodies are contaminated by pesti-cides (Gilliom *et al.*, 2007). Many pesticides have been withdrawn from use, due to their persistent, bioaccumulative and toxic proper-ties such as DDT (dichlorodiphenyltrichlor-oethane); DDT is still, however, used in a number of African countries to control mosquitoes. Although pesticides can affect the health of the environment, consequences for human health are not so well defined. Acci-dental ingestion, inhalation or absorption through the skin of small amounts of pesti-cides during preparation or application can result in temporary ill-health, such as nausea or skin irritation; chronic exposure may lead to serious illness through increased risk of cancers, respiratory disease and neurological disorders (such as Parkinson's Disease) (Brown *et al.*, 2005). Exposure to the herbi-cides 2,4-D and 2,4,5-T, which constituted the defoliant Agent Orange used by the US mili-tary during the Vietnam War (1962–1971), led to elevated levels of ill-health and birth defects, mainly due to contamination by 2,3,7,8-tetrachlorodibenzo-*p*-dioxin (TCDD) (Erickson *et al.*, 1984). There are also con-cerns about the very widely used herbicide glyphosate, which often has much higher water quality limits than other pesticides (Table 7.3) and yet is now deemed by the IARC (International Agency for Research on

Cancer) to be 'probably carcinogenic to humans' (IARC, 2017). The consequences to human health through consumption of low-level contaminated water supplies are not well understood but are largely prevented, at least in the developed world, through strict drinking water quality limits, treatment regimes and pesticide control regulations.

A range of organic compounds associated with human activities can be discharged into the environment in wastewater if appropriate treatment processes do not exist. Most substances are associated with industrial effluent, particularly organic solvents used in manufacturing, including petrochemicals. Different organic chemicals impact upon the environment, and ultimately on human health, in different ways depending on the chemical, its concentration, whether or not it is discharged in a mixture with other chemicals, and the chemistry of the receiving waters. A common symptom of water pollution by organic substances is fish death, which usually results from oxygen depletion (hypoxia); hypoxia is also a water quality indicator for sewage/wastewater pollution, which may also be discharged with effluent containing organic pollutants. Many organic compounds are priority pollutants, and hence their levels in water are closely monitored (Box 7.3). Effective removal during wastewater treatment, particularly using tertiary treatment such as ozonation and ultraviolet radiation (see Chapter 8), should prevent entry into receiving waters; water treatment for drinking water supplies offers a secondary treatment process. In developed countries, environmental exposure will be through accidental release, which will result in heavy fines and the enforcement of 'polluter pays' principles. While human exposure will be unlikely, recreational exposure to pollution incidences may result in gastrointestinal or neurological illness. In developing countries, where wastewater and water

treatments are less effective, environmental and health consequences can be more extreme. Reports of high rates of birth defects in the Iraqi city of Fallujah have been suggested to be linked to chemical contamination of the city's water supply, although other studies suggest a link to depleted uranium (Alaani et al., 2011). Previously, it has been suggested in both developed and developing countries that perceived pockets of illness and disease, including birth defects, cancer clusters and increased infant mortality rates, may be linked to water and soil contaminated by waste disposal sites and other associated industrial activities, but for many cases the epidemiological evidence is inconclusive (Elliott et al., 2009). The most high-profile of all such cases is Love Canal in the USA (Vianna and Polan, 1984; Gensburg et al., 2009; Austin et al., 2011), where a community living above a closed chemical waste landfill in the 1970s reported high levels of illness and birth defects. This incident resulted in new federal legislation in 1980 focusing on hazardous waste landfills (Comprehensive Environmental Response, Compensation, and Liability Act (CERCLA)). Overall, although human health impacts from exposure to contaminated water are not well defined, environmental consequences are well documented and the toxicological profile of most chemicals is well described, hence the limits imposed on drinking water.

Heavy metal contamination of water can be a consequence of natural processes, such as weathering of rocks, or derived from human activities. Incidences of water contamination by heavy metals such as lead and chromium usually derive from human activities. Lead, a neurotoxin which has an adverse impact on the brain development of infants and young children, has been used to make water pipes for centuries, and in many older properties in the developed world (pre-1970 in the UK and much of Europe; pre-1986 for the USA) lead pipes may still be found. Generally, water utility companies have replaced older lead pipes with plastic pipes up to the point of entry to a household, but it is then generally the householder's responsibility to replace household lead water pipes with pipes made from copper or other materials. Households with lead pipes are recommended to run their taps to flush out standing water before using water for drinking or cooking, but should ultimately look to replace the pipework, particularly if young children are resident.

Chromium, unlike lead, is a water contaminant associated with industry. Runoff from contaminated land or poorly treated industrial effluent (particularly from dye works) can lead to elevated chromium levels in water sources, and ultimately in drinking water. In particular, hexavalent chromium (Cr(VI)) is a recognised human carcinogen, but its presence in drinking water is still common in many developed countries (Costa, 1997). With a US-EPA maximum contaminant level of 0.1 mg L^{-1} applicable to total chromium (including Cr(VI)), levels in tap water across the USA and other countries can sometimes exceed this for Cr(VI) alone. Removal through ion exchange, reverse osmosis and coagulation/flocculation treatment is effective.

Many metals and other substances can enter water as a consequence of water (or wastewater) treatment. Contamination incidences are rare and usually accidental, but have resulted in some significant public health issues. In the English town of Camelford in 1988, a large amount of aluminium sulfate, which was used in water treatment for coagulation/flocculation (see Chapter 8), entered the water supply and resulted in widespread health effects, which varied from urinary and dermal complaints, diarrhoea and nausea to more long-term health issues that have been attributed to the pollution incidence but are still disputed. Key among the long-term effects were osteodystrophy and damage to brain function, leading to dementia and Alzheimer's disease-like symptoms (Ward *et al.*, 1978). Incidences of chlorine leaks from water and wastewater treatment works have also been reported, often with considerable impact on the local environment.

7.4.3 Emerging contaminants

Groups of compounds have been identified over the last few decades which may pose a threat to human health and the health of the wider environment. Often, the evidence to support health links is still much disputed and it is far from clear what impact these contaminants have on health. Perhaps the most famous of these are endocrine-disrupting compounds (EDCs). These compounds can mimic hormones (e.g. the female hormone oestrogen), inhibit the action of hormones (e.g. antiandrogenic effects) or alter the normal regulatory effects of the endocrine system and, potentially, the nervous and immune systems. EDCs include a wide range of chemicals, most of which have been identified in natural waterways, derived from a wide range of sources such as:

- dichlorodiphenyltrichloroethane (DDT) and other pesticides
- polychlorinated biphenyls (PCBs)
- bisphenol A
- polybrominated diphenyl ethers (PBDEs)
- phthalates
- alkylphenols
- perfluorooctanoic acid (PFOA)
- dioxins and furans
- various pharmaceuticals
- complex chemical mixtures such as industrial effluents, landfill leachate and municipal sewage.

EDCs have been cited as being responsible for a range of reported environmental consequences, including declining fertility in fish, birds and mammals; reduced hatching success and offspring survival in fish, amphibians, birds and reptiles; demasculinisation/feminisation or defeminisation/masculinisation across all vertebrate groups, particularly aquatic animals; and thyroid function and development problems in fish and birds. Laboratory analyses in particular, demonstrate the endocrine disruption behaviour of individual chemicals or mixtures. The role of the toxic effect is still uncertain, with suggestions that EDCs have a direct impact on endocrine function or that the impact results indirectly from other stresses on other physiological functions. Evidence of impact on human health is less substantiated, with no causal relationship established between EDC exposure and health effects, despite suggestions that exposure to EDCs may increase rates of breast cancer and endometriosis in women, and testicular and prostate cancers in men; result in abnormal sexual development; reduce male fertility; alter pituitary and thyroid gland function; suppress the immune system; and have neurological effects. While levels of individual EDCs in the environment, and specifically in drinking water, may be very small and hence may easily be buffered by a healthy adult (Kuch and Ballschmiter, 2001), the impact of mixtures of these chemicals and the response of foetuses and newborn babies is still unclear.

Certain pharmaceuticals and veterinary medicines demonstrate endocrine-disrupting behaviour, which is essential for their therapeutic purpose, such as the female contraceptive 17-α-ethinyloestradiol. Chapter 6 describes some of the impacts of these chemicals observed on aquatic organisms. Low levels of pharmaceutical residues have long been detected in natural waterways and there is increasing evidence that these residues may also be found in drinking water, albeit at trace levels (see also Chapter 4). Such residues include a wide range of therapeutic drug categories, from analgesics to antibiotics and treatments for coronary and respiratory diseases. While levels are low, the health effects from long-term exposure to biologically active chemicals have not been fully realised (Kuemmerer, 2009).

Chlorine is the most common method of drinking water disinfection worldwide and it is very effective at killing pathogens that remain in the water following filtration. While chlorine is a relatively cheap and effective disinfectant, it can react with natural organic compounds in water to produce the disinfection by-products trihalomethanes (THMs) and haloacetic acids (HAAs). THMs and HAAs are classified as carcinogenic agents, so exposure to these compounds in drinking water may increase cancer risk (Boorman et al., 1999). To reduce the amounts of THMs and HAAs produced in disinfected water and comply with regulatory requirements, water companies attempt to remove as much of the organic content of water (particularly dissolved organic carbon (DOC)) as possible. Removing high levels of DOC from water can be a costly (financially and energy-wise) process and many water utility companies in the developed world are involved in research aimed at reducing DOC prior to water treatment. While ozonation and UV disinfection do not lead to the generation of harmful residues, they tend not to be as effective, or are suitable for use only in certain situations.

Nanomaterials are increasingly used in consumer products such as clothes and cosmetics (e.g. silver nanoparticles added to socks to prevent foot odour and zinc oxide in sunscreen), as well as in water and wastewater treatment processes. The properties of nanomaterials offer very cost-effective methods of improving health and well-being through cleaner water. However, the environmental fate and behaviour of nano-materials in treated sewage effluent and septic tanks and as airborne particles requires investigation to ensure that environmental exposure does not pose a health risk. Two factors in particular have raised concerns about the increasing use of nanoparticles: their chemical properties and size. At nanoscales, the properties of many chemicals change, due to alterations to the surface area-to-volume ratio, increasing their solubility in water. Due to their small size, nanoparticles have the potential to be taken up by organisms via dermal absorption, inhalation or ingestion. However, there is no evidence that yet suggests that nanomaterials commonly used in consumer products and in the treatment of water or wastewater have a negative impact on environmental and human health (Moore, 2006; Nowack and Bucheli, 2007). Work is ongoing in this area.

Microplastics and nanoplastics are an area of increasing environmental concern with many studies finding evidence of plastics in many different environmental compartments, from terrestrial ecosystems through to rivers, estuaries, seas and oceans: plastics are now considered to be truly global contaminants (Andrady, 2011). Various different sources are being identified including wastewater treatment works (through fibres released from washing of clothes among other human activities), landfill sites and waste disposal at sea (Kay *et al.*, 2018; Horton *et al.*, 2017; Browne *et al.*, 2011). The impacts on animal and human health, if any, are yet to be realised despite evidence that plastics can cross the gut barrier (Andrady, 2011): research is actively being carried out in this area.

> ## REFLECTIVE QUESTIONS
>
> What is the difference between naturally occurring chemical contaminants and pollutants created by anthropogenic activities? Do these two categories differ in the level of risk they pose to human health?
>
> What do you think are the main chemical contaminants of concern today and why are they problematic?
>
> What contaminants require further research to identify their health impacts?

7.5 TOO MUCH WATER VERSUS TOO LITTLE

The previous sections have largely focused on water quality, including microbiological load and chemical contamination; water quantity has not been considered except with regard to water-washed disease and water availability for personal hygiene. Yet, water availability is vital to support health – directly through the provision of drinking water, but also for food production. This section evaluates some of the complex issues that can be considered to be physical risks to health from water, from flooding to drought. Between 2006 and 2015 almost 2 billion people were affected by over 3000 water- or weather-related natural disasters, leading to over 340,000 deaths (IFRC, 2016).

7.5.1 Flooding

The physical process of flooding is described in Chapter 3. Flooding can lead to direct loss of life through drowning, but can exert

BOX 7.4

Water-related health impacts of hurricanes, earthquakes and tsunamis

Location: New Orleans, USA.

Date: 29 August 2005.

Event: Hurricane Katrina, Category 3 storm (landfall).

Result: The city of New Orleans is built on marshlands of the Mississippi River delta, with much of the city below sea level. Hurricane Katrina caused a storm surge of up to 9 metres in height along a 32-kilometre stretch of the Mississippi coastline, with New Orleans hit by a 6-metre-high surge. The New Orleans surge caused over-topping of a number of levees built to protect the city, while other concrete floodwalls collapsed as safety margins were exceeded. Approximately 80% of the city was flooded.

Health impacts: While 1836 deaths were associated with Hurricane Katrina, 700 were a result of the levee breaches in New Orleans. Over 1 million people from the central Gulf coast were relocated across the USA. About 300,000 homes were destroyed or made uninhabitable. Looting, including of food and water supplies, increased and crime rates generally rose.

Aftermath: New Orleans's population plummeted in the years following the flooding, from over 480,000 in the year 2000 to around 230,000 in July 2006. By July 2015, the population had recovered to almost 387,000. Unemployment doubled in the region, with economic loss as oil production shut down for six months following the hurricane, and crop loss from flood waters. It is estimated to have cost the US government in excess of $120 billion in emergency relief alone, and an estimated $160 billion in damage.

Location: Tohoku, Japan.

Date: 11 March 2011.

Event: Earthquake, magnitude 9.1.

Result: The earthquake resulted in a large tsunami wave which hit the north-east coast of Japan. The wave, in excess of 10 metres high, overtopped many of the existing sea defences. This was exacerbated by a drop in land level of up to 1 metre in places, due to ground-level movements resulting from the earthquake (Universitat Politècnica de Catalunya, 2011). Large areas of the coastline were flooded and infrastructure, including houses and roads, was severely damaged as the wave travelled up to 10 kilometres inland. At Fukushima I Nuclear Power Plant, three nuclear reactors went into meltdown following flooding of emergency generators and loss of power.

Health impacts: Of the 15,821 deaths recorded by April 2011, 93% were as a result of drowning: estimates by 2018 suggest over 20,000 deaths with 2,500 missing. A number of towns were almost completely destroyed and key services were lost for many more – 1.5 million households lost water supplies. Following the Fukushima nuclear disaster, residents within 20 kilometres of the plant were evacuated, with many unlikely to return to their homes, due to the high levels of radioactivity released during the meltdown.

Aftermath: With 300,000 people displaced and living in temporary accommodation, the area has suffered considerably, both socially and economically. Industrial production was suspended following the earthquake and tsunami. Material damage resulting from the earthquake and tsunami was estimated at $300 billion (25 trillion yen).

Location: Texas, USA.

Date: 25 August 2017.

Event: Hurricane Harvey, Category 4 storm (landfall).

Result: The storm resulted in the highest rainfall levels ever recorded in the USA. Large areas were flooded (Figure 7.8), including the fourth largest city in the USA, Houston (Figure 7.8).

Figure 7.8 Some of the flooding caused by Hurricane Harvey, August 2017, League City, Texas.
Source: Shutterstock/Cire notrevo.

Health impacts: 13 million were affected by the storm, 135,000 homes damaged or destroyed, an estimated 30,000 people required emergency shelter and 88 people lost their lives.

Aftermath: Unemployment surged to a two-year high in Texas after the storm, which is estimated to have caused $125 billion in damage, second only in estimated cost to Hurricane Katrina. Recovery was slowed as Hurricanes Irma and Maria resulted in resources being diverted.

Location: South-eastern USA and the Caribbean.

Date: September 2017.

Event: Hurricanes Irma and Maria, both Category 4/5 storms (landfall).

Result: Hurricane Irma swept through the Caribbean from 6 September 2017 as a Category 5 storm before moving to the west coast of Florida as a Category 4 storm on 9 September. Puerto Rico and other Caribbean islands were hit again on 20 September 2017 by Hurricane Maria. Both storms caused extensive flooding as well as high winds.

Health impacts: Hundreds of thousands of homes were damaged or destroyed from Antigua and Barbuda to Cuba, Puerto Rico and the Bahamas as Irma swept through the Caribbean, and left a trail of destruction in Florida and the southern states of the USA. Despite millions of people having been affected, the death toll is not estimated to be more than 200. Maria, however, is estimated to be responsible for 2975 deaths on Puerto Rico alone (Figure 7.9), although only 64 are directly attributable to the storm itself. The remaining deaths occurred in the six months after the storm due to lack of power, medicine and other essentials. On Dominica, which suffered damage to 85% of its housing stock, there was considerable concern regarding the availability of clean drinking water for at least a month after the storm.

Aftermath: Irma is estimated to have cost the USA approximately $50–70 billion (the fifth costliest storm) but estimates exceed $100 billion when other Caribbean countries are factored into the damage. Maria is currently the third costliest storm affecting the US, with estimates in excess of $100 billion. A year after Maria, Puerto Rico was still feeling the impact of the storm, with many schools still closed, chronic health effects at an all-time high and high levels of migration to mainland USA.

Figure 7.9 Satellite views of a part of Puerto Rico a year before (a, 23 September 2016) and a few days after (b, 26 September 2017) Hurricane Maria, showing areas of extensive flood waters post-hurricane.
Source: NASA's Earth Observatory

Location: Philippines and South China; south-east coast, USA.
Date: 14/15–16/18 September 2018.
Event: Typhoon Mangkhut, estimated Category 4 super typhoon (landfall Baggao, Philippines); Hurricane Florence, Category 1 storm (landfall North Carolina, USA).
Result: The simultaneous development of large tropical storms, called a hurricane if forming in the North Atlantic or north-east Pacific Oceans or typhoons if forming in the north-west Pacific Ocean, in September 2018 resulted in considerable damage and flooding along the east coast of the USA (Hurricane Florence) and in the Philippines and South China (Typhoon Mangkhut). While Hurricane Florence was downgraded from a Category 4 storm on landfall, its slow movement inland resulted in extremely heavy rainfall and extensive flooding. Typhoon Mangkhut caused extensive damage to buildings in the Philippines, Hong Kong and Guangdong.
Health impacts: Landslides in the Philippines were responsible for most loss of life: early warning allowed many of the estimated 4 million people in the path of the storm to seek shelter in

emergency shelters and leave coastal areas. Flooding following both storms posed health threats to all residents.

Aftermath: Extensive damage to property and farm land will, however, have a long-lasting impact on the region: the northern Philippines is considered the 'bread basket' of the country and, despite many farmers attempting to harvest as much of the rice crop as possible before the storm hit, much was not ripe. Similarly, US farms were hit hard by the storm, with stock lost and crops destroyed.

Final word: With the exception of the Caribbean islands and Typhoon Mangkhut in the Philippines, the above case studies feature incidents that affected areas of the developed world for which expensive flood prevention measures or emergency procedures were in place (although not necessarily fit for purpose). Similar incidences in the developing world can result in greater levels of mortality and morbidity, with longer-term health impacts resulting from lack of food, water and shelter as has been seen in Puerto Rico following Hurricane Maria. The regular monsoon floods which occur June–September almost every year in Bangladesh kill thousands and displace millions. The Federation of Red Cross and Red Crescent Societies estimate that 1200 died and 41 million were displaced in September 2017 across South Asia due to flooding when a third of Bangladesh was under water. Among those affected were the Rohingya refugee camps on the Myanmar border in Bangladesh, many of which had been located on flood plains. Which event in September 2017 garnered more press coverage in the developed world: the US hurricanes or South Asian monsoon floods?

a greater influence on health through disease, loss of financial security and social cohesion, and famine. Flood waters can easily become contaminated with sewage, leading to the spread of waterborne diseases such as cholera and typhoid. For instance, during the monsoon season in India and Bangladesh, levels of cholera peak rapidly (Koelle *et al.*, 2005; Akanda *et al.*, 2011). In many countries with a well-developed water supply and wastewater network, this risk is well recognised and can be mitigated against; countries lacking this infrastructure are placed at a greater risk. An increase in standing water, and emergency housing with little or no mosquito protection, may also result in an increase in vector-borne diseases such as malaria, dengue fever and West Nile fever.

The greatest impact that flooding has on health is associated with loss of shelter, civil unrest, crop failure, loss of transport links for movement of food and emergency aid, reduced access to health care professionals, loss of power, radiation leaks from damaged nuclear power plants, etc. (Watson *et al.*, 2007). Disaster and post-disaster recovery situations often place women and children at greater threat of violence and economic distress (Alston, 2012). Physical and mental health consequences are equally likely for those impacted by flooding. Some, if not all, of these effects were seen in recent natural disasters including Hurricane Harvey in the USA in 2017 and the Japanese tsunami following the earthquake of 2011 (Box 7.4). The impact of floods can be reduced through the construction of suitable barriers (coastal and riverine), maintaining floodplains (free from development), development of flood strategies including improved

BOX 7.5

Disease burden and socio-economic loss

Premature death, ill-health or disability can have significant impacts on the livelihoods of individuals, families, communities and populations. There are various measures that can be used to calculate the cost, both financial and non-financial, of disease, but increasingly most disease burden studies are using DALYs.

A DALY is a **d**isability **a**djusted **l**ife **y**ear and is defined by the World Health Organization as 'the sum of years of potential life lost due to premature mortality (years of life lost, YLL) and years of productive life lost through living in states of less than full health due to disease/injury (years of life lost due to poor health/disability, YLD)', so representing a common metric for mortality (death) and morbidity (poor health/disability) (see Equation [7.1]: the weight factor reflects the severity of the disease on a scale from 0 (perfect health) to 1 (death)). First developed by the World Bank, the DALY is now widely used as a disease burden measurement by a number of international organisations: one DALY represents the loss of the equivalent of one year of full health:

$$DALY = YLL + YLD \qquad [7.1]$$

where YLD = number of incident cases in that period × average duration of the disease × weight factor.

While most measures of disease burden (where 'disease' encompasses all factors influencing physical and mental health, including infectious agents, non-communicable illness, and physical factors such as flooding and earthquakes) consider mortality only, the DALY seeks to evaluate the consequences of a lifetime spent with poor health or a disability that may influence social and economic well-being, including, for example, the ability to earn money to feed and house a family. In 2004, the global average burden of disease was 237 DALYs per 1000 population, with about 60% due to premature death and 40% due to non-fatal health outcomes. For particular diseases, the difference between mortality and DALY may be small, while others show a considerable difference (Table 7.4). For water-related diseases, drowning always results in death (100% mortality), while diarrhoeal disease, the second-largest cause of global disease burden, has a mortality to morbidity ratio of 77%, and malaria a ratio of 68% (Table 7.4). Neuropsychiatric disorders, however, result in higher levels of morbidity (mortality:morbidity ratio of 16%), a significant disease burden that would rank much lower if considering mortality alone (Table 7.4). Figures for 2016 reveal that mortality and morbidity from diarrhoeal disease both declined (death:DALY=3.03:3.12), while mortality associated with malaria declined with morbidity increasing (death:DALY=1.32:2.36) as a percentage of global disease burden (Global Health Data Exchange, 2018).

Table 7.4 Global burden of disease for selected factors (based on data presented by WHO, 2004)

Disease	Deaths (% of global total)	DALYs (% of global total)
Diarrhoeal disease	3.7	4.8
Malaria	1.5	2.2
Drowning	0.7	0.7
Neuropsychiatric disorder	2.1	13.1
(Unipolar depressive disorder)	(0)	(4.3)

DALYs provide an indication of where the greatest social, economic and health needs are, helping to prioritise intervention strategies such as improved sanitation or the provision of mosquito nets. Projections comparing 2004 global DALYs to 2030 suggest that diarrhoeal disease will fall from 4.8% of global DALYs, the second leading cause of burden of disease, to 1.6%, a ranking of 18 in the burden list. Early indications suggest that these projections are relatively accurate, with 2016 figures suggesting a global DALY for diarrhoeal disease of 3.12%. This suggests that efforts to address **Sustainable Development Goals** (SDGs) (SDG 6 addresses access to safe drinking water and basic sanitation for all) are beginning to make a real difference but will only be sustained with the continued economic development of low-income countries.

DALYs are usually calculated on the basis of age and sex only, although geographic location may also be considered. However, not all ages are considered equally, with age weightings and time discounting frequently applied to the calculation of YLL and YLD. The World Health Organization applies non-uniform age weights that give less weight to years lived at young and older ages (**Figure 7.10**) and a 3% discounting rate so that future years lost are not considered equally as future uncertainty increases (a 3% discount suggests that a weighted year of life saved next year is worth 97% of a year of life saved this year). Using discounting and age weights, a death in infancy corresponds to 33 DALYs, and deaths at ages 5–20 years to around 36 DALYs.

Figure 7.10 Age weightings applied to calculation of DALYs.

housing standards, and increasing awareness of populations living in flood-prone areas.

7.5.2 Drought

Where water is scarce, low-quality water resources may be consumed, increasing the likelihood of waterborne disease or illness from water contaminated by chemical residues. Drought is also linked to famine, conflict and civil unrest, and is a key factor in many social and economic disasters around the world, from the Dust Bowl of the Midwestern states of the United States (Schubert *et al.*, 2004) and Canada during the 1930s (see **Box 2.2** in Chapter 2) to famines and war in sub-Saharan Africa in more recent years. In many drought-prone countries, women are at particular risk, a factor which is explored further in Section 7.6. Reducing the impact of drought on human health and the environment can include the construction of water-storage ponds or containers (although this will also increase the number of mosquito breeding sites, and so elevate malaria risk), planting of drought-resistant crops, development of drought management strategies, water-sharing mechanisms, land management and aid programmes.

7.5.3 Food production

Food production is reliant on the provision of water for irrigation of crops, drinking water/ food for cattle and the processing/cooking of food. Particularly in sub-Saharan Africa, drought is the main factor behind crop failure, but too much water can also have a catastrophic effect (as seen in the Irish potato famine in the 1840s, when wet and warm conditions provided the ideal environment for the spread of potato blight, *Phytophthora infestans*). Malnourishment further aids infection by waterborne diseases,

increasing the risk of death, as compared to well-nourished individuals.

Many countries around the world are already suffering considerable water stress, which will be exacerbated by climate change and increasing population. To meet increasing demands for food production, and also for the production of biofuels for energy, more land will need to be taken into production and a greater reliance will be placed on irrigation. Agriculture is already responsible for the largest abstraction of water, with a global average of just under 70% of abstracted freshwater used directly in agriculture (see **Figure 9.1** in Chapter 9), but the total climbs to 86% of abstracted freshwater if we include the processing of agricultural products (Hoekstra and Chapagain, 2008). As rainfall becomes less predictable in more regions, a shift in agricultural practices towards more drought-tolerant plant varieties is needed to ensure a resilient food supply.

REFLECTIVE QUESTIONS

What are the key health risks associated with flooding and droughts?

What steps might be taken to help reduce health risks associated with too much or too little water?

7.6 SOCIAL AND ECONOMIC WATER RISK

Clean, safe water and good sanitation may be available, but not accessible to all in any given population. The barriers may be physical, for instance, not being connected to a water pipe

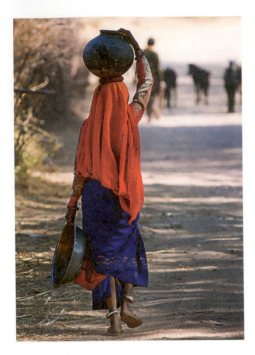

Figure 7.11 Collecting water can often involve a walk of several kilometres. For children, this may mean they do not have time to attend school, while for women such walks can expose them to gender-based violence.
Source: Shutterstock/Tim Booth.

or sewer, but will largely result from social and/or economic factors. There is a well-defined difference between water accessibility in developed countries as compared to developing countries.

In the developed world, all permanent homes have a water supply (either public or private) and sewerage connection (sewer network or septic tank), and householders pay for clean water and to dispose of wastewater. Non-payment of water bills will rarely result in a home being disconnected from a public water supply and sewerage network, due to the health issues that would result. Social and economic exclusion from accessibility to water and sanitation is therefore rare. However, as water metering is increasingly adopted in many developed countries, poorer households may seek to reduce their water use, at a risk to hygiene and sanitation practices.

Developing countries may be at greater risk from social and financial barriers to water supply and sanitation. People living in slum areas or in rural locations will be particularly vulnerable, owing to poor water/wastewater infrastructure (social exclusion). Similarly, householders may be unwilling or unable to pay for connections to existing networks (economic exclusion). Civil unrest, corruption or lack of political will may also hamper access to water supply and sanitation, ultimately leading to impacts upon health and well-being. Cultural or religious beliefs might also act as a barrier, with sanitation provision within the home regarded as impure in some faith communities, leading to a preference for open defecation even when latrines are provided (George, 2015). Death or chronic diseases resulting from waterborne infection or other water-related issues, such as crop failure due to too much/too little water, can also have a significant impact on the social and economic status of an individual, community or population (Box 7.5). Social and financial barriers to water accessibility are crucial to the consideration of water and health in developing countries, and Chapters 10 and 11 discuss these issues in more detail.

Gender plays a significant role in water, sanitation and health. It is women in many communities who source water for use in the home, often travelling long distances to procure water for cooking and washing. This travel affects the time they have available to fulfil other roles within society and also places them at greater risk of violence as they travel through often remote areas, with greater distances needing to be covered in times of drought. Water collection might also be delegated to the children of the household (Figure

7.11); while this allows women to have a greater role in the family and society, it often prevents children from completing their education and places them at risk of violence, accidents and other hazards. Poor sanitation provision also poses a greater risk for women, creating opportunities for violence, stress and personal hygiene during menstruation where open defecation is the only available option (Khanna and Das, 2015; Caruso *et al.*, 2017). It is increasingly recognised that provision of clean and safe toilets not only reduces environmental pathogen load but also contributes significantly to the health, well-being and safety of women.

REFLECTIVE QUESTIONS

What are the main differences in socio-economic water-related health risks in the developed and developing worlds?

Are there particular groups in society for whom inadequate water, sanitation and hygiene (WASH) pose greater health risks?

What other barriers exist to improving WASH for as many people as possible, as defined by the UN's Sustainable Development Goal 6?

7.7 HOW SAFE IS YOUR WATER?

Water and sewerage distribution networks and treatment systems are a staple of all public health schemes in the developed world, and are discussed in more detail in Chapter 8. Stringent drinking water quality standards have been implemented in all industrialised countries, which usually adopt World Health Organization guideline values (Box 7.3). There is an increasing focus on ecological water quality, as demonstrated by the EU's Water Framework Directive, which seeks to improve the chemical, biological and geomorphological status of all European water bodies (see also Chapters 4 and 6). With advanced treatment facilities available for the last 100 years for the removal of microorganisms from water, there has been a greater emphasis on improving the chemical quality of water (e.g. pesticide pollution, emerging pollutants) and the quality of water in the environment, particularly with regard to freshwater biodiversity and recreational water use, such as for swimming. Many water-related diseases in developed countries have been eliminated through eradication programmes (particularly if vector-related), improvements in water supply and sanitation, and hygiene education (Box 7.6). The high quality of drinking water means that the main water-related health concern in many developed countries is exposure during recreation, when diseases such as leptospirosis and cryptosporidiosis may be contracted.

In developing countries, major economic investment in public water supply and sanitation has not been as common as in developed countries. Combined with more favourable climates for many water-related diseases, greater water resource pressures (e.g. lower levels of rainfall), and poorly implemented legislation regulating the standards of water quality, as well as a range of social and economic factors affecting access to clean water and good sanitation, the health impacts associated with water-related diseases are greater. While improvement to water access will help to improve health, measures should also be adopted that prevent the further spread of diseases such as malaria and yellow fever, which are responsible for thousands of deaths per year. Reducing illness associated with poor quality (or quantities) of water will help

BOX 7.6

Minimising disease

There are a number of strategies involving water that can be followed to improve health. Provision of clean water and sanitation facilities considerably reduces the transmission of water-related disease. However, hand-washing is one of the most important ways of controlling the spread of infections, especially those that cause diarrhoea and vomiting, and respiratory disease. The recommended method is the use of soap, water and paper towels. Health care professionals are taught a standard procedure, as shown in Figure 7.12 (Ayliffe *et al.*, 1992), for maximising the effectiveness of hand-washing to ensure that the whole hand is washed for the removal of transient microorganisms. If adopted by all, particularly after defecation and before food preparation, the spread of many infectious diseases will be considerably reduced.

Hand washing technique

1 Palm to palm

2 Left palm over right dorsum and right palm over left dorsum

3 Palm to palm fingers interlaced

4 Backs of fingers to opposing palms with fingers interlocked

5 Rotational rubbing of left thumb clasped in right and vice versa

6 Rotational rubbing, backwards and forwards with clasped fingers of right hand in left palm and vice versa

Figure 7.12 An appropriate hand-washing technique to minimise the spread of disease.

to improve the economic well-being of poor communities, allowing money that would otherwise be spent on treatment to be shifted to education, food and improving household services.

Currently, the safety of your water is very much dependent on your geographical location. In industrialised and developed countries, high-quality public water supplies contribute to healthy societies; in many developing countries, public water supplies may not exist and a range of other factors (greater variety of pathogens, water scarcity, lack of political drivers, etc.) contribute to a greater health risk from water.

> **REFLECTIVE QUESTION**
>
> Why is it important to thoroughly wash hands?

7.8 SUMMARY

Both water quantity and water quality are important when considering the impact of water on human health and the health of the environment. Droughts and periods of water scarcity can lead to famine and death; too much water can also lead directly to death by drowning but can have larger-scale impacts through the spread of disease, food shortages, loss of shelter and livelihood disruption. Many organisms need water for some or all of their life cycle, and some cause disease in humans. Infectious diseases associated with water can be divided into four categories, based on their infection pathways:

- waterborne diseases – diseases spread by water, such as cholera and typhoid, including those that are inhaled in water droplets (e.g. Legionnaires' disease);
- water-washed diseases – infections spread by lack of water (for personal hygiene), and insanitary conditions that may not directly involve water as a means of infection transfer (e.g. scabies and typhus);
- water-based diseases – infections transmitted by aquatic invertebrates (e.g. schistosomiasis);

- vector-based water-related diseases – infections, of which malaria is the most common, spread by insects that need water to complete their life cycle.

Chemicals in water can also affect health. Three causes of non-communicable diseases can be considered, including:

- environmentally available chemicals such as arsenic;
- anthropogenic chemicals – contaminants from agriculture and other 'traditional' human activities, such as pesticides and fertiliser runoff;
- anthropogenic chemicals – emerging contaminants such as endocrine-disrupting compounds and disinfection by-products.

Treating water before use, through filtration and disinfection, can reduce infection rates. Separation of water and wastewater, with sanitation provision and facilities for good personal hygiene, also improves health by reducing the levels of many water-related communicable diseases. Water-based and vector-based disease levels do not respond as well to improvements in water quality; alternative strategies that prevent contact with the transmitting organisms are therefore required.

The abundance and scarcity of water will become an increasing feature of daily life for billions of people around the world as extreme weather events increasingly impact on lives and livelihoods. Recognising these risks and planning for them will help significantly in mitigating these impacts and so minimising consequences for health, both nationally and internationally through world food markets.

Developed and developing countries face different water-related health challenges. Developed countries are better able to adapt to water quantity issues, such as flooding and drought, and to biological threats, through well-developed treatment and distribution networks, meaning that the chemicals in water are of greatest concern. In developing countries, quantity and biological issues prevail, although there are incidences of extensive chemical contamination of water supplies, such as arsenic in the Indian subcontinent. There might also be a disparity within countries between the water available to the economically deprived compared to their wealthier counterparts. Gender issues, culture and faith are factors increasingly recognised as having an impact on health outcomes of communities, particularly with regard to sanitation.

FURTHER READING

World Health Organization factsheets: www.who.int/topics/infectious_diseases/factsheets/en

> The World Health Organization regularly update their factsheets on various water-related infectious diseases as well as providing information on other-related health impacts such as flooding and impact of extreme weather events.

Bennett, J.E., Dolin, R. and Blaser, M.J. (eds). 2015. *Mandell, Douglas and Bennett's Principles and Practice of Infectious Diseases (Volume 2)* (8th edition). Elsevier Churchill Livingstone; Philadelphia, PA.

> A key text providing further information on water-related diseases caused by pathogens.

> Fascinating research into infectious disease and water associations over the last 20 years is presented in the following four texts:

Fewtrell, L. and Bartram, J. (eds). 2001. *Water Quality Guidelines, Standards and – Assessment of Risk and Risk Management for Water-related Infectious Disease.* IWA Publishing; London.

Hunter, P.R., Waite, M. and Ronchi, E. (eds). 2002. *Drinking Water and Infectious Disease: Establishing the Links.* CRC Press; London.

Mara, D. and Horan, N.J. (eds). 2003. *Handbook of Water and Wastewater Microbiology.* Academic Press; New York.

Selendy, J.M.H. (ed.). 2011. *Water and Sanitation Related Diseases and the Environment: Challenges, Interventions and Preventive Measures.* Wiley-Blackwell; Chichester

Classic papers

Carson, R. 1962. *Silent Spring.* Houghton Mifflin; New York.

> The first report of concerns between widespread use of chemicals, particularly pesticides, and their potential impact upon the health of the environment.

Fraser, D.W. *et al.* 1977. Legionnaires disease – description of an epidemic of pneumonia. *New England Journal of Medicine* 297: 1189–1197.

> A paper describing the early stages of identifying the aetiology of a newly identified water-related disease.

Kolpin, D.W., Furlong, E.T., Meyer, M.T., Thurman, E.M., Zaugg. S.D., Barber, L.B. and Buxton, H.T. 2002. Pharmaceuticals, hormones, and other organic wastewater contaminants in US streams, 1999–2000: a national reconnaissance. *Environmental Science & Technology* 36: 1202–1211.

> A comprehensive assessment of emerging contaminants in natural waterways.

Murray, C.J.L. and Lopez, A.D. 1997. Global mortality, disability, and the contribution of risk factors: global burden of disease study. *Lancet* 349: 1436–1442.

> A demonstration of the use of DALYs in disease burden studies by the authors credited with the development of DALYs.

PROJECT IDEAS

- Globally, different approaches are being adopted to reduce the impact of water-mediated diseases. The different approaches range from policy and regulation (e.g. an aim/legal requirement to ensure that a certain proportion of a population receives clean and safe drinking water), through engineered solutions (e.g. water treatment systems, construction of reservoirs, etc.) and eradication programmes to public education campaigns. All possess an inherent but often very different economic cost. Research the benefits and disadvantages of the different approaches adopted to help understand how successful different schemes can be in different financial, social and environmental situations.

- During disasters, diseases such as cholera and malaria may rise to epidemic proportions. Review existing guidelines, together with an evaluation of current understanding of disease transmission and control. What recommendations can you make following this review that could contribute significantly to improved disaster planning?

- Concern regarding different aquatic contaminants tends to ebb and flow: in the late 1990s there was considerable concern about the health impacts of potential endocrine-disrupting chemicals, while more recently it is microplastics that are the focus, despite research into these areas generally beginning at the same time. Look at the different reasons that might raise the profile of different contaminants at any one time and reflect on the wider influences of scientific research.

REFERENCES

Akanda, A.S., Jutla, A.S., Alam, M., de Magny, G. C., Siddique, A.K., Sack, R.B., Huq, A., Colwell, R.R. and Islam, S. 2011. Hydroclimatic influences on seasonal and spatial cholera transmission cycles: implications for public health intervention in the Bengal Delta. *Water Resources Research* 47: W00H07. doi:10.1029/2010WR009914.

Alaani, S., Tafash, M., Busby, C., Hamdan, M. and Blaurock-Busch, E. 2011. Uranium and other contaminants in hair from the parents of children with congenital anomalies in Fallujah, Iraq. *Conflict and Health* 5: 15. doi:10.1186/1752-1505-5-15.

Alston, M. 2012. Gender-based-violence in post-disaster recovery situations: an emerging public health issue. In: Nakray, K. (ed.), *Gender Based Violence and Its Impact on Health*. Routledge; New York.

Andrady, A.L. 2011. Microplastics in the marine environment. *Marine Pollution Bulletin* 62: 1596–1605.

Austin, A.A., Fitzgerald, E.F., Pantea, C.I., Gensburg, L.J., Kim, N.K., Stark, A.D. and Hwang, S-A. 2011. Reproductive outcomes among former Love Canal residents, Niagara Falls, New York. *Environmental Research* 111: 693–701.

Ayliffe, G.A.J., Lowbury, E.J.C., Geddes, A.M. and Williams, J.D. 1992. *Control of Hospital Infection; A Practical Handbook* (3rd edition). Chapman and Hall; London.

Bath, J.L., Eneh, P.N., Bakken, A.J., Knox, M.E., Schiedt, M.D. and Campbell, J.M. 2010. The impact of perception and knowledge on the treatment and prevention of intestinal worms in the Manikganj District of Bangladesh. *Yale Journal of Biology and Medicine* 83: 171–184.

Black, R.E., Deromana, G.L., Brown, K.H., Bravo, N., Bazalar, O.G. and Kanashiro, H.C. 1989. Incidence and etiology of infantile diarrhea and major routes of transmission in Huascar, Peru. *American Journal of Epidemiology* 129: 785–799.

Boorman, G.A., Dellarco, V., Dunnick, J.K., Chapin, R.E., Hunter, S., Hauchman, F. Gardner, H., Cox, M. and Sills, R.S. 1999. Drinking water disinfection byproducts: review and approach to toxicity evaluation. *Environmental Health Perspectives* 107: 207–217.

Brown, R.C., Lockwood, A.H. and Sonawane, B. R. 2005. Neurodegenerative diseases: an overview of environmental risk factors. *Environmental Health Perspectives* 113: 1250–1256.

Browne, M.A., Crump, P., Niven, S.J., Teuten, E.L., Tonkin, A., Galloway, T. and Thompson, R.C. 2011. Accumulations of microplastic on shorelines worldwide: sources and sinks. *Environmental Science & Technology* 45: 9175–9179.

Burnham, G. 1998. Onchocerciasis. *Lancet* 351: 1341–1346.

Carson, R. 1962. *Silent Spring.* Houghton Mifflin; New York.

Caruso, B.A., Clasen, T.F., Hadley, C., Yount, K.M., Haardörfer, R., Rout, M., Dasmohapatra, M. and Cooper, H.L. 2017. Understanding and defining sanitation insecurity: women's gendered experiences of urination, defecation and menstruation in rural Odisha, India. *BMJ Global Health* 2: e000414.

CDC. 2018. *Legionella (Legionnaires' Disease and Pontiac Fever).* Centers for Disease Control and Prevention [online, accessed April 2019]. Available at: www.cdc.gov/legionella/index.html.

Chorus, I. and Bartram, J. 1999. *Toxic Cyanobacteria in Water: A Guide to Their Public Health Consequences, Monitoring and Management.* E and FN Spon; London.

Costa, M. 1997. Toxicity and carcinogenicity of Cr(VI) in animal models and humans. *Critical Reviews in Toxicology* 27: 431–442.

Crump, J.A., Luby, S.P. and Mintz, E.D. 2004. The global burden of typhoid fever. *Bulletin of the World Health Organization* 82: 346–353.

Elliott, P., Richardson, S., Abellan, J.J., Thomson, A., de Hoogh, C., Jarup, L. and Briggs, D.J. 2009. Geographic density of landfill sites and risk of congenital anomalies in England. *Occupational and Environmental Medicine* 66: 81–89.

Ellis, J., Oyston, P.C.F., Green, M. and Titball, R. W. 2002. Tularemia. *Clinical Microbiology Reviews* 15: 631–646.

Erickson, J.D., Mulinare, J., McClain, P.W., Fitch, T.G., James, L.M., McClearn, A.B. and Adams, M.J. 1984. Vietnam veterans risks for fathering babies with birth-defects. *JAMA – Journal of the American Medical Association* 252: 903–912.

Esrey, S.A., Potash, J.B., Roberts, L. and Shiff, C. 1991. Effects of improved water-supply and sanitation on ascariasis, diarrhea, dracunculiasis, hookworm infection, schistosomiasis, and trachoma. *Bulletin of the World Health Organization* 69: 609–621.

Fayer, R., Morgan, U. and Upton, S.J. 2000. Epidemiology of cryptosporidium: transmission, detection and identification. *International Journal for Parasitology* 30: 1305–1322.

Fewtrell, L. and Bartram, J. (eds). 2001. *Water Quality: Guidelines, Standards and Health – Assessment of Risk and Risk Management for Water-related Infectious Disease.* IWA Publishing; London.

Fraser, D.W., Tsai, T.R., Orenstein, W., Parkin, W. E., Beecham, H.J., Sharrar, R.G., Harris, J., Mallison, G.F., Martin, S.M., McDade, J.E., Shepard, C.C. and Brachman, P.S. 1977. Legionnaires disease – description of an epidemic of pneumonia. *New England Journal of Medicine* 297: 1189–1197.

Gensburg, L.J., Pantea, C., Fitzgerald, E., Stark, A., Hwang, S-A. and Kim, N. 2009. Mortality among former Love Canal residents. *Environmental Health Perspectives* 117: 209–216.

George, R. 2015. Faith and toilets. *Scientific American* 313: 19 November 2015.

Gillion, R.J., Barbash, J.E., Crawford, C.G., Hamilton, P.A. 2007. *Pesticides in the Nation's Streams and Ground Water, 1992–2001.* Reston: USGS. U.S. Geological Survey Scientific Investigations Report 2009–5189.

Global Health Data Exchange. 2016. *GBD Results Tool.* Available at: http://ghdx.healthdata.org/gbd-results-tool. Institute for Health Metrics and Evaluation.

Gryseels, B., Polman, K., Clerinx, J. and Kestens, L. 2006. Human schistosomiasis. *Lancet* 368: 1106–1118.

Gubler, D.J. 1998. Dengue and dengue hemorrhagic fever. *Clinical Microbiology Reviews* 11: 480–496.

Gupta, S.K., Gupta, R.C., Seth, A.K., Gupta, A.B., Bassin, J.K. and Gupta, A. 2000. Methaemoglobinaemia in areas with high nitrate concentration in drinking water. *National Medical Journal of India* 13: 58–61.

Hall, T.F., Penfound, W.T. and Hess, A.D. 1946. Water level relationships of plants in the Tennessee Valley with particular reference to malaria control. *Journal of the Tennessee Academy of Science* 21: 18–59.

Halliday, S. 2001. *The Great Stink of London: Sir Joseph Bazalgette and the Cleansing of the Victorian Metropolis.* History Press; Stroud.

Hempel, S. 2007. *The Medical Detective: John Snow, Cholera and the Mystery of the Broad Street Pump.* Granta Books; London.

Hoekstra, A.Y. and Chapagain, A.K. 2008. *Globalization of Water: Sharing the Planet's Freshwater Resources.* Blackwell; Oxford.

Horton, A.A., Walton, A., Spurgeon, D.J., Lahive, E. and Svendsen, C. 2017. Microplastics in freshwater and terrestrial environments: evaluating the current understanding to identify the knowledge gaps and future research priorities. *Science of the Total Environment* 586: 127–141.

Hunter, P.R., Waite, M. and Ronchi, E. (eds). 2002. *Drinking Water and Infectious Disease: Establishing the Links.* CRC Press; London.

IARC (2017) *Some Organophosphate Insecticides and Herbicides.* IARC Monographs on the Evaluation of Carcinogenic Risks to Humans Volume 112. Available at: http://publications.iarc.fr/Book-And-Report-Series/Iarc-Monographs-On-The-Evaluation-Of-Carcinogenic-Risks-To-Humans/Some-Organophosphate-Insecticides-And-Herbicides-2017.

IFRC (International Federation of Red Cross and Red Crescent Societies). 2016. *World Disasters Report 2016: Resilience: Saving Lives Today, Investing for Tomorrow.* IFRC; Geneva.

Kagbadouno, M.S., Camara, M., Bouyer, J., Courtin, F., Onikoyamou, M.F., Schofield, C.J. and Solano, P. 2011. Progress towards the eradication of tsetse from the Loos Islands, Guinea. *Parasites & Vectors* 4: 18.

Kavcar, P., Sofuoglu, A. and Sofuoglu, S.C. 2009. A health risk assessment for exposure to trace metals via drinking water ingestion pathway. *International Journal of Hygiene and Environmental Health* 212: 216–227.

Kay, P., Hiscoe, R., Moberley, I., Bajic, L. and McKenna, N. 2018. Wastewater treatment plants as a source of microplastics in river catchments. *Environmental Science and Pollution Research* 25: 20264–20267.

Khanna, T. and Das, M. 2016. Why gender matters in the solution towards safe sanitation? Reflections from rural India. *Global Public Health* 11: 1185–1201.

Koelle, K., Rodo, X., Pascual, M., Yunus, M. and Mostafa, G. 2005. Refractory periods and climate forcing in cholera dynamics. *Nature* 436: 696–700.

Kolpin, D.W., Furlong, E.T., Meyer, M.T., Thurman, E.M., Zaugg, S.D., Barber, L.B. and Buxton, H.T. 2002. Pharmaceuticals, hormones, and other organic wastewater contaminants in US streams, 1999–2000: a national reconnaissance. *Environmental Science & Technology* 36: 1202–1211.

Kotloff, K.L., Winickoff, J.P., Ivanoff, B., Clemens, J.D., Swerdlow, D.L., Sansonetti, P.J., Adak, G.K. and Levine, M.M. 1999. Global burden of shigella infections: implications for vaccine development and implementation of control strategies. *Bulletin of the World Health Organization* 77: 651–666.

Kuch, H.M. and Ballschmiter, K. 2001. Determination of endocrine-disrupting phenolic compounds and estrogens in surface and drinking water by HRGC-(NCI)-MS in the picogram per liter range. *Environmental Science & Technology* 35: 3201–3206.

Kuemmerer, K. 2009. The presence of pharmaceuticals in the environment due to human use – present knowledge and future challenges. *Journal of Environmental Management* 90: 2354–2366.

Kyrou, K., Hammond, A.M., Galizi, R., Kranjc, N., Burt, A., Beaghton, A.K., Nolan, T. and Crisanti, A. 2018. A CRISPR-Cas9 gene drive targeting doublesex causes complete population suppression in caged *Anopheles gambiae* mosquitoes. *Nature Biotechnology* 35: 1062–1066.

Levett, P.N. 2001. Leptospirosis. *Clinical Microbiology Reviews* 14: 296–326.

Mackintosh, C.A. and Hoffman, P.N. 1984. An extended model for transfer of microorganisms via the hands – differences between organisms and the effect of alcohol disinfection. *Journal of Hygiene* 92: 345–355.

Mandal, B.K., Wilkins, E.G.L., Dunbar, E.M. and Mayon-White, R.T. 2005. *Lecture Notes on Infectious Diseases* (6th edition). Blackwell Publishing; London.

Mandell, G.L., Bennett, J.E. and Dolin, R. (eds). 2010. *Principles and Practice of Infectious Diseases (Volume 2)* (7th edition). Elsevier Churchill Livingstone; Philadelphia, PA.

Mara, D. and Horan, N.J. (eds). 2003. *Handbook of Water and Wastewater Microbiology*. Academic Press; New York.

Marr, J.S. and Cathey, J.T. 2010. New hypothesis for cause of an epidemic among Native Americans, New England, 1616–1619. *Emerging Infectious Diseases* 16: 281–286.

Moore, M.N. 2006. Do nanoparticles present ecotoxicological risks for the health of the aquatic environment? *Environment International* 32: 967–976.

Murray, C.J.L. and Lopez, A.D. 1997. Global mortality, disability, and the contribution of risk factors: global burden of disease study. *Lancet* 349: 1436–1442.

Nowack, B. and Bucheli, T.D. 2007. Occurrence, behavior and effects of nanoparticles in the environment. *Environmental Pollution* 150: 5–22.

Pruss, A. 1998. Review of epidemiological studies on health effects from exposure to recreational water. *International Journal of Epidemiology* 27: 1–9.

Rasheed, H., Kay, P., Slack, R. and Gong, Y.Y. 2017. The effect of association between inefficient arsenic methylation capacity and demographic characteristics on the risk of skin lesions. *Toxicology and Applied Pharmacology* 339: 42–51.

Regli, S., Rose, J.B., Haas, C.N. and Gerba, C.P. 1991. Modeling the risk from giardia and viruses in drinking-water. *Journal American Water Works Association* 83: 76–84.

Resnikoff, S., Pascolini, D., Etya'ale, D., Kocur, I., Pararajasegaram, R., Pokharel, G.P. and Mariotti, S.P. 2004. Global data on visual impairment in the year 2002. *Bulletin of the World Health Organization* 82: 844–851.

Robertson, S.E., Hull, B.P., Tomori, O., Bele, O., Leduc, J.W. and Esteves, K. 1996. Yellow fever – a decade of reemergence. *JAMA – Journal of the American Medical Association* 276: 1157–1162.

Schubert, S.D., Suarez, M.J., Pegion, P.J., Koster, R.D. and Bacmeister, J.T. 2004. On the cause of the 1930s dust bowl. *Science* 303: 1855–1859.

Selendy, J.M.H. (ed.). 2011. *Water and Sanitation Related Diseases and the Environment: Challenges, Interventions and Preventive Measures*. Wiley-Blackwell; Chichester.

Smedley, P.L. and Kinniburgh, D.G. 2002. A review of the source, behaviour and distribution of arsenic in natural waters. *Applied Geochemistry* 17: 517–568.

Smith D. (ed.). 1999. *Water-Supply and Public Health Engineering: Studies in the History of Civil Engineering (Volume 5)*. Ashgate Publishing; Farnham.

Smith, G.D. 2002. Commentary: behind the Broad Street pump: aetiology, epidemiology and prevention of cholera in mid-nineteenth century Britain. *International Journal of Epidemiology* 31: 920–932.

Universitat Politècnica de Catalunya. (2011). Japan earthquake caused a displacement of about two meters. *ScienceDaily*, 8 April [online, accessed January 2013]. Available at: www.sciencedaily.com/releases/2011/04/110407121640.htm.

Vianna, N.J. and Polan, A.K. 1984. Incidence of low-birth-weight among Love Canal residents. *Science* 226: 1217–1219.

Vinten-Johansen, P., Brody, H., Paneth, N., Rachman, S. and Rip, M. 2003. *Cholera, Chloroform and the Science of Medicine: A Life of John Snow.* Oxford University Press; Oxford.

Ward, M.K., Feest, T.G., Ellis, H.A., Parkinson, I.S., Kerr, D.N.S., Herrington, J. and Goode, G.L. 1978. Osteomalacic dialysis osteodystrophy – evidence for a waterborne etiological agent, probably aluminum. *Lancet* 1: 841–845.

Watson, J.T., Gayer, M. and Connolly, M.A. 2007. Epidemics after natural disasters. *Emerging Infectious Diseases* 13: 1–5.

WHO. 2004. *The World Health Report 2004 – Changing History.* World Health Organization; Geneva.

WHO. 2010a. Cholera vaccines: WHO position paper. *Weekly Epidemiological Record* No. 13, 85: 117–128 [online, accessed April 2019]. Available at: www.who.int/wer.

WHO. 2010b. *Malaria Fact Sheet No. 94* [online, accessed August 2011]. Available at: www.who.int/mediacentre/factsheets/fs094.

WHO. 2011a. *Global Alert and Response (GAR) – Cholera* [online, accessed August 2011]. Available at: www.who.int/csr.

WHO. 2011b. *Cholera Fact Sheet No. 107* [online, accessed August 2011]. Available at: www.who.int/mediacentre/factsheets/fs107.

WHO. 2017. *Guidelines for Drinking Water Quality* (4th edition incorporating the first addendum). World Health Organization; Geneva.

WHO. 2018. *Zika virus* [online, accessed October 2018]. Available at: www.who.int/newsroom/fact-sheets/detail/zika-virus.

WHO/UNICEF. 2003. Basic needs and the right to health (chapter 5). In: UNESCO (ed.), *The First UN World Water Development Report: Water for People, Water for Life.* UNESCO Publishing and Berghahn Books; Paris, Oxford and New York.

CHAPTER EIGHT

Potable water and wastewater treatment

8

Nigel J. Horan

LEARNING OUTCOMES

After reading this chapter you should be able to:

- understand the requirements for potable water quality standards in order to ensure water is safe and wholesome;
- explain how potable water is produced to meet these standards, through a series of individual treatment stages;
- describe the desalination processes used to produce potable water from saline or brackish sources;
- recognise major pollutants present in wastewater and their contribution to aquatic pollution;
- understand the importance of legally enforced standards in achieving improvements to aquatic waters;
- appreciate our changing attitudes that view wastewater (and indeed all wastes) as a resource to be recovered and recycled and be aware of some of the technical advances in this area;
- understand how wastewater treatment technologies are now driven by energy saving, greenhouse gas reduction and resource recovery measures, as well as meeting legally enforced discharge standards;
- recognise those situations when a number of small, decentralised treatment systems might be more appropriate than a single, large centralised system.

8.1 INTRODUCTION

Previous chapters have identified the major sources of **potable water**, namely surface waters (rivers and lakes), groundwater (including springs) or engineered storage reservoirs and dams. Ensuring an adequate supply of water that will serve the population and which will not run out during times of drought is a basic necessity. But, in addition to adequacy, the water supplied must be clean, wholesome and palatable in order to protect the health of the consumer. Achieving this requires treatment, and the nature of the treatment is a function of the quality of the source water. Whereas groundwaters are *relatively* clean and need fewer treatment steps, surface waters are open to direct environmental input and so multiple treatment stages are generally needed to achieve the required water quality. In addition to the supply of safe drinking water there is the need to deal with wastewater. Wherever human populations settle, whether in villages, towns or cities, they are faced with the problem of disposing of their excreta in a safe manner. As populations have traditionally developed close to a source of potable water (typically a river or lake), this has usually been the first point of disposal. But as populations have increased this route has become increasingly polluted. The history of sanitary engineering is therefore the history of our attempts to remove waste from its point of origin and dispose of it safely, with minimal impact on the environment. This chapter deals with the steps taken to treat water so that it is safe to drink and then examines the principles of wastewater treatment and how we go about producing effluent that has a minimal impact on the aquatic environment. As discussed in Chapter 7, in the mid-nineteenth century, Dr John Snow was able to show that an epidemic of cholera in London was a result of a contaminated water supply, even though the causative agent of cholera itself (*Vibrio cholerae*) had not been discovered. By the beginning of the twentieth century, filtration of water followed by chlorination was widely practised in Europe and North America to ensure complete removal of **pathogenic** microorganisms. This single innovation is recognised as being one of the most important contributions to the 50% increase in life expectancy that we have witnessed in the developed world since the start of the twentieth century. The basic treatment flow scheme has not changed significantly since that time, but there have been new unit processes added to combat perceived threats to health such as those from nitrates and pesticides.

8.2 ENSURING A SAFE SUPPLY OF POTABLE WATER

Water treatment aims to remove three types of contaminant: microbial, inorganic and organic. Most importantly, water treatment should remove microbial contaminants, for instance, the **enteric pathogens** discussed in Chapter 7 such as the bacterium *Vibrio cholera*; protozoa such as *Cryptosporidium parvum* and *Giardia lamblia*; and viruses such as rotavirus and enterovirus. If present in water, onset of infection from these organisms is rapid, often happening in as little as a few hours after consumption. By contrast, inorganic chemical contaminants such as arsenic, fluoride and lead may require many years of ingestion before symptoms of illness develop, though this is a major concern in many regions (see Sections 7.4.1 and 7.4.2 of Chapter 7 for a discussion of the health impacts). Individual toxic doses of these compounds are unlikely as they would render the water unpalatable. Organic contaminants, such as pesticides and chlorinated hydrocarbons which may be present at remarkably low concentrations, are removed as a precaution, as they have been shown to induce symptoms in laboratory animals (Klaunig *et al.*, 1986), although such symptoms have not necessarily been observed in human populations. In all cases, contaminants follow a **dose–response relationship**, which was recognised as early as the sixteenth century by the Swiss alchemist Paracelsus, who noted that 'the dose makes the poison'. More recently, the Royal Society of Chemistry (RSC, 2009) has rephrased this well-known expression and stated that:

> While there is no such thing as a safe chemical, it must be realised there is no

chemical that cannot be used safely by limiting the dose or exposure. Poisons can be safely used and be of benefit to society when used appropriately.

The safe dose or exposure of a given chemical that may be ingested through drinking water is provided by the World Health Organization Guidelines for Drinking Water Quality (WHO, 2017). These guidelines provide a framework for drinking-water safety, covering the management of emergencies and unforeseen events, as well as fact sheets on individual organisms, inorganic and organic chemicals. Water quality standards for most of the world's countries are based on the recommendations of the World Health Organization (Table 8.1). It is the role of the water process engineer to ensure that these standards are met at all times, and there is a wide range of treatment alternatives available to achieve this. The underlying principle behind water treatment is to provide multiple barriers, and not simply to rely on a single protection. The first barrier comes with the protection of the supply itself. Whether this supply is a simple spring or a large dam, preventing the ingress of faecal contamination, fertilisers and pesticides through river basin management can significantly improve the quality of the supply (see Chapter 4). Additional barriers are then provided by treatment and disinfection.

Table 8.1 Water quality guidelines for safe drinking water for some key parameters as recommended by the World Health Organization (see also Table 7.2 in Chapter 7 for a more complete list of chemical parameters and Table 7.3 for pesticides and other organic pollutants)

Parameter	Maximum concentration or value
Microbiological	**number per 100 mL**
Escherichia coli	0
Enterococci	0
Chemical	**µg L^{-1}**
Antimony	20
Arsenic	10
Benzene	10
Boron	2400
Bromate	10
Cadmium	3
Chromium	50
Copper	2000
Fluoride	1500
Lead	10
Mercury	6
Pesticides	
Aldrin	0.03
Dieldrin	0.03
Heptachlor	0.03

Note: One microgram (µg) is one millionth of a gram (g) and is equivalent to 1000 milligrams (mg). 1 µg L^{-1} is equivalent to one part per billion (ppb) or one thousand parts per million (ppm).

8.2.1 Flow train for water treatment

Depending on the quality of the raw water source, several unit processes are assembled to form a flow train through which the water passes, becoming cleaner with each stage. A complete flow train will include one or more of the following processes: screening, clarification, filtration, activated carbon, ozone and chlorination (Figure 8.1). The major characteristics and applications of each of these processes are considered in more detail in the following sections.

Figure 8.1 Typical process flow train for producing water of potable quality that meets the World Health Organization guidelines summarised in **Table 8.1**.

8.2.1.1 Preliminary treatment processes

Prior to entering the water treatment works, surface water is subjected to a number of simple processes to remove larger physical contaminants. These processes are not required for groundwater, as they are mimicked by the action of the water percolating through the soil to the aquifer. The first treatment stage involves screening, which removes floating material such as leaves and twigs. This is accomplished by passing the water through a mesh or screen that retains particles larger than the mesh size. This material can then be removed by scraping or flushing the mesh. The screened water is stored for several days in a tank or reservoir, which acts both to balance the flow and settle out larger suspended particles. This in turn will reduce the number of pathogenic organisms both by settlement and by die-off. For certain waters that are acidic and which contain high concentrations of iron and manganese in solution, an aeration stage is used to drive off dissolved carbon dioxide and replace it with oxygen. This process also oxidises soluble ferrous (Fe^{2+}) and manganese (Mn^{2+}) into their insoluble forms (Fe^{3+} and Mn^{3+}), which can now be removed easily by settlement.

8.2.1.2 Coagulation and flocculation

After storage, the water will contain only finely suspended particles such as silt and mud of <1 mm diameter, as well as colloidal material. The particles can be removed by clarification, which helps to reduce colour and also helps to remove much of the microbial contamination that is associated with the solid fraction. Colloids in water have a net negative electrostatic charge and so electrostatic force prevents them from agglomerating (collecting together into a large mass). Brownian motion in the water prevents these colloids from settling. The negative charge is removed by adding ions to the water with an opposite, positive charge. A number of inorganic and organic compounds are used with ferric hydroxide, aluminium hydroxide (alum) and polyaluminium chloride being popular multivalent, inorganic 'coagulants'. More recently, owing to public worries as to the possible health impacts of aluminium in drinking water, ferric compounds are now proving more popular. Adding positively charged ions allows the colloids to come together and aggregate in size, a process termed **coagulation**. However, coagulation does not produce a strong material and so coagulant aids are necessary in order to provide additional strength by a process of **flocculation**. High molecular-weight synthetic polymers or polyelectrolytes such as polyquaternary amines are widely used for flocculation. They are long-chain organic molecules with charged groups along the chain length. They draw together the coagulated material to form a large and strong 'floc'. The flocculated material settles readily and is easily removed (**Figure 8.2**).

8.2.1.3 Solids removal

The flocs formed during coagulation and flocculation must now be removed from the water and this can be achieved either by allowing them to sink in a settlement tank or by encouraging them to float in a flotation tank. The former can be rectangular, with the flow passing horizontally along the tank; circular, where the flow radiates from the centre; or hopper-bottomed with upward flow. The latter is a very popular process option and it comprises a large tank shaped like an inverted cone with vertical sidewalls and a conical base. This arrangement permits solids to settle to the bottom of the tank and the sloping base encourages them to move towards a central hopper at the apex of the cone.

The density of the flocculated particles is not high and the buoyancy of these particles can be increased by passing air through the water. If this air is introduced at high pressure, then small bubbles are produced as the pressure reduces to atmospheric within the tank. The gas bubbles attach to the flocculated particles, reducing their density further so that they float. The big advantage of dissolved air flotation is that the smaller particles that would take a long time to be removed by sedimentation can be removed more rapidly and more effectively. Once at the surface, the floated particles can be removed by skimming.

The resultant sludge is one of the waste products from the water treatment process. It can be disposed of back to the reservoir from which the water was abstracted or, if sewerage is present at the treatment plant, it can be discharged to sewer.

Small particles in wastewater **Adsorption of polymer with particles** **Floc formed by bridges between particles**

Figure 8.2 Addition of a polymer destabilises small particles and allows them to aggregate into large, well settling particles in a process termed flocculation.

8.2.1.4 *Filtration*

The role of a filter is to sieve or strain suspended material from the water. The filter medium retains fine inorganic and organic particles while allowing the treated water to pass through. The filter medium may be comprised of sand or anthracite, and excellent results have been obtained using clean, crushed, recycled glass (Horan and Lowe, 2007).

There are two types of filter: slow sand and rapid gravity. The simplest of these is the slow sand filter (**Figure 8.3a**), which comprises a shallow tank of about 1 m deep with a layer of sand of at least 0.6 m and more typically 1 m, where the sand has a particle size in the range 0.15 to 0.35 mm, over which the water to be treated is allowed to flow. The water sits above the sand, taking several hours to percolate through to emerge at the base of the tank, where a piped collection system at its base takes away the treated water for discharge. These pipes are covered with a layer of gravel to a depth of 0.4 m. The raw feed water is replenished at the same rate that it percolates in order to maintain a constant head over the sand layer. In order to ensure the bed does not run dry in the event of a raw water supply problem, a weir is placed at the filter outlet and before the supply reservoir (**Figure 8.3a**). Thus the top layer of sand removes the larger particles from the flow. As this sand layer is exposed to the light, it permits the growth of algae that form a film on the surface. This film in turn attracts protozoa, bacteria and other microorganisms that form a gelatinous surface layer (also known as **schmutzdecke**). Thus, slow sand filters have a biological action, as the microorganisms break down organic compounds held within the water and the algae provide oxygen, which aerates the water. Over time the layer of **biofilm** slowly accumulates and blocks the flow of water through the bed. At this point the bed is taken out of service and partially drained and the layer of biofilm is removed for disposal. Depending on the turbidity of the feed water the filter will need cleaning at a frequency of between one and several months. Thus the raw water must have a low turbidity. It is because of the propensity of the biofilm to clog that slow sand filters are not used to treat coagulated water, as this can contain many fine, coagulated particles. Cleaning the filter results in the loss of sand and so the sand will need replenishing regularly. Slow sand filters operate at flow rates of 0.1–0.3 m hr^{-1} (see **Box 8.1** for an explanation of treatment flow rates) and this equates to 100 to 300 L hr^{-1} per m^2 of filter area. Although slow sand filters do not have a high capital or operating cost, they do need large land areas and close to cities and large towns such land is often not available, or very expensive.

By contrast, a rapid gravity filter (**Figure 8.3b**) is an intense and fully automated process that can treat flow velocities between 4 and 21 m hr^{-1} and so it needs only a fraction of the land space that is required for the larger, slow gravity filter. Rapid gravity filters are best suited for waters that have received pre-treatment and coagulation.

The main components of a rapid gravity sand filter are: the filter media, the gravel support layers and the filter underdrain. The sand used in a rapid gravity filter is of a larger grain size than that in a slow sand filter (typically >0.55 mm) and consequently its performance in removing microorganisms is poor. As a result, rapid sand filtration must always be followed by disinfection. Flow can pass either up or down through the rapid filter (upflow or downflow) and any suspended material in the water is retained on the filter by both **adsorption** and filtration. As a result, the flow rate through the bed will be reduced as

Figure 8.3 Schematic diagram to show the basic design of raw water gravity filters: a) a slow sand filter that operates under gravity; b) a rapid gravity filter operating under pressure.

the bed starts to clog. When the **head loss** through the bed reaches a certain point the medium will require cleaning and this is achieved by a process of backwashing. Air is introduced at a high flow rate, countercurrent to the flow of water, and this thoroughly scours the bed, releasing trapped particles. The bed is then washed with water to remove these particles, with this

backwash water taken away for disposal. The efficiency of the filter is dependent on the effectiveness of the backwash cycle, and this relies on a well-designed underdrain. An underdrain has many functions. The underdrain supports a gravel layer and the filter medium. The underdrain collects the filtered water from the base of the filter and it also distributes air and water evenly

BOX 8.1

Comparing process efficiencies

Velocity units (m hr^{-1}) are commonly used in both water and wastewater treatment to compare the flow rates that different unit processes can handle. They are obtained by dividing the flow rate (in m^3 hr^{-1}) by the area of treatment process (m^2) to give m^3 m^{-2} hr^{-1}, which is an areal loading rate. These units are exactly the same as m hr^{-1}, which is a flow velocity. The greater the areal loading rate or flow velocity, the greater the volume of water that a given process can treat. For instance, a primary settlement tank used in wastewater treatment is typically designed to ensure that the flow velocity does not exceed 2.5 m hr^{-1}. Thus, if a wastewater treatment plant has four primary settlement tanks, each of 30 m diameter, this provides a total area of 707 m^2. Hence the maximum flow that the plant can safely treat is 707 m^2 × 2.5 m hr^{-1}, or 1767.5 m^3 hr^{-1}.

Figure 8.4 Underdrain design for a rapid gravity filter showing the many roles this has to perform.

across the base of the filter during back-washing (**Figure 8.4**).

For many waters, filtration represents the final treatment before disinfection. However, for other water sources which may be contaminated with pesticides or have taste problems, additional polishing stages are required.

8.2.1.5 Activated carbon

The activated carbon stage is a specialised process that relies on the adsorption abilities of activated carbon to remove organic material from water. Activated carbon has a very porous structure and its small pores will absorb a wide range of contaminants. These contaminants include pesticides and pesticide by-products, trihalomethanes and other disinfection by-products, algal toxins and chlorinated hydrocarbons. The activated carbon is usually used in a granular form, known as GAC and contained in filters that are operated in a similar manner to a rapid gravity filter. Activated carbon will eventually become saturated with adsorbed products, at which stage it must be removed and replaced. The saturated material can be regenerated by heating to high temperatures >550°C, which will destroy the adsorbed organic contaminants.

8.2.1.6 Disinfection

Disinfection is the process that protects consumers from waterborne disease by eradicating any potential enteric pathogens in the treated water. It is not a sterilisation process and many bacteria will survive disinfection, but the disinfection choice and dose are intended to ensure that the more harmful enteric organisms are removed. A wide range of chemicals show disinfectant properties, but a key property for water treatment is that the disinfectant should not be harmful to humans, even at extremely low concentrations. Those most often used for treating potable water include chlorine (Cl_2), chlorine dioxide (ClO_2), hypochlorite (OCl^-) and ozone (O_3). In addition, ultraviolet (UV) light is often used, but this is a physical and not a chemical disinfectant. Disinfectants can be applied in two ways, either as a primary disinfectant to achieve the requisite degree of microbial kill or with the addition of a secondary disinfection stage that aims to maintain a residual concentration of disinfectant in the finished water in order to prevent either regrowth or ingress of pathogens in the distribution system. Physical disinfection such as UV cannot provide a residual effect that continues to disinfect the water down the supply chain. Chlorine is one of the most

effective disinfectants that is easy to apply and which provides a residual concentration in the source water. However, chlorine can also react with organic compounds in the water to form disinfectant by-products, in particular the halomethanes. These compounds substitute one or more of the hydrogen molecules on a carbon atom with chlorine to produce mono-, di- and tri-halomethanes. There are fears about the links between halomethanes and cancer (Pereira *et al.*, 1982), and a number of other health impacts (see Section 7.4.3 of Chapter 7) although it is widely accepted that the risks to health from drinking unchlorinated waters are many orders of magnitude greater than the risks of cancer from waterborne halomethanes. Nevertheless, water companies make great efforts to reduce the risks of these compounds, especially where the source waters are rich in organics, such as those flowing from peatlands (Chow *et al.*, 2003). Pre-treatment with GAC will help by reducing trace organic compounds in the water that might react with chlorine. In addition, other disinfectants might be employed, such as chloramines or ozone, which, although not as effective as chlorine, do not have by-products associated with them.

8.2.2 Membrane treatment

Membranes are being increasingly applied in the treatment of drinking waters because they can produce high-quality water in a single process with a relatively small footprint. A number of membrane types are used which are defined based on the membrane pore size. These are microfiltration (MF), ultrafiltration (UF), reverse osmosis (RO) and nanofiltration (NF) membranes. MF membranes have the largest pore size and will reject large particles as well as algae and protozoa. UF membranes can also reject bacteria, larger, soluble molecules such as proteins and many viruses. RO

Figure 8.5 Arrangement of membranes used in a range of water treatment applications.

membranes are effectively non-porous and will exclude particles, bacteria, viruses, soluble organics and even low molecular-weight inorganic ions such as salt. Membranes are generally made of synthetic organic polymers and used as flat sheets or in tubular arrangements (**Figure 8.5**). Membranes offer a good process option for small water treatment plants that are faced with removing a wide variety of contaminants. Where the source water is rich in particles, then prior filtration is essential to prevent membrane clogging. In addition, the material retained by the membrane, known as the 'concentrate', can comprise as much as 15% of the treated water, or 'permeate', and requires careful disposal.

8.2.3 Desalination

Global demand for water is doubling every 20 years and growing faster than population (see Chapter 9). Owing to changes in rainfall patterns, the movement of populations from rural to urban areas and increased demands from industry and agriculture, it is expected

that by 2025, demand for clean water will exceed supply by 56%. As the oceans make up 97% of the world's supply of water, it is understandable that engineers look to this as a source to bridge the gap between water demand and water supply. But seawater typically contains around 35,000 mg L^{-1} of salt, and brackish water between 2000 and 5000 mg L^{-1}, whereas the guidelines for drinking-water quality require that salinity levels are below 250 mg L^{-1} (**Table 8.1**). The process of removing this salt from water is known as desalination and it requires large amounts of energy (**Box 8.2**) and specialised, expensive infrastructure. As a result, desalinised water is much more expensive (typically double the price) than water from more conventional sources. However, desalination already plays a major role in the provision of water for drinking and **irrigation** in areas such as the Middle East, which are energy rich but water poor; the world's largest desalination plant is currently found in the United Arab Emirates. But now many other nations, including the USA, China, Spain and Australia, are having to utilise this

BOX 8.2

Energy required for desalination

Removing salt from seawater requires an energy input of between 3 and 5 kWh per m^3 of potable water produced. This energy requirement compares to <0.2 kWh per m^3 of water for treating conventional freshwater supplies. The technique is very popular in Middle Eastern countries, where freshwater is scarce but there is a ready supply of seawater and energy. Consequently, installed desalination capacity has doubled in the Gulf countries over the past decade. For example, Kuwait has an annual rainfall of just 110 mm yr^{-1} and no surface freshwater supplies. It has a water demand of more than 350 million m^3 yr^{-1} but only 6 million m^3 yr^{-1} of conventional freshwater. Therefore Kuwait uses desalination to meet the shortfall. To provide energy for desalination plants, oil and natural gas resources are used, but this consumes around 12% of Kuwait's annual oil supplies. In view of the country's climate, solar power would seem the obvious way forward and there are plans to construct a 280 MW solar system in Kuwait to help power desalination (Paleologos *et al.*, 2018).

technology to deal with their own water scarcity problems. Indeed, in 2010 the first desalination plant was opened in the UK, in East London, treating brackish water from the Thames Estuary, with the ability to produce 140,000 m^3 per day and serving 1 million people. However, the plant is intended to operate only during times of extended low rainfall and, to reduce its **carbon footprint**, it runs on recycled fats and oils recovered from London's restaurants.

Desalination can be achieved through distillation using heat, but reverse osmosis using membranes is now more popular and requires less energy. The RO membranes are used in a spiral-wound arrangement that rolls up many layers of flat membrane sheets around a central pipe, which delivers the saline water for treatment. The membranes are rated for their ability to reject compounds from contaminated water, and to achieve the guideline quality for chloride using a saline source water of 25,000 mg Cl^- L^{-1} requires a rejection rate of 99%. Polyamide thin-film composite membranes provide the necessary strength and durability and also offer the high rejection rates that are needed. Purification is achieved by applying a high pressure to the concentrated side of the membrane, which forces the purified water across the membrane to the dilute side. The impurities that remain on the concentrated side are taken away in the rejected water, and this can be as much as 50% of the total volume treated. The rejected water needs to be disposed of in such a way as to avoid environmental damage.

8.2.4 Wastewater as a potable water supply

One of the most environmentally benign sources of potable water is wastewater utilised through a process of direct reuse, thus completing the cycle and avoiding the need for water abstraction and wastewater discharge. Unplanned indirect reuse is practised daily throughout the world on rivers such as the Thames, the Yangtze, the Murray–Darling and Mississippi, where treated wastewater is discharged to watercourses, which are then abstracted further downstream, either for irrigation or as water intake for a water treatment works. Alternatively, direct reuse means that the treated wastewater is added directly to the normal drinking-water distribution system. With the standards of treatment now available for wastewaters and potable waters, this concept is perfectly feasible, although it may not appeal to the majority of consumers. It has been practised with some success at a limited number of facilities worldwide, with perhaps the best documented being that of Windhoek, the capital city of Namibia (du Pisani, 2006). This city is the driest city below the Sahara, with no freshwater resources within a 500 km radius. Due to severe water shortages in the 1960s, the city introduced a water reclamation scheme in 1968 to supplement the potable water available to the city with domestic wastewater. The scheme has been very closely monitored and it has demonstrated that water of acceptable quality can be produced from domestic wastewater and that, if fully informed and consulted, consumers will accept this alternative in order to maintain their potable supplies. A more recent scheme has been put in place in Essex, the driest county in the UK and one with no potential for new water resources. The scheme involves the diversion of 40,000 m^3 of treated wastewater from the Langford treatment works, where it discharges to an estuary. This effluent receives additional treatment and UV disinfection before discharge to an adjacent river, four miles upstream of a water treatment works. Here it is abstracted for treatment and put into potable supply. When discharged to the river, the treated wastewater was of better quality than the river itself. However, by using this route, instead of transferring directly to a reservoir, it was possible to satisfy public misgivings about the scheme. This simple scheme augmented the available water resources by 8%, and at a fraction of the cost of desalination.

> **REFLECTIVE QUESTION**
>
> Describe the flow train to treat a water supply that contains pathogenic bacteria and pesticides, so that it is potable, meeting the World Health Organization guidelines.

8.3 WASTEWATER TREATMENT

The ideal fate of excreted material is for it to be recycled to agricultural land, where it can complete the important nutrient cycles for nitrogen and phosphorus, as well as replenishing the soil with valuable organic material. Historically, this agricultural use was the traditional route for waste in the UK and Western Europe, with faecal material collected through chamber pots used indoors and pail closets or middens sited outdoors. The rapid growth of towns and cities during the Industrial Revolution of the eighteenth and nineteenth centuries meant that this agricultural route for disposal was overwhelmed and access to rural areas became limited. Thus, excreted material was either disposed of directly to watercourses or, in the poorer areas, directly onto the streets, where it would accumulate. As this material is a source of numerous pathogenic microorganisms and is responsible for many faeco-oral infections, it led to widespread epidemics, such as cholera, particularly among the poorer populations. It was in an attempt to improve the health of such people that the great Victorian reformers pioneered the widespread adoption of water-borne sewerage and the provision of a clean water supply. This measure is widely accepted as being the single largest contributor to improving the health and reducing the mortality of the population. Nevertheless, its uptake was slow, and even by 1970 in the UK a government report (HMSO, 1970) was critical of the 270,000 households that still lacked toilets connected to a sewerage system and which relied on earth and pail closets.

The water carriage system adopted in most developed countries serves the purpose of transporting excreta from its point of origin (the house, factory or office) to a point of ultimate disposal (the local watercourse). It has been in use for around 150 years and its major component is the sewerage network. The sewerage network comprises a system of drains, which connect individual properties to a larger sewer running close to the house. In turn, the individual sewers connect to a much larger interceptor sewer that carries the material to a sewage treatment works. Of course, this development is only made possible by the provision of mains water to a property, and it does come with a number of disadvantages. The sewerage network in many countries serves a dual function in that, as well as accepting sewage, it can also be used to convey surface water drainage, and such systems are termed combined sewers. This is to differentiate them from separate systems, where a single piped network carries the sewage and a separate piped network is used to convey drainage. Both combined and separate sewerage systems are now under stress as the population continues to rise, because not only does this increase the amount of sewage to the system but it also increases the amount of drainage as more and more land is covered with impermeable surfaces to provide housing, offices and industry. A final challenge comes from climate change, with more extreme storm events, which can overload the combined sewer network, leading to sewer flooding. The foul water can pour out onto streets, rivers or even properties when a sewer surcharges and floods.

Bringing together the sewage from large numbers of people to a single point source at the end of the main sewer generates a potentially huge pollution problem if it is not treated adequately prior to discharge to a watercourse. This section of the chapter will examine the principles of wastewater treatment and examine how an effluent can be produced that has minimal impact on the aquatic environment.

8.3.1 Components of wastewater

Wastewater comprises two fractions, one soluble and one particulate, with the latter measured by the amount of material retained on a 7 µm filter paper. The particulate component is termed the 'total suspended solids' or usually just the 'total solids' (TS). Both the soluble and particulate fraction will comprise organic and inorganic material. If the material retained on the filter is heated to a temperature of 550°C or more it volatilises, as all the organic carbon is oxidised to carbon dioxide. The organic material lost in this way is known as the volatile suspended solids (VS) and measures the organic fraction of the waste. The inert material, or ash, that remains is a measure of the inorganic fraction. The inorganic fraction will contain cations, including metals such as sodium, magnesium and phosphorus, and ammonia nitrogen. It will also contain anions such as chloride and sulfate.

Wastewater discharged to a receiving watercourse may cause pollution in four ways:

1. Lack of oxygen in the watercourse.

Figure 8.6 The oxygen sag curve that develops when wastewater is discharged to a watercourse with a concentration of organic material in excess of the available oxygen dissolved in the water.

2. Nutrients causing algal blooms or toxic effects.
3. Enhanced suspended solids.
4. Pathogen and chemical contamination.

1. The organic material in the wastewater acts as a food source for microorganisms in the receiving watercourse, which oxidise it to carbon dioxide and water as in this equation:

$$C_6H_{12}O_6 + 6O_2 \rightarrow 6CO_2 + 6H_2O$$

The oxygen for this reaction is derived from the watercourse itself and, as oxygen is sparingly soluble in water (at around 10 mg O_2 L^{-1} at 10°C), this is quickly exhausted. As a result, there is no oxygen remaining for other aquatic life, which either dies or moves elsewhere. This effect is illustrated in Figure 8.6, where the dissolved oxygen in the river declines below the point at which organic material is discharged. If there is a single discharge, then this figure shows clearly that a river is able to recover from organic pollution once the oxygen demand is reduced. If discharge contains a large amount of organic material (for instance, a major industrial spillage or a broken sewer that discharges raw sewage), then the river can be polluted for many km downstream. The oxygen sag curve illustrated in Figure 8.6 can be used to ensure that the organic material in a sewage effluent is reduced to such an extent that it causes minimal impact on the oxygen concentration in a watercourse. It can also be used to ensure that treatment plant discharges are not sited too close together, thus ensuring that river water has time to recover from an oxygen deficit along its course.

The amount of oxygen needed to oxidise all of the organic material in wastewater, as shown in the equation above, is expressed in terms of its oxygen demand. This can be undertaken either biologically as the **biochemical oxygen demand** (**BOD**), which is measured using naturally occurring microorganisms (see Section 4.2.5 in Chapter 4), or chemically as the **chemical oxygen demand** (**COD**), which uses strong chemical oxidising agents. An individual excretes about 45 g BOD per day, and where a sewerage network is installed it has a per capita water usage that ranges from around 100 L in areas of water shortage (such as the Middle East), 150 L in the UK and 320 L in Spain to as high as 575 L in the USA. Thus, the strength of sewage would be strong in countries with sparse water usage (45 g BOD in 100 L water gives a concentration of 450 mg BOD L^{-1}), medium in the UK at 300 mg BOD L^{-1} and weak in the USA at 78 mg BOD L^{-1}. During wet periods, drainage in the sewer can often account for more than six times the per capita water usage and so the BOD concentration can be as low as 50 mg L^{-1} in the UK, although the amount (or BOD load) is unchanged. Usually if the BOD in the treated wastewater can be kept below 20 mg L^{-1} and, if there is adequate dilution in the receiving watercourse (at least eightfold), it can be discharged with no detrimental impact to the oxygen concentration and, thus, aquatic life in the watercourse. Where rivers are much larger than this and provide much more dilution, a higher BOD concentration in the treated effluent can be tolerated without harm. Thus, the performance expected from a treatment plant depends very much on the characteristics of the receiving watercourse.

2. The nutrients ammonia and phosphorus can encourage blooms of algae, in particular the blue-green algae or cyanophyta. This process is known as **eutrophication** (see Section 4.2.7 of Chapter 4), which is more common in mid- to low-latitude countries, where sunlight is stronger, and also more common in lakes than in rivers, as the nutrients in lakes are not as easily removed by upstream flows. In addition, ammonia itself can be toxic to fish life, as the un-ionised form of ammonia (NH_3) interferes with oxygen transport across the gills of fish. It is apparent from the equation below that as the pH increases, so the reaction will be driven to the left, with the formation of more un-ionised ammonia:

$$NH_3 + H_2O \rightarrow NH_4^+ + OH^-$$

Only at pH values greater than 8.5 does the un-ionised ammonia exceed 10% of the total ammonia fraction. If the concentration of ammonia-N in treated wastewater is kept below 1 to 4 mg L^{-1}, then such toxicity is prevented. In a similar way, if eutrophication is likely to be a problem, it can normally be averted by achieving a concentration of <1 mg P L^{-1} in the treated wastewater.

3. Suspended solids contain organic material and thus contribute to a BOD demand. Ten mg TS can contribute up to 6 mg BOD. Thus, if a treatment plant is to achieve an effluent quality of <20 mg L^{-1} it must also ensure that the solids are removed, and generally <30 mg L^{-1} can be achieved with ease through simple settlement. Pollution from TS is not clearly defined and it has a number of effects. It will make the watercourse more turbid and thus prevent sunlight penetrating; it will also settle on the **benthic layer** and can make it difficult for invertebrate species to feed and reproduce (see Chapter 6).

In order to protect the receiving water quality, wastewater must be treated to reduce these contaminants to concentrations that will pose no pollution threat. The specific values are very much a function of the climate, the watercourse and the size of the effluent discharge. It is the role of the regulator (such as the US Environmental Protection Agency) to set standards that are appropriate, and these standards are known as consents to discharge. As wastewater treatment is an expensive process that requires energy and emits greenhouse gases, there is no environmental benefit in setting a consent higher than is necessary. It is the role of the engineer to design and build a treatment plant that will meet these consents. If the consents are not met, the watercourse is not protected, and so money spent on the facility has been wasted. Experience in many countries has shown that the best way to ensure protection of a watercourse is to have an effective monitoring procedure with random sampling, and powers to prosecute if the consent is not met. It is sometimes said that people 'do what you inspect, not what you expect'. It is necessary, therefore, to be very clear in the legislation as to exactly how consent will be measured. It is not adequate simply to state that the treatment plant must achieve an effluent of 20 mg BOD L^{-1}. It is far more useful to state that (for instance): 'the treatment plant must achieve a 95-percentile compliance with an effluent standard of 20 mg BOD L^{-1} based on a random spot sample analysed in a quality assured laboratory'.

4. In addition to the above three parameters, there are other contaminants that are of importance only in specific situations. For instance, excreted enteric pathogens do not impact on the water quality in terms of the aquatic organisms, but, if the water is used for recreational purposes such as bathing or canoeing, they pose a risk to human health. In addition, as the quality of watercourses improves and fish return to them, we are now recognising impacts from other pollutants that were not previously apparent. For instance, women who take the contraceptive pill or receive hormone replacement therapy excrete oestrogens in their urine. Oestrogen will pass through a wastewater treatment process with very little removal. When present in the receiving watercourse, it can cause many problems in fish which breed below the discharge point, such as birth defects and hermaphrodite characteristics. All of these problems are dealt with as they arise, new effluent standards are promulgated and additional treatment processes are added to the flow train. Other examples of emerging contaminants are provided in Chapter 4.

8.3.2 Types of wastewater treatment system

There are many different process options for treating wastewater, but a feature of all of them is that, as the effluent consent becomes tighter and the land available to construct a facility reduces, more highly engineered systems are needed. Thus, simple treatment options such as pond systems and reed beds are relatively low in capital and operating costs but require an extremely large land area. They can also provide reasonable treatment, and, where the consent is based on a composite sample with compliance based on an average, they can meet a standard of 20 mg BOD L^{-1} and 30 mg TS L^{-1}. However, they are not able to meet tighter standards where effluent quality is based on achieving a 95-percentile compliance from a spot sample, and neither can they meet consents for ammonia or phosphorus. For example, an effluent that achieves an average BOD of 20 mg L^{-1} would imply that on many occasions the effluent quality was in excess of 20 mg L^{-1}. In order to meet a 95-percentile effluent compliance, the average must be half this 95-percentile value (Figure 8.7). So to achieve a 95-percentile compliance with a BOD of 20 mg L^{-1} means that an average

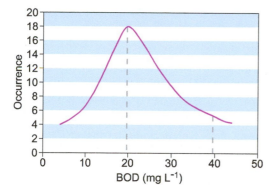

Figure 8.7 Annual distribution of biochemical oxygen demand (BOD) in the effluent from a wastewater treatment plant, illustrating the influence of compliance at 50 and 95 percentiles.

of 10 mg L^{-1} must be achieved, and this cannot be met by simple treatment systems.

Larger, more complex options, such as trickling filters, activated sludge or membrane systems, can routinely achieve effluent quality of 10 mg L^{-1} BOD, 10 mg L^{-1} TS and <1 mg L^{-1} ammonia with a 95-percentile compliance (although this is not always necessary for environmental protection). They also have a very small site footprint. However, their capital costs and operating costs are higher than those for simple systems. They also have a large

energy demand, and the current challenge for engineers is to reduce this demand through process efficiency and energy recovery from the wastewater (Table 8.2).

For those areas where wastewater infrastructure is not yet in place, there is also a great debate as to whether to build several smaller treatment facilities close to the point of production or a single large facility treating all the wastes from a river basin (Verstraete *et al.*, 2009). The arguments in favour of the smaller, decentralised treatment are that they do not require such a large sewerage network and this reduces the pumping (and thus energy) costs. It currently takes about 0.5 to 1 kWh m^{-3} for water to reach the home, and a similar amount to collect it from the home at a wastewater treatment works. The arguments in favour of centralised treatment are that it reduces the number of discharge points to the environment (and thus the risk of pollution); also economies of scale mean cheaper per capita treatment. The large volumes of wastewater to be processed permit the introduction of novel technologies for energy and resource recovery.

Wastewater treatment up until the end of the twentieth century focused simply on producing an effluent that met its discharge consent. This was very effective but has left us with processes that have a large

Table 8.2 The performance of a range of wastewater treatment systems, showing improving performance as capital and operating cost increases

Treatment process	Indicative effluent quality at a 95-percentile compliance				
	BOD (mg L^{-1})	TS (mg L^{-1})	Ammonia-N (mg L^{-1})	Phosphorus (mg L^{-1})	Faecal coliform (log removal)*
Waste stabilisation ponds	50	50	12	12	>6
Reed beds	40	40	12	12	>4
Trickling filters	20	30	<4	12	<3
Activated sludge	15	15	<1	<1	<3
Membrane bioreactors	<4	<4	<1	<1	>6

Note: * A 1-log removal means that 90% of faecal coliforms are removed, 2-log removal is 99%, etc.

BOX 8.3

Energy from warm sewers

The wastewater in the sewer has a huge embedded energy. In many temperate countries water entering the house has an average temperature of 10°C, but in the sewer it has an average temperature of 10 to 15°C in winter and 20 to 25°C in summer. This temperature rise is due to warming from hot taps, heating systems and warm body temperatures. To achieve this temperature rise has required an energy input of around 47 kWh m^{-3}, which represents a staggering 30% of the thermal energy input to a building. Sewer heat recovery is now being practised in a number of locations to utilise this energy (Meggers and Leibundgut, 2011), and heat exchangers in the sewer can provide space or hot water heating to local residential and commercial buildings.

energy demand and are wasteful of resources. As a result, a new approach has emerged in the twenty-first century, one that does not see wastewater as a problem but, rather, as a resource. This resource can provide energy (e.g. see Box 8.3) and it can permit the recovery of useful materials and the essential nutrients nitrogen and phosphorus. Such an approach is only possible at large works where the value of the recovered material will cover the capital costs of the processes required to achieve the recovery. For this reason, much of the rest of this chapter will focus on large, centralised treatment systems, and consider how we now treat wastewater as a resource and what a resource recovery centre of the future may look like. The potential opportunities for smaller decentralised systems are then discussed in the context of those environments where they may offer cost savings and environmental enhancements.

8.3.3 Siting wastewater treatment plants

The effluent that leaves a treatment plant must be discharged to the aquatic environment and thus treatment plants are always sited close to a watercourse. This may be a river or estuary, the sea or a lake. In order to ensure that the impact of this discharge is minimised it is important that there is adequate dilution at all times of year to reduce the concentration of the treated effluent. In the UK, for a plant that achieves a final effluent BOD of 20 mg L^{-1} m, a dilution of at least eight times is required. This value was determined from a mass balance of the upstream flow and BOD and the required value of downstream BOD to ensure the watercourse would remain unpolluted. Thus this value for dilution is specific to the UK. A similar calculation exercise would be required for each site or country of assessment in order to show the selected site has an acceptable receiving watercourse. As far as possible the site is normally selected to be on the lowest lying land but above the point at which flooding might be expected. This is to allow the sewage to flow by gravity from its point of origin and to arrive at the wastewater treatment works with adequate head to flow through the first stage of treatment. In hilly areas the sewage arriving at the works may have a large head that can be exploited for energy (see Box 8.4).

8.3.4 Stages in the treatment process

Wastewater treatment is focused on a solids removal process. Wastewater passes through a number of treatment stages, each of which removes smaller solid particles, until ultimately only colloidal and soluble materials remain. At this point a biological treatment stage is needed in which microorganisms are utilised that are able to oxidise this soluble and colloidal material, using it both as a source of energy and to provide cellular carbon and nitrogen. The new microorganisms that grow can then, in turn, be removed through a solids removal process. The initial stages of treatment are all physical solids removal processes and are designed exclusively based on the daily flow rate that they receive. They are termed preliminary and primary treatment. The biological stage is known as secondary treatment and is designed based on the daily load of organic material (as BOD) and ammonia that must be treated. These three stages are able to produce an effluent with a BOD <20 mg L^{-1}, a TS <30 mg L^{-1} and an ammonia <1 mg L^{-1}, all with a 95-percentile compliance. If additional treatment is required, for instance to reduce the BOD concentration for discharge to a very sensitive watercourse, or to remove enteric pathogens for discharge to recreational waters, this is termed tertiary treatment.

8.3.5 Physical treatment processes

The flow arriving at a wastewater treatment plant comprises: a) a domestic contribution, b) flow from commerce and industry, and c) infiltration of water into the sewerage network. These three components make up what is known as the dry weather flow (DWF) and represent the flow in the absence of rainfall, traditionally expressed with units of either m^3 per day or L s^{-1}. Rainfall will add significantly to this flow in the form of drainage and will serve to dilute the wastewater. To ensure that the sewerage network does not flood during times of heavy and extended rainfall, it contains a number of overflows whenever the sewer passes close to a watercourse. These are meant to operate intermittently during

Figure 8.8 a) A combined sewer overflow serves to protect the sewer from flooding, although it does discharge untreated sewage to the environment during storm events; b) innovation processes are now available to remove coarse material from combined sewer overflows to prevent aesthetic pollution.

Figure 8.9 Screens are the first stage of treatment, and remove solid material which is larger than the holes in the screen, in this case 6 mm.

extreme storm events and they discharge untreated, but very dilute, sewage to water-courses (Figure 8.8). The overflow is known as a combined sewer overflow (CSO). At the sewage works itself all the incoming flow passes by a weir that is set to receive any flows above 6 × DWF. This is known as a 'Formula A' weir, and again protects the works from flooding. Flows above 'Formula A' pass over this weir untreated to the watercourse.

The first stage in the treatment process is termed a 'screen', in which the wastewater passes through a moving drum or plate that is perforated with holes, typically of 6 mm diameter (Figure 8.9). Thus, all solid material >6 mm, known as 'screenings', is removed from the flow on this screen. The

screened material is then compressed to remove water, and traditionally is sent to landfill for ultimate disposal. In the UK more than 1000 tonnes of screenings are produced each day, and their carbon foot-print is high at 4.6 tonnes CO_{2Eq} per tonne of screenings. Thus, water companies are now looking at the potential for reducing this volume by, for instance, composting or anaerobic digestion, both of which also gen-erate valuable end-products.

The screened wastewater is subject to grit removal, usually through settlement in tanks or long channels. If the velocity of the waste-water is reduced to around 0.3 m s^{-1}, then the heavy grit particles will settle out and can be removed by scraping the base of the tank or channel. The resultant grit can be recovered and reused. The grit can be used around site for repair work to the roads, or it can be further processed and used as a building material or as valuable horticultural grit.

The screened and degritted wastewater now passes on to the primary treatment stage to receive additional or full treatment. However, not all the sewage will receive full treatment and the capacity of the primary and secondary stage is restricted to $3 \times$ DWF, what is known as flow to full treatment (FFT). For the majority of the time, the flow to the works is less than 3 DWF, but on those occasions when it exceeds this figure, flows >3 DWF will pass to large storm tanks for storage. Typically the tanks can provide storage for about two hours, but if the storm event is longer than this any additional flow to the storm tanks overflows directly to watercourses. Once the storm has ended, the storm tank contents are pumped back to the inlet, where they receive full treatment. It may appear from the above that a lot of wastewater goes directly to watercourses without any form of treatment, and this is quite true as it involves flows from CSOs, 'Formula A' weirs and storm tank discharges. However, this happens only when the flow is greater than 6 DWF (and often much greater). Therefore the sewage is diluted at least fivefold with rainwater. In addition, if the storm is of a magnitude to produce this volume of drainage, then the river also must have a very high flow rate. This provides a lot of dilution for the wastewater, and also ensures that it is carried away rapidly by the river. Nevertheless, this source of pollution remains a concern that will inevitably receive more attention in the future.

Primary treatment is a very simple process whereby the wastewater is held in a large tank under quiescent (still) conditions for a minimum of two hours. Under these conditions the majority of the solids in the wastewater will settle to the bottom of the tank to form a primary or raw sludge. Screened, degritted sewage entering the primary tank will typically contain around 140 mg L^{-1} solids, and between 40 and 60% of this is removed. The solids that settle in the tanks are also rich in organic material, and so primary tanks can achieve removals of up to 30% of the incoming BOD. Thus, the settled sewage leaving the primary tanks might have a TS of around 70 mg L^{-1} and a BOD of 110 mg L^{-1}. The raw sludge at the base of the tank is removed at a TS of about 15,000 mg L^{-1}, showing how effective this process is at thickening the sludge.

Primary sludge is an extremely valuable resource. Over 80% of the solids are volatile, and so it is rich in energy and nitrogen, and it passes on to the sludge handling process to recover these resources. The settled sewage passes forward to receive biological treatment.

8.3.6 Biological treatment processes

There is a wide range of process options that provide biological treatment, but they all have the same aim. Their aim is to maintain a large population of microorganisms in contact with the settled sewage and to provide adequate oxygen to meet the BOD demand of this wastewater within the time it is retained in the basin. An attached growth process uses an inert support material to which the microorganisms attach, creating a biofilm, which builds up to quite a substantial thickness. The inert supports range from natural materials such as crushed granite or basalt, to blast furnace slag or synthetic materials such as plastic. Attached growth processes include trickling filters, reed beds and rotating biological contactors (**Figure 8.10**).

Figure 8.10 Attached growth processes maintain a viable population of microorganisms by the provision of an inert support such as plastic or mineral media. Shown here are rotating biological contactors.

Source: D. Pimborough.

In all these processes oxygen is supplied naturally from the atmosphere and the process efficiency depends on the rate of oxygen transfer through the biofilm. Introduction of mechanical aeration can improve process efficiency and thus shrink the process footprint. Aerated attached growth processes include biological aerated flooded filters (BAFF) and submerged aerated filters (SAF).

In a suspended growth process the microorganisms are suspended within the settled sewage and air is introduced either by surface aeration using motor-driven impellers, or from the base of the tank through diffuser systems. The suspension of microorganisms is known as the mixed liquor suspended solids (MLSS) and is usually maintained at a concentration between 3000 and 4000 mg L^{-1}. In order to maintain the concentration of organisms at this level it is necessary to recycle a fraction of MLSS from the base of the final settlement tank back to the aeration basin. This is known as the return sludge. A system that utilises suspended growth with return sludge is known as the activated sludge process, and it has many process variants (**Figure 8.11**). Thus activated sludge uses energy for two processes, introducing air into the mixed liquor and pumping the return sludge from the final settlement tank. The challenge with wastewater treatment is to deliver the high-quality effluent associated with activated sludge, but at a reduced energy input, and this need is driving much process innovation (see **Box 8.4**).

In both attached and suspended growth processes it is necessary to remove a fraction of the biomass each day. This promotes an actively growing population of microorganisms and reduces the overall demand for oxygen. In attached growth systems the microorganisms slough off naturally from the inert support as the biofilm becomes too large to support the amount of BOD and oxygen available and it is sheared by the velocity of wastewater passing the support media. This is known as humus sludge. In suspended growth systems a fraction of the MLSS is pumped (or 'wasted') from the system each day. This is known as the waste or surplus activated sludge. There is a relationship between the amount of humus or waste activated sludge and the amount of BOD removed in the process. This relationship is termed the yield. Typically between 0.7 and 1.0 kg of sludge is produced for each kg of BOD removed.

8.3.7 Final or secondary settlement

The effluent leaving an attached or suspended growth process has a high TS concentration, which represents new microorganisms that have grown on the BOD in the settled sewage (as expressed by the yield). It is necessary to remove these solids before the treated wastewater is discharged to the watercourse. Fortunately, these microorganisms settle very well. he large particles of microorganisms have a strong negative charge of up to

Figure 8.11 The activated sludge process with air introduced into the tank. It is one of the most popular treatment processes worldwide, owing to the quality of the final effluent that it can produce.

−70 mV and are able to interact with positively charged material such as cations to form large agglomerates or flocs. They can have settlement velocities as high as 3 m hr^{-1}, and these solids are removed in settlement tanks that are very similar in appearance and behaviour to primary tanks. A final settlement tank (FST) is able to produce an effluent with a TS of 10 mg L^{-1} or less, which represents a solids removal efficiency of 99.75% for an MLSS of 4000 mg L^{-1}. As with a primary tank, these solids settle to the bottom of the tank to form a sludge with a solids concentration of around 1.2% or 12,000 mg L^{-1}. For a humus tank serving attached growth processes, all of this sludge is removed from the

tank, or wasted. If the tanks serve the activated sludge process, then a fraction of the solids are returned to the aeration basin as return sludge, and the rest are wasted.

This secondary sludge, like primary sludge, is also a good source of energy and comprises more than 75% volatile solids. However, it is more difficult to recover this energy because the microorganisms that make up the waste sludge have strong cell walls that resist destruction. The secondary sludge now passes forward for further treatment. The final effluent leaving the tank is generally discharged directly to the watercourse, but in certain situations it may pass forward for additional or tertiary treatment, such as disinfection for

Energy-neutral wastewater treatment

Esholt wastewater treatment works, owned by Yorkshire Water, treats the wastewater from Bradford in the UK. The plant deals with a population equivalent of 700,000 with a flow of up to 300 L s^{-1}. It was upgraded in 2008 to meet tighter discharge consents for BOD, TS and ammonia, and energy saving was central to the plant design philosophy. The influent to the works has a large available head with a fall of 10 m between the inlet works and the primary tank. Energy is recovered along this steep gradient by passing the flow through two 90 kW screw generators or hydro turbines, installed in series. A flow of 267 L s^{-1} passes sequentially through the two screws over the total fall of around 10 m and this generates 180 kW, around 5% of the site's energy needs. Better design and control of aeration and oxygen recovery by denitrification in the activated sludge process has resulted in an average energy efficiency of 4.5 kg O_2 per kWh, compared to older installations, which routinely achieve around 2.7 kg O_2 per kWh. The above processes result in 7.3 GWh per year of renewable electricity, about 30% of the site's needs.

In 2014 a thermal hydrolysis plant was opened at the site which heats up sludge to 165°C to kill off pathogens. This then enables the biodigestion process to generate more methane gas per unit of sludge. The methane gas is used for electricity generation, making the site self-sufficient, with excess electricity feeding back to the grid.

pathogen removal, or sand filtration to achieve a better removal of TS.

8.3.8 Membrane treatment

The final settlement stage of activated sludge is recognised as the weak link in the process, as it is expensive to construct and can frequently experience problems with the settlement of the sludge, which leads to a large loss of solids in the final effluent. It is also expensive to return sludge from the tank to the aeration basin. An alternative for removing solids from the aeration stage is to use a membrane with a small pore size, typically around 0.45 μm, that allows the treated effluent to pass but which restricts the passage of particles larger than the pore size. Inclusion of a membrane means that it is not necessary to recirculate solids back for aeration, as the membrane itself separates the solids retention time (as they are held in the basin) from the liquid retention time (as the liquid passes through the membrane and is discharged). It also means that effluent quality is extremely good, with TS <1 mg L^{-1} and BOD frequently <4 mg L^{-1}. The small pore size also completely removes bacterial pathogens, which are generally >1 μm in size, as well as many enteric viruses. As a result the treated wastewater can be directly reused. For instance, King County in the US state of Washington operates a large membrane plant serving 1.5 million people, including the city of Seattle. It produces 26,500 m^3 per day of high-quality water for non-potable reuse that is available for use by golf courses, farms and vineyards. This water reuse reduces withdrawals from the river, thus making more flow available for salmon.

8.3.9 Sludge handling

Primary and secondary sludge requires further treatment before it can be sent to its final destination. As it is produced with a solids concentration of just 1 or 2%, this means that up to 99% of the sludge is water. Thus, the first stage is to remove this water by a process of thickening. Thickening is achieved by the addition of a large molecular-weight organic compound that contains a lot of highly charged molecules, known as polyelectrolyte. These can be cationic or anionic, depending on their charge. They are used in conjunction with an inorganic cation such as iron or aluminium to provide additional charged material. Together these compounds cause the sludge to coagulate and then flocculate into large clumps that help to release the water from the sludge. The water is then removed either by **centrifuge** (spinning) or by running the sludge in a thin layer onto a belt with small holes that allows the water to pass through (**Figure 8.12**). In this way a sludge of up to 6% DS can be achieved. The water that is recovered is recycled back to the head of the works and is known as sludge return liquor.

Figure 8.12 A belt thickener allows the water contained in primary and waste activated sludge to drain away to produce a thickened sludge with a TS of up to 6%. The sludge requires conditioning first, to allow the particles to agglomerate and release the trapped water.

REFLECTIVE QUESTION

Sketch a flow scheme for a treatment plant using a blue line to represent the wastewater flow and a brown line to represent the solids flow.

8.4 ENERGY AND AGRICULTURAL USE OF SLUDGE

8.4.1 Energy reuse

Historically, the process of sludge treatment and handling has been one of the most expensive aspects of the treatment process, accounting for up to 60% of the total costs of treatment. Now, because of the rising cost of energy, this sludge is one of the most valuable aspects of the treatment process. Each tonne of sludge dry solids produced from wastewater treatment contains between 4 and 7 MWh of energy depending on its organic content. Indeed, sludge handling now generates surplus energy and this extra energy can be used in upstream processes, thus reducing the energy demand of the whole process.

The key to energy recovery is the process of anaerobic digestion. In many developed countries between 60 and 90% of sludge is treated by anaerobic digestion. An anaerobic digester is a very simple reactor in engineering terms but one that houses a very complex microbiology and biochemistry. The digestion process is undertaken in the absence of oxygen (hence the term 'anaerobic'). Microbial energy-yielding reactions always involve oxidative processes in which electrons are removed from an energy source and passed to an electron acceptor,

which becomes reduced. Oxygen is the traditional electron acceptor, accepting electrons to become reduced to water. In the absence of oxygen this causes a problem for the microorganisms, which are forced to use carbon in place of oxygen as an electron acceptor. Carbon can be reduced to many forms, often of commercial importance, such as ethanol, butanol or propanol. In the anaerobic digestion of sludge the desired end-product is the most reduced form of carbon, methane, often termed 'biogas'. Methane is energy rich, with 1 m^3 methane releasing 9.97 kWh energy on burning.

In order to generate the required end-product a four-stage biochemical transformation must be accomplished. Initially the large polymeric molecules of protein, carbohydrate and lipids are broken down to their component **monomers** of amino acids, simple sugars, glycerol and fatty acid by a process known as hydrolysis, which involves the insertion of a water molecule. Primary sludge undergoes rapid hydrolysis because the major components of a primary sludge have already undergone a preliminary digestion in the gut. By contrast, hydrolysis of secondary sludge is slow, as the microbial cell wall is very resistant to microbial attack. Thus the anaerobic digestion process is slow and the sludge must be retained for between 15 and 20 days at 35°C to fully degrade the sludge. As a result it is now common at larger wastewater treatment works (e.g. Blue Plains, Washington, DC; Gaontun, Beijing) to include an additional process to speed up hydrolysis. For instance, thermal hydrolysis (see **Box 8.4**) subjects the sludge to a temperature of 165°C and a pressure of 6 bar for up to 30 minutes. The pressure is then rapidly released, causing rupture of the microbial cell wall and the release of the cell contents. This means that much more energy can be recovered from the sludge in subsequent digestion. But such expensive and complex options as thermal hydrolysis are viable only for large works generating large quantities of sludge. The simple monomers released can be broken down further by oxidation to a number of volatile fatty acids (VFAs), such as acetic, propionic and butyric acid, by a process known as acidogenesis. These VFAs are then converted to acetic acid through acetogenesis and the acetic acid is cleaved to produce carbon dioxide and methane by methanogenesis.

Additional methane is also produced by hydrogenotrophic methanogens that are able to reduce carbon dioxide to methane with hydrogen. Each dry tonne of sludge can generate between 80 to 150 m^3 methane through anaerobic digestion, which has an energy value of up to 1.50 MWh, although typical yields in practice are around 90 m^3. This biogas contains other contaminants, such as carbon dioxide, hydrogen sulfide, ammonia and siloxanes, and will require cleaning up before it is used to generate energy. However, there are many uses for the biogas. It can be injected directly into the gas grid, or compressed for use as a vehicle fuel or sold as bottled gas. Its more widespread usage at the moment involves burning in a combined heat and power (CHP) engine. Rather like the engine of a car, a CHP engine, known as the prime mover, burns biogas, which turns an engine. This in turn drives a generator that produces electricity. This electricity can then be used on site to provide aeration and other activities that require electricity. The exhaust gases from the CHP, together with the engine cooling water, are then passed through heat exchangers to recover the heat, which is used to heat the digester itself, to provide heat for thermal hydrolysis and for space heating. Large sludge treatment centres fitted with thermal hydrolysis are now operated with a surplus of energy that can be exported to the grid.

8.4.2 Agricultural reuse

The digestion process destroys up to 60% of the VS content of the sludge, converting it to biogas. This still leaves the inert material and the remaining VS that is not destroyed, together with the new organisms that were produced during the digestion process. This material is known as 'biosolids' and it is produced at between 3 and 4% DS. A feature of the digestion process is that it renders the solids very easy to dewater and, using a similar system of belt presses or centrifuges that were applied during thickening, the biosolids can be dewatered to 30% DS or more to produce a sludge cake. The ideal route for this material is to recycle it to agricultural land, where it provides a valuable soil conditioner and source of nutrients and trace elements (see **Box 8.5**). During the digestion process, the breakdown of large polymers and bacterial cell walls also releases a lot of ammonia and phosphorus. These soluble materials are recovered during dewatering of the biosolids, and the dewatering liquor can contain ammonia as high as 1400 mg L^{-1} and phosphorus concentrations as high as 100 mg L^{-1}. Both of these nutrients can be precipitated in the form of magnesium ammonium phosphate or struvite, which is a valuable slow-release phosphate fertiliser. A number of commercial processes achieve this by dosing the dewatering liquor with either magnesium chloride or magnesium hydroxide and mixing this with air introduced in a rolling motion to form pellets of struvite that settle rapidly and can be recovered and dried. Thus, the end-products recovered from sludge handling are biosolids recycled to land, struvite sold as fertiliser and energy in the form of biogas.

Milorganite – a valuable resource from sewage sludge

Since 1926, the Milwaukee Sewerage Commission has been marketing and selling Milwaukee organic nitrogen (Milorganite) fertiliser derived from their waste activated sludge. They have been pioneers in devising appropriate marketing strategies and developing new markets for their product. Based on sound science involving field growth trials undertaken by agronomists, their early markets were fertiliser manufacturers (who used sludge as a component of mixed fertilisers), golf courses and flower growers. By 1932 it was the most widely used golf course fertiliser in the US. The process currently produces around 45,000 tonnes of product each year from a mixture of waste activated and anaerobically digested sludge, which is heat dried to 650° in order to remove water and ensure pathogen destruction. The sludge is then pelletised in a range of sizes to meet the needs of a range of final applications. It is notable how Milorganite has successfully faced a number of marketing challenges by developing new markets, and has tackled public prejudices against sludge-based products by sound and practical research and ensuring compliance with the most rigid standards.

Although the product has never been produced and sold profitably, the income it generates means it is cheaper than any alternative disposal option, for instance incineration, and does ensure full land recycling of the essential nutrients in wastewater.

8.4.3 Novel products from sludge

With many wastewater treatment plants now practising recovery of both energy and nutrients, these sites now more resemble resource recovery factories. This trend is likely to continue as the world tries to reduce its reliance on oil-based products and in particular non-biodegradable plastics. There are several technologies involving sludge, at various stages of development, that may contribute to further resource recovery (Smith, 2014). Many of these technologies are based around the manipulation of anaerobic digester biochemistry and microbiology to convert organic material in the waste stream into a wide range of high-value bioproducts such as acetate, caproate and butanol. This sort of production is sometimes referred to as the biorefinery concept or the VFA factory (**Figure 8.13**). Other approaches use an aerobic route for synthesis; for instance, polyhydroxyalkanoate (PHA) is a microbial intracellular storage compound stored by certain bacteria when a growth-limiting compound, for instance nitrogen, restricts their growth but there is an excess of organic material available to them. In this case the organic material is acetate and the bacteria can accumulate up to 70% of their cell mass as PHA (Akaraonye et al., 2010). PHAs are a precursor for a range of bioplastics and with a very high value. Work on digestate liquor suggests it is an ideal growth media for PHA accumulating bacteria.

REFLECTIVE QUESTION

Describe how the energy in wastewater can be recovered in a usable form and the products from this process successfully recycled to the environment.

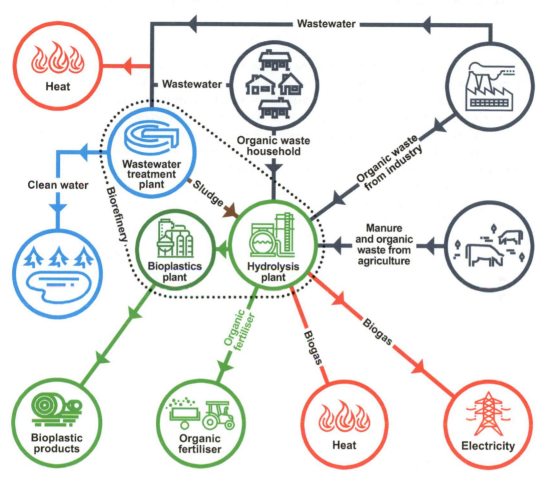

Figure 8.13 Converting the sludge produced in wastewater treatment together with other biodegradable organic wastes, into valuable products through anaerobic digestion via the biorefinery.

8.5 ALTERNATIVES TO CONVENTIONAL WASTEWATER TREATMENT

8.5.1 Decentralised sanitation

The centralised treatment processes described in the preceding sections are ideally suited for larger communities with populations >20,000 and that have adequate

supplies of potable water for toilet flushing. Centralised treatment has many advantages. There are cost savings in constructing a single, large facility and it permits capital investment in technologies to recover energy and nutrients from the wastewater that could not be justified at smaller plants. In addition, there is just a single point of discharge for the treated effluent and so not only are sampling and analysis costs reduced, the risks of consent failure (and thus prosecution) are also reduced. There are also cost savings to be made in the labour and sludge transport requirements that are needed for operating multiple, small treatment plants. However, centralised treatment systems require a large and expensive sewerage network to convey the wastewater from its site of production to the point of treatment; this is often many kilometres away. The sewage will also require pumping in areas where the gradient is flat, in order to convey it to the treatment plant, and this requires the expenditure of a lot of energy. There are also environmental implications as the upstream flow of the watercourse is reduced as a result of diverting many smaller wastewater flows downstream.

As a consequence of the above limitations, there are many situations where a decentralised approach to treatment may be more appropriate and in particular for new developments where there is no existing sewerage infrastructure. Choice of the most appropriate system is generally based on economics and feasibility. For instance, at present a new 'super sewer' is under construction in London and due for completion in 2023. Known as the Thames Tideway Tunnel, it is intended to intercept sewage flowing into the Thames from 34 overloaded CSOs (e.g. Figure 8.8a). It will be 7.2 m in diameter, 25 km long, constructed up to 65 m below ground and at an estimated cost of £4.2 billion. The validity of the super sewer is still being debated (National Audit Office, 2017) and many people considered that decentralised greener alternatives were available to prevent runoff entering the sewerage network and thus reducing spills from the CSOs. This green infrastructure involves the application of sustainable urban drainage (SUDS), in which runoff is dealt with at, or close to, the surface where the rainfall lands.

The main features of a decentralised treatment process are: a) the separation of the relatively unpolluted storm water or drainage from household generated wastewater; b) collection and storage of the former to allow it to percolate to groundwater; c) collection and treatment of the latter with discharge close to its point of origin. The first two of these are usually achieved by some form of SUDS (Figure 8.14). Surface water runoff from larger areas, such as part of a housing estate, major roads or business parks, can be channelled to detention ponds or wetlands for storage and pollution attenuation. Overflow from these can safely discharge to a local watercourse. Treatment of the wastewater from a decentralised scheme should, in principle, be much simpler than a conventional process because the drainage has been removed and consequently it does not experience wide fluctuations in flow and load. Simple options include a septic tank system to remove the solids (which then require tankering away at appropriate intervals) followed by constructed wetlands to oxidise the remaining colloidal or soluble pollutants. Alternatively, where large

Figure 8.14 Tanner Springs Park, Portland, Oregon, is a constructed urban wetland that collects surface runoff for storage and treatment.
Source: Shutterstock/Rachel Moon

tracts of land are available, a waste stabilisation pond system can receive and treat all of the flow without any need for a solids removal stage (Table 8.2). In order to ensure the success of decentralised systems an appropriate enforcement system must be in place, which routinely monitors the treated water quality and undertakes appropriate enforcement if the relevant discharge standards are not achieved.

REFLECTIVE QUESTION

Provide a list of the advantages for using a decentralised sanitation system and compare this to the advantages of centralised sanitation. What would be the most suitable approach for your own home town?

8.6 SUMMARY

The world's freshwater supplies are a finite resource but are rarely regarded and treated as such. Technology is available to produce palatable and safe drinking water from a wide range of source supplies, but this comes at a large cost and with its own pollution problems, particularly in relation to climate change and energy use. The flow train for water treatment includes screening and settlement, coagulation and flocculation, solids removal and filtration, and disinfection. Activated carbon may be used to remove organic contaminants, while desalination is used in some countries to treat seawater where water resources are scarce, but desalination is very energy intensive.

Water and wastewater management now seeks to safeguard and manage the world's water resources in a more sustainable manner while providing for increasing populations and economies. Water is a resource that can be used and then reused. In certain parts of the world treated wastewater is now recycled directly to the water treatment works for potable reuse. Introducing wastewater into water use means that less is released back into the environment, and wastewater treatment becomes important not only as a pollution prevention measure but also as a strategic tool in water resource conservation.

Many of the world's rivers, lakes, estuaries and seas are polluted as a result of the discharge of untreated human and industrial effluent. Reducing pollution from wastewater is not technically challenging, using physical and biological treatment processes, but it is expensive and, if not operated correctly, with proper monitoring, the technology will not achieve its aim. Wastewater is now viewed as a resource that can be recovered, and treatment plants can now be built that are energy neutral and consequently with much reduced operating costs. Over a period of less than a decade, wastewater treatment has gone from being an expensive, energy-intensive process with a number of waste outputs, to being an almost energy-neutral treatment process with negligible waste outputs and almost complete recycling of all input materials. Although the capital costs remain high, the money is well spent, as it increases both the amenity and the commercial value of the aquatic environment, providing enjoyment for both present and future generations. Indeed, the additional cost to the consumer to fund the development of wastewater treatment facilities has been small and demonstrates how it might be feasible to move towards a zero-waste economy operating largely on renewable energy, with a negligible carbon footprint.

FURTHER READING

Binnie, C. and Kimber, M. 2017. *Basic water treatment*. Thomas Telford; London.

Now in its sixth edition, this is a good practical guide to the range of unit operations that comprise a modern water treatment works, with clear explanations of their role and mode of operation.

Horan, N.J. 1990. *Biological wastewater treatment systems: Theory and operation*. John Wiley and Sons; Chichester.

A concise guide to understanding the biological principles which guide the design and operation of wastewater treatment facilities.

Mara, D.D. 2011. *Good practice in water and environmental management: Natural wastewater treatment*. CIWEM; London.

Natural wastewater treatment systems find many applications in small communities and this handbook reviews the design principles for such treatment processes.

Mara, D.D. and Horan, N.J. (eds). 2003. *The handbook of water and wastewater microbiology*. Academic Press; London.

A selection of articles from international experts introducing all aspects of water and wastewater microbiology.

Metcalf & Eddy Inc. 2013. *Wastewater engineering: Treatment and resource recovery* (5th edition). McGraw Hill; New York.

The classic textbook, comprehensive in its scope and coverage of the design of processes used for the treatment of wastewater and sludge.

MWH. 2012. *Water treatment: Principles and design* (3rd edition). John Wiley and Sons; Chichester.

This is becoming the classic textbook for water treatment, with a broad and comprehensive coverage of the treatment options.

Schneider, D. 2011. *Hybrid nature. Sewage treatment and contradictions of the industrial ecosystem*. MIT Press; London.

A detailed and scholarly account of the history of wastewater treatment and sludge recycling from the perspective of the changes made to natural ecosystems through human intervention, in order to deliver effective engineered treatment processes.

Classic papers

Anderson, J. 2003. The environmental benefits of water recycling and reuse. *Water Science and Technology: Water Supply* 3: 1–10.

One of the first contributions that outlined the possibilities for utilising treated effluents through recycling and reuse.

Ardern, E. and Lockett, W.T. 1914. Experiments on the oxidation of sewage without the aid of filters. *Journal of the Society of the Chemical Industry* 33: 523.

This paper, presented in Manchester to the Society of Chemical Industry, is acknowledged as the invention of the activated sludge process, which is now the most successful wastewater treatment process option.

Barnard, J.L. 1976. A review of biological phosphorus removal in the activated sludge process. *Water SA* 2: 136–144.

The discovery by Barnard of the mechanism for biological P-removal has led to huge improvements in the quality of treated effluents and helped to reduce the impact of eutrophication wherever it is practised.

Marais, G.V.R. and Ekama, G.A. 1976. The activated sludge process Part 1 – steady-state behaviour. *Water SA* 2: 164–200.

This work was the start of a unified model of the activated sludge process, which led to computer-based design and optimisation of treatment plants.

PROJECT IDEAS

- Design a poster that explains the steps of water treatment for potable supply and another that outlines wastewater treatment steps.
- Use Google Earth to identify the wastewater treatment plant that treats your own sewage. See if you can identify the individual unit processes from the aerial view.
- Find out how much sludge material from wastewater treatment is reused in your local area and perform a study to outline the potential for use of a larger proportion of the sludge. What markets are there for sludge products?
- Design an educational video that outlines how wastewater treatment could be an almost energy-neutral process.

REFERENCES

Agler, M.T., Wrenn, B.A., Zinder, S.H. and Angenent, L.T. 2011. Waste to bioproduct conversion with undefined mixed cultures: the carboxylate platform. *Trends in Biotechnology* 29: 70–78.

Akaraonye, E., Keshavarz, T. and Roy, I. 2010. Production of polyhydroxyalkanoates: the future green materials of choice. *Chemical Technology and Biotechnology* 85(6), 732–743.

Anderson, J. 2003. The environmental benefits of water recycling and reuse. *Water Science and Technology: Water Supply* 3: 1–10.

Ardern, E. and Lockett, W.T. 1914. Experiments on the oxidation of sewage without the aid of filters. *Journal of the Society of the Chemical Industry* 33: 523–539.

Barnard, J.L. 1976. A review of biological phosphorus removal in the activated sludge process. *Water SA2*: 136–144.

Binnie, C. and Kimber, M. 2009. *Basic water treatment*. Thomas Telford; London.

Chow, A.T., Tanji, K.K. and Gao, S. 2003. Production of dissolved organic carbon (DOC) and trihalomethane (THM) precursor from peat soils. *Water Research* 37: 4475–4485.

Cutler, D. and Miller, G. 2005. The role of public health improvements in health advances: the twentieth-century United States. *Demography* 42(1): 1–22.

du Pisani, P.L. 2006. Direct reclamation of potable water at Windhoek's Goreangab reclamation plant. *Desalination* 188: 79–88.

Eikelboom, D.H. 2000. *Process control of activated sludge plants by microscopic investigation*. IWA Publishing; London.

Henze, M., van Loosdrecht, M.C.M., Ekama, G. and Brdjanovic, D. 2008. *Biological wastewater treatment: Principles, modelling and design*. IWA Publishing; London.

HMSO. 1970. *Working party on sewage disposal (The Jeger Report). Taken for granted*. HMSO; London.

Horan, N.J. 1990. *Biological wastewater treatment systems: Theory and operation*. John Wiley and Sons; Chichester.

Horan, N.J. and Lowe, M. 2007. Full-scale trials of recycled glass as tertiary filter medium for wastewater treatment. *Water Research* 41: 253–259.

Judd, S. 2006. *The MBR book: Principles and applications of membrane bioreactors in water and wastewater treatment*. Elsevier; Oxford.

Klaunig, J.E., Ruch, R.J. and Pereira, M.A. 1986. Carcinogenicity of chlorinated methane and ethane compounds administered in drinking water to mice. *Environmental Health Perspectives* 69: 89–95.

Mara, D.D. 2011. *Good practice in water and environmental management: Natural wastewater treatment.* CIWEM; London.

Mara, D.D. and Horan, N.J. (eds). 2003. *The handbook of water and wastewater microbiology.* Academic Press; London.

Marais, G.V.R. and Ekama, G.A. 1976. The activated sludge process Part 1 – steady-state behaviour. *Water SA* 2: 164–200.

Meggers, F. and Leibundgut, H. 2011. The potential of wastewater heat and energy: decentralized high-temperature recovery with a heat pump. *Energy and Buildings* 43: 879–886.

Metcalf & Eddy Inc. 2003. *Wastewater engineering: Treatment and re-use* (4th edition). McGraw Hill; New York.

MWH. 2005. *Water treatment: Principles and design* (2nd edition). John Wiley and Sons; Chichester.

NAO National AuditOffice. 2017. *Review of the Thames Tideway Tunnel.* Report by the Comptroller and Auditor General. HC 783 Session 2016-17 2 March 2017.

Paleologos, E. K., Al Nahyan, M.T. and Farouk, S. 2018. Risks and threats of desalination in the Arabian Gulf. 4th International Conference on Water Resource and Environment. *IOP Conference Series: Earth and Environmental Science* 191: 012008.

Pereira, M.A., Lin, L.H., Lippitt, J.M. and Herren, S.L. 1982. Trihalomethanes as initiators and promotors of carcinogenesis. *Environmental Health Perspectives* 46: 151–156.

Poustie, M.S., Deletic, A., Brown, R., Wong, T., De Haan, F.J. and Skinner, R. 2015. Sustainable urban water futures in developing countries: the centralised, decentralised or hybrid dilemma. *Urban Water Journal* 12: 543–558.

RSC. 2009. *What is a poison?* Note produced by a Working Party of the Environment, Health and Safety Committee [EHSC] of the Royal Society of Chemistry [online]. Available at: www.rsc.org/images/PoisonV2_tcm18-37982 .pdf.

Schneider, D. 2012. Purification or profit: Milwaukee and the contradiction of sludge. *In*: Foote, S. and Mazzolini, E. (eds) *Histories of the Dustheap: Waste, material cultures, social justice.* MIT Press; Cambridge, MA.

Smith, S.R. 2014. How activated sludge has been transformed from a waste to a resource and the implications of this for the future of the activated sludge process. *In*: Horan, N.J. (ed.) *Activated sludge, past, present and future.* p. 267. Aqua Enviro; Wakefield.

Verstraete, W., Van de Caveye, P. and Diamantis, V. 2009. Maximum use of resources present in domestic 'used water'. *Bioresource Technology* 100: 5537–5545.

WHO. 2017. *Guidelines for drinking-water quality* (4th edition incorporating the first addendum). World Health Organization; Geneva.

CHAPTER NINE

Water demand planning and management

9

Adrian T. McDonald
and Gordon Mitchell

LEARNING OUTCOMES

After reading this chapter you should be able to:

- describe the key methods for measuring and estimating water demand;
- explain the key drivers of water demand and trends in water demand;
- outline methods of water management through resource development;
- outline methods of water management through efficiency.

9.1 INTRODUCTION

This chapter concerns water demand and the challenges of the water gap, which arises where there is a greater demand for water than available supply in the forms in which it is demanded. Demand for water grows because of demographic change (rising population, changing average household size, more affluent lifestyles), increased needs for agricultural produce, urbanisation, and the supply of manufactured products. The Middle East and North Africa is an area which, broadly, is considered to have poor water resources and is also where there have been very large reductions in water resources over the past 60 years (Jones, 2010). In other parts of the world, however, technological advances, drives to reduce demand through water efficiency, and reduced activity of water-consuming industry have resulted in declining demand (e.g. in Germany). Water demand management not only involves large-scale water resource inventories (groundwater and surface water accounting, usage rates, replenishment rates, etc.) in order to try to secure resources to meet demand but, importantly, it also involves working with people to alter attitudes and behaviour so as to reduce water demand.

9.2 SOURCES OF DEMAND

Just as water is not a simple, single product (floodwaters are very different from **potable water**, which in turn differs from urban wastewater effluents), so the demand for water varies, depending on the nature of the water involved. The demand for raw water is quite different from the demand for potable water. Indeed, raw water for potable consumption is simply one of the many – often minor – raw water demands.

Water can be used both in-stream and off-stream. In-stream water use refers to direct use of water in rivers or lakes where there is no withdrawal of the water. For example, the uses of water for navigation, hydropower, tourism and fisheries are all in-stream water uses. Off-stream water use involves withdrawal of the water from the source. Where water is withdrawn and then not returned to

the resource that provided it, this is known as **consumptive water use**. Examples might include manufacturing, where water is abstracted for cooling, which results in evaporation of the water, or water used in food preparation, where the water is not returned to the stream. Agricultural consumptive water use includes the water transpired by plants plus losses from evaporation from the soil surface and leaves in the crop area. Good farming practice attempts to reduce crop water consumption to its most efficient minimum.

The major demand for water globally arises from **irrigation** requirements for agriculture (**Figure 9.1**). Thus, this demand will

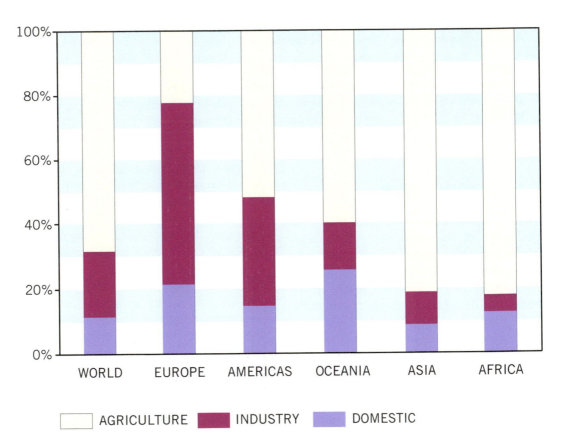

Figure 9.1 Off-stream water use (water withdrawal) by sector.
Source: Data from FAO Aquastat database, 2016.

be influenced by moisture deficit and by the market for agricultural products. Soil moisture alters in the short term with weather variability, in the medium term with seasonal crop cycles and inter-annual variability (e.g. El Niño Southern Oscillation; see Chapter 2), and over longer periods with the influences of climate change. The demand for agricultural products is influenced by a range of factors, such as population growth, lifestyle expectations, world market prices, food and energy sourcing policies and more. So a key point is to understand how the demand for water interacts with energy and food requirements and circumstances. For example, if more land is taken to grow crops that are used as biofuels, that is an additional water demand to that required for food provision. Box 9.1 presents some interactions which have been characterised as giving the possibility of the 'perfect storm'.

Demand for water abstraction (off-stream use) arises from industry, agriculture and domestic water supply requirements, but water is also demanded for in-stream use. It is needed in river channels for power generation – in past times primarily for direct mechanical power but today also for various types of hydropower – head fall, run-of-the-river, micro-hydro. Water is also demanded within channels for navigation for shipping. The maintenance of water habitats and water quality is also an in-stream requirement. In Europe there is an explicit requirement for sufficient water to maintain or achieve 'good ecological status' in all reaches of the river, as required by the EU Water Framework Directive (see Chapters 4 and 6). However, there has also been an *implicit* requirement in many countries that there must be sufficient water to dilute

BOX 9.1

The perfect storm

There have been warnings to water managers and scientists, policymakers and politicians, to the possibility of a 'perfect storm', wherein demand for energy and food in the BRIC economies (Brazil, Russia, India and China), driven by lifestyle and demographic change, when allied to climate change, drives up prices on global markets (Beddington, 2009). This may encourage some countries which are currently net importers of food to try to grow more food locally. For example, global price increases would force increased irrigation-driven food and energy cropping in the UK. Should this arise in a period of drought it would be a challenge to provide the water and prioritise supply.

It has been found that groundwater resources in China for agricultural irrigation are unsustainable, especially in the northern plain (Cao *et al.*, 2013). China could consider reducing food production in the region as it would be more economical to import than to keep trying to resolve the problem via major infrastructure and engineering programmes (see Chapter 12). However, such a move in itself could cause major shifts in global food production and prices.

See the project ideas at the end of the chapter for some work you could undertake on this topic.

pollutants – summarised in the phrase 'the solution to pollution is dilution'. In the UK this was formalised in the 1912 Royal Commission on Sewage Disposal, where it was suggested that there should be a river flow of eight times the volume of any pollutant discharge (and that such river flow should be of fair quality at least – an often forgotten requirement). It follows, then, that the demand for water in-stream is interconnected with the multiple off-stream uses via abstractions and discharges that occur on many river systems. Such demand therefore has to be evaluated for every river reach. This lies at the heart of abstraction licensing in those countries where legal processes exist to deal with licensing for abstraction (e.g. the USA, Australia, Germany and the UK). Even in those countries some rivers are now over-licensed for abstraction, in order to satisfy demand (e.g. Bobbie *et al.*, 2014; Santato *et al.*, 2016). While it has been argued that actual abstraction is frequently much less than licensed abstraction limits, it is now recognised that abstraction at the maximum rate (a value strongly dependent on the wording of the licence) is likely to occur at the time of greatest stress on the river environment. As a result, regulatory agencies are more frequently not renewing licences and water companies themselves are not seeking renewal of some abstraction licences. Such changes are often contained in phrases such as 'environmental efficiencies' or 'river husbandry'. It is an appropriate and correct course of action, but will require both water demand reductions and new alternative resource developments, which almost inevitably requires exploiting more distant water resources, presenting challenges on grounds of property rights, cost and institutional cooperation (see also Box 9.3).

9.3 MEASURING CURRENT POTABLE DEMAND

9.3.1 Water metering

At its core, demand is an economic concept (see Chapter 10) – how much of a particular good a population is willing to buy for a set price. Water is a rather more problematic resource to consider in that style, for three reasons; first, because it is not a matter of pure choice, as water is essential to life; second, because demand is not always directly measured but must often be estimated; and, third, because significantly more water must be abstracted than indicated by end-use demand, so as to accommodate water used in treatment or operational uses, or lost to leakage and theft. If a country has universal water metering, it is in theory a simple matter of aggregating some millions of meter readings to provide a measure of potable water demand. However, these meters will most likely be 'dumb' meters, read at different times and themselves representing an aggregation of different time periods (e.g. perhaps read once a year or once every six months), and so deriving a credible and useful figure for water demand is still not trivial. Box 9.2 provides details of more innovative and useful smart meters.

BOX 9.2

Intelligent metering

Water meters have typically been installed in a meter box, often outside the property. They have to be physically read. The set-up in Figure 9.2a is typical. However, the gas and, particularly, the electricity industries in the developed world have adopted smart meters that measure continuously and so can allow smart tariffs and provide excellent customer information.

Automatic meter reading can be carried out remotely. It often uses wireless technology to connect to a local station, which might be a van passing down the street to read all the meters. The frequency of the reading is the key to whether it is just a more convenient dumb meter or a much smarter device. A true smart meter reads the flows 'continuously' (e.g. Figure 9.2b), reports the information to the water company, analyses the information and presents various summaries to the consumer. So, in addition to the meter there is likely to be a consumer wireless display unit. A smart meter may be able to provide extremely valuable water demand information which enables water providers to deliver services more efficiently. For example, knowing if there are areas or households with particular patterns in water demand (e.g. religious or cultural patterns) may enable water pressures to be altered during the day to suit those areas, thereby also reducing any volumes of water leaking from underground pipes at other times of the day.

Water consumption in a dwelling can also be monitored using smart acoustic meters. These detect the noise water makes on passing through the supply pipe, and following calibration, are able to differentiate between specific end uses, such as a toilet flush, shower, washing machine or a running tap. These are not used for billing purposes, but provide insights into 'micro-component' water use that helps improve estimates of demand in unmetered properties.

Figure 9.2 a) A dumb meter, which simply records accumulated usage; b) a smart meter, which provides continuous data on usage through time.

In some countries, however, not all homes have a water meter and there may be an annual tariff (perhaps related to the size of the house or number of residents) no matter how much water is used (which is not useful for driving efficient behaviour). The problem remains as to how you estimate demand, and thus plan the supply needs, when there are no or few meter records. Surprisingly, this is not just restricted to the developing world. The UK is one example, with less than 50% of homes metered. This figure also varies markedly in different regions – in 2018 ~80% of the customers of Anglian Water, in the east of England, were metered, a figure unchanged from 2009, but the company has ambitions to raise this meter penetration to 95%. In contrast, Scottish Water had a mere 530 metered customers. This remarkably low number, 0.02% of the 2.3 million customers, was a reduction of over 10% from the previous year's figure. This is unsurprising as Scottish Water requires customers to pay all survey, adaptation and installation costs plus additional ongoing charges.

9.3.2 Demand estimation

There are two main approaches to demand estimation based on either 'domestic consumption monitors' or district meters (known in the industry as the bottom-up and top-down approaches, respectively). The bottom-up method aggregates estimates of demand made at the individual household level, while the top-down method disaggregates from an area measurement.

9.3.2.1 Bottom-up

A domestic consumption monitor or domestic water-use survey is a sample of about 2000 households (at most, but often many fewer) who pay on a standard tariff (i.e. based on not having a meter).

Establishing, and thereafter maintaining, a monitor is difficult, although the information that such a monitor provides can be invaluable to the industry. The sample of customers is usually drawn at 'random' from the company's billing computer, but that does not mean just entering the list of customers at random and printing the next several thousand names would be correct – start at K and you might get too many Khans, start at M and there might be too many Macs, and these could have implications for properly representing cultural water-use differences (see below). But let us assume for the moment that the initial sample was near random. The people on the list are contacted and invited to join the survey, but only a small proportion of those invited will choose to enter the survey and so they are self-selected, and this starts to introduce bias. People may decide to join the survey because they think they are low-water users and hope to become a metered customer if the survey bears out their perception; others may be reticent to join a survey because of their own real or perceived educational limitations; while yet others may simply wish to avoid contact with bodies with 'authority'. Table 9.1 lists the range of known biases. All domestic consumption monitors contain bias, some of which are present from the outset while others build up during the maintenance and upkeep of the monitor (Parker and Wilby, 2013). These biases can be partially counteracted by careful correction measures in the recruitment of the sample and the interpretation of the results, but such surveys will never give a precise measure and when used to report demand to fractions of a litre, as is sometimes required by regulators, give a false impression of precision. The challenge for demand analysis is to determine

BOX 9.3

Integrated water resources management

IWRM is a high-level framework which aims to foster integration:

- of all parts of the water cycle (water sources, flows, and their manipulation for human use) to produce more efficient, productive and environmentally benign use;
- of uses, including agricultural, household, industrial and environmental, to efficiently allocate water to maximise benefits;
- of water institutions, stakeholders and users – IWRM recognises the complexity of water governance, and the need for enhanced communication, collaboration, information sharing and participation, and the need to bring together disciplines and skillsets;
- across time, recognising short-, medium- and long-term needs and pressures, and the need for flexible adaptive systems able to cope with uncertainty and future change;
- across space, from a single site to the regional scale, so water management decisions upstream and downstream recognise each other, and institutions working within a basin or across multiple basins, are effectively coordinated;
- of innovation to improve productivity and efficiency. Innovations may comprise unconventional water sources, new infrastructure whether hard (technology) or soft (ecological), as well as non-structural tools such as regulation, education and application of market-based instruments.

Some related systems approaches to water management are listed below.

Water management hierarchy – argues for water conservation via elimination of use, reduction through demand management, direct reuse (e.g. grey water recycling) and outsourcing (e.g. rainwater harvesting) and finally recycling. Only once these options have been explored should new freshwater resources be developed.

Water soft path – this approach, developed from the energy sector, is oriented towards meeting user needs, rather than simply supplying water. Conventional supply-side systems remain relevant, but the water soft path fosters consideration of demand management, use of water of different quality for different needs, decentralised infrastructure (e.g. rainwater harvesting), and work at local and community scale to identify better alternatives to meeting water service needs.

Water efficiency trading and investment offsets – these facilitate delivery of more efficient water allocation across sectors. For example, water conservation may be more cheaply achieved in a river basin within the agriculture sector than in the domestic sector. Cities can thus support farmers to conserve water (e.g. via investing in drip irrigation), thus freeing water for urban use, in return for financial compensation, but at lower cost than for urban water conservation.

Virtual water trading – virtual water (see Chapter 12) is the total water required to produce a commodity. Water-intensive products can intensify water scarcity. Using water footprinting (Chapter 12) and life cycle analysis, virtual water can be quantified and it may be possible to adjust trade (e.g. switch to less water-intensive crops) to reduce pressure from virtual water in water scarce areas.

Water neutral development – specifies that development should only occur if no new water is needed, with investment made in water efficiency and productivity elsewhere, equivalent to the projected water consumption in the new development.

the overall direction of each bias and to allocate an estimate of the extent of bias so that a correction can be entered into the final demand estimate.

If the sample is really representative of the household population as a whole we can simply take the average. Unfortunately, in many countries water supply bodies are required by legislation to report not just the household water demand but the average per capita demand, and this introduces significant complexities. In particular, it requires that the size of every household in the sample be known. Unfortunately, household sizes change frequently, due to births and deaths, lodgers, students away from home and so on, and households typically move in their entirety every few years on average, so in, say, just three years, the sample information on which to calculate per capita demand could have drifted at least 25% into error. Thus, for

the bottom-up approach water supply bodies must resurvey on at least an annual basis, with both cost and enhanced bias impacts.

Although the water industry seeks a large sample (e.g. 2000 in a domestic water-use survey), for many supply bodies that figure falls significantly, as data returns may be incomplete and recruitment is difficult because people frequently leave the survey. Whatever the sample size, the water-demand analyst eventually finishes with a table similar to that shown in Table 9.2. Each row provides the result showing the water demand for a household. However, Table 9.2 shows two alternative values for the final calculation. The first is derived from the averaging of the individual household per capita demands, and the second is arrived at by summing all the daily demands and dividing by the aggregate population.

Table 9.1 Biases in domestic consumption monitors (DCM)

Bias	Description
Self-selection	When a water company wishes to create a DCM the people who choose to join are self-selected. They may join because they believe they are low-water users, or because they are 'good citizens', or because they like being part of an experiment.
Staff	To get a DCM started it is possible that a company may encourage its own staff to join – both to get it underway and to iron out initial difficulties. However, such members will be more informed and are more likely to behave 'correctly'.
Exclusion	Not everyone feels able or is allowed to join a survey. In the early years of UK private water companies, local authorities in some areas, in a mistaken view of their own authority and because they had political disputes with the concept of 'privatisation', told council house tenants that membership of a water survey was not allowed. The inevitable bias which carries through to type of property, family size and several other factors has been difficult to eradicate. People can also self-exclude if, for example, they have concerns about language and competence.
Financial advantage	People are compensated for the time it takes to complete forms and record information. For some this is an income source, but it is likely to be biased towards those with time available and need for supplemental income.
Hawthorne effect	When people are surveyed they form an opinion about the purpose of the survey and tend to make an (unknown to them) effort to perform in the direction expected. If using less water or being more water efficient is perceived as being an appropriate outcome they will be biased in that direction.

Table 9.2 A very simplified domestic consumption monitor table

	Household consumption (litres per day)	Number of residents	Water demand per capita (litres per day)
	400	4	100
	150	1	150
	280	2	140
	360	3	120
Total	1190	10	119 or 127.5

Note: In a real table there are likely to be 1000 rows (households) and perhaps hundreds of columns, with each column holding a weekly or monthly average demand (each derived, in many cases, from 15-minute demand data), together with data on location, socio-economic status and property characteristics. One outcome (first number, bottom right) is derived by dividing the sum of column 1 by the sum of column 2. The second outcome is derived by averaging the per capita demand of the sample households (second number, bottom right).

Starting from an even more fundamental base, the demand for water can be calculated from the product of the population of interest multiplied by the per person demand. The total population can be segmented by household characteristic to perhaps gain a more sophisticated measure. The per person demand can be determined by the use of OVF (ownership, volume and frequency) calculations, sometimes known as micro-components (Williamson *et al.*, 2002). Components of demand are such things as toilet flushing, garden watering, tap use, dishwasher use, etc. For each component the average ownership is multiplied by the average volume per use and by the average frequency of use, and all component use is summed to derive the use per person or per household. This is a very mechanistic process, the averaging hides a great deal of variability, and the values required are very difficult to determine with confidence (see also Box 9.2). Clearly, the ownership of three baths does not result in the owner having triple the average number of baths, and so working with OVF determinations requires considered views of both the data and the outcomes. However, in some countries this is a mandatory approach to demand measurement. The measurement of micro-components is outlined in Ball *et al.* (2003). Data from surveys of domestic water use can be analysed using multivariate

statistical methods (predominantly regression) to relate demand to property and household characteristics. This has some advantages but is also more opaque and so less available to water customer representatives. Multivariate methods assume the sample is unbiased, which, as we have seen, may not be the case.

9.3.2.2 Top-down

Water in a distribution system is monitored by a system of area meters. Top-down demand estimation is derived from these meters. Each meter covers, typically, a minimum of several hundred households, but the number is often much larger. The meters measure total water passing through the pipe, irrespective of final use. Area meter-based determination of the household demand is therefore more complex and less direct. An area meter gives the water use in a relatively large area. It is therefore the sum of many demands, few of which are known. Initially the nightline is determined. This is the water demand in the early hours of the morning when it is assumed that household use is at a minimum. From the nightline an allowance for legitimate night use is removed, as is metered domestic usage and commercial and industrial use (also metered) for the night period; the remainder is deemed leakage. From the average demand on the area meter, commercial and metered demand is subtracted, as is the now 'known' leakage. Further allowances for estimated operational use, estimated theft, estimated emergency use and the lowered leakage under the lower-pressure daytime regime in the metered area are also entered into what is simply a mass balance. The result is average demand of the unmetered population. This can now be divided by the number of households to derive per household average unmetered water demand and by the population (although this too is often not a readily available, up-to-date, accurate value) to gain average per capita unmetered domestic demand.

The two values of bottom-up and top-down demand estimation will not be the same, but if they lie within 5% of each other little further action is deemed necessary. If they lie outside this band, then some form of outcome reconciliation is required, which may be either an arithmetical or an investigative correction.

REFLECTIVE QUESTIONS

What methods can be used to estimate domestic and industrial water demand?

How can we estimate leakage rates?

9.4 TRENDS IN WATER DEMAND

Water demand is driven by population, lifestyle and economy. Of course, lifestyle is dependent on age, affluence, education, culture, family size and several other more minor factors, few of which are directly measurable. Aggregate water demand is trending upwards both globally and in most countries, and will continue to do so for the next 20 years because the population forecasts for that period are well understood. Thereafter the water demand is less certain, reflecting the uncertainty in the population forecasts. The components of the trend also change over time. In less developed countries, irrigation is the major water demand, but, as countries become more industrialised, so commercial and domestic water requirement becomes more important. The trend depends on how much water is already demanded, and expected population growth. Globally, water demand is increasing at a 2% compound annual growth rate, and by 2030 a major gap between water resources demanded and sustainable resource availability is expected. This is illustrated in Figure 9.3, taken from a report by the global consulting firm McKinsey & Company, which aggregates the views and research of several major global companies.

The implication of this water-balance forecast is that, without significant interventions, there will be water shortages. In some countries, typically in the developed world, this might be addressed by demand reduction *and* resource development, but in many less developed nations resource development is the only option, assuming that the water resource is available for exploitation, a situation that does not always exist. The deeper implications of this last statement are troubling: drought, crop failure, inability to develop, water-related diseases and ultimately conflict (see Chapter 11) over access to water and, if the conflicts are across national boundaries, war. In the words of Ban Ki-moon,

All are places where shortages of water contribute to poverty. They cause social hardship and impede development. They create tensions in conflict-prone regions.

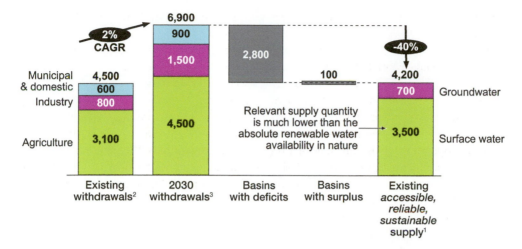

Figure 9.3 Global water demand and resource availability for 2010 as a baseline compared to a forecast for 2030.

Source: Charting our water future, November 2009, McKinsey & Company, www.mckinsey.com. © 2019 McKinsey & Company. All rights reserved. Reprinted by permission.

Notes: [1] Existing supply which can be provided at 90% reliability. [2] Based on 2010 agricultural production analysis. [3] Based on GDP, population and agricultural projections and considers no water productivity gains to 2030. CAGR is compound annual growth rate.

Too often, where we need water we find guns. … There is still enough water for all of us – but only so long as we keep it clean, use it more wisely, and share it fairly.

Ban Ki-moon, UN secretary-general
(United Nations, 2012)

The OECD has forecast that the population suffering from *severe* water stress will rise by a further 1 billion to 3.9 billion by 2030 (**Figure 9.4**). Water stress is the least severe of the categories of reduced water availability. The OECD suggests that on a quantitative basis **water stress** occurs when the water resource falls below 1700 m^3 per person per year, **water scarcity** occurs when the resource drops below 1000 m^3 per person per year and **absolute water scarcity** occurs when the water resource falls below 500 m^3 per person per year. The UN Water for Life decade also provides a qualitative definition of water scarcity:

Water scarcity is the point at which aggregate impact of all users impacts on the supply or quality of water under prevailing institutional arrangements to the extent that demand cannot be satisfied fully.

(United Nations Department of Economic and Social Affairs, 2014)

In the above definition, scarcity occurs when the water resource availability is significantly below that which we have become used to and which we have anticipated being available in the future. Mapping (**Figure 9.5**) shows that water scarcity, based on a quantitative measure that addresses the balance between water resources and population, is not just focused on the desert countries.

Demand varies greatly in different countries (**Figure 9.6**). Of particular note are the very significant differences between countries at broadly the same state of economic development. The USA per capita water

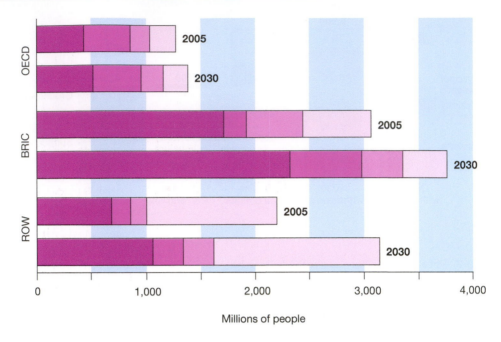

Figure 9.4 Populations living in areas of water stress. Colours denote, from left to right: severe, medium, low and no water stress.

Source: Kindly reproduced with permission from OECD (2008), *OECD Environmental Outlook to 2030*, OECD Publishing, doi: 10.1787/9789264040519-en.

Notes: OECD countries = 34 developed countries; BRIC countries = Brazil, Russia, India and China; ROW = rest of the world.

demand is quadruple that of the UK. How much faith to place in such aggregate figures is, however, open to question; Germany is recorded in **Figure 9.6** as using more water than the UK but is quoted elsewhere as being more water efficient. Alternative tabulations of water use per head, for example, as reported by the EEA (n.d.), show Germany using 125 litres per person per day. There are reasons why countries at the same level of economic development can have different water demands. Differences in average household size, propensity to live in flats in Europe and in grassed lots in the USA and differences in the way the data is collected – including or excluding leakage are examples. Price drivers are also partly responsible for the differences. **Figure 9.7** shows household water use indexed against water prices. At lower

water prices a range of water demands exist: a Canadian household using triple the water of a Hungarian household. But, as price increases above $3 per m^3, the gap between highest and lowest water use diminishes, and demand is fairly consistent around the 50 m^3 per capita per year.

Water demand by industry can be for both raw and potable water. For example, drinks, canning or food processing companies will require potable supplies, while steel making will require primarily raw water. However, many industries, such as energy processing and refining, that might be expected to require raw water in fact use considerable quantities of potable water.

Almost all potable water demands made by industry and commercial premises are metered in the developed world. Some companies have a diverse range of industrial users,

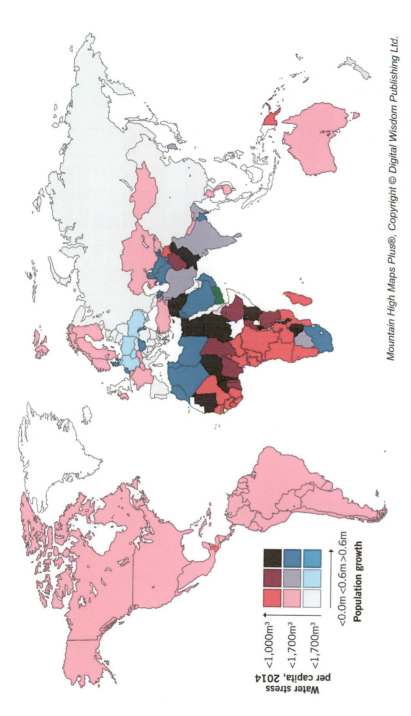

Figure 9.5 The interaction between current water availability per person and forecast population growth to 2050. Data based on UN population forecasts and the FAO Aquastat database.

Source: Damania *et al.* (2017). World Bank. Licence: Creative Commons Attribution CC BY 3.0 IGO.

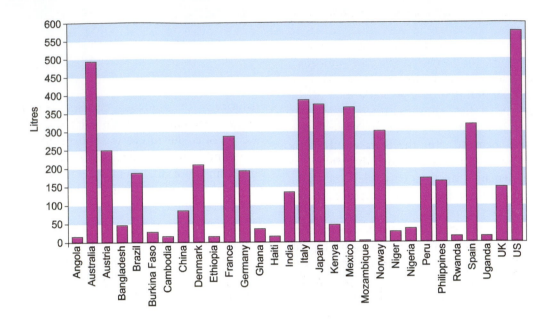

Figure 9.6 Water demand in different countries – average water use per person per day.
Source: Data from UNDP, 2006.

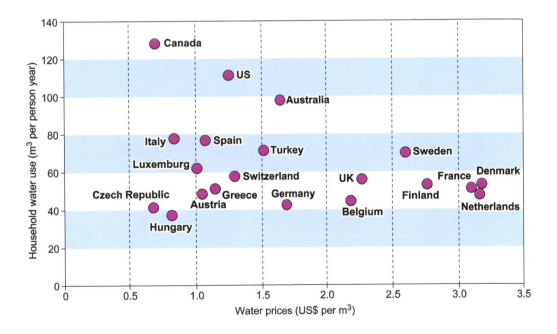

Figure 9.7 Household water use indexed against water prices across a range of countries.
Source: Data based on values reported in OECD, 1999.

and in times of economic 'tranquillity' it is possible simply to use a time trend to forecast future industrial water demand. However, in some cases the industrial demand is focused, for the most part, on a single industry, and so variations in the activity of that industry caused by markets or policy can cause dramatic changes in water demand. In these cases there is a need to track the factors that influence the demand for that particular industry's products or services. The water demand of a specific industry is driven by outputs, and these can be measured either by material finished product metrics, by sales or by the labour force employed (albeit an input measure). This is modified slightly by water price (industry appears to be more responsive to small price changes than is domestic metered demand) and by water stress, as measured by regional soil moisture deficit. Since these drivers cannot be determined with certainty for the future, forecasts have to be made on the basis of a range of scenarios. Additionally, as we go into the future, the water-saving strategies of each industry need to be determined. These will depend on the technologies thought to be available and the rate of uptake of the technology. This in turn is likely to depend on the size of the individual company and the cost of water used, relative to other inputs (Mitchell *et al.*, 2000).

REFLECTIVE QUESTIONS

What is the water availability in your region in m^3 per person per day?

How does your region's water availability compare to water stress/water scarcity measures?

9.5 TACKLING RISING DEMAND THROUGH WATER RESOURCE PLANNING AND WATER EFFICIENCY

Historically, the approach to water demand management has been to 'predict and provide' (*predict* the future water demand and devise and implement plans to *provide* sufficient supply for that demand). To a great extent this remains true today. Water supply organisations predict the major elements of demand for some decades ahead and present the forecasts of available water across the same time period, including any planned resource developments, to identify when resource availability will be insufficient to meet demand. This is then used as the basis of proposals to achieve a secure water balance with sufficient surplus to deal with supply perturbations. Historically, this would simply be proposals to develop new resources. However, more recently proposals by water planners would also be likely to involve measures to drive down demand by enhancing water efficiency.

9.5.1 Water resource development

What constitutes water demand is very dependent on the role of the person assessing the demand. A demand planner will focus only on the demand made by domestic and commercial customers. The resource planner, on the other hand, needs to determine all the other uses and losses of water from abstraction onwards such that, after leakage, theft, operational and emergency use, there will be sufficient water to provide for the needs of the customer demand. The scale of the difference in these two interpretations depends on the nature of the water resource system and how it has been developed.

Water systems start with abstractions from ground or surface water (there are small exceptions, for example rainwater harvesting). Seawater is effectively unlimited, but for all other sources the rate of exploitation determines whether the operation is sustainable. If the rate of exploitation of groundwater far exceeds recharge rate it is effectively the mining of a finite resource and should be analysed from that perspective. A surface water, say a river, could be exploited, to the level informed by a) single mass curve analysis, b) minimum acceptable flows in the river and c) the duration of any abstraction cessation that the system and customers can tolerate, assuming there is no storage. To exploit further, to the long-term yield of a river, will require storage, usually in the form of a dam and reservoir (although other options, such as infiltration galleries, are possible), so that periods of above-average yield are stored to satisfy demand in periods of below-average yield. A single mass curve is a basic water resource analysis tool (Figure 9.8). It is a plot of the cumulative flow at a point in a river over time. It can be derived from daily data, but weekly or monthly total flow volumes are usually satisfactory for water resources purposes. The longer the period of record, the more faith the analyst can have in the outcomes. A straight line drawn from the final point to the origin is the long-term yield (LTY) of the river. In a temperate river the cumulative plot will depart only modestly from the LTY, but in a markedly seasonal river subject to Mediterranean, monsoonal or snowmelt drivers, the annual departures above and below the average line will be more marked. The point of lowest slope in the curve is the reliable yield without storage. As storage is increased, so the river can be used in greater volume until the LTY is reached. The volume of storage

Figure 9.8 A stylised single mass curve. Regions with drought periods or strong seasonality will have larger 'curves' than temperate regions. Some of the 'yield' will be used for ecology support. Stable data from a gauge is vital and can be verified using double mass curves.

(the capacity of the reservoir) required is determined as the largest departure of the actual volume trend line from the LTY line, but of course this assumes that the period is representative of the longer-term conditions. Additional storage volume beyond that point will not add to the river yield, although it may have value in providing a safety margin, a recreational resource, or the ability to deliver above-average volumes if required at critical periods – say, to generate power in high energy-demand periods.

Dams are by far the most common method of providing storage. The International Commission on Large Dams (ICOLD) records 59,000 large dams (those >15 m high or 5–15 m high and impounding >30 million m^3 of water), of which about half are in China and two-thirds in just four countries (China, Iran, Japan and Turkey). Half the single-function dams and reservoirs are for irrigation, 20% for hydropower and 11% for water supply (others provide for recreation, river navigation, fisheries and river regulation to reduce flooding and maintain dry season flow). In the twentieth century about $2 trillion was spent on dams, which display a range of structural types. Embankment dams comprise over three-quarters of all large dams (the largest example is the 300-metre-high Nurek dam in Tajikistan; Figure 9.9a). These are suited to wide valleys, and are of mounded earth with a central core of impermeable clay, or rock fill with an exterior cover of concrete. The gravity dam (Figure 9.9b) is weighted to the underlying rock by its mass of concrete or masonry which prevents the dam being toppled by the weight of stored water. Arch dams (Figure 9.9c) are thinner and so cheaper and easier to construct in remote areas, although only suited to steep narrow valleys with sound geology. The curve resists the pressure of stored water that presses the dam into its foundations. Buttress dams

(Figure 9.9d) may be an alternative in steep valleys. These are constructed like gravity dams, but use of downstream supporting buttresses mean less concrete or masonry is needed than in gravity dams. The ICOLD database provides extensive statistics on many aspects of dams, which you can interrogate at: www.icold-cigb.org/GB/world_reg ister/world_register_of_dams.asp.

The map provided in Figure 6.13 in Chapter 6 shows how river discharges around the world are regulated by dams. Impoundment and flow modification means that environmental impacts of dams can be significant. Dams cause ecological fragmentation by creating a barrier to river migration (fish ladders are an attempt to mitigate this effect for economically important species like salmon), while changes to river flow, temperature and quality have further impacts on ecology with subsequent impacts on public health particularly in tropical regions where changes in distribution and intensity of diseases such as river blindness occur (Chapter 7). Dams also trap sediment, causing downstream riverbank erosion and starving lowland agriculture, deltas and coastal beaches of deposition (see Chapter 3, Section 3.5.4). Owing to these effects, loss of capacity from reservoir sedimentation, or concerns over safety and maintenance costs, dams may be removed. Thousands of small dams have been removed across the USA and Europe in recent years, while the largest dam removal project to date, for ecological restoration, has been on the Elwha River in Washington (dams of 33 m and 64 m).

Raw waters are transferred to a treatment works by aqueducts, pipes and siphons, depending on local circumstances and the age of the system. Following treatment and disinfection (see Chapter 8 for the techniques by which this is achieved), which will vary according to the raw water source (upland waters tend to be cleaner than

Figure 9.9 Examples of the main types of dam: a) embankment dam – Nurek Dam on the Vakhsh River, Tajikistan, is the tallest embankment dam in the world; b) gravity dam – Grand Coulee dam on Columbia River, USA; c) arch dam – Kariba Dam on the Zambezi River, Zimbabwe; d) buttress dam – Roselend dam, Alps, south-east France.
Source: Shutterstock – a) Lucas Bischoff; b) Nadia Yong; c) Lynn Yeh; d) Nichola Auckland.

lowland river abstractions because of the long storage in a large reservoir and the maintenance of a relatively pristine catchment), the now-potable water is measured into the water distribution network system to be held in service reservoirs, from which it is distributed by a bifurcating pipe system to individual homes and businesses. Area meters will record flows to different areas. The last few metres of the water supply pipe tend to be owned by the customer and water meters may be located at the property boundary, external or internal to the house.

Abstracting, storing, treating and distributing water requires major investment in infrastructure and has impacts on ecology, landscape and energy. In the early industrial countries of Western Europe, the infrastructure is now ageing and requires a major intervention and replacement programme. Reliance on an old legacy, no matter how well built, has a finite time limit. Investment in infrastructure replacement across most of Europe has accelerated markedly in the last 20 years, but in some countries it will still take about 100 years to replace all the water assets and thus, at the end of the period, assets replaced today will again have become legacy assets. With climate change, however, it will be necessary to consider additional assets if periods of extreme flood and drought are to be managed successfully.

A water company seeking to develop new resources to address a potential future water shortage will have an array of supply options. Typically, the least-cost option that satisfies the supply gap and does not cause damage has been selected. This has sometimes been referred to, somewhat flippantly, as 'CATNAP' – cheapest available technology narrowly avoiding prosecution. In some countries, the least-cost approach has now been superseded by 'best value'. This new approach has many merits – it is more positive, explicitly incorporates further, more diverse measures and is congruent with current social and environmental expectations. The challenge for the water company is to identify what constitutes best value, determine how it can be measured and whether appropriate, credible and reliable data are available and how information derived from these data should be presented to decision influencers and decision makers. Complexity is increased where best value solutions might involve shared development of a resource between several companies, perhaps located outside the supply areas of all these partners and the longevity and flexibility of any solution. Core determinants needed for a best value solution are likely to be: cost, resilience, adaptability, customer preference, community preference, environmental damage, environmental gains, added value, feasibility, intergenerational equity and resilience. Thus for each of these factors, specific metrics (indicators) and associated data are required to support decision-making.

In many low-income developing countries, rapid urbanisation has resulted in shanty towns. Here infrastructure, as perceived in the Western world, is non-existent, and investment in infrastructure development is not sufficient to match the rate of urban expansion. It may well be that water must be supplied to the shanty perimeter and informal local systems are encouraged to distribute water locally.

All of the approaches (reservoirs, groundwater and surface water abstraction) discussed above are, to a greater or lesser extent, conventional. There are many less widely used resource development techniques such as cloud seeding, desalination and iceberg sourcing. The latter two suffer from relatively high energy costs, as the water is provided at sea level and therefore has to be pumped up into the distribution system. Given that, even with more energy-efficient desalination systems and energy-efficient large iceberg towing, considerable energy has gone into the provision of the water. It is unlikely, therefore, that such sources will be widely exploited until water prices have risen considerably. Cloud seeding is rather different. Initially explored in the 1950s, cloud seeding promotes rainfall release from existing clouds (Bruintjes, 1999). However, it is fraught with legal and ethical problems: capturing water from clouds that had customarily deposited rain elsewhere, creating major flood-producing rainfalls, and generating rain in the wrong place are all concerns that may limit cloud-seeding practice – and of course there have to be clouds to seed in the first instance.

A more viable alternative source is wastewater reuse (see Chapter 8). Conventionally, the ambition is to collect and treat all sewage before discharge to river or sea. However, wastewater can be recycled and reused, so increasing water availability and reducing the need to draw on groundwater or other sources. Wastewater reuse is most common in arid countries, where treated sewage is predominantly used for urban and peri-urban agricultural irrigation. Initially use was restricted to non-food crops, but, as treated effluent quality has risen, a wider range of wastewater uses have been found. In some countries, recycled water is also used for municipal functions, such as irrigation of parks and golf courses, for industrial processes and toilet flushing, and some for maintaining river flows. About 21% of Israel's water is derived from recycled sewage, and over 85% of wastewater is recycled (Israel Water Authority, 2015). To achieve this, a dual clean/recycled network infrastructure has been developed, with reclaimed water piped to the Negev, which, as a consequence, has seen a reversal of desertification. In arid areas, recycled water offers a particular advantage, in that it is a reliable and predictable supply, unlike rainfall. However, agricultural irrigation demand is seasonal, so storage infrastructure is required. Treatment relies on technology, which has energy and financial costs. Natural remediation may also be part of the process, making use of wetlands, or filtering water through sand deposits laying over aquifers that can then be recharged, if separate from those used for potable water. In some cases recycled sewage water is also used for drinking water, either via direct supply or more usually mixing with reservoir waters. In Singapore, recycled wastewater – so-called 'NEWater' – which has a high public acceptance rate, meets about 30% of demand, with plans to increase this significantly (Tan and Rawat, 2018). While some may feel disgust at the thought of drinking recycled water, evidence shows that the principal health risk comes not from organic contamination but pharmaceutical residues and other persistent household chemicals. Tackling these requires effective monitoring and high-level treatment, such as reverse osmosis.

Finally, additional water resources can also be made available at point of use through improved network management. Considerable effort goes into sourcing, cleaning and distributing high-quality water, so unaccounted-for water, primarily leakage that does not reach the customer, is unwelcome. Many systems leak. Electricity leaks as heat from transmission lines. Universities 'leak' students who give up, and prisons 'leak' prisoners. While the very best rates of leakage are around 5% (Germany), leakage rates are typically much higher, at 10–35% (Denmark, Finland, the UK, Sweden, Italy), with others with rates of 40–80% (Slovenia, Bulgaria, Nigeria). Leakage occurs as systems age and decay, as pressure increases and as ice, traffic, roots and drought shift the ground. Finding and fixing leaks is not trivial. Major bursts that reach the surface are reported by the public, but the more modest, yet more numerous, leaks have to be traced by discrepancies in the local water balance, by geophysical techniques (e.g. Crocco et al., 2009) or locating the leak by the noise it makes, using either a skilled operator with a 'listening stick' or a correlator, in essence a software-supported microphone, to do the same task. Once a leak is located the ground can be excavated and the leak made good – but this is highly disruptive to traffic and residents, and so many companies now seek 'no-dig' solutions, by lining existing pipes with an inner plastic pipe, for example.

Water conservation options, like leakage control and recycling, form part of a water management hierarchy, itself part of a widely recognised systems approach to water management, known as the integrated water resources management approach (IWRM). This seeks to address all parts of the water cycle, with different sectors viewed as parts of an integrated physical and institutional system, which, if managed collectively, can better meet goals of water efficiency, environmental protection, economic development and democratic participation in water governance. IWRM, and some associated system approaches, including the water management hierarchy, and water soft path are described further in **Box 9.3**. A critical systems approach to meeting water need is the consideration of both supply-side development, discussed above, and demand reduction, through increasing efficiency of water use, which we discuss next.

9.5.2 Water efficiency

Today, in some countries, the approach to water resource planning is different and is called 'twin track', reflecting that achieving a water balance requires considering resource development *and* demand reduction. However, this twin track approach introduces uncertainty into what is already an uncertain equation. Demand reduction is ultimately in the hands of the individual consumer, so there can be only limited confidence in: a) the scale of demand reduction achievable; b) the longevity of that reduction and c) the sustainability of the reduction in the face of changing societal circumstances and attitudes. Of course, there is uncertainty in resource development as well – the water industry has no contract with its major supplier – rainfall – and so holds a reserve of water, called '**headroom**', in its water-balance forecast accounts. In more extreme droughts, headroom can prove insufficient and water supply organisations may ask customers to be more economical with their water use. If they respond, the reduced demand helps to stave off more aggressive water controls such as hosepipe bans, closure of water-using commercial companies

such as car-wash firms, suspension of abstraction licences, rotational cutoffs, the *threat* of standpipes in the street and, in the final step, the cessation of household supply substituted by standpipes in the streets.

Water supply organisations may have already promoted water efficiency and choose to use (or have been so instructed by a government regulator) lower water use simply to defer investment on major water development projects, and perhaps to plan for a permanently different demand. In such cases restraint and efficiency cannot now be called upon in time of drought. So the availability of this informal headroom (the ability of the public to be more water efficient in times of crisis) is important in offsetting climate, and thus resource, uncertainty. We therefore need to think through water efficiency in a more analytical way. In particular, we need to differentiate between embedded water efficiencies and behavioural efficiencies. The former should be encouraged, perhaps through building regulations and the equipment and industry standards regimes. The latter might best be retained as informal headroom until such time as there is a credible national plan to share headroom in the event of drought. Both domestic and industrial water users are being encouraged to use less water. Groups such as 'Water Use It Wisely' in the USA or 'Waterwise' in the UK have been formed to encourage such reductions in water use.

9.5.2.1 Industry

Industry, since its water use is metered, will be persuaded by economic arguments and will invest in new technologies once the saving is sufficient to overcome the many initial barriers and inertia that exist in all organisations. If a commercial company can gain some assurance that it will be less likely to be called upon to curtail operations because it is more water efficient, it is probable that risk reduction will be a greater driver than simple savings. However, in many countries water efficiency has risen up the political agenda. In the UK, for example, water efficiency can mean a company pays less tax *in a single year* (as opposed to the norm of a small amount per year over several years).

All companies seek to minimise waste in their inputs and raw materials. However, unless a company is a large water user, the focus of waste minimisation will be on energy and raw materials. This is perhaps changing as companies, particularly in retail and companies with a direct link to customers, seek to market their good environmental credentials (Hartmann *et al.*, 2017). A larger water-using company will seek to adapt its product design and manufacturing processes to use less water, and the particular approaches adopted, while often supported by environmental auditing and life cycle assessment of products and processes, will be driven by least-cost analysis with water simply as one cost.

9.5.2.2 Agriculture

Most water for agriculture comes from rainfall stored in the soil profile and only 15% is provided through irrigation. However, of the 3800 km^3 of water abstracted from rivers and groundwater, 70% is used for agricultural irrigation (Molden *et al.*, 2007). Additional water required for agriculture under forecasted population growth will strain ecosystems and increase competition for water resources. As shown in **Figure 9.1**, efforts to reduce water consumption in Europe and the Americas for industry and domestic use are sensible, but for the rest of the world the focus needs to be on using water more efficiently in agricultural processes. Here, water conservation relies on minimising evaporation and runoff. This can be achieved through a variety of practices, including replacing flood irrigation with more targeted overhead irrigation (sprinkler) systems or, more efficient, drip irrigation. These systems are costly to install, however, so conservation seeks to ensure maximum productivity by improving soil fertility, creating furrows or reducing soil compaction to reduce runoff and enhance soil water holding capacity (Holden *et al.*, 2019). In arid environments, up to 90% of rainfall evaporates, leaving just 10% for productive transpiration by plants. Breeding crops that quickly expand their leaves to shade the ground to reduce evaporation and increase productive transpiration may be beneficial, as would more precisely understanding crop water requirements, through soil–water–plant interaction studies. As the climate changes, water stress may lead to reduced agricultural water use through changing of crops, possibly including use of more drought-tolerant plants (Varshney *et al.*, 2011). While there has been a lot of research into breeding wheat, rice and maize to become more water efficient and resistant to disease, further future water efficiency improvements in these grains may be moderate. However, grains such as millet and sorghum have received less attention and so there may be benefit in targeted work to enhance water efficiency and subsequent enhanced use of those grains (Molden, 2007).

9.5.2.3 Domestic

Reducing domestic demand is much more challenging than for industry and agriculture. Raising prices is not an option for unmetered customers, while increasing prices for metered customers often raises concerns for low-income households, who may struggle to pay the higher prices, and so reduce usage to levels which are potentially detrimental to health and well-being. For these reasons so-called 'lifeline tariffs' (inclining block tariffs with a low fee for the first block of use and a successively higher unit charge for added blocks of consumption) appear attractive. However, there is great difficulty in setting efficient block fees and ranges, and in communicating higher charge to users in time to reduce use, plus concerns over suppressing demand via cost not scarcity, sending the wrong price signal to low users, and inadequate protection for large low-income households and cost recovery.

Table 9.3 shows the water savings that are possible through behaviour change;

Table 9.3 Potential water savings through change of habit proposed by Northumbrian Water and Essex & Suffolk Water, UK (reproduced with kind permission)

Change of habit	Saving per property per day, litres
Turn off the tap when brushing your teeth	30.7
Shower rather than bathe	25.0
Use a bowl when hand-washing dishes	17.5
Take shorter showers	17.0
Use a basin when personal washing	11.8
Repair dripping taps	8.9
Use full loads in washing machines	8.3
Use a bowl when washing food	6.5
Use full loads in dishwashers	2.0
Use water resourcefully in the garden	3.8

these represent significant potential savings but are difficult to achieve, as water companies must influence the attitudes and behaviours of the many different types of consumer that make up the population of water-using device owners. Technological devices can be built into the home to make them more efficient, a process that may be encouraged for new homes by building regulations. Technology, assumed to 'embed' savings more firmly than encouraging behavioural change through education and conservation campaigns, includes pressure-limiting valves, spray taps, low-flow showers, smaller baths and toilet cisterns and dual-flush systems. However, home owners might not choose to replace a faulty spray tap with a similar device if they have found it inconvenient, a process known as regression. In older homes people can be encouraged to install water-capture devices such as the 'drought plug' (which facilitates using bath water for garden irrigation) or displacement devices such as 'hippo bags' (which ensure that each toilet flush is reduced by about a litre). Consumers now have more choice of water-efficient domestic appliances, although they must still choose the more water-efficient option. Influencing the behaviour of the consumer to install, retain and use water-saving devices and to modify lifestyles to use less water remains the key.

Local rainwater harvesting by capturing roof water and the recycling of grey water from baths and showers for use in toilet flushing are much more costly than any of the options discussed earlier and appear not to be cost-effective, particularly if retrofitted. More problematic are the potential health impacts associated with toilet aerosols containing potential pathogens from bird droppings or the accidental filling of, say, a child's paddling pool from a tap providing grey or roof water.

9.5.3 Resilience

Population growth and, particularly, concentration in urban areas, expectations of improving environmental quality, climate change and a more risk-averse public are fuelling expectations of resilient water systems that recognise risk and uncertainty and that can adapt to unexpected alterations in the supply–demand balance, whether this arises through system failure, rapid demographic change or drought. Traditionally this has been achieved by holding sufficient headroom, a strategic reserve between expected supply and demand. Such a reserve is, however, at a cost and is planned never to be fully used. It is therefore likely that increasing the size of any reserve in an individual water supply organisation in many areas of the developed world, even in the face of growing uncertainty, will be difficult. There is therefore a need to share headroom between adjacent water supply companies where possible. In effect this is a form of increased centralisation that can be achieved by informal agreements, formal water transfer agreements, regulator insistence or the acquisition of adjacent companies.

Altering the supply side of the supply–demand balance has greater certainty than managing down demand. Further, there is an additional loss of resilience in the simplistic reduction in demand. When demand is constrained by education, tariffs, information or habit, this could be used to increase the gap between supply and demand, effectively increasing headroom and making the system more resilient to outside pressures, such as a prolonged drought. However, such a course of action is often unlikely and instead investment in new resources will be deferred. However, when an inevitable drought does occur the water managers can no longer call on behavioural changes and simple technical adaptations to constrain demand because these have already been employed. Demand management, naively pursued, can reduce resilience. What is required is, first, a more sophisticated selection of demand-management techniques so as to implement those efficiencies that cannot be delivered by voluntary behaviour change, so retaining for use in drought those adaptations that can be delivered quickly by customers – in effect, to retain the informal distributed headroom that is customer-driven reduced demand in response to supply problems. Second, more robust decision-making, as advised by the RAND think tank below, is required:

> Consider large ensembles of plausible futures, not a single best estimate forecast. Seek robust strategies that perform well across many plausible futures, not a strategy optimal for one particular view. Employ adaptive strategies, ones that evolve over time in response to changing conditions. Use the computer as a 'prosthesis for the imagination,' not a calculator.
>
> (Lempert *et al.*, 2003)

This robust decision-making might direct strategy towards (what some might see as) a sub-optimal solution, but it is a solution that is flexible across several futures and so is resilient (see Lempert *et al.* (2003, 2009) for more on resilient methods). This is important, as most water conservation opportunities are geographically decentralised and exist across a diverse range of uses, and, hence, water companies tend to have relatively little direct control over demand (leakage control being an obvious exception). Under such circumstances, it is no surprise that water resource planners have traditionally sought to bridge the water gap through supply augmentation, rather than demand management. However, the impacts of increasing supply (economically, socially, environmentally) are now widely recognised in water resource planning, and 'supply-fix' is no longer automatically seen as the solution. Indeed, often the opportunities to increase supply through increasing storage, or further abstraction, do not exist. Thus integrated, adaptable strategies that may seek to increase supply, reduce demand, or both, but which are sensitive to place and water users in those places, and which recognise the limits and uncertainties associated with the various water management instruments, are required. In adopting this more adaptable approach water companies are more likely to continue to achieve promised standards of service in the face of the uncertainty inherent on both sides of the supply–demand equation.

> **REFLECTIVE QUESTIONS**
>
> What mechanisms can be used to tackle rising demand?
>
> Why might a focus on water efficiency in agriculture be important?

9.6 SUMMARY

Measuring and understanding the precise nature of existing water demand is challenging. Metering, top-down and bottom-up processes need to be adopted. The demand for water is rising globally. Balancing water demand and supply, and protecting or enhancing environmental quality at the same time, is challenging under conditions of growing uncertainty. Water resource development (e.g. infrastructure for desalinisation, reservoirs, groundwater and surface water abstraction) takes a long time, and is shifting away from 'least cost' to 'best value' approaches. Demand forecasting over a relatively long term is required. Such forecasting requires a strong demographic element, and the information base on water demand fluctuations at the household level is weak. Water supply also needs to account for a range of other demands on resources, including leakage, theft, operational use and emergency use. Globally, agriculture is the biggest off-stream user of water and so enhanced agricultural water efficiency is required. Water efficiency in both domestic and industrial arenas has a significant role, but sophistication in precisely what efficiencies are promoted is vital. Decisions need to be robust and to reflect uncertainty, if they are to contribute to a resilient system.

FURTHER READING

Butler, D. and Memon, F.A. (eds). 2006. *Water demand management*. International Water Association; London.

    A detailed review of water demand management.

Jones, J.A.A. (ed.). 2010. *Water sustainability: a global perspective*. Hodder Education; London.

    Describes the drivers of demand, and a range of technical and institutional approaches to close the water gap.

Loucks, D.P., van Beek, E., Stedinger, J.R., Dijkman, J.P.M. and Villars, M.T. 2017. *Water resources systems planning and management: an introduction to methods, models and applications*. Springer; Deltares.

A higher-level text for those who want to know more about practical water resource system planning and management.

Maksimovic, C., Butler, D. and Memon, A. (eds). 2003. *Advances in water supply management*. Balkema; Lisse/Abington/Exton/Tokyo.

An edited book covering most aspects of water supply, with every chapter jointly authored by practitioners and scientists.

Postel, S. 1992. *Last oasis: facing water scarcity*. Worldwatch Institute Series. W.W. Norton; London.

Describes the causes of water scarcity, the consequences of shortages, techniques for promoting water efficiency, and the policies and laws required to promote sustainable water use.

Classic papers

Bowden, C. 1977. *Killing the hidden waters*. University of Texas Press; Austin, TX.

This book charts the slow destruction of the water resources of the American south-west.

It has an indigenous people perspective and is relevant to other parts of the world today.

Clarke, G., Williamson, P., McDonald, A. and Kashti, A. 1997. Estimating small area demand for water: a new methodology. *Journal of the Institution of Water and Environmental Management* 11: 186–192.

Herrington, P. 1996. *Climate change and the demand for water*. HMSO; London.

Both of the above papers were very influential in supporting policy or water company best practice.

Howe, C.W. and Linaweaver, F.P. 1967. The impact of price on residential water demand and its relation to system design and price structure. *Water Resources Research* 3: 13–32.

An early study into the role of water price, geography and water-use habits on water demand.

PROJECT IDEAS

- Draw up additional evidence for the perfect storm on energy, food and water supply by searching for information on the internet. Possible questions to look up include: How many days' food supply does the world hold in reserve? How many droughts has your country suffered in the last 20 years? What proportion of US cropping is for energy?

- Draw a simple flow diagram to show how water flows into different uses from the water source to the domestic potable supply. Do some research to try to quantify these components for a case study region of your choice. Remember to include leakage.

- Try to obtain some water meter data at area and household level and see if you can estimate the level of leakage which might be occurring in your area. If you have your own meter or can ask colleagues to keep records from theirs you could try plotting usage rates over time to identify patterns (daily, weekly, seasonal, etc.).

- Search on the internet for household water use and water price information for different countries and add this to the graph on **Figure 9.7**; update data for countries already on the figure and add new countries to it. Are there any interesting patterns?

REFERENCES

Ball, A., Styles, M., Stimson, K. and Kowalski, M. 2003. Measuring microcomponents for demand forecasting. *In*: Maksimovic, C., Butler, D. and Memon, A. (eds) *Advances in water supply management*. Balkema; Lisse/Abington/Exton/Tokyo, 673–682.

Beddington, J. 2009. Food, energy, water and the climate: a perfect storm of global events? Government office for science. webarchive.nationalarchives.gov.uk/20121206120858/www.bis.gov.uk/assets/goscience/docs/p/perfect-storm-paper.pdf.

Bobbie, C.J., Warfe, D.M. and Hardie, S.A. 2014. Implementing environmental flows in semi-regulated and unregulated rivers using a flexible framework: case studies from Tasmania, Australia. *River Research and Applications* 30: 578–592.

Bowden, C. 1977. *Killing the hidden waters*. University of Texas Press; Texas.

Bruintjes, R.T. 1999. A review of cloud seeding experiments to enhance precipitation and some new prospects. *Bulletin of the American Meteorological Society* 80: 805–820.

Butler, D. and Memon, F.A. (eds). 2006. *Water demand management*. International Water Association; London.

Cao, G., Zheng, C., Scanlon, B.R., Liu, J. and Li, W. 2013. Use of flow modeling to assess sustainability of groundwater resources in the North China Plain. *Water Resources Research* 49: 159–175.

Clarke, G., Williamson, P., McDonald, A. and Kashti, A. 1997. Estimating small area demand for water: a new methodology. *Journal of the Institution of Water and Environmental Management* 11: 186–192.

Crocco, L., Prisco, G. and Cassidy, N.J. 2009. Early-stage leaking pipes GPR monitoring via microwave tomographic inversion. *Journal of Applied Geophysics* 67: 270–277.

Damania, R., Desbureaux, S., Hyland, M., Islam, A., Moore, S., Rodella, A., Russ, J. and Zaveri, E. 2017. Uncharted Waters: the new economics of water scarcity and variability. World Bank; Washington, DC. doi:10.1596/978-1-4648-

1179-1. Licence: Creative Commons Attribution CC BY 3.0 IGO.

EEA. n.d. *Household water use*. European Environment Agency Website [online, accessed April 2019]. Available from: www.eea.europa.eu/data-and-maps/figures/household-water-use-1.

Hartmann, P., Apaolaza, V., D'Souza, S., Barrutia, J.M. and Echebarria, C. 2017. Corporate environmental responsibility communication: implications from CSR and green advertising research. *In*: Diehl, S., Karmasin, M., Mueller, B., Terlutter, R. and Weder, F. (eds) *Handbook of integrated CSR communication*. Springer; Cham.

Herrington, P. 1996. *Climate change and the demand for water*. HMSO; London. HMRC [online, accessed March 2013]. Available from: www.hmrc.gov.uk/capital-allowances/fya/water.htm.

Holden, J., Grayson, R., Berdeni, D., Bird, S., Chapman, P.J., Edmondson, J.L., Firbank, L.G., Helgason, T., Hodson, M.E., Hunt, S., Jones, D.T., Lappage, M., Marshall-Harries, E., Nelson, M., Prendergast-Miller, M., Shaw, H., Wade, R. and Leake, J.R. 2019. The role of hedgerows in soil functioning within *agricultural landscapes*. *Agriculture, Ecosystems and Environment* 273: 1–12, /doi.org/10.1016/j.agee.2018.11.027.

Howe, C.W. and Linaweaver, F.P. 1967. The impact of price on residential water demand and its relation to system design and price structure. *Water Resources Research* 3: 13–32.

Israel Water Authority. 2015. The wastewater and treated effluents infrastructure development in Israel [online, accessed April 2019]. Available from: www.water.gov.il/Hebrew/ProfessionalInfoAndData/2012/03-The%20Wastewater%20and%20Treated%20Effluents%20Infrastructure%20Development%20in%20Israel.pdf.

Jones, J.A.A. (ed.) 2010. *Water sustainability: a global perspective*. Hodder Education; London.

Lempert, R., Popper, S. and Bankes, C. 2003. *Shaping the next one hundred years: new methods for quantitative, long-term policy analysis*. RAND report MR 1626 [online, accessed April 2019]. Available from: www.rand.org/international_programs/pardee/methods.html.

Lempert, R., Scheffran, J. and Sprinz, D. 2009. Methods for long-term environmental policy challenges. *Global Environmental Politics* 9: 106–133.

Loucks, D.P., van Beek, E., Stedinger, J.R., Dijkman, J.P.M. and Villars, M.T. 2017. *Water resources systems planning and management: an introduction to methods, models and applications*. Springer; Deltares.

Maksimovic, C., Butler, D. and Memon, A. (eds). 2003. *Advances in water supply management*. Balkema; Lisse/Abington/Exton/Tokyo.

Mitchell, G., McDonald, A., Williamson, P. and Wattage, P. 2000. A 'Standard Industrial Classification' coded strategic planning model of non-household water demand for UK regions. *Journal of the Chartered Institution of Water and Environmental Management* 14: 226–232.

Molden, D., Oweis, T.Y., Pasquale, S., Kijne, J.W., Hanjra, M.A., Bindraban, P.S., Bouman, B.A.M., Cook, S., Erenstein, O., Farahani, H., Hachum, A., Hoogeveen, J., Mahoo, H., Nangia, V., Peden, D., Sikka, A., Silva, P., Turral, H., Upadhyaya, A. and Zwart, S. 2007. Pathways for increasing agricultural water productivity. *In:* Molden, D. (ed.) *Water for food, water for life: a comprehensive assessment of water management in agriculture*. Earthscan; London and International Water Management Institute; Colombo, 279–310.

OECD. 1999. *Household water pricing in OECD countries* [online, accessed June 2012]. Available from: http://tinyurl.com/af4tg45.

OECD. 2008. *Environmental outlook to 2030* [online, accessed June 2012]. Available from: www.oecd.org/dataoecd/29/33/40200582.pdf.

Parker, J.M. and Wilby, R.M. 2013. Quantifying household water demand: a review of theory and practice in the UK. *Water Resources Management* 27: 981–1011.

Postel, S. 1992. *Last oasis: facing water scarcity*. Worldwatch Institute Series. W.W. Norton; London.

Santato, S., Mysiak, J. and Pérez-Blanco, C.D. 2016. The water abstraction license regime in Italy: a case for reform? *Water* 8: 103.

Tan, P.T. and Rawat, S. 2018. NEWater in Singapore. *Global Water Forum* [online, accessed April 2019]. Available from: www.globalwaterforum.org/2018/01/15/newater-in-singapore.

UNDP. 2006. *Human development report. Beyond scarcity: power, poverty and the global water crisis*. UNDP; New York.

United Nations. 2012. *UN Department of Economic and Social Affairs* [online, accessed June 2012]. Available from: www.un.org/waterforlifedecade/scarcity.shtml. World Resources Institute (WRI) in collaboration with United Nations Development Programme.

United Nations Department of Economic and Social Affairs. 2014. *Water for life decade: scarcity* [online, accessed August 2019]. Available from: https://www.un.org/waterforlifedecade/scarcity.shtml.

Varshney, R.K., Bansal, K.C., Aggarwal, P.K., Datta, S.K. and Craufurd, P.Q. 2011. Agricultural biotechnology for crop improvement in a variable climate: hope or hype? *Trends in Plant Science* 16: 363–371.

Williamson, P., Mitchell, G. and McDonald, A.T. 2002. Domestic water demand forecasting: a static microsimulation approach. *Water and Environment Journal* 16: 243–248.

WWF. 2012. *Dam facts and figures* [online, accessed June 2012]. Available from: http://wwf.panda.org/what_we_do/footprint/water/dams_initiative/quick_facts/.

CHAPTER TEN

Water economics

Sonja S. Teelucksingh,
Nesha C. Beharry-Borg,
Yixin Hu, Zhao Zeng
and Dabo Guan

LEARNING OUTCOMES

After reading this chapter you should be able to:

- understand the role of water in an economic system;
- explain the total economic value of water resources;
- understand the different types of economic valuation techniques that can be applied to water resources;
- describe the methodologies of the Millennium Ecosystem Assessment and how these methodologies can be applied to different ecosystems.

10.1 INTRODUCTION

The fluctuations in supply, coupled with a myriad of competing demands, make the study of water resources perfectly suited to the discipline of environmental economics. This chapter uses concepts from environmental economics but with a focus on water as the environmental component of concern. Water is a core environmental component and is crucial for economic development, and it is therefore essential to understand water economics if we are to develop an integrated approach to understanding both water and development.

Since its much-cited use in the Brundtland Report (United Nations, 1987), plenty of attention has been paid to the concept of sustainable development. The notion of development itself can be seen through the perspective of human needs rather than solely in terms of economic growth. This enables us to pay more attention to the way growth is achieved, how it is distributed and how livelihoods are affected in the development process. Sustainability has frequently been defined as 'the maintenance of social well-being' for both current and future generations. That definition is an elusive phrase that can be interpreted in a number of different ways. However, despite contradictory interpretations, a common international consensus is that sustainable development is something for which economies should strive. This is made clear by the list of eight Millennium Development Goals derived by the United Nations in 2000 and signed up to by 189 nations (UN, 2015). One of the goals is to 'ensure environmental sustainability'. Later, in 2012, the UN Conference on Sustainable Development in Rio proposed a new set of universal goals to meet the urgent global challenges that were raised from environmental, political and economic systems to replace the Millennium Development Goals. In 2015, these 17 goals were named the **Sustainable Development Goals** and came into effect in 2016 (UN, 2018). While water is important for many of the Sustainable Development Goals (see Chapter 13), two of them in particular are most directly associated with water challenges. The first is about water scarcity, in which the objective is to 'ensure availability and sustainable management of water and sanitation for all by 2030' (Goal 6); and the other is about water management, as demonstrated in Goal 12, to 'ensure sustainable consumption and production patterns'. In order to achieve economic growth and sustainable development, it requires us to reduce our ecological footprint through changing the production and consumption of natural resources. Water availability can represent a significant constraint on the sustainable development of an economy. This, coupled with increasing population, places water resources under growing demands and leads to the need for improved water management as a critical element in the achievement of sustainable development. Water resource management affects every aspect of an economy. At the same time, the patterns of water demand relate directly to economic structures and sectors. Water stress and water insecurity have particular implications for developing countries, in the context of those dominated by rural subsistence-based communities heavily dependent on agriculture and characterised by a lack of water infrastructure.

There has been a recent global research effort to both provide a snapshot of current linkages between biodiversity and human well-being and to suggest ways for its future conservation (TEEB, 2010). This global study, known as 'The Economics of Ecosystems and Biodiversity' (TEEB), came up with an interesting approach to the concept of the value of an environmental resource, called the three-tiered approach (**Figure 10.1**). The methodology suggests that values need to be recognised, while understanding that different people in different situations can apply different values to the same resources. Once recognised, the value needs to be demonstrated in economic terms by

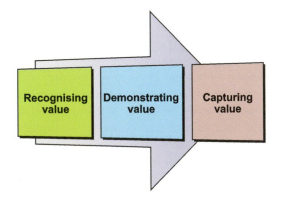

Figure 10.1 The three-tiered approach to economic value.

the use of the appropriate economic valuation tools. Finally, once demonstrated, these ecosystem values then need to be embedded into the decision-making processes by the management of the governance frameworks and institutional structures which oversee the resource, and via the use of economic instruments such as payments for protecting the environment so as to enhance water quality.

This chapter follows the three-tiered principles to provide an overview of the main issues in water economics. First, we *recognise value* by discussing the role of water within an economic system, in terms of both consumer and producer demands. Coupled with demand must come a supply to meet this demand. How natural systems supply water, and how these systems can be threatened by human activities, are then discussed. The intersection of demand and supply inevitably raises the question of the pricing of water. Given the role of water in an economic system, we then ask the question – what is the *value* of water? Different philosophical conceptions exist around the notion of value. Environmental economists speak of the 'total economic value' of an environmental good, which deconstructs into a series of sub-classifications that combine to form this total value. The Millennium Ecosystem Assessment (2005) offered an alternative approach, classifying environmental goods into what has become known as 'ecosystem goods and services', with a focus on the flow of benefits from ecosystems, via these goods and services, to human well-being. Whichever typology is adopted, there exists a range of economic valuation techniques that can be applied to water resources to *demonstrate value*. Once values are estimated, the question then becomes how to integrate these values into decision-making processes, or how to *capture value*, where the complex issue of water

management is addressed. Complexity is to be found primarily in a governance sense, where water resources transcend political boundaries, therefore requiring a collective approach to the overall management of water.

10.2 WATER WITHIN THE ECONOMY

10.2.1 Supply and demand

Water by its very nature is defined as a **common property resource**. It is widely accepted that those resources upon which poor rural households from developing countries depend for their daily livelihoods are open access or common property, with a major problem facing developing countries being the degradation of these common resources. It is inevitable that a high dependence on any open access or common property resource, together with a lack of, or improperly designed/enforced, **property rights**, can lead to conflicts over resource use and ownership, leading to an increasing need for comprehensive management strategies. Freshwater resources are characterised by uneven spatial distribution across the globe. Water is also mobile across political entities (e.g. the River Danube flows through ten European countries) and, as discussed in Chapter 1, there are 44 countries which depend on other countries for over 50% of their renewable water resources, and so this increases the risk of conflict over water.

Like many environmental resources, freshwater is characterised by a limited supply, together with an array of competing wants by a range of **stakeholders**. When these stakeholder demands inevitably conflict, the issue becomes more complicated. Pollution that crosses political boundaries is known as **transboundary pollution**. This can occur, for example, with marine pollution in small island states that are geographically close to one another. Transboundary pollution is a problem that results from the lack of clearly defined property rights in areas where environmental resources are shared or impacted upon. This brings into the analysis the question of governance (see Chapter 11 for a more detailed treatment). Constitutional settings in many developing countries are characteristically weak. This has direct implications for environmental resource use and management. Institutional and government failures are one of the reasons identified for environmental destruction, through environmentally adverse policies or the inability to resolve competing objectives (for example, where expansion of industrial activity in the pursuit of economic development can involve practices that pollute groundwater). This is further compounded by the different spatial scales being considered. It is inevitable that trade-offs in use, impacts upon supply and factors affecting demand can occur at local, regional and international levels.

The supply of freshwater originates from two main sources: surface water from rivers and lakes, and groundwater. Each of these sources has characteristics of both renewable and non-renewable resources (Dalhuisen *et al.*, 2000). Surface water and groundwater is to some extent renewable, through seasonal rainfall, as well as through inputs back into the surface water system of treated wastewater, via water management and purification systems (see Chapter 8). Surface water is easily accessible but can be highly vulnerable to agricultural, industrial and domestic pollution; it is seasonally variable; and its supply is susceptible to periods of drought. As described in Chapter 5, groundwater in some places is considered to be non-renewable due its age and slow rates of recharge. Groundwater needs to be extracted, and at an optimal

rate, but could be available at times of drought. The question in any geographical setting then becomes the extent to which surface and groundwater can be simultaneously and optimally used (conjunctive use) to meet economic demand.

The main distinction in the economic demand for water is the types of activities for which the water is demanded. We can distinguish between productive purposes (where water is used as an input resource in the production of goods and services) and consumptive purposes (where water is used by households for uses such as drinking and sanitation). Though varying across countries, a significant component of water demand is the demand of domestic households (see Figure 9.1 in Chapter 9). Households are not, however, homogeneous units, and the extent of residential water demand is impacted upon by a host of characteristics such as size of household, level of income and type of accommodation (see Chapter 9). On the productive side – which is by far the largest component of water demand globally – water demand varies according to type of industry, with industrial demands for water in manufacturing processes and agricultural demands for irrigation and the watering of crops and animals dominating the demand.

The differences in use of water for consumption and for production raise the possibility of using water of different qualities for different purposes. In this way water could be classed as a mixed resource rather than as a uniform one. Water with a quality suitable for drinking (potable water) may not be demanded for many residential and industrial purposes. For example, there is no need to flush a toilet or wash a car with potable water. However, in many developed countries water for toilet flushing is supplied at the same quality as water for all other household purposes. The largest part of water demand is for activities that do not require potable water (Dalhuisen *et al.*, 2000). However, water is usually supplied as a uniform good, in particular in developing countries where no distinction is made as to the type of water provided for these different uses. Little is known about consumer demand for different qualities of water. It is feasible to assume that, for different aspects of water quality, there exists a range of supply and demand characteristics that must be examined in their own right (Dalhuisen *et al.*, 2000).

10.2.2 Water industry

The water industry forms the bridge between the supply and demand for water, through its activities of collecting, treating, storing, transporting and distributing water. By definition this is an industry that requires a significant level of investment and that is also associated with high costs. Because of this, water supply in many developing countries is characterised by a monopoly and is under some sort of central government control. However, the water supply chain can itself be disaggregated and various elements of it can be privatised. For example, the collection of water, the treatment of water and the transport and distribution of water can all be undertaken by different organisations that come from the private sector. An interesting issue then becomes the extent to which private versus public organisations exist in the water industry, and which mixture is in the best interests of water security.

Given the role of the water industry in harnessing water supply and meeting water demands, the question now becomes the optimum price at which this is achieved. The water industry traditionally passes on the costs of physical supply (e.g. collecting, treating, pumping and piping the water) to

homes and businesses. However, the water itself as a publicly provided good is free, with no royalties implemented to extract the resource. Cheap water therefore reflects the lower capital costs of the water industry rather than an abundance of the resource. This means that the economic value of water does not enter the transaction which takes place on the market, and therefore is not captured in the price. Consumers of water are therefore unaware of the actual cost of the resource to society at large, or of its relative scarcity. Economists therefore apply a range of methodological tools to the valuation of water resources. The history of non-market valuation in the United States is in fact closely linked with water projects (Hanemann, 2006).

10.2.3 Water, climate and the economy

Water resources affect, and have always affected, every aspect of economic life (e.g. see Chapter 1). There exists a delicate balance between water demand and water supply at all levels, from the household itself, the community, the country and on the global scale. Forecasts of economic growth and population tell us that in the next few years global demand for water will outstrip its supply. There are already many countries where demand outstrips supply (see Table 1.2 in Chapter 1) and industrial and agricultural water demand is set to substantially rise further (e.g. Wada and Bierkens, 2014). There is therefore an urgent need to address the challenges of water security, for both current and future livelihoods.

As a part of natural variability, precipitation, and thus the availability of water resources throughout time and space, varies within bounds determined by given climate conditions. In this context, climate change has a direct impact on water supply (see Chapter 2). Many of the anticipated impacts of climate change upon human livelihoods will operate through water. Changes in the global hydrological cycle as a result of changing temperatures can lead to changes in water quantity and quality through accelerated glacier melt, changes in precipitation, temperature and evaporative demand, runoff and recharge patterns and rates, and saltwater intrusion into coastal aquifers (Bates *et al.*, 2008). There is also the increased risk of extreme meteorological events such as floods, droughts and tropical cyclones (see Chapter 2). These impact upon human livelihoods, as humans depend on the provisioning, regulating, cultural and supporting ecosystem services provided by freshwater ecosystems. Such impacts may be felt more strongly in some places than in others. Water resources in small island developing states have been identified as highly vulnerable to climate change, as described in Box 10.1.

The predictions of the Fifth Intergovernmental Panel on Climate Change indicated that water resources will be one of the sectors most affected by changes in climate (Jiménez *et al.*, 2014; see Chapter 2). Water scarcity is estimated to increase as climate change effects are felt. The growing world population, unsustainable practices and inefficient allocation of water threaten to induce and/or intensify water scarcity, with disastrous consequences for the environment and societies. Health issues are also relevant, where lack of access to potable water and proper sanitation, in particular, can cause the persistence of associated ailments (see Chapter 7). It therefore becomes necessary for society to adapt to the range of water-related hazards that accompany climate change.

The negative environmental effects have the potential for disrupting human social and economic activities everywhere on Earth, but are particularly threatening to coastal communities (e.g. Box 10.1). Healthy ecosystems are at the front line of defence against climate change, through both a) protection from natural hazards and b) the ability to reduce economic vulnerabilities through their contribution to human well-being and sustainable livelihoods. There has therefore been a shift from a focus on natural hazards in particular to a focus on the series of factors that can change the coping capacities (resilience) of social systems, and the ways in which these capacities can be affected by policy decisions at community, national, regional and international levels.

Water resource management affects every aspect of the economy. At the same time, the patterns of water demand relate directly to economic structures and sectors. Increasing pressure of demand on water resources, coupled with climate-induced changes in supply, will have different effects, depending on geographical, economic and social criteria. While climate change can directly affect the supply of freshwater, it is non-climatic drivers that most impact upon the demand. Effective water management and principles for adaptation with a goal of achieving water security therefore require an analysis of both the supply and demand patterns of a country and the climatic, economic and other non-climatic drivers that determine them.

REFLECTIVE QUESTIONS

When you pay for your water what are you actually paying for?

Why are water resources on small island developing states particularly vulnerable to climate change?

BOX 10.1

Areas at risk: small island developing states

Small island developing states are found in the Caribbean and the Pacific and close to the African continent. Recognised by the Sustainable Development Goals, the Millennium Ecosystem Assessment and the IPCC reports as a unique category of analysis, small island developing states face a particular set of economic, environmental and developmental challenges. Their key underlying characteristic is that of the vulnerability of local livelihoods to environmental and economic changes. Small populations are coupled with high population densities, concentrated in coastal zones that comprise much of the land area. An inevitably high ratio of coastal to total land area means that island ecosystems are characterised by highly coupled terrestrial and marine ecosystems (McElroy *et al.*, 1990). There is a heavy reliance on natural resource exploitation, with tourism (and, increasingly, ecotourism) dominating island economies (Mimura *et al.*, 2007). They also exhibit a high degree of economic vulnerability to the world economy, due to their dependence on international trade. These characteristics make small island developing states especially vulnerable to the effects of climate change, sea-level rise and extreme events (Nurse *et al.*, 2014). Due to the heavy reliance on natural resource exploitation for economic livelihoods at both micro and macro levels and the high prevalence of coastal zone areas, environmental shifts such as ecosystem changes, natural disasters and climate change impacts can have extreme economic and welfare effects in small island developing states. In particular, there is an increasing focus on the adaptation of coastal communities to climate impacts (Robinson, 2017).

Small island developing states are most at risk from sea-level rises, increased average global temperatures affecting the normal seasonal cycles, increased incidences and intensity of storms and hurricanes, along with the consequent increased risks of storm surges, other floods and droughts. These impacts can have a direct effect in the area of water resources and water services where, under most climate scenarios, water resources in small island developing states are forecast to be highly compromised (Mimura *et al.*, 2007; Bates *et al.*, 2008; Karnauskas *et al.*, 2016). The IPCC reports explicitly identify small island developing states as a 'hotspot' area where climate change effects are present or imminent and where urgent action is required in the water sector (Nurse *et al.*, 2014).

The characteristics of small island developing states constrain effective management of the water sector; parallel to this is the pattern of demand for water resources in small island developing states. Water demand can be intensified by the large economic reliance of many small island developing states on water-intensive industries such as the tourism sector (e.g. **Figure 10.2**), and to a lesser extent the agricultural sector, as their main developmental option. At the community level, many coastal livelihoods can be closely linked to these industries. IPCC scenario analyses confirm that for both of these sectors climate change impacts will be negative (Nurse *et al.*, 2014).

Figure 10.2 Luxury swimming pools are common in Barbados to support the tourist industry and yet freshwater is very scarce on the island state, with a higher annual water demand than available renewable freshwater.
Source: Shutterstock/Chris Sargent

10.3 WATER PRICING

In *The Wealth of Nations*, published in 1776, Adam Smith pointed out a well-known paradox regarding the usefulness of water and its price: 'Nothing is more useful than water, but it will purchase scarce anything; scarce anything can be had in exchange for it.' Water, like many other goods we might want, is scarce. Under drought conditions it becomes a bit scarcer than normal. For any other good, we would expect supply reductions to correspond to increasing prices. When frosts damage citrus crops, orange juice prices rise. When hurricanes knock out oil refineries, gasoline prices shoot upwards. However, water is one of the exceptional goods where price is not an indicator of its scarcity.

10.3.1 Water as an economic good in private and public ownership

Increasing urbanisation, population and industrial development have increased demand for water, resulting in a considerable decrease in annual renewable water resource per capita. As the price of water is kept very low, the consumer takes this low price to be the real value of water and so increases consumption. Water has become scarce, but its use is rampant. It has become important to sort out an efficient pricing structure for water.

In the Dublin conference of 1992, for the first time in a United Nations setting, water was declared an economic good. However, water had been treated as an economic good for many centuries before this. In Europe and the early period of the United States, private water supply companies thrived in a wide variety of settings. However, the nineteenth-century 'sanitary revolution' saw the demand for public ownership and management of most of these companies in the name of public welfare. Heavy emphasis on the nature of water as a public good led to the development of heavily subsidised public systems. However, by the 1980s the World Bank and other institutions started stressing the need for privatisation in supplying water. For the World Bank, privatisation increases efficiency and brings about proper allocation.

Water privatisation refers to private-sector participation in the provision of water and sanitation. Over the past three decades, there have been heated debates over whether water should be supplied by the public or private sector. Advocates for privatisation claim that the private sector delivers services more efficiently than the public sector, as, first, the private sector has stronger commercial incentives to dominate the market by improving services and market competition would drive water resources to flow to the most capable supplier with lowest costs; second, the public sector, particularly in developing countries, makes slow progress in extending water services owing to a lack of finance and existence of corruption. However, in some cases water privatisation has led to a violation of basic human rights. Natural monopoly, a common type of market structure that accompanies water privatisation, tends to overprice and underproduce water services, making clean water unaffordable or inaccessible to the poor people. This thus constitutes a major reason against water privatisation.

After England and Wales sold off water systems in 1989, water-related investment increased and water quality improved significantly, while in the meantime water tariffs and profits went up. In the wake of globalisation and neo-liberalism, which stresses the role of the free market and dislikes governmental intervention, water privatisation was pushed forward in the developing world by international institutions (e.g. the World Bank and IMF). Massive funds were donated to help developing countries improve their water services through privatisation. Success can be witnessed in the cases of Manila, Guayaquil in Ecuador, Bucharest and some other cities, but failures also exist. One of the best-known examples of failure was Cochabamba, Bolivia, where protests against high tariffs and monopoly of water resources finally forced out the water company in operation at that time. The outcome of water privatisation varies case by case, largely depending on the way privatisation is developed and the local context (Budds and McGranahan, 2003). The bulk of water privatisation research has been focused on the tap/piped water system (as above); however, there is another type of water privatisation – bottled water (**Box 10.2**).

10.3.2 Pricing water: from benefit to cost

Pricing water is one of the ways to increase efficiency, equity and sustainability in the water sector. For a proper price, government intervention is typically required so that equity and other important issues are properly addressed. Through the Sustainable Development Goals the main concern is to eradicate poverty from the global map by giving the poor more access to the resources they need

BOX 10.2

The bottled water industry

The consumption of bottled water has increased dramatically over the past decades and is now worth over $200 billion per year globally. Its growth was against the context of shifting lifestyles that demand for greater convenience and public fears about tap water quality. Profit margins for bottled water are typically higher than tap water, ranging from 25% to 35%, which stimulates the capital to accumulate in the bottled water industry. However, the energy needed to produce and transport bottled water is many more times higher than that for tap water, causing more carbon emissions accordingly. Other environmental issues related to the bottled water industry include plastic bottle disposal and recycling, pollution in the bottling plant locations, and removal of water for other uses in the catchment (Jaffee and Newman, 2013).

In most regions, water rights are a prerequisite for private companies to access water resources. Nestlé Waters, one of the world's largest bottled water companies, has been in continuing dispute with authorities of California, in the USA, for several years over water rights of the Arrowhead Mountain Spring. The dispute is mainly focused on the validity and scope of a contract signed by the predecessors of Nestlé Waters and local residents dating back more than a century. As the contract was ruled to be still valid, the water rights remained within the company. Further updates have been made to clarify the scope of the rights in the contract, and the company was charged an annual permit fee of only $624 by the Forest Service. However, some local residents are dissatisfied with the judgement and have protested with strong concern for the local environment and ecosystems.

to live sustainably. It is here that water pricing, by encouraging judicious use of resources, can contribute to environmental protection and should help (but not always does) promote economic efficiency and social equity.

In a competitive market, price is determined by the demand and supply of goods and services. Water resources invoke an economic demand as they are used by people in the production of goods and services, such as agricultural output, human health, recreation and quality of life. Supplying water to production or residential usage is usually associated with costs for transmission, treatment and maintenance.

Providing or protecting water resources involves active employment of capital, labour and other scarce resources. Utilisation of these resources to provide water supplies means that they are not available to be used for other purposes. The economic concept of the 'value' of water is thus couched in terms of society's willingness to make trade-offs between competing uses of limited resources, and in terms of aggregating over individuals' willingness to make these trade-offs. The economist's task of estimating the benefits or loss of benefits resulting from a policy intervention is easiest when the benefits and costs are revealed explicitly through prices in established markets. When it comes to measuring environmental and some other impacts, however, valuing benefits is more difficult, and requires indirect methods (see Section 10.4).

On the other hand, the economist's notion of cost, or, more precisely, of opportunity cost, is linked with – but distinct from – everyday usage of the word. Opportunity cost is an indication of what must be sacrificed in order to obtain something. In the water resources context, it is a measure of the value of whatever must be sacrificed to make those resources available. It has been observed over and over again in diverse markets for goods and services of various kinds that the incremental costs of providing an additional unit increase, as the total quantity supplied increases.

In the language of economics, there are increasing (or upward-sloping) **marginal costs**. The costs of a cubic metre of water flowing out of a kitchen tap include the costs of transmission, treatment and distribution; some portion of the capital cost of reservoirs

and treatment systems, both those in existence today and those future facilities necessitated by current patterns of use; and the opportunity cost in both use and non-use value of that cubic metre of water in other potential functions. This is the long-run marginal cost (LRMC) of supplying water.

In a competitive market – which, as we have explained above, is not the context for most water resources – the quantity of a good or service provided and its price are jointly determined by the forces of supply and demand, which are closely linked with costs and benefits, as described here. In fact, the downward-sloping marginal benefit curve is the demand curve, and the upward-sloping marginal cost curve is the supply curve. Where these intersect, where demand and supply balance one another, markets achieve equilibrium, determining the

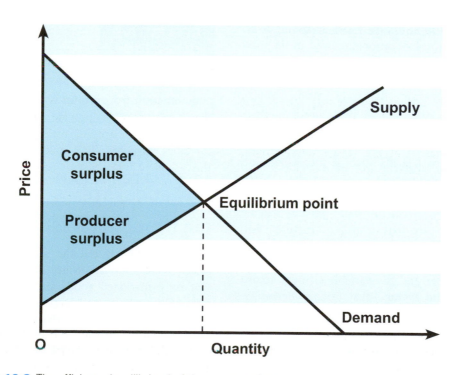

Figure 10.3 The efficiency (equilibrium) of the water market.

quantity provided and the price in the process. That particular combination of price and quantity maximises the difference between benefits and costs; that is, it maximises what economists call net benefits (the sum of consumer surplus and producer surplus). This is the definition of economic efficiency, and the efficient quantity and the efficient price of any good or service (Figure 10.3).

10.3.3 Types of water tariffs

Since water is not traded in markets, we would not expect prices to adjust automatically to reflect periods of scarcity, as they do for other goods and services. Instead, most water pricing is regulated by public institutions – city councils, public utility commissions, water boards and other entities.

A set of procedural rules used to determine the condition of services and charges among various classes is known as a water tariff. Tariff structure varies according to the state as the institution involved in its provision also differs. The water charges can be in the form of non-volumetric water tax, non-volumetric flat rate tariff, uniform metered tariff and metered block tariffs or can be a combination of the above charges. The most popular tariffs in OECD countries are two-part tariffs and the increasing block tariff. Countries such as Australia, Austria, Denmark, Finland and the United Kingdom that have successful water pricing schemes use a two-part tariff structure with a variable part and a fixed part. The fixed part varies according to some characteristic of the user and the variable element according to the average cost pricing. Sometimes the variable part is replaced by an 'increasing block tariff'. This increases the tariff structure. Advantages of the two-part tariff structure are that it brings in a stable revenue. The fixed element encourages conservation, for it charges the consumer according to their consumption.

In the increasing block tariff there are two or more blocks, for which prices are charged differently, and with each successive block the price increases. The utility must decide on the number of blocks, volume of water use associated with each block, and prices to be charged for each block, when designing an increasing block tariff structure (Boland and Whittington, 1998). The increasing block tariff is a progressive tariff. There is a lifeline block for the poor at below-cost rate, and beyond this minimum volume the charge increases. The lifeline block works as a subsidy for the poor and promotes welfare. Thus the increasing block tariff encourages equity. The system enables poorer households to get access to low-rate water since they possess fewer water-consuming appliances, and also allows for rich-to-poor and industrial-to-household subsidies (Boland and Whittington, 1998).

> ### REFLECTIVE QUESTION
>
> What factors ought to be involved when setting a price for water?

10.4 WATER AS AN ECOSYSTEM SERVICE

10.4.1 Total economic value

Economic value is not the same as market price. Were they the same, it would imply that only marketed goods and services had value. Given the essentialness of water for all life, the notion of the *value* of water (as opposed to its price) has occupied great minds for centuries. Adam Smith, in *The Wealth of Nations*, observed that the concept of value itself was subject to a myriad of sometimes paradoxical interpretations, where 'the things which have the greatest value in use have frequently little or no value in exchange; and, on the contrary, those which have the greatest value in exchange have frequently little or no value in use'. Smith used the example of diamonds and water as the illustration for this concept, where diamonds could command a great many other goods in exchange but had little use value, whereas water, of which 'nothing is more useful', could not be used in exchange. Benjamin Franklin attributed value to scarcity, with the famous proverbial saying 'when the well's dry, we know the worth of water'.

Economists, in considering the value of nature and its components, have attempted to demystify the concept of value by defining what has become known as **total economic value**, and disaggregating this into various categories and sub-categories. The total economic value of environmental goods consists of *use value* and *non-use value* (**Figure 10.4**).

The use value is a value related to the present or future use by individuals. It can be subdivided into direct use values and indirect use values. Direct use values are derived from the actual use of a resource (such as water consumption); indirect use values refer to the benefits derived from ecosystem functions (such as river basin protection). A separate category is made up of *option values*, which are values attributed by individuals who have knowledge that a resource will be available for future use. Thus it can be considered as an assurance that a resource will be able to supply benefits in the future. For example, the continued availability of a certain amount of forest area to sequester carbon and mitigate the impacts of climate change would provide future benefits to society. The quasi-option use value shown in **Figure 10.4** (sometimes also classified as a non-use value) represents the value derived from the preservation of the future potential use of the resource, given some expectation of an increase in knowledge. For example, the continued availability of certain species may have value to humanity, but this value may not be obvious now. However, with increased scientific knowledge, these species may come to play a significant role in human welfare in the future.

The non-use values (**Figure 10.4**) are associated with the benefits derived simply from the knowledge that a natural resource – such as a river basin – is maintained

Figure 10.4 A typology of economic value.

(Young, 2005). By definition, such a value is not associated with the use of the resource or the tangible benefits deriving from its use. It can be subdivided into two parts that overlap, according to its definition. First, there are *existence values*, which are not connected to the real or potential use of the good but reflect a value that is inherent in the fact that it will continue to exist independently from any possible present or future use by individuals. Second, *bequest values* are associated with the benefits to individuals derived from the awareness that future generations may benefit from the use of the resource. These can also be altruistic values when the resource in question should, in principle, be available to other individuals in the current generation.

10.4.2 Ecosystem services and goods

The Millennium Ecosystem Assessment (MEA), in its survey of the world's natural resources, provided an innovative methodological framework for analysing environmental resources. The emphasis was placed strongly on the concept of human well-being, the contributions made by ecosystems through their supply of ecosystem goods and services, and the drivers of change of ecosystems that would then impact upon human well-being (Figure 10.5). The MEA approach provided a framework that addressed the many mechanisms through which nature contributes to human well-being at different levels of spatial scale, and the factors (both direct and indirect) that can impact upon these mechanisms. The MEA methodology has resulted in a growing body of literature on ecosystem services.

The MEA identified four categories of ecosystem services, namely provisioning, regulating, cultural and supporting services (Millennium Ecosystem Assessment, 2005). These contribute in different ways to the

constituents of human well-being. The notion of human well-being and livelihood is a complex one. The MEA presents a comprehensive overview of the factors that both comprise and affect the state of human well-being. Furthermore, the MEA methodology hypothesises that drivers of change impact upon ecosystems and the goods and services they provide, thus affecting human well-being. If we are able to quantitatively or qualitatively define these linkages, the impacts upon human well-being of decisions that affect ecosystems at any spatial scale can be mapped. In addition, we can theorise a number of feedback loops: the supply of ecosystem services itself can impact upon the drivers of change; the state of human well-being can affect ecosystem services via increased or decreased demand placed on these services; and human well-being is directly linked to the drivers of change.

As an example of the ecosystem service framework we can examine wetland ecosystems. Wetland ecosystems are areas such as swamps or peatlands. Wetland ecosystems are at the source of many water resource services. A major supply of renewable freshwater comes from inland wetland ecosystems (e.g. Xu *et al.*, 2018). Many wetlands are responsible for water purification, treating and detoxifying a host of different waste products through their natural processes. Against this background, the MEA approach can be used to categorise the wide range of ecosystem service benefits of wetland ecosystems to local, regional and international communities (Millennium Ecosystem Assessment, 2005).

By identifying such services, economic tools can then enable us to trace the benefit flows from wetland ecosystems to human beneficiaries (see e.g. Box 10.3). Figure 10.7 shows some of the main services that wetland ecosystems can provide. As Figures 10.7 and 10.8 show, water resources flowing from wetland ecosystems are considered both

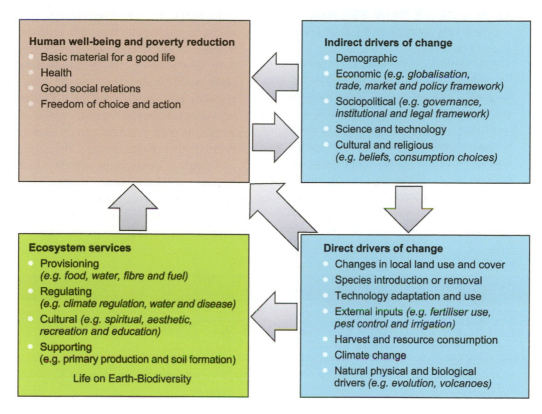

Figure 10.5 Linkages between ecosystem services, human well-being and drivers of change.
Source: After Millennium Ecosystem Assessment, 2005.

provisioning and regulating ecosystem services. The concept of the total economic value of an environmental resource has given way to the concept of goods and services flowing from ecosystems upon which human well-being depends (**Figure 10.8**). However, these two concepts are related. Water resources are considered mainly in the provisioning and regulating categories of the MEA. Freshwater for human consumption represents a provisioning service of wetland ecosystems, which also coincides with direct use values and option values. Water regulation and purification are considered regulating services of wetland ecosystems, and these coincide with indirect use values and option

values. Whichever methodological approach is adopted, the components that make up the value of the resource need to be identified before economic valuation can be undertaken.

REFLECTIVE QUESTIONS

What is the difference between use value and non-use value?

How might an ecosystem services approach allow new decisions to be made about managing a water resource?

10.5 VALUATION TECHNIQUES FOR WATER RESOURCES

There exists a methodological toolbox for the valuation of non-marketed environmental goods and services such as wetland ecosystems and water resources. Some of the techniques are more capable of revealing the values of some types of ecosystem goods and services than others. Furthermore, it is undeniable that, no matter the technique, some of these values themselves, in the context of human welfare, are by definition notoriously difficult to reveal. Whatever the

outcome of the valuation exercise, it is now frequently the case that the resulting estimates are considered as baseline estimates only. Furthermore, the **precautionary principle** will dictate that, in the absence of rigorous economic values, and despite pressing immediate economic needs, policy should lean in the direction of conservation rather than exploitation.

There are three main approaches to economic valuation: direct market valuation, revealed preference and stated preference. Direct market valuation approaches use data from actual markets. In revealed

---

BOX 10.3

Koshi Tappu Wildlife Reserve, Nepal

There are many examples of the monetisation of services provided by wetlands. One such example is that carried out for the Koshi Tappu Wildlife Reserve, located in eastern Nepal (**Figure 10.6**), where the economic benefits generated by the ecosystem services were estimated through value transfer and market-based approaches (Sharma *et al.*, 2015). The Koshi Tappu Reserve was established in 1976 and declared a Ramsar wetland site (the Ramsar classification indicates a wetland of international importance) in 1987, owing to it supporting at least 480 notable birds species and 30 species of mammals, such as the Bengal falcon, white-tailed eagle and panther. The reserve covers 175 km^2 of wetlands in the Sunsari, Saptari and Udayapur Districts, and, thanks to the Sapta Kosi River, it consists of extensive mudflats, reed beds and freshwater marshes. The study by Sharma *et al.* (2015) estimated that the Koshi Tappu Reserve can create an annual value at least US$16 million, which equates to US$982 per household. In particular, provisioning services (food, timber, fodder, fisheries, thatch and water) accounted for nearly 85% of the value. Services such as flood protection, carbon sequestration and ecotourism were also important. Although some non-use values and regulatory services are not taken into account, the results clearly emphasised that local people can achieve significant economic benefit from the wetland area. However, the low development level, high poverty pressure, and few alternative livelihood options for local people within the areas surrounding the wetland have led to increasing stress on natural resources and insufficient reserve management investment. If this continues, the result will be degradation and poorer quality of life. Thus implementation of an improved management programme which takes account of the ecosystem service value of the wetland will be important for improving the system and deriving benefits for society. In other words, the costs of implementing a new management plan will easily be paid back by the value of the ecosystem services that are provided in return. Some of the techniques used to determine the valuation of different ecosystem services are provided in Section 10.5.

Figure 10.6 Some birds on the Koshi Tappu Wildlife Reserve, Nepal. *Source:* Shutterstock/Yatra Rai.

Provisioning	Regulating
Freshwater	Water regulation
Food	Water purification
Fibre and fuel	Climate regulation
Genetic materials	Erosion regulation
Biochemical	Natural hazard regulation
	Pollination

WETLAND ECOSYSTEMS

Cultural	Supporting
Spiritual and inspirational	Soil formation
Recreational	Nutrient cycling
Aesthetic	
Educational	

Figure 10.7 The goods and services provided by wetland ecosystems.

preference approaches, people 'reveal' their preferences through their choices. Finally, in stated preference approaches, a market and demand for ecosystem services are simulated through conducting surveys that include some hypothetical changes in ecosystem services originating from policy changes. A number of studies have used stated preference methods when market information has been unavailable. In converse situations, where such data exist, revealed preference methods are applied. Moreover, stated preference methods are frequently used to elicit non-use or cultural values, as these values do not have recorded behaviour, unlike use values,

Group	Service	Direct use	Indirect use	Option	Non-use
Provisioning	Freshwater Food Fibre and fuel Biochemical	✓	NA	✓	NA
Regulation	Water regulation Water purification Climate regulation Erosion regulation Natural hazard regulation Pollination	NA	✓	✓	NA
Cultural	Spiritual and inspirational Recreational Aesthetic Educational	✓	NA	✓	✓
Habitat/ supporting	Soil formation Nutrient cycling	Valued through other ES categories			

Figure 10.8 Total economic value versus ecosystem services of wetland ecosystems.

which can be better measured by revealed preference methods. These approaches are described below. **Figure 10.9** suggests a classification of economic values related to freshwater resources and the various market and non-market approaches used to capture the magnitude of the respective benefits.

10.5.1 Direct market valuation

A good part of the value of ecosystems is represented by commercial and financial gains and losses through the market that can be measured by market-based methods of economic valuation. Some of the market valuation approaches (**Figure 10.9**) include: averting behaviour, replacement/restoration, the production factor method and dose–response. The averting behaviour (or 'preventive expenditure') technique measures the expenditure incurred in order to avert damage to the natural environment, human infrastructure or human health. Averting

behaviour should be seen as a minimum estimate of impact costs, since it does not measure the consumer surplus (i.e. the additional amount above actual expenditure that consumers would be willing to pay in order to protect a particular good or service). The replacement/restoration cost technique can be used to measure the costs incurred in restoring or replacing productive assets or restoring the natural environment or human health as a result of the impacts of environmental degradation. As with preventive expenditure, restoration costing is a relatively simple technique to use, and has the added advantage over preventive expenditure of being an objective valuation of an impact – i.e. the impact has occurred, or at least is known. Use of the replacement costs method relies on replacement or restoration measures being available and the costs of those measures being known. As such, the method is unlikely to be appropriate for costing the impacts on irreplaceable assets. Another shortcoming of

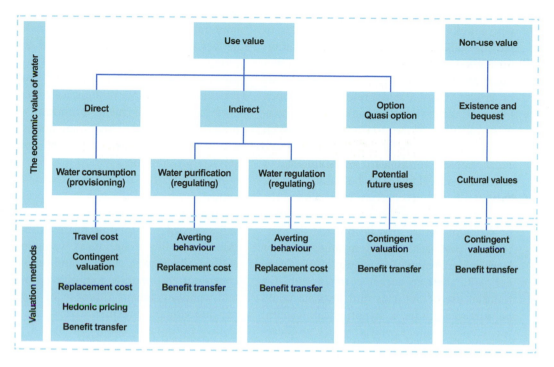

Figure 10.9 Economic valuation techniques for water resources. See text for explanation of each method listed.

the technique is that actual replacement or restoration costs do not necessarily bear any relationship to the willingness of individuals to pay to replace or restore something.

The production factor method estimates the economic value of an environmental commodity through an 'impact-pathway' approach, in which a change in the environmental attribute is linked to impacts on 'end-points' that are relevant for human well-being. For example, the benefits of tree planting, through reduced erosion, are measured first by the link between soil cover and erosion rates and then by the link between erosion rates and agricultural productivity. Such methods can be very useful for valuing many services provided by ecosystems, including forestry (timber and non-timber), agriculture (value of diversity in crops and use of genetic material) and marine systems (losses from overfishing, species invasion).

10.5.2 Stated preference valuation

Market prices and costs can provide estimates of the increase in the value of commercial activities. However, some environmental goods and services do not affect markets, or market data are not available to value them. The development of non-market valuation has been closely linked with water projects and the development of cost–benefit analysis. Hanemann (2006) noted that the positive and negative environmental impacts of water projects have been frequently assessed using non-market valuation since the 1980s. In such cases methods have been developed to derive consumers' preferences. These are revealed preference and stated preference valuation methods.

Contingent valuation is currently the most-used stated preference technique for the valuation of environmental goods, where individuals state their *willingness to pay* or *willingness to accept* compensation for a good or service. This valuation is often done during survey work with individuals or groups. Only stated preference methods like contingent valuation can elicit the monetary valuation of the non-use values, which typically leave no 'behavioural market trace'. Furthermore, contingent valuation allows environmental changes to be valued even if they have not yet occurred. It allows the specification of hypothetical policy scenarios or states of nature that lie outside the current or past institutional arrangements or levels of provision. For example, the farmers' willingness to accept compensation for reduced use of pesticide and fertiliser could be surveyed under a hypothetical scenario for a new water project which requires higher standards of water quality downstream. The resulting aggregates of individual farmers' willingness would reveal the indirect use value of water purification. Finally, contingent valuation allows a richer debate to occur as the process of valuation is part of a public discussion. Against this is the criticism that the values are hypothetical (payments are not actually made or cash paid out), and so it might be that the method is subject to many biases. Box 10.4 provides an important example of contingent valuation for water management in the USA.

The 'conjoint choice', 'choice experiment' or 'attribute-based' method (these are all the same) is also a commonly used stated preference method. This method elicits information on values by asking individuals to choose between alternatives (e.g. six compensation payments ranging from £2–28 per hectare annually may be presented to the farmers who are surveyed to examine their minimum willingness to accept requirements of reducing use of fertiliser to protect water quality). This can be done using conjoint ranking, where individuals rank alternatives in order of preference, and conjoint rating, which indicates their strength of preference on a cardinal scale. Box 10.5 provides an example of a choice experiment used to measure farmers' minimum willingness to accept requirements under a possible local payment for an ecosystem services programme.

According to Bateman *et al.* (2002a), choosing which of the two stated preference approaches to use depends on: the kind of value needed (i.e. total or relative), information availability (contingent valuation has a larger literature), welfare and/or welfare-consistent estimates, cognitive processing and sampling means (number of responses per individual). Freeman (2003) was 'cautiously optimistic' about the stated preference method and reported that others are attracted to stated preference methods because of the 'relatively easy and inexpensive way to get usable values for environmental resources'. In a similar vein, Whittington (2002) concluded that stated preference is vital to a developing country's policy application, but it is far from being a high-quality option at a low cost. However, almost two decades later, we now have more tools with which to judge the accuracy or validity of stated preference estimates and an emerging consensus on the criteria we should use to do so (Kling *et al.*, 2012). Johnston *et al.* (2017) provided contemporary guidance for stated preference studies, grounded in evidence from past literature and experience. Recommendations have been made from the process of survey development and implementation, to data analysis and to validity assessment, so as to raise the quality of stated preference studies and promote research that will further enhance the practice of these studies worldwide.

10.5.3 Revealed preference valuation

The revealed preference methods include hedonic pricing, the travel cost method and benefit transfer. Hedonic pricing estimates the economic value of an environmental commodity, such as clean water, clean air or an attractive view, by studying the relation between such attributes and house prices (Palmquist, 1991). House prices are often higher if the house has a nice view or the watercourse nearby is clean. Hedonic price estimation has been applied to elicit environmental/ecosystem values associated with recreation, landscape values and genetic and species diversity. Hedonic techniques are particularly employed in valuing visual amenity, quality of soil assets and exposure to air pollution. The travel cost method estimates the economic value of recreational sites by looking at the consumer surplus (or economic benefit of recreational users) of using these sites; the valuation is then based on deriving a demand curve relevant to the travel costs, for the site in question, through the use of various economic and statistical models. Normally, a rise/drop in travel costs would cause the recreational demand to decrease/increase along the demand curve, which indicates a down-sloping demand curve; and enhancement/degradation in ecosystem service delivery, such as water-quality regulation, would cause the demand curve to shift to the right/left, which continues to result in an increase/decrease in consumer surplus. The difference in the economic benefits that recreational users obtain from activities before and after changes in ecosystem service delivery is then considered the economic values of the environmental commodity in question. Where the individual makes a choice involving more than one site, the discrete choice models have used the random utility theory framework to value not only visits to different sites but also the attributes of sites, such as water quality. The travel cost technique has been widely applied, especially in North America. Choe et al. (1996) studied the economic value of clean surface water in Davao, a coastal urban area in the Philippines, using a simple single-site travel cost model. Their analysis was focused on one popular recreational site – Times Beach. The sea at the beach was polluted by the nearby discharge of the Davao River, which carries silt, storm water drainage, and household and industrial wastewater from Davao. Warned by the city health department about the health risks of swimming at Times Beach, most people stopped using Times Beach for recreational activities (e.g. picnicking and swimming), which implies a decrease in recreational demand at each level of travel costs, and the consequent loss of consumer surplus is an indication of the economic value of the clean surface water in Davao.

The method of benefit transfer consists of exporting previous benefit estimates (either from stated preference or revealed preference) from one site to another, at one point in time, with regard to the researcher's area of interest. In benefit transfer estimates there are three possible forms of transfers: transfer of an average of willingness to accept compensation estimates from one primary study, transfer of the willingness to accept compensation function, and transfer of willingness to accept compensation estimates by aggregating other estimates employing meta-analyses (Bateman et al., 2002b). Benefit transfer involves the use of economic values from one specific area with a known resource and policy conditions, and applying these to another site in similar circumstances. Generally, the first site is known as the 'study site' and the second as the 'policy site'. Sites differ in characteristics, and one has to be cautious when applying values from one site to another. On the one hand, this method reduces the cost of starting

BOX 10.4

Contingent valuation – economic effects of regulating flow releases from Glen Canyon dam, USA

The Glen Canyon Dam is upstream from the Grand Canyon National Park and the Glen Canyon National Recreation Area. The hydropower operations in the dam were thought to be impacting upon fisheries and rafting, particularly as water was stored for times of peak energy demand rather than for good baseflow in the river network. The US Bureau of Reclamation decided to use contingent valuation to establish how much the recreation was worth and therefore to understand whether providing more baseflows would be worth more than the water stored for the energy required at peak times. Visitor surveys were conducted in the 1980s and results suggested that a change in practice might be worth $2 million per year (Bishop *et al.*, 1989). However, the study also led to a more nuanced outcome; rather than the debate being about recreation versus power generation, it was realised that flows could be optimised so as to maximise economic value across all uses and that society was also concerned with other impacts, such as native vegetation and birds, erosion and endangered fish. A willingness-to-pay survey was conducted for households across the USA, which showed strong support for promoting a more natural flow regime for the river network, including large spring flows (Loomis, 1997).

BOX 10.5

A choice experiment to support water-quality management in UK upland drinking water

A project was undertaken to understand whether farmers are willing to accept privately funded compensation payments for protecting water-quality supplies in the UK uplands. This study, by Beharry-Borg *et al.* (2012), examined the feasibility of using payments financed by a water company to pay farmers to adjust agricultural land management practices to protect water quality. The choice experiment method was used to measure farmers' minimum willingness to accept requirements under a possible local payment for an ecosystem services programme to adjust agricultural land management practices in Nidderdale and the Washburn Valley (Yorkshire, UK). Choice data for the study were drawn from three geographical locations in the Nidderdale catchment in the north of England. These were Fearby, Middlesmoor, Pateley Bridge and the Washburn area.

Each land management attribute was chosen based on water-quality monitoring data and hydrological modelling results for Nidderdale, together with feedback on an early version of the questionnaire from a pilot group of farmers. An example of the choice card used in this study is illustrated in Figure 10.10. The results of monitoring of water-quality indicators and hydrological modelling showed that changes in the input of nitrogen and ditch blocking were the land management practices that were expected to have the greatest impact on water quality in the Nidderdale area, affecting nitrate levels and discoloration (i.e. reducing the colour) of drinking water, respectively. The ditches are found in organic upland soils and peatlands. The ditches result in a deepening of the water table and cause the organic matter to decompose, releasing brown dissolved organic carbon into watercourses. Each land management attribute and its levels were chosen to satisfy scientific understanding of the impact of such measures on water quality, and also to represent a feasible range of land-management changes that could be considered by farmers.

	Inorganic fertiliser use	Farmyard manure use	Grip blocking	Agreement length	Compensation payment	My choice (choose only one option)
Program A	Apply your current level of fertiliser	Apply your current level of manure	Maintain existing grips	3 years	£2 per acre	
Program B	Apply 25% less fertiliser	Apply 50% less manure	Allow 50% of grips to be blocked	5 years	£4 per acre	
Program C	Apply 50% less fertiliser	Apply 25% less manure	Allow 100% of grips to be blocked	10 years	£10 per acre	
Do nothing	I do not wish to participate in any of the programs					

Figure 10.10 Example choice card. On the card the word 'grips' is a local term used by the farmer to refer to drainage ditches.

Some mathematical models were used to examine the data derived from the choice experiment with farmers. Multinomial logit and latent class models were used to analyse 97 farmers' responses to establish their preferences, quantify willingness to accept compensation values and explain the presence of unobserved heterogeneity. Results from the multinomial logit model showed that, on the whole, farmers had significant aversions to reducing applications of artificial fertiliser and farmyard manure and, in general, the aversion to reducing farmyard manure application was stronger than the aversion to reducing applications of artificial fertiliser. Somewhat surprisingly, on average across all the farmers surveyed, there appeared to be no particular preference for, or aversion towards, blocking drains in the upland landscape, and no particular preference for, or aversion towards, the length of management agreement offered. This overall lack of preference/aversion to drain blocking may arise because decision rights over drain blocking are typically held by large landowners rather than by tenant farmers. The compensation payment analysis showed that, provided adequate levels of compensation payments were offered, farmers could be encouraged to adopt land management practices that have the potential to enhance water quality. In other words, farmers could be encouraged to reduce applications of artificial fertiliser and farmyard manure and to block drains that had previously resulted in deterioration of water quality.

The findings suggested that there is likely to be considerable variation across farmers in their strength of preference for, or strength of aversion to, different elements of land management programmes which aim to protect drinking-water quality. In addition, the results suggested that farmers can be encouraged to voluntarily adopt more environmentally sound land management practices and thereby enhance delivery of ecosystem services such as drinking-water provision or water-quality regulation, through the use of incentive payments. However, these initial results provide only a broad indication of the levels of incentive payments necessary to achieve partici-pation. The analysis tested could be used in a cost-effectiveness framework to address the spe-cific costs of water treatment, or the predicted cost of implementing payment programmes to farmers to change their land management practices, to achieve specified requirements for water colour and nitrate levels at a sub-catchment level. Within such a framework, private-sector water companies would be able to use these estimates to consider a number of policy scenarios and estimate the costs of each, so as to identify their preferred policy option.

a completely new valuation study, whereas, on the other hand, the compilation of a comprehensive database often proves costly.

10.5.4 Other techniques

There also exists a host of non-monetary tech-niques that have attempted to assess the importance of ecosystem services; a compre-hensive discussion of these is provided by Christie *et al.* (2008). Such methods include consultative and participatory approaches such as focus groups, citizens' juries, health-based approaches, Q-methodologies and Delphi surveys. While these methods do not yield monetary indicators, as do the eco-nomic techniques outlined above, they can provide useful insights into how water resources are perceived and utilised. The pos-sibility also exists that these non-economic methods can serve to complement the eco-nomic methods, which can then, with added non-economic insights, yield more accurate, rigorous and robust monetary estimates.

REFLECTIVE QUESTIONS

What are the main valuation techniques for water?

Under what sorts of circumstances would you need to ask people's opinions to undertake a water valuation?

10.6 ECONOMIC MANAGEMENT OF WATER FOR THE PRESENT AND FUTURE

The essential nature of water to all life requires a coherent and consistent management of water resources at a variety of stakeholder and spatial scales, with trade-offs among competing wants and uses. This brings up the need for the creation, strengthening or maintenance of the relevant institutions and instruments in order to manage these different interests. Underlying all of this is the need for the recognition, demonstration and capture of values of water resources. The public-good and transboundary nature of water resources brings the critical need for water governance. In recognition of this growing need, there has been a contemporary shift towards diverse governance mechanisms in the water sector, with a focus on multiple stakeholders at different levels of spatial and political scale. In Western Europe in particular, regional governance is playing an increasing role in water governance and management (Lieberherr, 2011).

There exist several ways in which the government can intervene in the water sector (Chapter 11 discusses this issue in detail). By the provision of supply, public institutions hold the responsibility for water infrastructure and water pricing as well as the management of the upstream or downstream private-sector institutions that may have developed along the water chain. Governments can apply taxes, grant subsidies and provide information (Dalhuisen *et al.*, 2000). They can also regulate or intervene in the demand chain by, for example, encouraging water conservation among residential users or imposing restrictions on industrial or agricultural activities that may impact upon water supply and water quality. An increasingly popular scheme for the management of ecosystem services across competing wants, using economic incentives, is that of payments for ecosystem services (e.g. Box 10.4). Much current work is dealing with establishing the design and potential operation of such payment systems from theoretical and empirical perspectives, considering the efficiency of alignment between payments and outcome, and estimating public willingness to pay for the ecosystem services so secured (Martin-Ortega *et al.*, 2015).

It is important to note explicitly that this type of economic management framework for water is by definition interdisciplinary. In particular, the disciplines of humanities, social sciences and natural sciences all play overlapping roles. Analyses of natural systems, traditionally linked to ecology and the natural sciences, can now be complemented by a coordinated interdisciplinary approach including humanities and social science disciplines such as economics, sociology, anthropology and governance. Via such a holistic approach, it is possible to better undertake a coherent analysis of the impacts of changes in water resources on human welfare, within the context of sustainable livelihoods, and the institutional and policy settings within which such changes occur. Ultimately, research towards these goals will inform policy on a progression towards poverty reduction, sustainable development and the achievement of the Sustainable Development Goals (UN, 2014).

REFLECTIVE QUESTION

Why is the economic management of water a complex interdisciplinary process?

10.7 SUMMARY

Water availability can represent a significant constraint on the sustainable development of an economy. Water by its very nature is defined as a common property resource.

Resource allocation issues, coupled with increasing levels of economic growth and populations that place water resources under increasing demands, lead to the need for improved water management as a critical element in the achievement of sustainable development. Value and price hardly equate in the context of environmental goods and services. To this end, economists have produced a methodology of total economic value in which the use and non-use components of environmental goods and services can be disaggregated. The Millennium Ecosystem Assessment took this methodology one step further by categorising ecosystem goods and services into provisioning, regulating, cultural and supporting. No matter what methodology is used, there exist a range of economic valuation tools that can be used to monetise ecosystem goods and services. Valuation techniques can include market-based and non-market-based methods. In fact, the history of non-market valuation is linked to water sector evaluations. In terms of water resources, we can for each category of value or for each category of ecosystem services apply particular techniques to produce economic values.

Non-monetary techniques are also becoming increasingly popular, in an effort to complement the range of available economic techniques. Once values are established, the work then turns to how these values can be embedded into the decision-making processes. This brings forth the crucial concept of 'water governance', where a contemporary shift has taken place towards diverse governance mechanisms in the water sector, with a focus on multiple actors at different levels of spatial and political scale. An increasingly popular scheme for the management of ecosystem services across competing wants, using economic incentives, is that of payments for ecosystem services. Ultimately, what is urgently required is a coherent analysis of the impacts of changes in water resources on human welfare, within the context of sustainable livelihoods and the institutional and policy settings within which such changes occur, with movements towards water security placing the world economy firmly on a path to poverty reduction, sustainable development and the achievement of the Sustainable Development Goals.

FURTHER READING

Dinar, A., Pochat, V. and Albiac, J. 2015. *Water pricing experiences and innovations.* Global Issues in Water Policy 9. Springer; Basel.

A collection of chapters from different parts of the world giving examples of water pricing successes and failures.

Hanemann, W.H. 2006. The economic conception of water. *In*: Rogers, P.P., Llamas, M.R. and Martinez-Cortina, L. (eds) *Water crisis: myth or reality?* Taylor & Francis; London.

This paper presents the economists' view of water and how water can be managed from the economists' perspective. By appealing to various economic thoughts and concepts, the article provides a history of economic thinking from the perspective of water issues.

Martin-Ortega, J., Ferrier, R.C., Gordon, I.J. and Khan, S. 2015. *Water ecosystem services: a global perspective.* Cambridge University Press; Cambridge.

This free-to-download book provides a wide-ranging set of examples from different regions and different parts of the water cycle, of approaches to water economics through adopting an ecosystem services approach.

Millennium Ecosystem Assessment. 2005. *Ecosystems and human well-being: wetlands and water synthesis*. World Resources Institute; Washington, DC.

This synthesis report of the Millennium Ecosystem Assessment for the Ramsar Convention uses the ecosystem services methodology and provides key findings on the range of ecosystem services provided by global wetlands (including water provision, regulation and purification). The report includes current status and trends, direct and indirect drivers of change, future scenarios and impacts on human well-being.

Young, R.A. 2005. *Determining the economic value of water: concepts and methods*. Resources for the Future; Washington, DC.

A detailed book with plentiful examples of water economics approaches.

Classic papers

Arbués, F., García-Valiñas, M.Á. and Martínez-Espiñeira, R. 2003. Estimation of residential water demand: a state-of-the-art review. *The Journal of Socio-Economics* 32: 81–102.

This paper surveys the main issues in the literature on residential water demand. Several tariff types and their objectives are analysed. The main contributions to the literature on residential water demand estimation are reviewed, with particular attention to variables, specification model, dataset, and the most common econometric problems. The paper concludes with comments on future trends and a summary of the contents of the study.

Barbier, E.B., Acreman, M. and Knowler, D. 1997. *Economic valuation of wetlands: a guide for policy makers and planners*. Ramsar Convention Bureau; Switzerland.

This handbook was one of the seminal contributions to the economic valuation literature and wetlands analysis. Though this came before the Millennium Ecosystem Assessment and its methodologies on ecosystem services and human well-being, this handbook provides a clear discussion of the concept of total economic value, the outputs, values and benefits associated with wetlands, and corresponding valuation techniques.

Dalhuisen, J.M., de Groot, H.L.F. and Nijkamp, P. 2000. The economics of water: a survey of issues. *International Journal of Development Planning Literature* 15: 3–20.

This paper surveys the central issues of the economics of water, including demand and supply characteristics, institutional capacities, pricing and rate structures of water, and provides a case for government intervention due to market failures.

Rogers, P., De Silva, R. and Bhatia, R. 2002. Water is an economic good: how to use prices to promote equity, efficiency, and sustainability. *Water Policy* 4: 1–17.

This paper focuses on the role of prices in the water sector and how they can be used to promote equity, efficiency and sustainability.

PROJECT IDEAS

■ Design a project to work out the economic value of water-quality improvements in a local lake. What factors would you need to consider and what data are required?

■ Investigate whether the price you or your organisation (e.g. university) currently pay for your water is fair.

■ Test the different techniques for valuing water in your local area and see whether they provide very different results.

REFERENCES

Barbier, E.B., Acreman, M. and Knowler, D. 1997. *Economic valuation of wetlands: a guide for policy makers and planners.* Ramsar Convention Bureau; Switzerland.

Bateman, I.J., Carson, I.T., Day, B., Hanemann, W.H., Hanley, N., Hett, T., Lee, M.J., Loomes, G., Mourato, S., Özdemiroglu, E. and Pearce, D.W. 2002a. *Economic valuation with stated preference techniques: a manual.* Edward Elgar Publishing; Northampton, MA.

Bateman, I.J., Lovett, A.A. and Brainard, J.S. 2002b. Developing a GIS-based methodology for benefit transfers. *In*: Florax, R.J., Nijkamp, P. and Willis, K.G. (eds) *Comparative environmental economic assessment.* Edward Elgar; Cheltenham.

Bates, B.C., Kundzewicz, Z.W., Wu, S. and Palutikof, J.P. (eds). 2008. *Climate change and water.* Technical Paper of the Intergovernmental Panel on Climate Change, IPCC Secretariat; Geneva.

Beharry-Borg, N., Smart, J.C.R., Termansen, M. and Hubacek, K. 2012. Evaluating farmers' likely participation in a payment program for water quality protection in the UK uplands. *Regional Environmental Change.* doi:10.1007/s10113-012-0282-9.

Bishop, R., Brown, C., Welsh, M. and Boyle, K. 1989. Grand Canyon and Glen Canyon Dam operations: an economic evaluation. *In*: Boyle, K. and Heekin, T. (eds) W-133, *Benefits and costs in natural resources planning, Interim Report #2.* Department of Agricultural and Resource Economics, University of Maine; Orono, ME.

Boland, J.J. and Whittington, D. 1998. *The political economy of increasing block tariffs for water in developing countries.* EEPSEA Special and Technical Paper sp199801t3, Economy and Environment Program for Southeast Asia (EEPSEA).

Budds, J. and McGranahan, G. 2003. Are the debates on water privatization missing the point? Experiences from Africa, Asia and Latin America. *Environment and Urbanization* 15: 87–114.

Christie, M., Fazey, I., Cooper, R., Hyde, A., Deri, A., Hughes, E., Bush, G., Brander, L.

Nahman, A., de Lange, W. and Reyers, B. 2008. *An evaluation of economic and non-economic techniques for assessing the importance of biodiversity to people in developing countries* [online, accessed April 2019]. DEFRA; London. Available at http://science search.defra.gov.uk/Document.aspx?Document=WC0709_7562_FRP.pdf.

Choe, K., Whittington, D. and Lauria, D.T. 1996. The economic benefits of surface water quality improvements in developing countries: a case study of Davao, Philippines. *Land Economics*: 519–537.

Dalhuisen, J.M., de Groot, H.L.F. and Nijkamp, P. 2000. The economics of water: a survey of issues. *International Journal of Development Planning Literature* 15: 3–20.

Dinar, A., Pochat, V. and Albiac, J. 2015. *Water pricing experiences and innovations.* Global Issues in Water Policy 9. Springer; Basel.

Freeman, A.M. 2003. *The measurement of environmental and resource values: theory and methods* (2nd edition). Resources for the Future; Washington, DC.

Hanemann, W.H. 2006. The economic conception of water. *In*: Rogers, P.P., Llamas, M.R. and Martinez-Cortina, L. (eds) *Water crisis: myth or reality?* Taylor & Francis; London.

Jiménez Cisneros, B.E., Oki, T., Arnell, N.W., Benito, G., Cogley, J.G., Döll, P., Jiang, T. and Mwakalila, S.S. 2014. Freshwater resources. *In*: Field, C.B. *et al.* (eds) *Climate change 2014: impacts, adaptation, and vulnerability. Part A: global and sectoral aspects. Contribution of Working Group II to the Fifth Assessment Report of the Intergovernmental Panel on Climate Change.* Cambridge University Press; Cambridge and New York, pp. 229–269.

Jaffee, D. and Newman, S. 2013. A bottle half empty: bottled water, commodification, and contestation. *Organization & Environment* 26: 318–335.

Johnston, R.J., Boyle, K.J., Adamowicz, W., Bennett, J., Brouwer, R., Cameron, T.A., Hanemann, W.M., Hanley, N., Ryan, M., Scarpa, R., Tourangeau, R. and Vossler, C.A. 2017. Contemporary guidance for stated preference studies. *Journal of the Association of Environmental and Resource Economists* 4: 319–405.

Karnauskas, K.B., Donnelly, J.P. and Anchukaitis, K.J. 2016. Future freshwater stress for island populations. *Nature Climate Change* 6: 720–725.

Kling, C.L., Phaneuf, D.J. and Zhao, J. 2012. From Exxon to BP: has some number become better than no number? *Journal of Economic Perspectives* 26: 3–26.

Lieberherr, E. 2011. Regionalization and water governance: a case study of a Swiss wastewater utility. *Procedia Social and Behavioral Sciences* 14: 73–89.

Loomis, J.B. 1997. Use of non-market valuation studies in water resource management assessments. *Water Resources Update* 109: 5–9.

Martin-Ortega, J., Ferrier, R.C., Gordon, I.J. and Khan, S. 2015. *Water ecosystem services: a global perspective.* Cambridge University Press; Cambridge.

McElroy, J.L., Potter, B. and Towle, E. 1990. Challenges for sustainable development in small Caribbean islands. *In*: Beller, W., d'Ayala, P. and Hein, P. (eds) *Sustainable development and environmental management of small islands.* Parthenon/UNESCO; Paris, 299–316.

Millennium Ecosystem Assessment. 2005. *Ecosystems and human well-being: wetlands and water synthesis.* World Resources Institute; Washington, DC.

Nan Do, T. and Bennett, J. 2010. Using choice experiments to estimate wetland values in Viet Nam: implementation and practical issues. *In*: Bennett, J. and Birol, E. (eds) *Choice experiments in developing countries, implementation, challenges and policy implementation.* Edward Elgar; Cheltenham.

Nurse, L.A., McLean, R.F., Agard, J., Briguglio, L.P., Duvat-Magnan, V., Pelesikoti, N., Tompkins, E. and Webb A. 2014. Small islands. *In*: Barros, V.R., Field, C.B., Dokken, D.J., Mastrandrea, M.D., Mach, K.J., Bilir, T.E., Chatterjee, M., Ebi, K.L., Estrada, Y.O., Genova, R.C., Girma, B., Kissel, E.S., Levy, A.N., MacCracken, S., Mastrandrea, P.R. and White, L.L. (eds) *Climate change 2014: impacts, adaptation and vulnerability. Part B: regional aspects, Chapter 29 (Small islands).* Contribution of Working Group II to the Fifth Assessment Report of the Intergovernmental Panel on Climate Change, pp. 1613–1654.

Palmquist, R.B. 1991. Hedonic methods. *In*: Braden, J.B. and Kolstad, C.D. (eds) *Measuring the demand for environmental quality.* North Holland; Amsterdam.

Robinson, S. 2017. Climate change adaptation trends in small island developing states. *Mitigation and Adaptation Strategies for Global Change* 22: 669–691.

Sharma, B., Rasul, G. and Chettri, N. 2015. The economic value of wetland ecosystem services: evidence from the Koshi Tappu Wildlife Reserve, Nepal. *Ecosystem Services* 12: 84–93.

Smith, A. 1776. *The wealth of nations: an inquiry into the nature and causes of the wealth of nations.* W. Strahan and T. Cadell; London.

TEEB. 2010. *The economics of ecosystems and biodiversity: mainstreaming the economics of nature: a synthesis of the approach, conclusions and recommendations of TEEB.* United Nations Environment Programme.

United Nations. 1987. *Our common future.* Oxford University Press; Oxford.

United Nations. 2015. Millennium Development Goals. Official website of United Nations [online, accessed April 2019]. Available at www.un.org/millenniumgoals.

United Nations. 2018. 17 goals to transform our world. Official website of United Nations [online, accessed April 2019]. Available at www.un.org/sustainabledevelopment.

Wada, Y. and Bierkens, M.F.P. 2014. Sustainability of global water use: past reconstruction and future projections. *Environmental Research Letters* 9: 104003.

Whittington, D. 2002. Improving the performance of contingent valuation studies in developing countries. *Environmental and Resource Economics* 22: 323–367.

Xu, J., Morris, P.J., Liu, J. and Holden, J. 2018. Hotspots of peatland-derived potable water use identified by global analysis. *Nature Sustainability* 1: 246–253.

CHAPTER ELEVEN

Water conflict, law and governance

11

Anamika Barua and Kitriphar Tongper

LEARNING OUTCOMES

After reading this chapter you should be able to:

- distinguish between various types of water conflicts and their causes;
- understand what water rights are and how they are related to water access;
- understand how interdependencies of countries on shared waters can also create grounds for cooperation;
- outline major international laws and institutions dealing with transboundary water conflict and to what extent they have been successful in resolving conflicts.

11.1 INTRODUCTION

Like many other resources that sustain and support life, water fit for human consumption and use has become a scarce resource even though it is supposedly abundant in supply. As long as water was abundantly available for use among people, there were no problems related to sharing or **property rights**. While the availability of freshwater in the world has roughly remained the same, the demand for water from agricultural, domestic and industrial users has increased. Water resources have, therefore, become increasingly under stress and the issue of water rights and sharing water has become a very important one. Since rivers, lakes and groundwater often flow across national, political boundaries, the conflicts related to water sharing are often **transboundary**, involving two or more nations. More than 90% of the world's population lives in states that share watercourses and almost every country with land borders shares some water resources with its neighbours. The River Danube flows through 18 different nations, while the Congo, Niger, Nile, Rhine and Zambezi are each shared between nine to 11 countries. The resultant interdependencies (political, economic, social and environmental) on shared waters make transboundary water management inherently a political process and an issue of state sovereignty. While various laws, schemes and plans have been designed to ensure that everyone gets their fair share of water, these plans, however, are not always perfect and they often do not address a number of issues. The lack of proper sharing mechanisms between, as well as within, countries has often been the cause of conflict situations related to water sharing and access.

However, the interdependencies of countries on shared waters can also create grounds for cooperation. There is a huge and growing body of literature that highlights, in transboundary waters, that conflict and cooperation can coexist. Research indicates that, despite the potential for dispute in international basins, there are many more instances of cooperation over shared water resources too. The evidence from different transboundary river basins across the world, such as the Nile or Jordan, also shows that water, by its very nature, tends to induce even hostile neighbours to cooperate, even as disputes rage over other issues.

This chapter deals with the types and causes of water conflicts, both within and between states, and distinguishes between various types of water rights and how these rights determine access to water. The chapter also discusses how water can act as a catalyst for cooperation and enables countries to cooperate. Through various case studies, the chapter discusses the role of international treaties, laws and institutions in resolving water conflicts both within and between nations.

11.2 WATER CONFLICTS

Water conflict is a term describing a conflict between countries, states or groups over access to water resources. As pressure on freshwater supplies rises, due to population growth, economic development and pollution, access to water and its allocation and use are becoming increasingly critical concerns that can drive conflict and which may have profound consequences on societal stability (Haftendorn, 2000). While water has not been a major cause of violent conflicts historically, water-related tensions can emerge between and within states (Zetoun and Mirumachi, 2008). These tensions occur on three interdependent levels:

- the local level: such as between societal groups over access to a water point, or between the state and people affected by the construction of a dam;
- the national level: such as between different interest groups (farmers, industry, tourism, environmentalists) in relation to national policies affecting water management, for example over the reallocation of water between economic sectors;
- the international level: such as between upstream and downstream states over the use of shared rivers.

Water-related tensions can occur when water is scarce, but, even when the resource is not severely limited, its allocation and use can still be hotly contested. The coexistence of a variety of uses and users – such as agriculture, industry, different clans or ethnic groups, and rural and urban users – increases the likelihood of conflicting interests over water. These can drive or exacerbate existing threats of conflict on intra- and inter-state levels. Many regions around the world deal with shortages of water. However, some areas deal more with conflicts over inadequate water supplies and disputes over shared water supplies. In regions where countries compete for access to water, the relations between countries are likely to be unstable. In the absence of strong institutions and agreements, changes within a basin can lead to transboundary tensions. Strife over water is plaguing states, with disputes in both the developed and developing world (Kornfeld, 2019). Notably there is considerable dispute within the Middle East, South Asia and South East Asia, regions where the amount of water available per person within several countries is scarce (<1000 m^3 per person per year) or where there is **absolute water scarcity** (<500 m^3 per person per year) (see Section 1.5.1 in Chapter 1 and Section 9.4 in Chapter 9). Although conflicts arise mainly because of the use of common water resources, a distinction can be made between conflicts arising through **consumptive water use**, pollution and distribution (Table 11.1).

11.2.1 Conflict through use

A conflict through consumptive use could be found in a situation where one state using a river clashes with another state, citing environmental concern or loss of water due to the other state's activities. **Box 11.1** provides an example. Most frequently, activities such as the construction of a dam or the channelling of the river flow lead to international conflict.

11.2.2 Conflict through pollution

The possibility of conflict through pollution arises when industry, urbanisation or mechanised agriculture in the upstream area leads to harmful consequences for water quality for the lower-lying states. An example of the conflicts on the Rhine from the 1970s to 1990s is provided in **Box 11.2**.

11.2.3 Relative distribution conflict

A relative conflict of distribution presents itself where a disparity over the use of water exists between upper- and lower-lying states. There are many examples of this type of conflict around the world, often on large rivers. These often occur due to large-scale dam construction or water diversion projects. One of the world's largest rivers is the Ganges and this is used as an example of a relative distribution conflict in **Box 11.3**.

11.2.4 Absolute distribution conflict

An absolute conflict of distribution exists when there simply is not enough water to meet all the legitimate needs of the contesting states sharing water resources. **Box 11.4** provides a case study.

Table 11.1 Causes of conflicts

Conflict type	Conflict through consumptive use	Conflict through pollution	Relative distribution conflict	Absolute distribution conflict
Conflict cause	Water use	Water quality	Water distribution	Water distribution and availability
Examples	Cauvery water dispute, India (see **Box 11.1**)	River Rhine	Ganga basin	Jordan River basin

Source: After Haftendorn, 2000.

BOX 11.1

Conflict through consumptive use: Cauvery water dispute, India

Cauvery is the name of the 802-kilometre river which originates from Talacauvery in Kodagu district in the southern Indian state of Karnataka and flows mainly through Karnataka and another state, Tamil Nadu. Its basin also covers areas in the other southern state of Kerala, and some areas of the Union Territory of Pondicherry. Both Karnataka and Tamil Nadu are heavily dependent on the river to meet all their water needs, especially agricultural needs. The use of the river waters was regulated by agreements dating back to 1892 and 1924, when India was a colony of the British Empire and the two states of Karnataka and Tamil Nadu were known as the Princely State of Mysore and the Madras Presidency, respectively. The River Cauvery is dependent on rains, and during periods of heavy monsoon rains there is excessive flow of water, even leading to floods. But when the monsoon fails there is severe shortage of water, leading to a drought situation: the water needs of farmers are not met and tension and conflict arise in the two states.

According to the 1892 and 1924 agreements the approximate river water allotments were 75% to Tamil Nadu and Pondicherry, 23% to Karnataka, and the rest to Kerala. The state of Karnataka has always indicated that it feels it did not get its due share of water in comparison with Tamil Nadu, and it demanded a renegotiation of the settlement. Tamil Nadu, on the other hand, has pleaded that it has already developed 12,000 km^2 of land and depends on the existing pattern of water usage.

On 2 June 1990 a tribunal known as the Cauvery Water Disputes Tribunal (CWDT) was constituted by the government of India to adjudicate the water dispute regarding the inter-state River Cauvery and its river valley. The tribunal gave an interim award on 25 June 1991 of 5.8 billion m^3, which Karnataka had to ensure reached Tamil Nadu in a water year. The award also stipulated the weekly and monthly flows to be ensured by Karnataka for each month of the water year. The tribunal further directed Karnataka not to increase its irrigated land area. Violence followed the interim order and Karnataka witnessed the worst ever anti-Tamil riots. In 1992, 1993 and 1994 the rain was sufficient to pacify the dispute between the two states. But in 1995 the monsoons failed badly in Karnataka and the state found itself unable to fulfil the interim order (Rahman, 2006). The resulting problem was solved only when the Supreme Court recommended the intervention of the prime minister to provide a political solution.

The CWDT submitted its reports and decision under Section 5(2) of the Inter-State River Water Disputes Act, 1956, on 5 February 2007, allocating 11.9 billion m^3 of water annually to Tamil Nadu, 7.6 billion m^3 to Karnataka, 0.8 billion m^3 to Kerala and 0.2 billion m^3 to the Union Territory of Pondicherry (MoWR, 2007). The tribunal also reserved 0.3 billion m^3 for environmental protection and 0.1 billion m^3 for inevitable escape into the sea. However, all four parties to the dispute were not satisfied with the verdict and they have decided to file review petitions seeking a possible renegotiation.

BOX 11.2

Conflict through pollution: the twentieth-century Rhine dispute

Although the Rhine is a relatively small river, it is one of Europe's best-known and most important rivers. Its length is 1320 km, of which 880 km is navigable. From its source in Switzerland, the Rhine flows via France, Germany and the Netherlands into the North Sea. The catchment area of 170,000 km^2 also covers parts of Italy, Austria, Liechtenstein, Luxembourg and Belgium. Hence, the Rhine is a river whose drainage area falls in a rich, highly industrialised area of Europe. Furthermore, the interests of the upper- and lower-lying areas around the river are particularly distinct, as a result of their different uses of the Rhine. The primary cause of pollution of the Rhine was from the chemical industry in the upper-lying states of Switzerland and Germany (Figure 11.1), as well as from French potassium mines in Alsace and the German coal works in the Ruhr and Lippe (Frijters and Leentvaar, 2002). Many of the costs of this pollution were borne by the lower-lying states such as the Netherlands, whose primary use of the Rhine was for drinking water and agriculture. Rising toxic sediment in the Rhine forced costs upon the city of Rotterdam (in the Netherlands), which had to remove the toxic mud from Rotterdam harbour to special waste depots.

Figure 11.1 Heavy industry next to the River Rhine near Leverkusen, Germany.
Source: Shutterstock/360degreeAerial.

The conflicting interests of the upper- and lower-lying parties along the Rhine prevented a quick solution. In the late 1960s and early 1970s, the Netherlands, being one of the most negatively affected lower-lying states, called for continued measures to combat the increasing pollution. In 1976 the environment ministers of the Rhine states signed two agreements for the protection of the Rhine against chemical pollution and the reduction of salt loads.

Under the Chemical Agreement a financing proposal was put forward with regard to the clean-up costs, whereby France and Germany assumed 30% each, the Netherlands 34% and Switzerland 6% of the costs. In exchange, the Netherlands agreed not to sue Germany and Switzerland. However, in response to local resistance to the agreed measures, France declined to ratify the Chemical Agreement in 1979. It was only after the 1983 elections and a change in the French government that the process of ratification was once again taken up. An additional catalyst, a series of chemical accidents on the Rhine in 1986, increased public awareness (Haftendorn, 2000), and in 1987 the Chemical Agreement finally came into force. In 1991 it was replaced by a second agreement calling for an even greater reduction in pollution levels.

11.2.5 Future conflicts

As water becomes increasingly scarce, conflicts are expected to arise. First, there may be local conflicts arising from competition among users over access to a water point. Second, there will be sectoral conflicts where industry, agriculture and citizens are in competition for larger volumes of the resource on a much wider level. Third, there may be more international conflicts between countries over the use of shared rivers and water resources. The distribution of water resources throughout the world is increasingly becoming a political issue, and a lack of understanding among countries over what constitutes an equitable and fair share of available water resources and supply is a major problem for the future.

REFLECTIVE QUESTION

Are there any water conflict examples that are local to you – if so, what types of conflict are they?

11.3 RIGHTS TO ACCESS WATER

Having seen that conflicts arise over water, it is only natural to assume that such conflicts sometimes occur over claims and counter-claims over the right to use available water. This brings into focus the issue of rights over water in different situations and at different times. The term 'water right' is not easy to define, owing to the fact that there is no universally agreed definition. The term is used to mean different things in different contexts and different jurisdictions. In its simplest conception, a water right is frequently understood to be a legal right to abstract and use a quantity of water from a natural source such as a river, stream or aquifer.

11.3.1 History of water rights

A whole body of legal arrangements has grown up over many generations that is specifically related to the use of and right to use water. The history of water rights can be traced back to Roman law, one of the elements of which was to confer a privileged position on the owners of land adjacent to watercourses (Bannon, 2017).

BOX 11.3

Relative distribution conflict: the Ganges basin

The Ganges basin conflict is primarily a conflict of water sharing between India and Bangladesh. It can also be characterised as a conflict of interests between upstream and downstream river use. India, which manages the upper part of the drainage basin, developed plans for water diversions for its own irrigation, navigability, and water-supply interests. Bangladesh, however, had interests in protecting the historic flow of the river for its own downstream uses. The potential clash between upstream development and downstream historic use set the stage for very long-drawn-out attempts at conflict management (Priscoli and Wolf, 2009).

The attempts at conflict management started from the time when Bangladesh was a part of Pakistan and was known as East Pakistan. In 1960 an expert-level meeting was held between India and Pakistan, followed by three more meetings in 1962. While the meetings were going on, India informed Pakistan that it had started the construction of the Farakka Barrage, designed to divert 1100 m^3 s^{-1} from the Ganges basin into the Bhagirathi–Hooghly River during the dry season to flush out the accumulating silt, which would improve navigability for Calcutta city port and keep the city's water supply low in salts (Figure 11.2). The meetings continued with not much headway being made and the Farakka Barrage was completed in 1970 (but it was not yet in operation). Bangladesh came into being in 1971, after gaining independence from Pakistan. The governments of India and Bangladesh had agreed to establish the Indo-Bangladesh Joint Rivers Commission, but the question of the Ganges was excluded and was supposed to be handled only at the level of the prime ministers of the two countries.

The prime ministers of India and Bangladesh met in New Delhi in 1974 and made a declaration that, during the period of minimum flow in the Ganges, there might not be enough water for both an Indian diversion and Bangladeshi needs. They also expressed their determination to work for a mutually acceptable allocation of the water available during the periods of minimum flow before the opening of the Farakka Barrage. The problems, however, persisted and there was a series of five commission meetings between the two countries from June 1974 to January 1975, with one ministerial-level meeting in April 1975. In 1975 the two sides agreed to a limited trial operation of the barrage. However, without renewing or negotiating a new agreement with Bangladesh, India continued to divert the Ganges waters at Farakka after the trial run. This had serious consequences for Bangladesh and there were setbacks to agriculture, fisheries, navigation and industry. There were more meetings, but the conflict was unresolved and in January 1976 Bangladesh lodged a formal protest against India with the General Assembly of the United Nations. The United Nations encouraged the parties to meet urgently for negotiations and on 18 April 1977 an understanding was reached on fundamental issues, followed by the signing of the Ganges Water Agreement on 5 November 1977. This agreement broadly covered the sharing of the waters of the Ganges at Farakka and the finding of a long-term solution for augmentation of the dry season flows of the Ganges (Priscoli and Wolf, 2009).

The agreement was to cover a period of five years and could be extended by mutual agreement, but no solution had been reached by the end of the five-year period. Following this, both sides agreed not to extend the 1977 agreement but to initiate fresh attempts to arrive at a solution within 18 months. This was followed by an Indo-Bangladesh Memorandum of Understanding, signed on 22 November 1985, on the sharing of the Ganges dry season flow through 1988 and establishing a Joint Committee of Experts to help resolve development issues. On

12 December 1996 a new treaty was signed by the prime ministers of the two states and this agreement was based generally on the 1985 accord, which delineates a flow regime under varying conditions. The agreement also established a 30-year water-sharing arrangement and recognised the rights of Bangladesh as a downstream river user.

Figure 11.2 The diversion of flow caused by the Farakka Barrage, LANDSAT image, February 2019. The image is about 100 km north–south and 80 km east–west. The Ganges flows south to the barrage, where some of the flow is diverted into a long canal, which heads south to the Bhagirathi–Hooghly River. Immediately south of the barrage the image shows how the Ganges has a lower discharge with more exposed sediment on the floodplain.

Source: U.S. Geological Survey.

BOX 11.4

Absolute distribution conflict: the Jordan basin

The River Jordan supplies Israel and Jordan with the vast majority of their water. As the population of Israel grew, the demand for water and reliance on the Jordan River grew. In the early 1950s Israel created a system called the National Water Carrier to transport water from the River Jordan, which was not well received by the Arab Nations. In 1955, Syrian artillery units opened fire on the Israeli construction team. In an attempt to settle the water dispute, US president Eisenhower appointed Eric Johnston as mediator. Negotiations between Arab states and Israel on regional water-sharing agreements continued for more than two years, with no success beyond a ceasefire.

In response to the completion of the Israeli National Water Carrier, in 1964 the Arab League attempted to divert two tributaries of the Jordan – the Banias and the Hasbani – to the north. Israel in turn prevented this by military action. Meanwhile, with international financial aid, Jordan built a side canal on the eastern Jordan shores, the Abdullah Canal, for irrigation of cultivated land in north-western Jordan and to satisfy the water needs of Amman. In order to guarantee a continuous water flow, two dams were proposed for the Yarmuk River, the most important tributary of the Jordan, which forms the border with Syria. One of these was completed in 1967 and subsequently destroyed by Israeli forces in the Six-Day War. The Six-Day War of 1967 changed the entire water scenario in the region (Haftendorn, 2000). Israel now controlled the Jordan waters, along with unlimited access to the underwater reservoirs of west Jordan.

With the 1994 Israeli–Jordanian Peace Agreement, the division of the transboundary waters was contractually defined for the first time. According to the agreement, Jordan received the right to use the largest part of the Yarmuk, up to 45 million $m^3 yr^{-1}$, and Israel the largest part of the River Jordan discharge, up to 40 million $m^3 yr^{-1}$. To this end, the two parties declared their intention to jointly construct a dam on the Yarmuk River on the Israeli–Jordanian–Syrian border, in spite of Syrian objections. In addition, a joint water commission was established. However, to date there has been no agreement with Syria.

This, in turn, had a major influence on conceptions of water rights in the influential European legal tradition, prior to the introduction of modern water rights regimes. Indeed, some of these influences can still be observed. For example, Roman law denied the possibility of private ownership of running water. The Institutes of Justinian, published in AD 533–534, held that running water was a part of the 'negative community' of things that could not be owned, along with air, the seas and wildlife.

At the same time, it was recognised that things in the negative community could be used and that the 'usufruct' or right to use the advantage of the resource needed to be regulated so as to provide order and prevent over-exploitation.

Roman law also distinguished the more important, perennial streams and rivers from the less important seasonal water bodies. The former were considered to be common or public, while the latter were private. The right to use a public stream or river was open to all those who had access to them. Roman law, however, recognised the right of the government to prohibit the use of any public water and required an authorisation for taking water from navigable rivers.

11.3.2 Water rights in modern society

Societies in the past have devised their own ways of dealing with the issue of allocating water resources for use among their members, and these formed a system of water rights applicable within particular societies. Remnants of these systems can still be found in customary practices, which still wield some influence in the allocation of water in developing countries. Today, different countries have devised different arrangements related to water rights, and over the years these systems have become increasingly complex. Issues such as population growth creating water scarcity, pollution and so on have contributed to the complexity of legal systems and arrangements in different countries, and modern governments are increasingly challenged to make the allocation and apportioning of water usage fair and equitable. It has also been observed that European conceptions of water and water law have strongly influenced the development of formal water laws around the world, through the two principal European legal traditions: the civil law and the common law (Hodgson, 2006).

The civil law tradition, which is sometimes referred to as the Romano-Germanic family, is found in most European countries (including the former socialist countries of Eastern and Central Europe), nearly all countries in Latin America, large parts of Africa, Indonesia and Japan, as well as the countries of the former Soviet Union (Hodgson, 2006). In such countries, codified statutes and ordinances are given priority. The common-law tradition emerged

from the law of England. Countries in which the common-law tradition applies include Australia, Canada, India, New Zealand, Pakistan, Singapore and the United States of America, and the remaining African countries that are not in the civil law tradition, as well as other Commonwealth countries and a number of countries in the Middle East. In such countries, published judicial opinions and past rulings are important to determining a case.

11.3.3 Types of water rights

There are fundamental differences between the nature and source of water rights in different countries. In some cases water rights are linked to land ownership; this means that those who buy land are assured of having rights over the water resources available there. But there are systems where the rights to water and other resources are separate from the right of land ownership. Generally, there are two kinds of rights related to water: *riparian water rights* and *prior appropriation water rights*. Riparian rights refer to the system of allocating water among those who own the land at or near the source of the water. The origins of the system of riparian rights can be traced back to the English common law. The prior appropriation water rights, on the other hand, refer to the system of allocating water rights from a source of water on a first-use basis. Thus, prior appropriation rights are not linked to ownership of the land over which the water sources are present. Instead, the first person to use the water from a water source for a beneficial use will have the right to continue to use that quantity of water for that purpose. Other users may use the remaining water for their own beneficial purpose, provided that they do not impinge on the rights of previous users. Both types of water rights exist in the United States: riparian rights in the eastern states and prior appropriation rights in the western states.

Riparian rights are property rights that have come to be regarded as inherent in a riparian parcel of land. A simple definition of a riparian parcel of land is the land which borders a natural body of water. Ownership of a riparian parcel of land means that a person owns the land, the beach, the bottom land covered by water, the right to fish, hunt, swim and boat on the surface of the entire body of water, in common with all other riparian property owners. Riparian rights also include the right to make reasonable use of the water. However, it is also implied that, if the available water is not enough to meet the needs of all users, allotments are made in proportion to frontage on the water source (this is known as *correlative rights*). Riparian rights cannot be sold or transferred other than with the adjoining land.

The doctrine of prior appropriation was developed in the nineteenth century to serve the practical demands of water users in the western United States. It originated in the customs of miners on federal public lands, who accorded the best rights to those who first used water, just as they had accorded mining rights to those who first located ore deposits. The doctrine of prior appropriation was later extended to include farmers and other users, even on private lands. The flexibility of the common-law tradition is such that this new, more suitable water rights doctrine was accepted as the law in a number of states in the United States (Hodgson, 2006). All of the states in which the doctrine of prior appropriation applies have statutory administrative procedures to provide an orderly method for appropriating water and regulating established water rights (Getches, 1997). The doctrine of prior appropriation clearly indicated that water rights and land rights were different. Water rights under this doctrine are acquired on the basis of beneficial use rather than landownership, and those rights continue as long as the beneficial use is maintained. The right of individuals to use water under the prior appropriation system is based on the application of a quantity of water to a beneficial use.

Groundwater is managed in a similar way to surface water both by appropriation rights and by the principle of riparian rights (typically using correlative rights), in different places (e.g. different states of the USA), to restrict abstraction. However, good groundwater data are required on flow directions, recharge rates and so on (see Chapter 5) in order to support the legal framework and resolve or avoid disputes. Under the doctrine of appropriation, expansion of groundwater pumping operations would be illegal, as it could adversely affect either others who use the groundwater or, as groundwater often supplies river baseflow, other surface water users.

In some places water rights are not clearly defined. India, for example, does not have any specific law defining ownership and rights over water sources. The rights are derived from several pieces of legislation and customary practices and beliefs. In India water is a regional state-based subject rather than a national one and states have the exclusive power to regulate water supplies, irrigation and canals, drainage and embankments, water storage, hydropower and fisheries. There are nevertheless restrictions with regard to the use of inter-state rivers. The central government is entitled to legislate on certain issues. These include shipping and navigation on national waterways, as well as powers to regulate the use of tidal and territorial waters. The Indian constitution also provides that the central government can legislate with regard to adjudication of inter-state water disputes.

11.3.4 Water allocation

Following the above discussion of water rights, it will be understood that conditions leading to the increasing scarcity of water bring into focus another important issue, that of water allocation, or the manner in which water is distributed to those who have a right to it. Water allocation has been defined by Burchi and Andrea (2003) as 'the function of assigning water from a given source to a given user or number of users for abstracting it and applying it to a given use'. They also pointed out that, within a system where the state is responsible for a country's water resources, the decision of who should abstract water and for what use rests with a public authority.

The way in which water is allocated is no longer as simple as it was in early human societies. The term 'water use' today does not only refer to abstraction of water for basic human needs but extends to irrigation for agricultural purposes, industrial uses, hydroelectricity and so on. Modern civilisation has seen levels of development that have never been experienced before. Global population is around 8 billion. To compound the issue of scarcity there is the additional problem of pollution, which has also become a global menace (see Chapters 4 and 5). Pollution has (to a very large extent) had an impact on the availability of clean sources of water suitable for human use.

As the resource comes under greater pressure, better institutions are needed to allocate and reallocate water between uses and users. Both government agencies and user groups dealing with water have tended to be sectorally defined, focusing exclusively on irrigation, or domestic water supply, or environment. Even within the irrigation sector, allocation between irrigation systems has often been unclear. As long as there is surplus water in the river basin, this is not problematic, as different users do not interfere with each other. But, as overall water use increases, there is interference, not only between surface abstractions along a river but also between surface and groundwater use (Bruns and Meinzen-Dick, 2000). The allocation and use of water resources today falls under the control of governments. Governments in most countries today have a specific department monitoring water resources.

However, the manner in which water resources are allocated by governments differs from one country to another and there is no uniform method applicable to all countries. This creates challenges for social justice. According to Meinzen-Dick and Mendoza (1996) and Meinzen-Dick and Rosegrant (1997), institutional arrangements for water allocation, and particularly for reallocation, can be grouped into three broad categories. First, in *user-based allocation*, water users join together to coordinate their actions, managing water resources as a form of common property. Second, in *agency allocation*, water is treated as public property, with government agencies assuming authority for directing who does and does not receive water in accordance with policies and procedures. Third, in *market allocation*, which corresponds with private property held by individuals or organisations, water may be allocated and reallocated through private transactions, with users trading water through short- or long-term agreements and reallocating rights in response to prices. These three forms of water allocation institutions may be combined in various ways at different locations and levels of water management.

11.3.5 Water and social justice

Water interacts with people in several ways – as a vital resource for life, agriculture, sanitation, health and overall well-being. In spite of the fact that water plays a significant role in people's lives, figures suggest that at least 2.1 billion people across the world are living without access to safe, clean drinking water in the home, and over 4.5 billion are without access to safely managed sanitation (WHO and UNICEF, 2017). Over 2 million people die each year from water-related diseases, and there is a much bigger annual loss of meaningful productive healthy years – perhaps 80 million work years are lost each year (McDonald *et al.*, 2011). Hence, management of water resources has critical implications for people's lives and livelihoods, for overall economic development and for social prosperity (Rasul and Choudhury, 2010). Social theories of justice, equity and fairness underscore the need to ensure social justice in water resource management (Syme *et al.*, 1999; Tisdell, 2003). Social justice and water refers to fairness of access to water resources and the equality of burden from poor water quality and water hazards.

Competing water needs have triggered conflicts between disparate water users, such as commercial interests and subsistence livelihoods, the rich and the poor. The rural poor, who depend heavily on public water bodies such as rivers, streams, khals (natural channels of water) and beels (permanent backwater lakes in the floodplain) for sustenance and well-being, have often been marginalised and even deprived of access to public water bodies in the process of water resource development, such as the construction of dams, irrigation canals and flood control structures (Phansalkar, 2007). Competing needs have also been a bone of contention between different sectors and different regions, such as domestic and agriculture, agriculture and industry, agriculture and fisheries, upstream and downstream, rural and urban areas, and fisheries and flood control.

The movement towards the privatisation of water, which is gaining momentum worldwide, may make water access much more difficult and unaffordable for poor people in developing countries if it is not governed properly (see also Section 9.5.2.3 in Chapter 9 on pricing structures). On the other hand, it may work very well if a fair pricing structure is in place. Proponents say that such a privatised pricing system is the only way to distribute water to the world's thirsty. However, experience in some locations shows that selling water on the open market does not address the needs of poor thirsty people (e.g. Lima, Peru). On the contrary, privatised water is sometimes delivered only to those who can pay for it, such as wealthy cities and individuals and water-intensive industries, like agriculture and high-tech. Therefore, privatisation has to be associated with social justice so that we can avoid an outcome which suggests that those who can afford to pay for water are assured of a supply of clean and safe water, while those who cannot afford to do so have to manage with whatever is available. Such a situation would obviously go against the principle of social justice, which stands for the equitable treatment of all citizens.

On 28 July 2010 the United Nations General Assembly voted overwhelmingly to adopt a resolution recognising the human right to drinking water and sanitation; 122 countries voted in favour of the resolution, none opposed and 41 abstained. The General Assembly also voted to call on member states to provide financial resources and technology to help realise this right in poorer countries. This was lauded as a historic landmark in the fight for water justice, and, while the resolution itself was not binding, it demonstrated the intent of the General Assembly. Access to water is considered to be a prerequisite for the realisation of other human rights. After the resolution of 2010, states became bound to protect, fulfil and respect the human rights to water in their territories.

Hence now governments have a legal obligation to accomplish the human right to water and also it gives rights to the population to claim their legal right to access to water.

European Union legislation known as the Water Framework Directive places emphasis on environmental protection and makes it illegal for land managers or water users to deteriorate the ecological status of water bodies. Efforts must be made to improve the status of water bodies towards 'good ecological status', and groundwater use must be sustainable. This means that a river basin approach is required to water management, and water use across the entire basin needs to be considered so that rivers, lakes and estuaries are protected for the sake of the environment. Thus, water allocation for the environmental good must now be considered, monitored and acted upon within EU Member States and there must be clearly delineated responsibilities and regulatory agencies within each state (see Section 4.5 in Chapter 4 and Box 6.4 in Chapter 6 for further details).

> ### REFLECTIVE QUESTIONS
>
> Why do we need rights to water? What would be the benefits of such a right? What is the relevance of linking human rights and water?

11.4 WATER GOVERNANCE

'Water governance refers to the range of political, social, economic and administrative systems that are in place to develop and manage water resources, and the delivery of water services at different levels' (Rogers and Hall, 2003). As stated by the United Nations (2006), water governance has four dimensions:

- a social dimension concerned with 'equitable use';
- an economic dimension concerned with 'efficient use';
- an environmental dimension concerned with 'sustainable use';
- a political dimension concerned with 'equal democratic opportunities'.

Each of these dimensions is 'anchored in governance systems across three levels: government, civil society and the private sector'. To realise 'effective governance', the UN report *Water: A Shared Responsibility* (2006) proposes a checklist that includes the following:

- participation
- transparency
- equity
- effectiveness and efficiency
- rule of law
- accountability
- coherency
- responsiveness
- integration
- ethical considerations.

The absence of some or all of these practices has resulted in 'bad' or 'poor' governance, a simple definition of which is the inability and/or unwillingness to alter patterns of resource allocation, use and management, despite clear evidence of resource degradation, uneconomic behaviour and abiding poverty and social inequality (UN, 2006).

An effective water governance framework that would be universally applicable is very difficult to conceptualise, since each and every country has its own unique set of issues and problems related to water allocation and distribution. However, the generally accepted principle is that there should be a healthy balance between policy, law and institutions, which are considered to be the main components of effective water governance. Whatever its policies towards water management, a country needs to develop each of these areas – policy, law, institutions, regulations, contracts and compliance – in order to have effective water governance. The establishment of water governance capacity may follow different patterns in different countries. Not every country pursues the same sequence in terms of adopting policies on water, enacting the laws to realise the policies and establishing the institutions to implement the law (Isa and Stein, 2009).

The use of the word 'policy' here refers to a government's plan to address the issue of water sharing and use within its boundaries. This plan is usually reflected in legislation and executive orders. Law is an essential component of effective water governance because when a legal system is created it formalises the processes through law. It is also very important that institutions should be set up to carry out the laws that are created. Institutions may exist at the local, national and even international levels. When a balance is struck between policy, law and institutions, the establishment of an effective water governance capacity is a very strong possibility.

The case of South Africa is relevant to understanding how a process of policy and legal reform contributes to the progressive development of a country's water governance capacity. In 1994, the Minister of Water Affairs and Forestry appointed a Policy and Strategy Team that reviewed all South African water laws and published a document entitled *You and Your Water Rights*. This document sought to assist the public in making meaningful contributions to policy development, set out the main principles and provisions of the existing legal structure and also

contextualised these against their origin and historical development (Isa and Stein, 2009). As a result of the call for a public response to the review, simple and concise statements were published for comment, which would constitute a framework for the development of a new, detailed policy and a new national law. The 1997 White Paper which followed confirmed that water is an indivisible resource and a national asset, and abolished the system of riparian rights through which water ownership was tied to landownership along rivers. It recognised water to meet basic human needs and maintain ecosystems as a right. It also recognised the authority of the country to prioritise water uses to meet the requirements of neighbouring countries and promoted an integrated system for managing water quality, quantity and supply. These principles became law in subsequent years and a National Water Act included the following: a national strategy, a classification of water and reserves, a mechanism for allocating and regulating water uses through a licensing system, a system of prices, catchment and management agencies, and water users' associations.

REFLECTIVE QUESTION

According to the UN checklist, what are the elements of good governance?

11.5 INTERNATIONAL INSTITUTIONS AND LAWS TO RESOLVE WATER CONFLICT

'International water law' (also known as international watercourse law or international law of water resources) is a term used to identify those legal rules that regulate the use of water resources shared by two or more countries. The primary role of international water law is to determine a state's entitlement to the benefits of the watercourse (substantive rules) and to establish certain requirements for states' behaviour while developing the resource (procedural rules) (Vinogradov *et al.*, 2003). What distinguishes international law from domestic law is that the former is both created and enforced by states (at the international level) primarily in order to regulate inter-state relations in various areas of human activity, while the latter involves matters within a state's borders (Cosgrove, 2003). International law offers a series of means to resolve international disputes, both diplomatic (negotiations, consultation, good offices, mediation, fact-finding, inquiry, conciliation and the use of joint bodies and institutions) and legal process (arbitration and adjudication). Generally, water conflicts are settled through negotiations, with an agreement as the final outcome (Cosgrove, 2003). Few water laws are drafted to deal with all future water conflicts, and there are some water regulations which are drafted specifically to deal with a particular conflict. Two examples of general water regulations are the 1992 Helsinki Convention (Box 11.5), which seeks to protect the Baltic Sea, and the 1997 United Nations Convention on the Law of the Non-Navigational Uses of International Watercourses (see Box 11.5). The aim of both of these conventions is to establish general principles for the use of transboundary water resources.

11.5.1 International treaties

Most transboundary water resources are subject to a treaty regime. Around 3600 international agreements govern the use of most of the world's shared waters. Some of the agreements are watercourse specific (e.g. the Columbia River Treaty, see Box 11.5),

some are boundary agreements (e.g. the 1909 Canada–United States Boundary Waters Treaty, see Box 11.5) and some are umbrella agreements regulating all regional waters. For example, the Convention on the Protection and Use of Transboundary and International Lakes, also known as the Water Convention, is an international environmental agreement to improve national attempts and measures for the protection and management of transboundary surface waters and groundwaters. Parties are obliged to cooperate, create joint bodies and operate with mutual assistance and exchange of information. All these treaties act as instruments for dispute resolution to maintain cordial relations between neighbouring states.

11.5.2 The role of international institutions in conflict management

Water is a resource that transcends national boundaries and scarcity leads to conflict situations. The different case studies described in earlier parts of this chapter above have highlighted that water disputes leading to conflict have become part of modern civilisation. The parties to the dispute using negotiations and agreements, without any third-party involvement, usually resolve the disputes. However, there are many instances where parties to a dispute enter into negotiations with third-party involvement. There is no international tribunal or institution to govern the relations of states in the utilisation and protection of transboundary freshwater resources. Therefore, international law is the only instrument available that provides a legal framework while at the same time making it possible to determine respective legal rights and obligations, along with the mechanisms for ensuring compliance and resolving disputes between states. However, international treaty regimes should be

BOX 11.5

Examples of international water agreements

1 *Helsinki Convention on the Protection of the Marine Environment of the Baltic Sea Area (1992 Helsinki Convention)*

The 1992 Helsinki Convention came into force on 17 January 2000. The governing body of the convention is the Helsinki Commission – Baltic Marine Environment Protection Commission – also known as HELCOM. The present contracting parties to HELCOM are all nine Baltic Sea riparian states: Denmark, Estonia, Finland, Germany, Latvia, Lithuania, Poland, Russia and Sweden, plus the EU. While the earlier 1974 Helsinki Convention was concerned primarily with issues of technical pollution control and the pollution of the Baltic Sea, the 1992 Convention, which still operates, is concerned with the entire marine environment of the Baltic Sea area. Its purpose is to prevent and eliminate pollution in order to promote the ecological restoration of the Baltic Sea area and the preservation of its ecological balance. Its geographical scope covers not only the entire Baltic Sea, including the sea floor and coastal zones, but also its drainage area.

2 *United Nations Convention on the Law of the Non-Navigational Uses of International Watercourses*

The 1997 UN Water Convention, for which 103 member states voted in favour, is the only treaty governing shared freshwater resources that is of universal applicability. It was the first global water law. It is a framework convention, in the sense that it provides a framework of principles and rules that may be applied and adjusted to suit the characteristics of particular international watercourses. The convention applies to uses of international watercourses and of their waters for purposes other than navigation and to measures of protection, preservation and management related to the uses of those watercourses and their waters. A key principle is that all states bordering an 'international watercourse' can utilise the resource in an 'equitable and reasonable manner' in order to achieve 'optimal and sustainable utilisation'. However, the state is obligated to undertake all necessary measures to ensure that such utilisation does not lead to any of the other riparian states suffering 'significant harm'. States also need to tolerate some disadvantages caused by the legitimate use of the water resource by the other riparian states, as long as this damage is not 'considerable'. The convention represents an important contribution to the strengthening of the rule of law in this increasingly critical field of international relations and to the protection and preservation of international watercourses. In an era of increasing water scarcity, it is to be hoped that the convention's influence will continue to grow (McCaffrey, 2001).

3 *The Columbia River Treaty*

The 1964 Columbia River Treaty is an international agreement between Canada and the United States for the cooperative development and operation of water resources in the Columbia River basin. The treaty was signed to mandate the construction and operation of three water-storage dams in British Columbia (Canadian storage) for the purpose of providing flood control and optimum hydropower generation in the Columbia River basin in Canada and/or the United States. The treaty also authorised the construction of the Libby Dam on the Kootenai River in Montana for flood control and other purposes in the United States. The Libby Dam creates power and flood control benefits downstream in Canada and the United States, and these benefits have no payment requirements. All four dams were built. The treaty has no end date but has a minimum length of 60 years. The treaty provides that either Canada or the United States can unilaterally terminate the treaty by providing a minimum of ten years' advance, written notice. However, if both the Canadian and United States' federal governments agree, the treaty can be renegotiated or terminated at any time. Unless terminated or renegotiated, most of the treaty's current provisions will continue indefinitely (Vinogradov et al., 2003).

4 The 1909 Canada–United States Boundary Waters Treaty

The Boundary Waters Treaty of 1909 was a treaty between the United States and Great Britain relating to the use of boundary waters and to settle all questions pending between the United States and the Dominion of Canada involving the rights, obligations or interests of both countries along their common frontier. The treaty created the independent International Joint Commission (IJC) to prevent and resolve boundary waters disputes between Canada and the United States. The IJC makes decisions on applications for projects such as dams in boundary waters and regulates the operations of many of those projects. The treaty covers all the 'boundary waters which are defined as the waters from main shore to main shore of the lakes and rivers and connecting waterways, or the portions thereof, along with the international boundary between the United States and the Dominion of Canada passes, including all bays, arms, and inlets thereof, but not including tributary waters which in their natural channels would flow into such lakes, rivers, and waterways, or waters flowing from such lakes, rivers, and waterways, or the waters of rivers flowing across the boundary' (preliminary article – text of the treaty). The treaty was supposed to remain in force for five years from the day of ratification and until terminated by 12 months' written notice given by either party. The treaty still stands, more than 100 years after agreement.

flexible enough to reflect the constantly changing natural status of water resources, as well as the growing human impact and demands on them. Ideally, the regimes should contain built-in flexibility that would allow them to adapt to changing conditions such as fluctuations in precipitation, droughts, floods and other emergency situations (Vinogradov et al., 2003).

11.5.3 Conflicts and cooperation can coexist

In the case of transboundary water, while there are studies which focus on the conflict potential of international water and stress the dangers of violence over water resources, recent works emphasise that conflict and cooperation in transboundary waters are not mutually exclusive; they operate in continuum (Wolf, 2001; Zeitoun and Warner, 2006; Cascão, 2009; Zeitoun et al., 2011). For example, in the last 50 years, 157 treaties have been negotiated and signed, marking the initiation of institutional agreements between riparian nations to cooperate and to mitigate future conflicts over their shared water sources.

Benefit sharing presents an alternative method of cooperation over transboundary water resources, focusing on the needs, demands and concerns of each of the parties, through a proper channel of communication between the parties. Benefit sharing explicitly articulates the options for water sharing and the resulting trade-offs, with the ultimate goal of promoting the sustainable use of transboundary water resources. The UN (2008) suggested that sharing benefits and responsibilities for transboundary waters would enable better ecological management, underpin increased food and energy production, contribute to poverty reduction and help control migration from rural areas to urban centres. By developing transboundary early-warning systems, the UN (2008) also suggested that risk to life from flood hazards can be reduced. Political benefits may include easing of tensions and improved economic integration between states, directly or indirectly contributing to international trade, economic development, food security, political security, poverty alleviation and regional integration. Box 11.6 provides an example of cooperation through benefit sharing between two South Asian nations – India and Bhutan.

11.5.4 Water diplomacy and informal dialogue

On a transboundary level, water diplomacy is most often used to prevent conflict escalation and to improve cooperation (Huntjens and de Man, 2017). It helps in communicating at different levels and between different parties (representatives of states, civil society, academia and so on), in establishing connections, and in building trust. It may include formal or informal dialogues between high-level diplomatic delegations of states or developing people-to-people relationships at grassroots level. It is a dynamic process that

seeks to develop reasonable, sustainable and peaceful solutions to water allocation and management while promoting or influencing regional cooperation and collaboration (Huntjens *et al.*, 2015; Barua and Vij, 2018).

In transboundary contexts, owing to the complexity related to allocation and management of water resources, formal diplomacy or formal dialogue alone is not sufficient to bring about cooperation among sovereign states. Informal dialogue tries to include more people (influential academic, religious and non-governmental organisation leaders, and other civil society actors) rather than just the figureheads in governments. Such informal dialogue helps in re-establishing relationships between the parties to the conflict, rather than bargaining over specifics (Wahrenfennig, 2016). Informal discussions between government and non-governmental stakeholders can be facilitated by a neutral third party (Hanasz, 2017). There are a large number of dialogue projects trying to resolve different transboundary conflicts by communicating at different levels and between different parties, including the Brahmaputra Dialogue outlined in Box 11.7.

REFLECTIVE QUESTION

What contribution can water diplomacy and informal dialogue make to enhance water cooperation?

11.6 SUMMARY

Water for human consumption and use is becoming increasingly scarce on a global scale and 90% of the projected increase in population by 2050 is most likely to be in developing countries, where even at present levels the

BOX 11.6

India–Bhutan partnership on hydropower development

Bhutan has enjoyed a strong relationship with India, which has been sustained by hydropower cooperation that has proved to be mutually beneficial for both the countries. Bhutan, being a Himalayan country in the Brahmaputra basin, is endowed with high per capita water availability. Hydropower is the backbone of Bhutanese economic development.

India–Bhutan hydropower cooperation began in 1961 with the signing of the Jaldhaka Agreement, which perpetuates the so-called 'sweetheart deal' exhibiting an ideal win–win arrangement for these neighbours. Since then, power-deficit India has been buying power from Bhutan and revenue-deficit Bhutan has been selling power to India. Bhutan has a capacity to generate about 30,000 MW of hydroelectricity every year. However, Bhutan is not able to harness hydropower energy on its own owing to financial and technological constraints. For both, it depends on India. Major treaties on hydropower dam development between the two countries include the 1974 Chukha Agreement, the 1994 Kurichu Agreement and the 1996 Tala Agreement. In 2006, the two countries signed a cooperation agreement on hydroelectric power, with a further protocol signed in 2009. Accordingly, the Indian government agreed to assist Bhutan in installing a minimum of 10,000 MW of hydro capacity by 2020, and to purchase all of the surplus power. In April 2014, the two countries also signed the 'Framework Inter-Governmental Agreement' that would facilitate the commencement of four hydroelectric power projects and further strengthen hydropower cooperation between the two countries. During his visit to Bhutan in 2014, Prime Minister Narendra Modi laid the foundation stone of the 600 MW Kholongchu hydroelectric project, a joint venture between India and Bhutan.

Bhutan not only gets electricity for its own use but also satisfies India's growing energy demand, earning much needed revenue and foreign exchange. Hydropower exports contribute around 40% to Bhutan's revenue and 25% of its GDP. In order to mitigate climate change impacts, India is committed to shift 40% of its electricity capacity from the current heavy reliance on fossil fuel sources by 2030. Given strong resistance to building hydropower dams in its own territory, it is reasonable to consider India's incentive in investing in hydropower dams in Bhutan to be a way to shift its energy mix domestically.

majority of people do not have safe and sustainable access to drinking water and sanitation. Competing needs between the industrial and agricultural sectors, especially in developing countries, will lead to conflicts over water, with little possibility of arriving at solutions acceptable to all parties involved.

Conflict occurs due to consumptive use, pollution, relative distribution of water and absolute water scarcity. There is no universally acceptable formula for the sharing and use of water. Issues like water rights, water allocation and water access are treated in different ways, in different locations, depending on the available legal framework and traditional and customary practices. Developed countries have a way of managing water resources, which can be characterised as being streamlined and uniform. However, in developing countries there are different ways of dealing with the

BOX 11.7

Informal water diplomacy in Brahmaputra River basin

The Brahmaputra River basin is the principal arm of the Ganges–Brahmaputra–Meghna system. It is one of the most critical transboundary rivers, playing a significant strategic role in South Asia's water, energy and food security. Shared by four countries – China, India, Bangladesh and Bhutan – the Brahmaputra basin is one of the most densely populated river basins in the world, supporting an estimated population of 700 million people. In the Brahmaputra basin, the conflict between countries ranges from less intense conflicts like stereotypical perception and differences of opinion to more intense conflicts of interest and influence of power. However, the absence of integrated management of water resources and lack of coordination among neighbouring states poses a considerable threat to the future development plans within the basin. Since an integrated and coordinated Brahmaputra River management would play an important role in the development of the entire South Asia region, it was realised that there is a strong need for dialogue and discussion across multiple stakeholders (government, civil society organisations, funding agencies, academia, scientific community and the media). The project, titled 'Transboundary Policy Dialogue for Improved Water Governance in Brahmaputra River', is one such initiative that aims to create a platform for the discussion of the issues, challenges and opportunities towards improved co-management of the river basin. The project was started by a group of researchers from India and Bangladesh in 2013 as a bilateral initiative, which later became a multilateral dialogue programme with the involvement of China and Bhutan. The major achievement of the Brahmaputra Dialogue has been that it has generated willingness among different stakeholders to participate in the dialogue process (Barua and Vij, 2018). It has also created a platform for the exchange of ideas and concerns among these stakeholders. The overarching aim is to develop a basin level institutional framework for strengthening the co-management of the Brahmaputra basin.

issue, and often traditional and customary beliefs and practices play a very important role in managing and sharing water resources, and this may mean that the models adopted are not very equitable and fair.

The need for international institutions and laws to resolve water conflicts and to work towards a more equitable and fair distribution of water resources is an urgent one. Although it is clear that there is no single uniform formula that can provide a solution for all countries and regions, the presence of laws and institutions at the international level will undoubtedly go a long way towards working out a solution to the future and present water-sharing problem. However, informal mechanisms of creating dialogue can also be important. Presenting the debate in a more positive light such as using terms associated with the co-benefits of sharing resources and responsibilities across territories in integrated river basin management approaches may be very important in mitigating difficult future conflict.

FURTHER READING

Bruns, B.R. and Meinzen-Dick, R.S. (eds). 2000. *Negotiating water rights*. Intermediate Technology Publications; London.

Getches, D.H., Zellmer, S. and Amos, A. 2015. *Water law in a nutshell* (5th edition). West Academic Publishing; Saint Paul, MN.

Mirimachi, N. 2015. *Transboundary water politics in the developing world*. Routledge; New York.

Priscoli, D. and Wolf, A.T. 2009. *Managing and transforming water conflicts*. Cambridge University Press; New York.

The above four books provide a good overview on types of water rights, water laws and water conflicts and cooperation.

Kornfeld, I.E. 2019. *Transboundary water disputes: state conflict and the assessment of their adjudication*. Cambridge University Press; Cambridge.

This book focuses on international water disputes, with a number involving the USA, and provides rich detail on each case, assessing the effectiveness of the courts and tribunals.

Classic papers

Falkenmark, M. 1986. Fresh water – time for a modified approach. *Ambio* 15: 192–200.

Hodgson, S. 2006. *Modern water rights: theory and practice*. Food and Agriculture Organization of the United Nations; Rome.

The above two papers have been highly influential in understanding water rights and water laws.

Wolf, A. (ed.). 2001. *Conflict prevention and resolution in water systems*. Edward Elgar; Cheltenham.

A collection of classic papers on water conflict at various scales.

Zeitoun, M. and Mirumachi, N. 2008. Transboundary water interaction I: reconsidering conflict and cooperation. *International Environmental Agreements: Politics* 8: 297, doi: 10.1007/s10784-008-9083-5.

A paper that shows how water conflict and positive water cooperation can exist at the same time.

PROJECT IDEAS

- Perform a study in your own country to explain the economic and social barriers associated with water allocation and distribution. How can efficient management of water resources lead to overall economic development and social prosperity?

- Provide a report on a case study from a transboundary river basin showing the ways in which neighbouring nations cooperate to mitigate future conflicts over their shared water resources.

REFERENCES

Bannon, C. 2017. Fresh water in Roman law: rights and policy. *Journal of Roman Studies* 107: 60–89.

Barua, A. and Vij, S. 2018. Treaties can be a non-starter: a multi-track and multilateral dialogue approach in Brahmaputra river basin management. *Water Policy* 20: 1027–1041.

Bruns, B.R. and Meinzen-Dick, R.S. (eds). 2000. *Negotiating water rights*. Intermediate Technology Publications; London.

Burchi, S. and Andrea, A.D. 2003. *Preparing national regulations for water resources management: principles and practice.* Food and Agriculture Organization; Rome.

Cascão, A.E. 2009. Changing power relations in the Nile river basin: unilateralism vs. cooperation? *Water Alternatives* 2: 245–268.

Cosgrove, W. 2003. *Water security and peace – a synthesis of studies prepared under the PCCP – water for peace process.* UNESCO; France.

Falkenmark, M. 1986. Fresh water – time for a modified approach. *Ambio* 15: 192–200.

Frijters, I. and Leentvaar, J. 2002. *Rhine case study.* UNESCO; Paris.

Getches, D.H., Zellmer, S. and Amos, A. 2015. *Water law in a nutshell* (5th edition). West Academic Publishing; Saint Paul, MN.

Haftendorn, H. 2000. Water and international conflict. *Third World Quarterly* 21: 51–68.

Hanasz, P. 2017. A little less conversation? Track II dialogue and transboundary water governance. *Asia Pacific Policy Studies* 4: 296–309.

Hodgson, S. 2006. *Modern water rights: theory and practice.* Food and Agriculture Organization of the United Nations; Rome.

Huntjens, P., Lebel, L. and Furze, B. 2015. The effectiveness of multi-stakeholder dialogues on water reflections on experiences in the Rhine, Mekong, and Ganga-Brahmaputhra-Meghna river basins. *International Journal of Water Governance* 5: 39–60.

Huntjens, P. and Man de, R. 2017. Water diplomacy: making water cooperation work. Policy Brief [online, accessed March 2019]. Available at www.planetarysecurityinitiative.org/sites/default/files/2017-04/PB_Water_Diplomacy_WG_4.pdf.

Isa, A. and Stein, R. 2009. *Rule: reforming water governance.* International Union for Conservation of Nature; Gland.

Kornfeld, I.E. 2019. *Transboundary water disputes: state conflict and the assessment of their adjudication.* Cambridge University Press; Cambridge.

McCaffrey, S.C. 2001. *The law of international watercourses: non-navigational uses.* Oxford University Press; Oxford.

McDonald, A., Clarke, M., Boden, P. and Kay, D. 2011. Social justice and water. *In*: R. Hester and R.M. Harrison (eds) *Sustainable water.* Royal Society of Chemistry; London, 93–113.

Meinzen-Dick, R. and Mendoza, M. 1996. Alternative water allocation mechanisms; Indian and international experiences. *Economics and Political Weekly* 31: 25–30.

Meinzen-Dick, R. and Rosegrant, M.W. 1997. Alternative allocation mechanisms for intersectoral water management. *In*: J. Richter, P. Wolff, H. Franzen and F. Heim (eds) *Strategies for intersectoral water management in developing countries; challenges and consequences for agriculture.* Deutsche Stiftung für internationale Entwicklung; Feldafing, 256–273.

MoWR (Ministry of Water Resources, Government of India). 2007. *The Report of the Cauvery Water Disputes Tribunals.* Ministry of Water Resources, Government of India; New Delhi.

Phansalkar, S.J. 2007. Water equity and development. *International Journal of Rural Management* 3: 1–25.

Priscoli, D. and Wolf, A.T. 2009. *Managing and transforming water conflicts.* Cambridge University Press; New York.

Rahman, M. 2006. The Ganga water conflict. *Asterisko* 1/2: 195–208.

Rasul, G. and Choudhury, J.U. 2010. *Equity and social justice in water resource management in Bangladesh.* International Institute of Environment and Development; London.

Rogers, P. and Hall, A.W. 2003. *Effective water governance.* TEC background paper 7. Global Water Partnership; Stockholm, 37.

Syme, G.J., Nancarrow, B.E. and McCreddin, J.A. 1999. Defining the component of fairness in the allocation of water to environmental and human use. *Journal of Environmental Management* 57: 51–71.

Tisdell, J.G. 2003. Equity and social justice in water doctrines. *Social Justice Research* 16: 401–416.

United Nations. 2006. *Water: a shared responsibility.* UNESCO and Berghahn Books; Paris and New York.

United Nations. 2008. *Transboundary waters: sharing benefits, sharing responsibilities.* United Nations; Zaragoza.

Vinogradov, S., Wouters, P. and Jones, P. 2003. *Transforming potential conflict into cooperation potential: the role of international water law*. UNESCO; Paris.

Wehrenfennig, D. 2016. Multi-track diplomacy and human security. *Human Security Journal* 7: 80–89.

WHO and UNICEF. 2017. *Progress on drinking water, sanitation and hygiene: 2017 update and SDG baselines*. World Health Organization; Geneva.

Wolf, A. (ed.). 2001. *Conflict prevention and resolution in water systems*. Edward Elgar; Cheltenham.

Zeitoun, M. and Warner, J. 2006. Hydrohegemony: a framework for analysis of transboundary water conflicts. *Water Policy* 8: 435–460.

Zeitoun, M. and Mirumachi, N. 2008. Transboundary water interaction I: reconsidering conflict and cooperation. *International Environmental Agreements: Politics* 8: 297, doi: 10.1007/s10784-008-9083-5.

Zeitoun, M., Mirumachi, N. and Warner, J. 2011. Transboundary water interaction II: the influence of 'soft' power. *International Environmental Agreements: Politics, Law and Economics* 11: 159–178.

CHAPTER TWELVE

Virtual water

Martin Tillotson, Dabo Guan and Xu Zhao

12

LEARNING OUTCOMES

After reading this chapter you should be able to:

- ■ describe virtual water;
- ■ explain the water footprint concept and provide examples at different scales;
- ■ show an awareness of some global and regional issues associated with water footprints.

12.1 INTRODUCTION

This chapter introduces some emerging topics which influence how water may be evaluated on a global and regional scale in the future. In particular, it focuses on the concept of virtual water as a way of understanding how water is consumed in the goods and services we all rely on, and how, as a result, water flows around the world as a part of trade. Mapping these flows of water is a powerful tool that can be useful to inform water use efficiency and water-management plans. Some regions are already water scarce and yet are producing products that consume a lot of water and shipping them off to other regions, which may themselves have plentiful water. In other cases it may be possible to determine more efficient water use based on trade and in the knowledge that water use to produce the same product might be different in different places. In this chapter we also provide an in-depth case study examination of virtual water flows in China with its contrasting water endowment, huge and uneven population distribution and trading surplus. It is possible to compare the geographical and physical bulk water transfer projects in China to the transfer of virtual water in traded goods and services. Virtual water is often ignored when it comes to water resource planning, and many regions or countries with large export volumes of water-intensive products exhibit significant and worsening water stress. The problem is exacerbated by waterborne pollution, which often arises as a result of manufacturing processes and poor legislative or regulatory controls which, in effect, allow recipients of goods and services to export their own water pollution.

12.2 THE BASIC WATER FOOTPRINT CONCEPT

Traditional measures of water use simply rely on measuring the amount of physical water withdrawal from rivers, lakes and groundwater within a country or region; this may be termed *blue water*. However, these measures ignore indirect water uses. Indirect water uses arise from water incorporated into products or services or utilised in their production and supply to the consumer, the water used to deal with the pollution which arises from the production of these goods or services (termed *grey water*), or water which is assimilated into soils and vegetation (termed *green water*). The recognition of this *embedded* or *virtual* water into goods and services was first introduced by Allan (1996), in which he described a nation's ability to offset its own water demand through importation of goods and services. The virtual water concept was further developed into the idea of the water footprint by Hoekstra (2003), in which he described it in terms of a nation's total use of global water resources, which is the sum of domestic water plus net virtual water import. The net virtual import thus refers to the amount of water used in the production of a good or service which is imported into a country (Hoekstra, 2003). The term has become accepted in the field of water management (Hoekstra and Chapagain, 2008) but can be confusing since all water use is real rather than virtual.

The idea of calculating the water footprint of a country was based on the analogy of an ecological footprint, which calculates the amount of land needed for the production of goods and services to sustain a country but also the land needed to assimilate the waste produced by the population. For the water footprint the same concepts apply, including water use in treating waste. The concept of the water footprint is aimed at improving water management on a global level (Hoekstra, 2007). The water footprint can be viewed as an indicator of total freshwater use (Hoekstra *et al.*, 2011) and is governed ultimately by human consumption. This differentiates it from other measures, which associate water demand with production within a country rather than consumption by a country. Research on water footprinting aims to show links between global trade and water management, and between human consumption and water use (Hoekstra, 2007). However, the water footprint can also be calculated at other boundary levels as well as country levels – for example, regions, cities, industrial sectors, companies, products and even individuals.

The water footprint can be split into three categories – *blue*, *green* and *grey* (Hoekstra *et al.*, 2011). The *blue* water footprint involves 'clean' water consumption from freshwater sources such as surface and groundwater which does not return back to the catchment from which it was taken in a clean form. The *green* water footprint includes consumption of water from rainfall that is stored on or assimilated into the soil or plants. Green water may evaporate or be lost through plant transpiration/evaporation and can be made productive for crop use. The grey water footprint represents the water used in dealing with pollution created throughout a supply chain and is defined by the volume of freshwater that it takes to assimilate the load of pollutants and return the water source back to its original quality (Hoekstra *et al.*, 2011). For a particular good, product or service the total water footprint is therefore a summation of the combined green, blue and grey water footprints at a particular point in space and time. It is important to note that, in water footprint concepts, the definition of grey water is different from the definition used in traditional environmental management, where it refers to water used for things such as washing clothes and bathing but where the water can then be further used for other purposes such as for watering gardens. Furthermore, while green and blue water are quantity measures, grey water is effectively a quality measure and, as such, is somewhat controversial. Therefore, when you read about grey water in the literature or on websites you need to check exactly what it is referring to.

Figure 12.1 shows a schematic of some of the direct and indirect water uses for a home that contribute to a water footprint in a typical northern European household. The interesting point here is that the metrics of traditional water use are based entirely on blue water, which accounts for only about 3% of consumption. Of the 97% or so of virtual water consumed by a typical household, over three-quarters is derived from agricultural products (Chapagain and Orr, 2008). The water footprint of a range of products and goods has been calculated (Mekonnen and Hoekstra, 2011, 2012); **Table 12.1** presents a selection of these water footprints and it is interesting to note that:

1 The majority of water consumption is related to the green water footprint, the primary production of plants.

Figure 12.1 Examples of indirect and direct water use for a home.

Table 12.1 Water footprint composition data for a selection of products

Product	Water footprint (litres)	Unit per	Green %	Blue %	Grey %
Lettuce	237	kg	77	7	16
Milk	1020	L	85	8	7
Eggs	3300	kg	79	7	13
Cotton (fabric)	10,000	kg	54	33	13
Chocolate	17,196	kg	98	1	1
Coffee (roasted)	18,900	kg	96	1	3

REFLECTIVE QUESTIONS

What is a water footprint?

Based on your knowledge of the water footprint of a particular product, what socio-economic implications could you foresee?

2 More intensively manufactured products, or those involving more complex supply chains, have proportionately greater blue and/or grey water consumption.

3 The combined water footprints are presented as global averages. This is because environmental factors, farming, and manufacturing practices and technologies vary significantly around the world.

The water footprint gives a volumetric measure of freshwater use and water pollution that consumers and producers can relate to (Hoekstra *et al.*, 2011). However, it should be noted that the water footprint does not provide an environmental impact assessment, as the severity of impacts of water use depends on the vulnerability of the catchment. The use of the water footprint concept for policy-making must therefore include this caveat (Wichelns, 2010). There is no distinction in the water footprint between sustainable and unsustainable water use, as a certain volume of water used in one catchment may not have a detrimental effect, but elsewhere the case may be different (Hoekstra, 2007).

12.3 CALCULATING A WATER FOOTPRINT

There are several methods for calculating a water footprint, and for a full list and description readers are referred to Hoekstra *et al.* (2011). However, the main methods are the chain summation approach, the stepwise accumulative approach and the input–output model. A water footprint can be calculated for a number of different products and processes, for a group of consumers, a geographical area, a nation, a river basin and a business. The water footprint of a product is measured in the volume of water used per unit of product (see Box 12.1), whereas the water footprint of a consumer, nation or business is expressed as the volume of water used per year (Hoekstra, 2008).

Blue water data can be sourced from historical information such as water abstraction meter readings. Models are sometimes used to calculate the green water footprint based on data on soil moisture, plant type and growth processes, climate conditions and so on (e.g. Fader *et al.*, 2011). These models predict green water use based on our best understanding of the physical processes involved. Grey water data can be calculated from measuring the concentrations of certain pollutants in the flow that are disposed into natural systems, detected by measuring local ambient water quality conditions

BOX 12.1

Calculating the water footprint of a soft drink

The following example is based on the scientific literature (Ercin *et al.*, 2011) to illustrate the detailed accounting process involved in calculating the total water footprint of a product. In this case it is for a hypothetical sugar-containing carbonated soft drink. The calculation is based on green, blue and grey water consumption and includes the following elements:

1 At the factory in the *direct* production of the packaged product. This includes:

   • blue water incorporated into the product;
   • blue water consumed in the production process;
   • grey water produced as a result of the production process.

2 At the factory in the *indirect* support of production of the packaged product. This is deemed an overhead:

   • blue water used by factory employees, for example, for drinking, eating, washing;
   • grey water produced as a result of the above activities.

3 In the supply chain *directly* concerned with the production and supply of the ingredients and packaging which make up the final product. The direct supply chain inputs into the factory are:

   • sugar and other ingredients (caffeine, phosphoric acid, lemon oil, orange oil, vanilla extract, carbon dioxide);
   • packaging materials (plastic bottle, bottle cap, labelling/adhesive, other packaging).

4 In the supply chain *indirectly* associated with production, for example energy, transport, and construction materials.

Table 12.2 presents the high-level outcomes and a summation of the manufacturing and supply chain analysis described above.

Table 12.2 Water footprint (litres) of a hypothetical 0.5-litre plastic bottle containing a sugar carbonated soft drink (data from Ercin *et al.*, 2011)

		Green	**Blue**	**Grey**	**Total**
Direct production		0.00	0.50	0.00	0.50
Indirect production		0.00	0.00	0.00	0.00
Direct supply chain					
Sugar	Min	0.00	7.00	2.40	26.00
	Max	117.90	123.50	12.00	167.00
Other ingredients		133.51	0.30	0.00	133.81
Packaging		1.034	0.256	6.0694	7.3594

(Continued)

Table 12.2 (Cont.)

		Green	Blue	Grey	Total
Indirect supply chain		0.0012	0.005	0.7184	0.7246
Total	Min	134.55	8.06	9.19	168.39
	Max	252.45	124.56	18.79	309.39

A number of observations can be made:

- Direct blue water consumption equates to the volume of water incorporated into the soft drink, which in this case is 0.5 litres.
- In practice it was very difficult to derive accurate water footprint data for many of the elements involved. Values, based on other literature, or assumptions were made, particularly where the contribution of a particular element is small. In this case an assumption has been made that net indirect consumption of water is zero.
- The type of sugar (i.e. beet, cane or high-fructose maize syrup and its country of origin) used in the soft drink is the most significant variable in the overall water footprint calculation. A much more detailed water footprint analysis underpins this data and is summarised in Figure 12.2. This type of variability is extremely common in primary production, and mean values are often presented.
- While the agriculturally derived ingredients make up the majority of the water footprint, they constitute only a tiny fraction of the mass of the final product.
- The overwhelming majority of the water footprint (99.7–99.8%) is associated with the supply chain, rather than with on-site production. However, common practice in water-efficiency assessment is to focus purely on consumption on site.

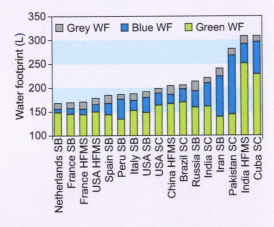

Figure 12.2 Green, blue and grey water footprint contributions for sugar beet (SB), sugar cane (SC) and high-fructose maize syrup (HFMS) sourced from different geographical locations. Volumes have been normalised for sugar content of a 0.5-litre soft drink.

Source: Reprinted by permission from Springer Science+Business Media: *Water Resources Management*, Corporate water footprint accounting and impact assessment: the case of the water footprint of a sugar-containing carbonated beverage, v25, 721–741, Ercin, A.E., Aldaya, M.M. and Hoekstra, A.Y. © 2011.

(Gerbens-Leenes and Hoekstra, 2008). All of these data on blue, green and grey water use can be very difficult to collect and highly location specific, and water footprint assessments in many cases are therefore in their early stages of development. In cases where data on grey and green water are not available, blue water must be used as the primary source of information, as this can be acquired from water bills, meter readings or regional water provision statistics (but see Chapter 9).

The chain summation approach is simple and is used when there is just one output product and when the process steps in the production system can be fully attributed to the production of that product. It simply involves working out how much water is used in the production of the product at each stage of the production process. The totals from each stage are then added together (Hoekstra *et al.*, 2011). The place where a product is produced may impact on the total water footprint, as similar processes may use different amounts of water in different locations. Therefore, it is important to note the locations of production and water use so that the final water footprint of a product can be mapped.

There are not many systems which have only one final output product or where the inputs can simply be ascribed to the product. The stepwise accumulative approach can be used when there is more than one product input to create the final output product. It takes the sum of the water footprints of the input products used in the last stage of processing to calculate the output product's water footprint. It may also be used when one input product creates a number of output products, and here the water footprint of the input product is distributed to the output products accordingly (Hoekstra *et al.*, 2011). This approach

is more realistic but is significantly more complicated, and it is more difficult to obtain water footprint data for the intermediate processing stages without carrying out water footprint investigations for each stage itself.

The environmental input–output analysis is another method to calculate a water footprint from a top-down accounting perspective. It is usually applied to calculate a water footprint for a large geographical entity such as a country, region or city. Sometimes it can be used to calculate the water footprint for a commercial/industrial sector or company in combination with the chain summation approach. The environmental input–output approach provides a complete description of the national and/or international supply chain. The method is based on final consumption and ensures that water used in production is assigned to the end-product consumed (and its location). For example, the water required for the production of work clothes provided by a car manufacturer is assigned to the final consumption of the car. This can, for instance, lead to considerable bias in the allocation of virtual water to different countries. A multi-regional environmental input–output model provides a methodological framework to comprehensively deal with international trade interlinkages across countries and sectors. Such models provide a description of the entire global supply chain and are able to trace water across any supply chain layer.

Environmental input–output analysis has a long history in water accounting studies. An early study undertaken by Hartman (1965) examined features of input–output models in terms of their usefulness for analysing regional water consumption and allocation. More recent studies of water-related input–output models include those in Table 12.3.

Table 12.3 Recent water accounting studies using environmental input–output models

Reference	Study
Guan and Hubacek (2008)	Advanced the work of Hubacek and Sun (2005) by taking the pollution absorption capacity into account, using North China as a case study.
Hubacek *et al.* (2009)	Examined environmental implications of urbanisation and lifestyle change in China by calculating ecological and water footprints, which was one of the first studies on water footprints using the input–output model.
Zhang *et al.* (2012)	Conducted an input–output analysis to investigate changes in Beijing's water footprint and used structural decomposition analysis to study the effects of contributing factors to the changes during 1997–2007.
Lenzen *et al.* (2013)	Incorporated water scarcity into an assessment of global virtual water flows combined with input–output analysis to include indirect virtual water flows.
Zhao *et al.* (2015)	Compiled a full inventory for physical water transfers and mapped virtual water flows between Chinese provinces in 2007 and 2030, and evaluated the associated impacts on provincial water stress.
Serrano *et al.* (2016)	Conducted a consumption-based assessment of virtual water flows in the European Union (EU27) using multi-region input–output analysis, and applied structural path analysis to identify major global routes conveying pressure on water resources from centres of consumption to regions of water scarcity.
Zhao *et al.* (2018)	Multiplied the water use in production with the water stress index to quantify scarce water, and assessed the scarce water saving/loss through inter-provincial trade within China using a multi-region input–output table.

The water footprint of a consumer is the total volume of freshwater consumed or polluted in the production and use of goods and services. For a consumer it is necessary to calculate their direct water footprint which is made up of the water use in the home. Interestingly, Crawford (2010) found that over 50 years the total operational water requirement for consumers living in a house (direct water use) equalled that of the water used in the original construction of the house. To calculate the total water footprint of a consumer the direct water use needs to be added to the indirect water footprint derived from the water used in producing the products (e.g. food, coffee, clothes, energy supply, furniture) that are used by the consumer (Godfrey and Chalmers, 2012). The water footprint of each product or service privately consumed is added to the proportion of shared water footprint of public goods and services (e.g. street lighting, local parks). It is therefore easy to see how factors such as GDP, personal habits and lifestyle, household size, etc. impact upon water footprint and vary dramatically across the globe.

The water footprint of a nation includes the internal water footprint, which is the total water in goods and services used by the population, minus the water used by the country in producing goods that are exported. The external water footprint of a nation must also be measured to determine the overall water footprint. The external water footprint of a nation is equal to the volume of water resources used in producing goods and services in other countries that are imported into the nation and used by that nation.

How would you calculate the water footprint of a packet of sugar?

How would you calculate your own water footprint?

12.4 THE GLOBAL WATER FOOTPRINT PICTURE

The water footprints of national consumption are shown in Figure 12.3 (average per person for domestic use and consumer goods). American citizens consume twice the global average of water, while Chinese citizens consume around half the global average per person. Trade between countries means that products consumed in many rich countries in Europe, the USA and Japan, for example, impact on water resources in developing countries. Many countries in Africa are net exporters of water and yet suffer from water stress. Figure 12.4 shows how much of the water footprint in different countries is dependent on virtual water imports. Much of Western Europe and the Middle East is highly dependent on virtual water imports (Hoekstra and Mekonnen, 2012). Through such trade a nation can save domestic water resources (the water is used elsewhere in the world to produce the products imported by that country). It may be that in some instances such an arrangement

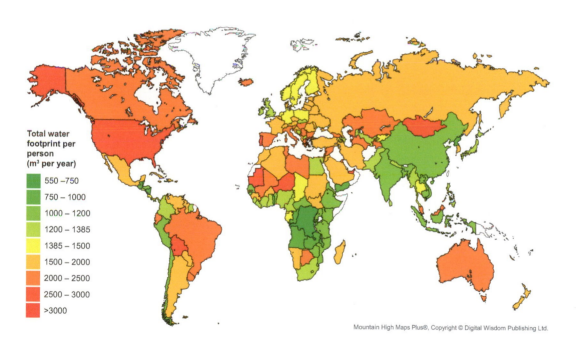

Total water footprint per person (m³ per year)

- 550 –750
- 750 – 1000
- 1000 – 1200
- 1200 – 1385
- 1385 – 1500
- 1500 – 2000
- 2000 – 2500
- 2500 – 3000
- >3000

Mountain High Maps Plus®, Copyright © Digital Wisdom Publishing Ltd.

Figure 12.3 The annual water footprint of national consumption averaged per person for drinking, washing, food consumption and consumption of other consumer goods. Countries shown in green have a water footprint that is smaller than the global average; countries shown in yellow-red have a water footprint larger than the global average.

Source: Hoekstra and Mekonnen (2012) under a CC-BY-SA 3.0 licence.

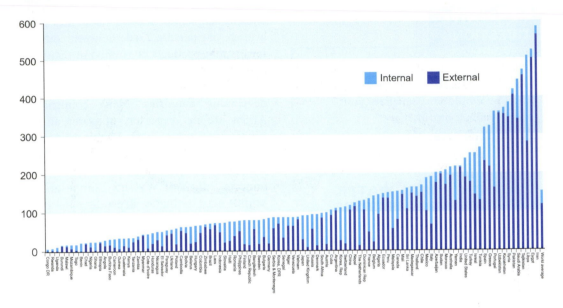

Figure 12.4 National blue water footprints dependent on imports (shown as external water footprint) compared to domestic provision (shown as internal water footprint). Data are in m³ per year per person between 1996 and 2005. Only countries with a population greater than 5 million are shown.
Source: Hoekstra and Mekonnen (2012) under a CC-BY-SA 3.0 licence.

is better for global water resources. If the virtual water flow occurs from regions which use water productively (e.g. little water required to grow a particular crop) to regions which are not very productive with water use (e.g. requiring lots of water to grow the same crop), that will benefit the global water footprint. It is estimated that 263–352 billion m³ of water is saved per year through such trade (Fader *et al.*, 2011; Hoekstra, 2010). Regional net virtual water balances are shown in **Figure 12.5** related to agricultural and industrial trade (which represents 80% of virtual water flows), with some of the largest flows shown by the arrows. Thus, global policymakers could look at virtual water flows and the water footprint (to include green and grey water use) to see how trade can be used to effectively reduce global water resource use and also support those areas that are water scarce.

In 2010 the UN told big business to prepare for legislation on water footprints, and many multinational corporations have committed to reducing their footprint. Data on water footprints for companies (especially large ones) is rather sketchy, and significant effort is required to pull together the information so as to enable full footprints to be established. However, evaluating the water footprint of a company may well benefit that company, allowing it to become more efficient and save costs, as well as establish best practice for the global good. It is also possible to think about how the water footprint concept could be extended to cover other water-related issues, which may be informative to economic planning (e.g. **Box 12.2**).

Figure 12.5 Virtual water balance and direction of gross virtual water flows related to trade in agricultural and industrial products between 1996 and 2005 for the very largest global gross flows (>15 billion m³ per year); the fatter the arrow, the bigger the virtual water flow.
Source: Hoekstra and Mekonnen (2012) under a CC-BY-SA 3.0 licence.

REFLECTIVE QUESTIONS

How does trade affect net virtual water balances across the world?

How dependent on water imports is your home country?

12.5 VIRTUAL WATER FLOW AND ITS APPLICATION TO CHINA

In order to illustrate how informative an assessment of virtual water flow might be, we turn to a case study which looks at the geographical regions of China. China's water resources are typified as being unevenly distributed and in severe shortage; with a temperate south and an arid north. China's annual per capita freshwater quantity is only 25% of the world average, yet China provides agricultural produce to over 21% of the world's population with only 6% of the world's water resources (Liu *et al.*, 2013). The geographical mismatch between freshwater demand and available freshwater resources is one of the largest threats to sustainable water supply in China. The North China Plain shows the greatest water scarcity, with per capita water availability under 150 m³ per year. At the same time, this area is home to 200 million people and provides more than one-half of China's wheat and

BOX 12.2

Flood footprint

It is possible to envisage other types of water footprint in addition to the main one described above which have excellent potential to support decision-making. 'Footprint' is a measure of impact, which can also be applied to risk and disaster management. Flooding in one location can impact upon the whole economy. Neglecting these knock-on costs (i.e. the true footprint of the flood) means that we might be ignoring the economic benefits and beneficiaries of flood risk management interventions. In 2015–16, for example, floods cost the UK economy about £1.6 billion directly, but the wider effect might actually add another 50 to 250% to that. Some of the effects of regional flooding might also extend internationally. If flooding caused damage to crops which could no longer be exported, the importing country would have to source its products elsewhere, which may influence its overall water footprint. The flood footprint is therefore a measure of the exclusive total socio-economic impact that is directly and indirectly caused by a flood event to the flooding region and wider economic systems and social networks.

one-third of its maize. The North China Plain is also the home to the megacities of Beijing and Tianjin, as well as many other major industrial cities. The water issues in northern China are constraining its development and threatening its water security, a problem compounded by its rapid economic development, urbanisation, population growth and climate change.

Scientists have used a water stress index (WSI) to evaluate China's provincial water stress (Zhao X. *et al.*, 2015). WSI here refers to water stress arising from the proportion of water withdrawal from available annual renewable local water resources. **Figure 12.6** shows that 23 out of 30 studied Chinese provinces had at least moderate water stress (WSI>0.2), with six provinces showing extreme water stress (WSI>1; i.e. they consumed more water than the available annual renewable amount). **Figure 12.6** also shows that water is scarcer in northern China, and that the Beijing, Tianjin and Hebei Provinces in the North China Plain suffer extreme water stress, along with Ningxia in northern

central China. While southern China exhibits less water stress, it is not spared from it. Six provinces in southern China show levels of moderate water stress, and two provinces show extreme water stress.

Recognising the above situation and paramount importance of water to the Chinese economy, the Chinese government has implemented multiple policies and actions designed to alleviate water stress. To an extent, these policies have been implemented through huge investment in water conservation infrastructure in northern China, such as irrigation technology, water transfer projects, seawater **desalination** and reclaimed water recycling (Liu *et al.*, 2013). **Figure 12.7** shows, using the latest data available (which unfortunately are from 2007), that provinces in northern China tend to utilise inter-provincial transferred water to replenish local water resource. Indeed, China has been developing over 20 major physical water transfer projects, with a combined length of over 7200 km (Aeschbach-Hertig and Gleeson, 2012), including

Figure 12.6 Water stress evaluation of China's provinces (2007 baseline). The colour coding of the regions distinguishes between moderate water stress, severe water stress, extreme water stress and no water stress. The size of the dots reflects the amount of physical water transfer.

Source: Zhao, X., Liu, J., Liu, Q., Tillotson, M.R., Guan, D. and Hubacek, K. 2015. Physical and virtual water transfers for regional water stress alleviation in China. *Proceedings of the National Academy of Sciences of the United States of America* 112: 1031–1035.

the world's largest – the South–North Water Transfer Project (SNWTP). Three routes are projected for the SNWTP, which will ultimately transfer 44.8 billion m^3 per year of water from the Yangtze River basin to the Huang–Huai–Hai River basin, of which 14.8 billion m^3 per year is for the East Route, 13 billion m^3 per year is for the Middle Route, and 17 billion m^3 per year is for the West Route (**Figure 12.7**). After completion of the three routes, the

transferred water is projected to amount to 30.5% of total water withdrawal in the Huang–Huai–Hai River basin.

Because northern China is not endowed with abundant water resources, scientists recommend the importation of more water-intensive goods and produce, as opposed to local production, from southern China. This may be countered through export of less water-intensive goods in order to maintain a trade balance. In other words, researchers

Figure 12.7 Routes and affected provinces of the South–North Water Transfer Project (SNWTP).
Source: Zhao, X., Liu, J., Liu, Q., Tillotson, M.R., Guan, D. and Hubacek, K. 2015. Physical and virtual water transfers for regional water stress alleviation in China. *Proceedings of the National Academy of Sciences of the United States of America* 112: 1031–1035.

recommend the flow of virtual water from southern China to northern China. Yang and Zehnder (2007) indicated that virtual water imports can be an additional measure in dealing with water scarcity in the North China Plain, compared with the conventional wisdom of 'opening up new sources' to meet growing water demand. Ma *et al.* (2006) were among the earliest to study virtual water trade between southern and northern China. They found water-abundant southern China

imported 52 km^3 of agricultural virtual water from water-scarce northern China. Guan and Hubacek (2007) also studied virtual water flows between northern and southern China using an input–output analysis. They drew a similar conclusion from their work, in that northern China, with water scarcity, virtually exported 5% of its available freshwater resources. In contrast, southern China imported a large amount of virtual water. Controversially, these researchers suggested that importing virtual water of agricultural products into the North China Plain may be more efficient than transferring physical water through the SNWTP in terms of water resources utilisation. However, the SNWTP is a result of trade-offs from multiple options, taking into consideration factors such as land resources and socio-economic and environmental benefits and costs. Importing virtual water of agricultural products for the North China Plain or for northern China seems like a promising solution, but it is difficult to realise. One of the notable reasons for this is the mismatch between freshwater resources and arable lands (Zhang and Anadon, 2014). The largest proportion of arable land is located in the North China Plain (Liu *et al.*, 2013), while southern China is more mountainous and has less potential to develop arable land.

While the above studies differentiate southern China from northern China in terms of virtual water flows from the water-scarce north to the water-abundant south, the differentiation is still too coarse to understand virtual water flows within China. The authors of this chapter previously studied inter-provincial physical water transfers and virtual water trade within China (Zhao *et al.*, 2015). Our study focused on 30 provincial-level administrative regions (provinces, autonomous regions and municipalities – for simplicity, they are referred to as provinces) in mainland China where data were available. The results showed that physical water transfer in China is incomparable to virtual water flows in terms of water quantity, although there were many water transfer projects in China. Physical water flows by water transfer projects amounted to 26.3 billion m^3 per year in 2007, accounting for 4.5% of national water supply in China. The magnitude of virtual water flows was found to be much larger than physical water transfers. The total volume of virtual water flows was 201 billion m^3 per year in 2007, meaning that 35% of the national water supply was used for inter-provincial virtual water trade.

More importantly, the results revealed a different spatial pattern of virtual water flows compared with previous studies focusing on north and south. As shown in Figure 12.8, a large amount of virtual water flowed from the economically poor and less populated west to the more affluent and densely populated coastal areas of the east, in which most of China's megacities are located. A small number of provinces were responsible for most of the net virtual water flows: the top five importing provinces (Shandong, Shanghai, Guangdong, Zhejiang and Tianjin, which are all coastal) accounted for 74% of net virtual water imports, whereas 78% of net virtual water exports were from five provinces (Xinjiang, Heilongjiang, Inner Mongolia, Guangxi and Hunan). These findings also highlighted that virtual water trade is closely linked to affluence (Feng *et al.*, 2014; Tamea *et al.*, 2014). Many developed provinces along China's east coast, such as Shanghai, Shandong, Beijing and Tianjin, prosper at the expense of water resource depletion in inland provinces, such as Xinjiang and Inner Mongolia. In effect, water scarcity is being shifted from coastal to inland regions.

The concept that developing countries and regions use large amounts of water to produce goods for export to developed regions is known as the burden-shifting of water stress. Using Shanghai as an example, we can see how a megacity has shifted the burden of both water quantity and quality stress to its trading partners. Combining water quantity stress evaluation with provincial net virtual water export to Shanghai, 60% of net virtual water imports to Shanghai were from 13 provinces experiencing extreme and severe water quantity stress (WSI>0.4). The top net virtual water-exporting provinces to Shanghai were all from

Figure 12.8 Virtual water balance for each economic region and the net direction of virtual water flows (2007 baseline). Only the largest net virtual water flows are shown (>2 billion m³ per year).

Source: Zhao, X., Liu, J., Liu, Q., Tillotson, M.R., Guan, D. and Hubacek, K. 2015. Physical and virtual water transfers for regional water stress alleviation in China. *Proceedings of the National Academy of Sciences of the United States of America* 112: 1031–1035.

northern China, and all experience different levels of water quantity stress (**Figure 12.9**). Hebei, for example, is among the most water-scarce provinces (**Figure 12.6**), however it virtually exports 6.2% of its annual renewable water resources to Shanghai. A common feature of the major virtual water-exporting provinces is that they are major producers of agricultural products. In the top six net virtual water-exporting provinces, 78–98% of virtual water export to Shanghai is embedded in food-related sectors.

The water pollution generated from exported goods will remain in the producing provinces, resulting in negative impacts on the local environment. Effectively,

Shanghai is exporting water pollution to other provinces. In other words, some provinces virtually accept waterborne pollution from Shanghai by exporting goods and services to Shanghai. Water bodies may become so polluted that they are unable to provide any beneficial use, termed 'pollutant induced water scarcity' (Liu *et al.*, 2016). The water quality stress index (WQSI) has therefore been designed to quantify the extent of pollutant-induced water scarcity. The WQSI can be calculated according to the ratio of grey water footprint to available local water resources.

Combining the provincial water quality stress evaluation with provincial pollution

Figure 12.9 a) Shanghai's net virtual water import from other provinces. The colours of the provinces indicate their water quantity stress status. The flows with arrows show the top net virtual water exporters to Shanghai. Only net flows larger than 500 million m³ are shown; b) chemical oxygen demand (COD) in other provinces due to the consumption of Shanghai. The direction of arrows from Shanghai to other provinces indicates that Shanghai outsources its COD pollution to other provinces. Only volumes larger than 35,000 tonnes are shown; c) ammonia nitrogen (NH₃-N) pollution in other provinces due to the consumption of Shanghai. Only volumes larger than 1000 tonnes are shown.
Source: Zhao *et al.* (2016).

outsourcing, we can study the burden-shifting of water quality stress between Shanghai and its trade partners. Nineteen provinces suffering water quality stress accounted for 79% of net chemical oxygen demand (COD) and 75.5% of net ammonia nitrogen (NH₃-N) outsourcing from Shanghai (**Figure 12.9**). The main provinces with

different levels of water quality stress suffering Shanghai's net COD outsourcing were Shandong, Hebei, Zhejiang and Henan, while the main provinces suffering Shanghai's net NH₃-N outsourcing were Anhui, Henan, Hebei, Zhejiang and Jiangsu. Hebei was found to suffer from the highest negative impact due to virtual water export and pollutant outsourcing from Shanghai, while also suffering from extreme water stress in terms of both water quality and water quantity.

The above analysis of inter-provincial virtual water trade in China shows that virtual water trade is closely correlated with regional water scarcity. Hence, it is important to incorporate virtual water assessment into water management at both regional and national levels.

REFLECTIVE QUESTIONS

Why does virtual water generally flow from economically poor regions to more affluent regions?

In which directions are the biggest virtual water flows in China?

12.6 SUMMARY

Virtual water is the water used in the production of a product or service. By consuming that product or service your indirect water use will increase by the amount of virtual water associated with that product. Therefore, your personal water footprint will increase beyond direct water use (e.g. drinking water, washing) to indirect uses. For example, if you bought a book there would be water use associated with making the paper and ink, ranging from water abstracted for manufacture (blue water), water used to grow the trees from which the paper was derived (green water) and pollution of water used in the printing process (grey water). Water footprints can be calculated for individuals, groups, businesses, regions and countries. Examining virtual water flows between these groups, and in particular mapping totals and net flows, can be highly informative in terms of optimising the efficiency of water use and examining where there are water-scarce regions which are highly vulnerable because they are net exporters of water. These sorts of tools can also be excellent for businesses and individuals to help them establish how we can be more efficient with our total water consumption (direct and indirect) and how trade can be used to reduce total global water consumption. This is a necessity as we move into a future with increasing population and growing concern over climate change impacts, both for ourselves as humans and for the wider, global ecosystem too.

FURTHER READING

Allan, J.A. 2011. *Virtual water: tackling the threat to our planet's most precious resource.* I.B. Tauris; London.

An important book by the person who came up with the term 'virtual water'.

Fader, M., Gerten, D., Thammer, M., Heinke, J., Lotze-Campen, H., Lucht, W. and Cramer, W. 2011. Internal and external green-blue agricultural water footprints of nations, and related water and land savings through trade. *Hydrology and Earth System Sciences* 15: 1641–1660.

An interesting modelling study to establish agricultural water flows through trade.

Hoekstra, A.Y. 2013. *The water footprint of modern consumer society.* Routledge; London.

Hoekstra, A.Y. and Chapagain, A.K. 2008. *Globalization of water: sharing the planet's freshwater resources*. Blackwell; Oxford.

Hoekstra, A.Y., Chapagain, A.K., Aldaya, M.M. and Mekonnen, M.M. 2011. *The water footprint assessment manual – setting the global standard*. Earthscan; London.

Classic papers

Allan, J.A. 1996. The political economy of water: reasons for optimism but long-term caution. *In*: Allan, J.A. (ed.), *Water, peace and the Middle East: negotiating resources in the Jordan basin*. Tauris Academic Studies; London.

One of the first articles to use the term 'virtual water'.

Hoekstra, A.Y. and Mekonnen, M.M. 2012. The water footprint of humanity. *Proceedings of the National Academy of Sciences of the United States of America* 109: 3232–3237.

This paper provides lots of interesting maps and data.

The two papers below provide mode detail on the regional virtual water flows in the Chinese case study:

Zhao, X., Liu, J., Liu, Q., Tillotson, M.R., Guan, D. and Hubacek, K. 2015. Physical and virtual water transfers for regional water stress alleviation in China. *Proceedings of the National Academy of Sciences of the United States of America* 112: 1031–1035.

Zhao, X., Liu, J., Yang, H., Duarte, R., Tillotson, M.R. and Hubacek, K. 2016. Burden shifting of water quantity and quality stress from megacity Shanghai. *Water Resources Research* 52: 6916–6927.

PROJECT IDEAS

- Quantify the amount of water used for producing exports in a country or region and trade to other places for final consumption. Discuss the findings in terms of impacts on domestic water stress and outsourcing of waterborne pollution to the producing country.

- Calculate the water footprint of different dietary lifestyles for individuals (different groups of population or across different countries). Discuss the findings within the context of current national and international sustainable development policies.

- Determine the possible impacts of water-related policies in your country or region for water efficiency, taking into account physical bulk water transfer projects as well as virtual water.

REFERENCES

Aeschbach-Hertig, W. and Gleeson, T. 2012. Regional strategies for the accelerating global problem of groundwater depletion. *Nature Geoscience* 5: 853–861.

Allan, J.A. 1996. The political economy of water: reasons for optimism but long-term caution. *In*: Allan, J.A. (ed.), *Water, peace and the Middle East: negotiating resources in the Jordan basin*. Tauris Academic Studies; London.

Allan, J.A. 2011. *Virtual water: tackling the threat to our planet's most precious resource*. I.B. Tauris; London.

Chapagain, A.K. and Orr, S. 2008. *UK water footprint: the impact of the UK's food and fibre consumption on global water resources*, Volume 1. WWF-UK; Godalming.

Chen, Z.M. and Chen, G.Q. 2013. Virtual water accounting for the globalized world economy: national water footprint and international virtual water trade. *Ecological Indicators* 28: 142–149.

Crawford, R.H. 2010. Life cycle water analysis of an Australian residential building and its occupants. *Proceedings of the Seventh Australian Conference on Life Cycle Assessment: revealing the secrets of a green market, Melbourne, 9–10 March 2011.*

de Villiers, M. 1999. *Water wars. Is the world's water running out?* Phoenix Press; London.

Ercin, A.E., Aldaya, M.M. and Hoekstra, A.Y. 2011. Corporate water footprint accounting and impact assessment: the case of the water footprint of a sugar-containing carbonated beverage. *Water Resources Management* 25: 721–741.

Fader, M., Gerten, D., Thammer, M., Heinke, J., Lotze-Campen, H., Lucht, W. and Cramer, W. 2011. Internal and external green-blue agricultural water footprints of nations, and related water and land savings through trade. *Hydrology and Earth System Sciences* 15: 1641–1660.

Feng, K., Hubacek, K., Pfister, S., Yu, Y. and Sun, L. 2014. Virtual scarce water in China. *Environmental Science & Technology* 48: 7704–7713.

Gerbens-Leenes, P.W. and Hoekstra, A.Y. 2008. *Business water footprint accounting: a tool to assess how production of goods and services impacts on freshwater resources worldwide* [online, accessed March 2019]. Available at www.waterfootprint.org/Reports/Report27-BusinessWaterFootprint.pdf.

Godfrey, J.M. and Chalmers, K. (eds). 2012. *Water accounting: international approaches to policy and decision-making*. Edward Elgar Publishing; Cheltenham.

Guan, D. and Hubacek, K. 2007. Assessment of regional trade and virtual water flows in China. *Ecological Economics* 61: 159–170.

Guan, D. and Hubacek, K. 2008. A new and integrated hydro-economic accounting and analytical framework for water resources: a case study for North China. *Journal of Environmental Management* 88: 1300–1313.

Hartman, L.M. 1965. The input–output model and regional water management. *Journal of Farm Economics* 47: 1583–1591.

Hoekstra, A.Y. 2003. *Virtual water trade*. IHE; Delft [online, accessed March 2019]. Available at www.waterfootprint.org/Reports/Report12.pdf.

Hoekstra, A.Y. 2007. Human appropriation of natural capital: comparing ecological footprint and water footprint analysis. *Ecological Economics* 68: 1963–1974.

Hoekstra, A.Y. 2008. *Water neutral: reducing and offsetting the impacts of water footprints* [online, accessed March 2019]. Available at www.waterfootprint.org/Reports/Report28-WaterNeutral.pdf.

Hoekstra, A.Y. 2010. The globalization of water. *In*: Jones, J.A. (ed.), *Water sustainability: a global perspective*. Hodder Education; London.

Hoekstra, A.Y. 2013. *The water footprint of modern consumer society*. Routledge; London.

Hoekstra, A.Y. and Chapagain, A.K. 2008. *Globalization of water: sharing the planet's freshwater resources*. Blackwell; Oxford.

Hoekstra, A.Y. and Mekonnen, M.M. 2012. The water footprint of humanity. *Proceedings of the National Academy of Sciences of the United States of America* 109: 3232–3237.

Hoekstra, A.Y., Chapagain, A.K., Aldaya, M.M. and Mekonnen, M.M. 2011. *The water footprint assessment manual – setting the global standard*. Earthscan; London.

Hubacek, K. and Sun, L. 2005. Economic and societal changes in China and its effects on water use. *Journal of Industrial Ecology* 9: 187–200.

Hubacek, K., Guan, D., Barrett, J. and Wiedmann, T. 2009. Environmental implications of urbanization and lifestyle change in China: ecological and water footprints. *Journal of Cleaner Production* 17: 1241–1248.

Lenzen, M., Moran, D., Bhaduri, A., Kanemoto, K., Bekchanov, M., Geschke, A. and Foran, B. 2013. International trade of scarce water. *Ecological Economics* 94: 78–85.

Liu, J., Liu, Q. and Yang H. 2016, Assessing water scarcity by simultaneously considering environmental flow requirements, water quantity, and water quality. *Ecological Indicators* 60: 434–441.

Liu, J., Zang, C., Tian, S., Liu, J., Yang, H., Jia, S., You, L., Liu, B. and Zhang, M. 2013. Water conservancy projects in China: achievements, challenges and way forward. *Global Environmental Change* 23: 633–643.

Ma, J., Hoekstra, A.Y., Wang, H., Chapagain, A.K. and Wang, D. 2006. Virtual versus real water transfers within China. *Philosophical Transactions of the Royal Society B – Biological Sciences* 361: 835–842.

Mekonnen, M.M. and Hoekstra, A.Y. 2011. The green, blue and grey water footprint of crops and derived crop products. *Hydrology and Earth System Sciences* 15: 1577–1600.

Mekonnen, M.M. and Hoekstra, A.Y. 2012a. A global assessment of the water footprint of farm animal products. *Ecosystems* 15: 401–415.

Mekonnen, M.M. and Hoekstra, A.Y. 2012b. *Water footprint* [online, accessed March 2019]. Available at www.waterfootprint.org/?page=files/productgallery.

Serrano, A., Guan, D., Duarte, R. and Paavola, J. 2016. Virtual water flows in the EU27: a consumption-based approach. *Journal of Industrial Ecology* 20: 547–558.

Tamea, S., Carr, J.A., Laio, F. and Ridolfi, L. 2014. Drivers of the virtual water trade. *Water Resources Research* 50: 17–28.

Velázquez, E. 2006. An input–output model of water consumption: analysing intersectoral water relationships in Andalusia. *Ecological Economics* 56: 226–240.

Wang, S. and Chen, B. 2016. Energy–water nexus of urban agglomeration based on multiregional input–output tables and ecological network analysis: a case study of the Beijing–Tianjin–Hebei region. *Applied Energy* 178: 773–783.

Wichelns, D. 2010. Virtual water and water footprints offer limited insight regarding important policy questions. *Water Resources Development* 26: 639–651.

Yang, H. and Zehnder, A. 2007. 'Virtual water': an unfolding concept in integrated water resources management. *Water Resources Research* 43: W12301.

Zhang, C. and Anadon, L.D. 2014. A multiregional input–output analysis of domestic virtual water trade and provincial water footprint in China. *Ecological Economics* 100: 159–172.

Zhang, Z., Shi, M. and Yang, H. 2012. Understanding Beijing's water challenge: a decomposition analysis of changes in Beijing's water footprint between 1997 and 2007. *Environmental Science and Technology* 46: 12373–12380.

Zhao, X., Liu, J., Liu, Q., Tillotson, M.R., Guan, D. and Hubacek, K. 2015. Physical and virtual water transfers for regional water stress alleviation in China. *Proceedings of the National Academy of Sciences of the United States of America* 112: 1031–1035.

Zhao, X., Liu, J., Yang, H., Duarte, R., Tillotson, M.R. and Hubacek, K. 2016. Burden shifting of water quantity and quality stress from megacity Shanghai. *Water Resources Research* 52: 6916–6927.

Zhao, X., Li, Y.P., Yang, H., Liu, W.F., Tillotson, M.R., Guan, D., Yi, Y. and Wang, H. 2018. Measuring scarce water saving from interregional virtual water flows in China. *Environmental Research Letters* 13: 054012.

CHAPTER THIRTEEN

The future of water

Joseph Holden

13

LEARNING OUTCOMES

After reading this chapter you should be able to:

- explain some of the ways water relates to most of the Sustainable Development Goals;
- describe some issues with predicting and responding to water challenges of the future;
- outline the features of integrated catchment management;
- consider why an interdisciplinary approach to water resources is important.

13.1 INTRODUCTION

Water is of fundamental importance to our existence and, in addition, has huge economic and cultural significance. Yet the world today faces massive water challenges. Given major drivers like population growth, economic development, climate change, sea-level rise, migration and forced displacement, the water challenges ahead will be even greater than those at present. The United Nations has adopted 17 **Sustainable Development Goals** (SDGs) to eradicate poverty by 2030. Two of these SDGs refer directly to water, while water must be considered in order to deliver most of the other SDGs. As members of the United Nations, countries must strive to deliver the SDGs.

However, while countries are ever more interconnected via trade, travel and social media the global geopolitical context at present is also one of increased nationalisation. This nationalist agenda enhances turbulence and uncertainty, which, when combined with the fact that water problems are also often non-linear with sudden catastrophic shifts in state driven by what at first appears to be slow, linear change, means that the exact nature of future water issues are more difficult to forecast. This chapter deals with the above issues, noting that many water solutions require integrated management and interdisciplinary approaches that deal with interconnectedness between environmental and social processes, at both river basin scale and at a global scale.

13.2 WATER AND THE SUSTAINABLE DEVELOPMENT GOALS

The SDGs were agreed in 2015, by the United Nations, as targets to be achieved by the year 2030. There are 17 such goals (Figure 13.1). Their purpose is to drive activity that ends all forms of poverty, calling for actions by all countries, whether rich or poor, to promote prosperity while providing environmental protection. Each SDG has a set of targets associated with it, resulting in 169 targets in total (United Nations, 2015). It is estimated that to achieve the SDGs will cost around $4 trillion per year. Many major multinational corporations, such as Unilever, have also understood that they can play a major role in supporting the SDGs and that, in turn, they will benefit from them with a more sustainable cycle of supply and demand, with reduced risks and better skills, in a world without poverty and hunger and that is resilient to climate change (e.g. see https://tinyurl.com/y55se3ce). While SDGs 6 and 14 have the word 'water' in their title (Figure 13.1), it should be understood that to achieve many of the other SDGs requires significant attention to be paid to water resource management and that there are both water-related trade-offs and water-related synergies across the SDGs that need to be worked through. In fact, some have argued that interrelationships between sectors and the various SDGs and targets need to be better mapped out so that we do not end up with perverse outcomes (Stafford-Smith *et al.*, 2017). For example, promoting increased consumption to alleviate poverty may lead to a failure in sustainably managing water.

To achieve 'zero hunger' will require efficient use of water for agriculture (Chapters 1 and 9) since the food sector is the largest water user globally. It will also require new forms of aquaculture and hydroponics (Chapter 6) to ensure that we protect and enhance life below water (SDG 14), rather than deplete our aquatic systems by over-harvesting or polluting them. Health and well-being (SDG 3) are reliant on dealing with physical and biological water risks (Chapter 7). In addition, the promotion of access to ecologically rich water features in urban settings can improve mental well-being (e.g. Völker and Kistemann, 2013), encourage exercise and support cultural and community engagement (Thomas, 2015; Higgins, 2019) (SDG 11). In Chapter 7 it was explained how gender inequalities (SDG 5) meant that in some poor communities women often faced much greater challenges than men in the collection of clean water and in sanitation-related burdens. Indeed, for SDG 6 there is a specific target that focuses on women and girls, which is 'by 2030, achieve access to adequate and equitable sanitation and hygiene for all and end open defecation, paying special attention to the needs of women and girls and those in vulnerable situations'.

One of the greatest water challenges is that of safe and clean sanitation systems (SDG 6) for large parts of the world where currently the systems are totally inadequate or non-existent. There truly is a global water crisis. It cannot be otherwise when so many people in the world have no access to safe water provided to their home (2.1 billion people) or adequate sanitation (4.5 billion people) (WHO and UNICEF, 2017) and while the population is set to increase by another 2 billion within the next three decades. Water supply and sanitation systems are crucial for health and well-being, and while it is possible to move towards carbon-neutral systems that recover resources from wastewater in the developed world (Chapter 8) there is still an enormous way to go before large proportions of the

Figure 13.1 The Sustainable Development Goals for 2030.

Source: From United Nations. 2015. *Transforming our world: the 2030 Agenda for Sustainable Development.* © 2015 United Nations. Reprinted with the permission of the United Nations.

world's population have adequate sanitation infrastructure (Chapter 7). Groundwater resources are being used unsustainably in many regions, which means that they will become more scarce and more costly to obtain (Chapter 5). However, it is no good sitting back and letting disaster happen. The purpose of the SDGs is to ensure countries and organisations take action. We need to be innovative with technology and look for new approaches. Plenty of examples have been mentioned in the other chapters of this book of recent innovation and discoveries which show that new ideas are emerging throughout the wide-ranging subject of water resources and novel ideas should be appropriately supported, and this requires political will. An example is provided in Box 13.1.

Hydropower is often thought to be an important way of generating affordable and clean energy (SDG 7), thereby delivering climate action (SDG 13). However, often hydropower schemes can cause conflict (Chapter 11; SDG 16) if they involve water diversion projects, reduce river flows downstream or damage aquatic ecosystems (Chapter 6). Nevertheless, as indicated in Box 11.6 in Chapter 11, hydropower can also be a way of bringing countries together, providing co-benefits. At a smaller scale, local community-driven hydropower schemes are possible with small in-river devices that rotate, generating electricity, driven by river flow (Figure 13.2). Using wastewater recovery techniques to generate power is also important and should be rolled out much more widely to enable self-sufficiency of the treatment process and the surplus energy to be provided back to communities (Chapter 8).

SDGs 8–12 each require an approach that utilises thinking around water demand planning (Chapter 9) and water economics (Chapter 10). Economic growth that is built around greater and greater water demand will be unsustainable. Thus, approaches that consider whole system economics and the true value of ecosystem services provided by life on land (SDG 15) and life below water (SDG 14) are required. In other words, we need to understand the true value of water, and this is not equivalent to the price we pay for water being supplied to a tap in our homes. Innovation and infrastructure are required that are water efficient, enabling sustainable cities. For example, researchers at the University of Waterloo (Cody et al., 2018) have developed technology to detect small leaks in water supply pipes by combining sophisticated signal processing techniques with artificial intelligence software to automatically detect the sound of leaks carried along water pipes. Artificial intelligence is a computing technique that learns the patterns in something (e.g. an image, sounds, text) from thousands of real examples. Once it has learned these patterns it can then be deployed to new situations and will give an output when it detects the same pattern occurring. The acoustic signatures are recorded by hydrophone sensors that can be installed in accessible places on the pipe network (such as fire hydrants). Detecting and fixing leaks early is important as water supply pipes in most urban areas have a high water leakage rate (typically 15–30%).

SDG 12 reminds us of the important analysis presented in Chapter 12 that showed how production, trade and consumption moves virtual water around the world. The term 'virtual water' can be somewhat misleading as the water use is real. Use of water in the production of goods (perhaps also causing some pollution) which are purchased and used in another region or country means that water use has been imported by that purchasing region. Including water footprint analysis in thinking about water resource management can enable enhanced water efficiencies and more responsible production and consumption. The same product can require more water to produce it in some places compared

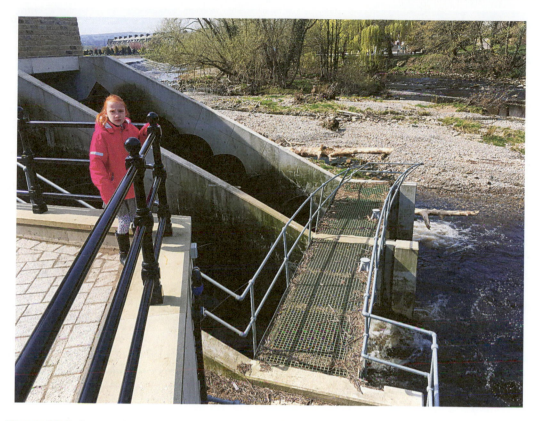

Figure 13.2 Community hydropower scheme on the River Wharfe, Otley, UK.

to others, or it may be more logical to support the production of certain products in water-rich regions, rather than have to develop hugely disruptive water diversion projects or create unnecessary water scarcity (and hence enhanced risk from climate change) in other regions.

SDGs 16 and 17 (Figure 13.1) remind us of the discussion in Chapter 11 on water conflict, law and governance. There are a huge number of international water conflicts around the world as a result of the fact that neighbouring countries share water resources such as lakes, rivers and groundwater, and that often water use in one country has an impact on the water availability in a neighbouring country. Such conflict can also occur within countries from region to region and landowner to landowner. Partnership working is required to tackle such issues.

REFLECTIVE QUESTIONS

What water-related actions is your own country undertaking to support the SDGs?

Do you take any personal actions that support the SDGs?

BOX 13.1

Bill and Melinda Gates Foundation: reinvent the toilet challenge

Since 2011 the Bill and Melinda Gates Foundation has supported a programme of work with engineers and inventors to reinvent the toilet for use in locations where there is no access to sanitation or energy grids. The main features should be that it:

- removes germs from human waste and recovers valuable resources such as energy, clean water and nutrients;
- operates 'off the grid' without connections to water, sewer or electrical lines;
- costs less than US$0.05 per user per day;
- promotes sustainable and financially profitable sanitation services and businesses that operate in poor, urban settings;
- is a truly aspirational next-generation product that everyone will want to use – in developed as well as developing nations.

In November 2018 an expo in Beijing showcased some of the inventions. Companies from China (Clear, CRRC, EcoSan), the United States (Sedron Technologies), India (Eram Scientific, Ankur Scientific, Tide Technocrats) and Thailand (SCG Chemicals) announced the world's first pathogen-killing toilets and small-scale waste treatment plants (called omni-processors). In addition, finance institutions at the expo such as the World Bank, the Asian Development Bank and the African Development Bank confirmed investments in city-scale projects that provide people in all areas, including the poorest, with safely managed sanitation services, supporting adoption of novel non-sewered sanitation solutions. UNICEF also announced a new market-shaping strategy to help scale and deploy sanitation product and service innovations and increase private-sector engagement. These examples from the same expo illustrate how inventions on their own are not enough, and a range of sectors have to come together to deliver the transformations needed.

13.3 WATER IN A TUNA WORLD

13.3.1 The globalisation–nationalisation paradox

Water resource planning requires making predictions about future demand requirements. However, the world is changing at a fast rate in many ways, which makes planning more difficult. Economic growth is accelerating, as is water use. Atmospheric CO_2 concentrations, energy use, the number of large dams, international tourism and urban population are among the indicators showing accelerating growth since 1950 (Steffen *et al.*, 2015). These are signals of what many researchers call the Anthropocene – the period of Earth's history where environmental change is dominated by human action. While the world is becoming more interconnected via trade, tourism, telecommunications and social media, most population growth since 1950 has been in non-OECD countries, yet the world's economy is still dominated by OECD countries (Steffen *et al.*, 2015). Thus, there is greater inequality between rich and poor. This inequality, a number of major events such as 9/11, the global financial crisis of 2007/08, and migration from countries with strife such as Syria and Libya have driven much more nationalism. The start of the third decade of the twenty-first century has quite a different feel to the one that appeared to be moving forward at the start of the century, when globalisation and inter-state cooperation seemed to be dominating the agenda. Inward-looking nationalism can drive countries to cooperate less with each other (e.g. the vote for the UK to leave the European Union, talk of border walls between Mexico and the USA, and tariff wars between the USA and China). The paradox is that, because of globalisation, countries are intricately connected to each other and rely on other countries for a huge array of products, from basic foodstuffs to electronics and clothing, often supplied in a 'just-in-time' manner (i.e. they are not stockpiled). This interconnectedness at a time of enhanced nationalism means that the risks of disruption can be greater. For example, extreme weather resulting from climate change (Chapter 2) in one region, a change in a government's policy on steel tariffs or some inappropriate words by a political leader on social media can result in a strong impact felt around the world on food prices and other sectors, since sectors are also strongly connected (Challinor *et al.*, 2018).

Some researchers talk about the future world being a TUNA world (e.g. Ramirez and Wilkinson, 2016). This is a world in which there is ever increasing *t*urbulence, *u*ncertainty, *n*ovelty and *a*mbiguity. Such a world may make partnership working more challenging and water demand forecasting and planning less reliable; what worked in the past may no longer work in the future. One approach to dealing with this problem is to develop a number of possible future scenarios to enable us to think through the implications of these scenarios for making effective decisions today. A scenario would be a made-up, but plausible, future context within which we could think through the water demand requirements. Such an approach would require people from different disciplines to come together to work out for each scenario what the situation might mean for different sectors or global connections. This process in itself would be beneficial because it would enable people to learn from one another as well as to think through the implications of actions taken now.

13.3.2 Linear and non-linear systems

In the water environment, change may be linear or non-linear. It can be reversible or irreversible. Slow, linear and reversible change on a small scale is easier to predict and deal with than fast, non-linear, irreversible change on a global scale.

A linear, reversible change might be the drawdown of the water table in groundwater (making it more costly to pump) through abstraction, in a location where the groundwater can be rapidly replenished by rainfall. By reducing the rate of abstraction to below the recharge rate we can recover groundwater levels. An irreversible change might occur when we pump groundwater out for crop irrigation in an arid or semi-arid region. Here groundwater may not be replenished so readily by rainfall. Therefore, each time the water is used for irrigation, because some is evaporated, the concentration of dissolved **solutes** in the groundwater may increase and the water table may get deeper. This concentration effect is exacerbated because water has to be pumped from deeper layers progressively over time and deeper groundwater is likely to be richer in solutes. Thus, slowly over time, the water becomes saltier with each irrigation event and eventually salinisation of soils occurs, making them unsuitable for crops (**Figure 13.3**). At this point the system has been damaged to the point where it can no longer be used for agriculture and hence a critical threshold has been reached beyond which the system can no longer function and the change may be irreversible. Around 20% of irrigated land globally is affected by salinisation.

At a global scale, sea-level rise can be viewed as a gradual linear change, where positive feedback effects compound the

changes which on human timescales are now not reversible (**Box 13.2**).

Non-linear change can mean that what at first seems like a slow transition can suddenly turn into a rapid shift of the system. Rivers can sometimes suddenly change the position of their channel (channel avulsion) on the floodplain following a slow and steady accumulation of sediment on a meander bend. Another example comes from studies of shallow lakes. Many shallow lakes have fairly clear water with aquatic plants that have roots which stabilise the bed sediments and shelter grazers that feed on phytoplankton. However, a lake may be subject to a slow, gradual increase in nutrient loading over time, from agricultural management in the river basin upstream, for example. Once nutrient concentrations increase to a sufficient level a sudden bloom of phytoplankton can occur. This can starve other

Figure 13.3 Soil salinisation at this site in southern India has made the land unsuitable for crops.
Source: Shutterstock/Maxsimilian.

BOX 13.2

Sea-level rise

Ice sheet melt, glacier melt and thermal expansion of water (Chapter 1) are contributing to global sea-level rise, which is currently around 3.2 mm per year (Church *et al.*, 2013). Climate projections suggest that, by 2100, sea levels will be 26 to 82 cm higher than at present. This would suggest that coastal flooding will become more common and erosion will increase as coastlines retreat. Even today 200 million people are currently exposed to coastal flooding during storm surges. If sea levels rise by 1 m then over 1 million km^2 of land will be inundated (Rowley *et al.*, 2007). At current population densities and locations this would displace around 150 million people. Deltas that are currently being starved of sediment due to upstream dams and water diversion projects (e.g. the Indus – see Box 3.4 in Chapter 3) will erode much more rapidly, affecting major population centres. Saline intrusion of groundwater sources (Chapter 5) will also be increased. This is important as around 600 million people live within 10 m of sea level and 21 megacities are coastal (e.g. Shanghai, Tokyo, Mumbai, Dhaka). Of further concern is the survival of low-lying small island states such as Vanuatu, Tuvalu, Fiji and the Maldives. These are at risk of inundation or severe coastal erosion.

aquatic plants of light, meaning that sediment on the lake floor is no longer held stable. This sediment can now be stirred up and smother the phytoplankton grazers. The lake has suddenly become a turbid one (Figure 13.4). Sometimes, even if farm management upstream is altered and nutrient loads to the lake decline remarkably, the lake can still stay in a stable, turbid state because there is a lack of phytoplankton feeders and a lack of plants to stabilise the lake floor. Thus, the lake jumped from one stable state to another in a non-linear way (Scheffer *et al.*, 1993). This is an example of a 'catastrophic' change. To get the lake to return to its clear state may require human interventions in the lake. Such sudden changes from one state to another may not be restricted to the physical environment. Sociopolitical and economic regimes may also undergo slow gradual change that then cause the system to reach a threshold beyond which it suddenly changes. The global financial crisis of 2007/08 is a good example.

REFLECTIVE QUESTIONS

Despite good population forecasts and climate change models, why is it challenging to predict water resource problems in the future?

How can slow linear changes cause sudden non-linear change?

13.4 GOING BEYOND INTEGRATED CATCHMENT MANAGEMENT

River basins tend to be the unit of study for hydrologists (Chapter 3). Inputs and outputs are often measured at a river basin (also known as a catchment) level. They are also

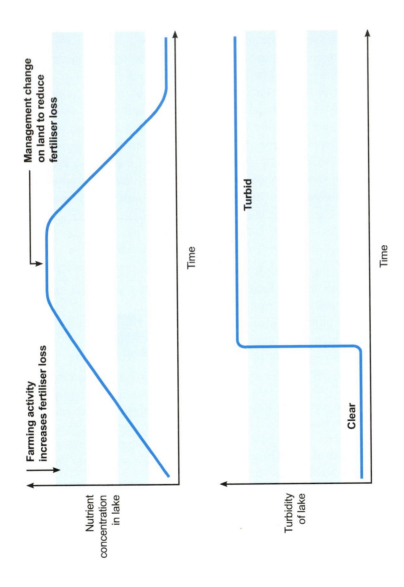

Figure 13.4 Sudden non-linear change in the turbidity of a shallow lake as a result of a steady linear change in nutrient loading to the lake.

a suitable unit for considering how management strategies can be used to manage water resources. Of course, river basins are nested within each other, but often it works well to consider the largest basin area for the system that flows to the sea or inland lake. Hydrological flowpaths influence water quality and river flows, and within a river basin there will be surface water – groundwater interactions. Land management (e.g. farming, industry, urban drainage, mining) impacts flowpaths for water within a river basin and influences water quality and aquatic ecosystems. Thus, rather than take a single factor approach (e.g. flood defences for a town), it may be more suitable to holistically consider catchment management to understand multiple land and water factors that interact. In doing so, a range of efforts need to be aligned including the efforts of those responsible for biodiversity, land management, and water management.

There are lots of examples, from all over the world, where integrated water management (Box 9.3 in Chapter 9) and integrated catchment management approaches have been adopted. For example, such an approach became state policy in Western Australia in 1988. An innovative approach in the UK, where a partnership between researchers and non-academic organisations is providing confidence in new catchment-based solutions, is described in Box 13.3. Where integrated catchment management approaches appear to be more successful is in situations where strong partnership approaches have been adopted involving not only government agencies, but also local community organisations and businesses (e.g. community engagement in Thailand; Box 1.1 in Chapter 1). For example, because much water pollution is diffuse, it is often challenging to trace its origin and so many landowners need to be encouraged to adopt

similar practices to reduce pollution. By joining forces and acting together, people can be encouraged to adopt strategies that they find workable on their land. An integrated catchment approach also indicates that rural and urban systems are considered holistically. Flood management might be served by replanting felled forests in the headwaters rather than urban defences for a town that could make the floods worse for the next town downstream. However, it is fair to say that much more research is needed to support such decision-making (Rogger *et al.*, 2017).

Taking a catchment-based approach, however, can sometimes mean socially inequitable solutions are most appropriate from a scientific perspective. The same land management practice (e.g. sheep farming) may have a different impact on water quality in one location (e.g. where overland flow is common) compared to the same activity in another location (where overland flow rarely occurs). This is because some parts of the catchment are more sensitive to a disturbance or to certain management practices than others (Burt, 2001). Thus, from a whole-catchment perspective it may be reasonable to allow the sheep farming in the second location and not in the first location. However, the farmer that runs the farm at the first location may feel very unfairly treated if just because of their location they are asked to do things differently. But that is the nature of taking an integrated catchment management approach as it seeks to identify where in the landscape activity is most effective from a joint land and water management perspective. It may be that economic or other incentives can be given to the first farmer who may, in turn, provide excellent financial benefits to water quality by changing land use on their land. Some ecosystem service economics approaches (Chapter 10) may help determine the effective payment schemes at a catchment level, and models have been developed that couple economics with bio-physical assessment at catchment scales (e.g. Kragt *et al.*, 2011).

Major water diversion projects often transfer water from one catchment to another. Therefore, these need to be borne in mind when taking a catchment approach. In addition, as we saw in Chapter 12, through trade, embedded water is moved from catchment to catchment and country to country. Hence, we also need to think about land and water management at much larger scales to determine water-efficient approaches to production and consumption. Thus, while integrated catchment management provides an excellent suitable starting point for tackling many freshwater issues, an even broader global management perspective is required. Many individual governments are unwilling to consider such a perspective but because of the interconnectedness of trade and the risks of food security it is important that they do so. Intergovernmental groups also need to come together to consider global water demand and big multinational corporations can play a vital role too as they have the ability to switch production to different locations or change practice across their entire supply chain to enhance water use efficiency to lessen the effects of water scarcity.

REFLECTIVE QUESTION

Can you think of any examples of integrated catchment management operating in your region?

13.5 INTERDISCIPLINARITY

This book has compiled materials from several disciplines applied to water resources. Water is not just a physical environmental subject. Ever since the beginning of civilisation there have been conflicts over water rights (Chapters 1 and Chapter 11), and access to water provides opportunities for economic advantage. Indeed, de Villiers (1999) argued that if you do not have enough freshwater you need to do one of the following: import it or make more of it (e.g. desalinisation), reduce demand and use less (typically by water conservation or pricing), or steal water from others! While some of the ways in which we currently manage our water resources make economic sense, others do not (Chapters 9 and 10). Thus, scientists, engineers, social scientists and humanities experts need to come together to develop new solutions for water. For example, evaluations of the basic economics of water do not necessarily get us around social justice issues (Chapter 11), but there are economic instruments that can help – through having tiered tariffs, for example (Chapter 7). The world is more globalised and interconnected than ever before, through trade and communications. These processes impact dramatically upon water resource use and allocation. As we have seen in this book, there are new ways of looking at water resources and water economics, such as water footprinting and virtual water flows (Chapter 12) and also the concept of ecosystem services (Chapter 10), which may help to redefine the way we make decisions on, govern, or at least accept water resource allocation and use on global to local scales, which means that we can avoid the last of de Villiers's (1999) strategies (the stealing of water).

Many of the chapters in this book deal directly with issues that are associated with

BOX 13.3

The Yorkshire Integrated Catchment Solutions Programme

The Yorkshire Integrated Catchment Solutions Programme (iCASP) is an ambitious and exciting programme to generate benefits by applying environmental science to catchment challenges. It brings together academic researchers working on water, climate and land-related issues with businesses, government bodies, charities and community partnerships. The focus is the Yorkshire Ouse drainage basin, home to 6.7% of the UK population and covering a third of the land area of northern England. This catchment area is facing some complex and costly challenges: flooding and drought, soil and water degradation, loss of agriculturally productive land and important ecosystems. There are major cities and towns in the region and huge potential for further environmentally sustainable growth. By taking a whole-catchment approach the programme seeks to understand where there are opportunities to maximise the return for investment. For example, if a water company seeks to improve water quality at its raw water intakes by investing in land management, it would be even better if that investment can be combined with planned investment in natural flood management, biodiversity or carbon sequestration by government authorities or community groups. That way we can seek to obtain many benefits from the same targeted solutions such as tree planting or peatland restoration. By integrating solutions and by integrating organisations, iCASP is helping to deliver change on the ground, change in policy, change within organisations and creating a network of catchment management experts. In making the urban and rural parts of the region more resilient to droughts and floods in the future, and embedding a strong understanding of water-related climate predictions, this gives regional planners and businesses more confidence to invest in new developments or change the way they operate, thereby supporting economic growth. In turn, iCASP can inform those new developments to ensure they include water-sensitive, water-efficient, climate-resilient designs. Examples of projects undertaken by iCASP are listed below:

- supporting business case developers and appraisers with green and blue infrastructure economic valuation and changing government guidelines on economic appraisal;
- piloting new climate change projections with stakeholders to ensure they can best use such datasets;
- agricultural land management and public goods delivery – review of evidence to support policy and practice;
- optimal peatland restoration in Yorkshire – spatial mapping and modelling tools to optimise restoration works, and guidance for economic investment appraisal;
- supporting the Yorkshire natural flood management pilots to deliver evidence and guidance that will influence national implementation and financing of natural flood management;
- integrating natural flood management into payment for outcomes schemes in the Yorkshire Dales;
- enhanced surface water flood forecasts for Yorkshire in rapid response catchments;
- supporting local authorities to reduce the spread of invasive non-native species.

A number of useful tools have been developed including summaries of evidence, economic valuation guidance documents, business case guidelines and user-friendly spatial planning models, which can also be more widely implemented. After only two years of operation the iCASP projects have been estimated to have delivered many millions of dollars of economic benefit, supported new business cases for investment worth hundreds of millions of dollars, and delivered a larger, better-connected network of practitioners and policymakers. For more information visit: https://icasp.org.uk.

sustainability. Here we refer to sustainability in the sense of meeting the needs of the present without compromising the ability of future generations to meet their own needs. To achieve sustainability requires people from different backgrounds and disciplines to work together. Abstracting groundwater that is ancient, and non-renewable in the short term, is not sustainable (Chapter 5). This is why most assessments of available freshwater resources for countries or per person per year are based on 'renewable' freshwater resources (Chapter 1). Managing and forecasting water demand is clearly an important approach to sustainability. This requires scientists and engineers who understand water-cycle processes to work with climate change modellers and with those who work on population dynamics, migration, land use, cropping systems and so on in order that the science base is best prepared for the challenges ahead and so that the policy community can be best informed. Chapter 9 outlined several strategies for demand management, noting that a focus on innovation in the agricultural sector was crucial for global water resources because that is where most consumptive water use takes place. Food waste and overconsumption of water will have to be reduced. The world is interconnected and trade means that, even if you live in an area that has abundant water, you might be contributing to water scarcity for other people in different parts of the world through your use of products from those water-scarce areas. However, at the same time this consumption provides income for the people in those regions, so assessments cannot be based solely on an analysis of water alone.

The misuse and pollution of water (e.g. Chapters 4–6) adds unnecessary pressure to the water resource system, makes water resources less sustainable, worsens risks to human health (Chapter 7), damages the economy owing to reduced provision of ecosystem services (Chapter 10), causes conflict (Chapter 11) and reduces resilience to climate change (Chapter 2). It also impacts upon the world's vast oceans. In Chapter 1 it was noted that only a tiny proportion of the water on Earth is available as an accessible freshwater resource for us to use. However, our pollution of this resource has entered the oceans and causes damage to marine life. For example, in Chapter 4 the nutrient enrichment of the Gulf of Mexico caused in large part by agricultural pollution across the US Midwest was noted. The United Nations estimates that 10% of the plastic we produce ends up in the oceans, with around 18,000 pieces per square kilometre. Such plastic can be harmful to sea creatures, not only due to entanglement but also when it is eaten. Many tiny pieces of plastic that have broken off larger pieces as a result of wave action and abrasion can be mistaken for plankton (in some parts of ocean centres there may be six times as many tonnes of plastic than plankton) and eaten, potentially causing the bioaccumulation (see Chapter 6) of toxins in sea creatures.

The ocean currents move the floating debris of plastic bags, bottles, cartons, caps and so on, which often later collect in concentrated areas in the centre of the surface ocean circulation systems. Some of these remote ocean locations can have more than 335,000 pieces of plastic per square kilometre, demonstrating the far-reaching impacts of humans upon water across the world.

Contemporary climate change is impacting upon the water cycle and will have major consequences for many parts of the world in terms of water availability and water hazards (Chapter 2), which in turn will impact upon economic performance. Climate change impacts need to be understood by people from different disciplinary backgrounds in the context of current and future projected trade patterns and virtual water flows so as to establish whether patterns are sustainable in the long term. Both climate and land-management change (through population growth, development, agricultural expansion, etc.) affect how water moves through river basins and impacts upon river flow (Chapter 3) and surface water quality (Chapter 4), which has knock-on impacts for aquatic ecosystems (Chapter 6) and associated costs in terms of economics, as well as costs for human health (Chapter 7) and wider ecosystem services (Chapter 11). These processes also impact upon groundwater, while the expanding need for water is also causing water quantity and quality problems for groundwater abstraction (Chapter 5).

It is therefore clear that we need to ensure environmental scientists, engineers, social and political scientists, businesses, policymakers and local communities work together to tackle the world's water resource issues. An integrated, interdisciplinary approach is essential because our water problems are multifaceted and interrelated; they are also vital to the survival of every person on the planet. Solving a technological problem on water may not be beneficial if there is no political or social will or the economic means to utilise that technological advance. An example of a major interdisciplinary water project that brings researchers from different disciplines together with government agencies, community groups and practitioners is provided in Box 13.4.

> **REFLECTIVE QUESTION**
>
> Can you think of a water resource problem that would benefit from different disciplines working together?

13.6 SUMMARY

The United Nations have adopted SDGs. Some of these SDGs are directly about water, while others must be underpinned by efficient water management and use. All countries have a responsibility to deliver actions to support the SDGs. However, anticipating what the future may hold for water and water cooperation is challenging because, while the world is currently in a phase of increasing globalisation and interconnectedness, it is also, paradoxically, in a phase of increasing nationalisation and self-interest. Water problems (both physical and sociopolitical) are also often non-linear, with sudden catastrophic shifts in state driven by what at first appears to be slow, linear change. To tackle the SDGs and to tackle major water issues therefore requires integrated management solutions that deal with interconnectedness and feedbacks

BOX 13.4

Interdisciplinary water research in East Africa

Availability of water is fundamental for development in the East African region. However, this vital resource is already under stress from land degradation, pollution and overfishing. Climate change adds to these problems, greatly increasing the vulnerability of the poorest people in the region. A project led by John Marsham, of water@leeds, called 'Integrating Hydro-Climate Science into Policy Decisions for Climate-Resilient Infrastructure and Livelihoods in East Africa', shortened to HyCRISTAL, is developing a new understanding of climate change and its impacts in the region, working with decision makers to manage water for a more climate-resilient future. The focus region includes Burundi, Kenya, Rwanda, Tanzania and Uganda, but the project addresses a wider region including Somalia and Ethiopia. Driven by East African priorities, the overarching goal of HyCRISTAL is to develop a new understanding of East African climate variability and change, and their impacts, and to work with regional decision makers to support effective long-term (five- to 40-year) decision-making in the face of a changing climate. In particular, it is designed to understand, quantify and reduce the uncertainty in the regional climate projections and, in collaboration with a range of stakeholders, co-develop climate change coping options that meet the region's societal needs in both urban and rural areas through a series of pilot demonstration projects, covering urban water sanitation and health, rural livelihoods, water management, tea production, transport and lake levels. As such, the project involves climate and atmospheric scientists, hydrologists and hydrogeologists, water and sanitation engineers and social scientists working together (Figure 13.5).

Figure 13.5 The HyCRISTAL research team includes researchers from many disciplines coming together to develop solutions to major climate-related water challenges in East Africa.
Source: Julio Araujo.

HyCRISTAL is making several active contributions to East African policymaking, explicitly based on climate science and impact pilot experiences within the project. HyCRISTAL outputs have been discussed with the city authorities of Kampala and Kisumu, the Ugandan Ministry of Water, tea producers and the World Bank. HyCRISTAL has been funded by the UK Natural Environment Research Council (NERC), the UK Department for International Development (DFID) through the Future Climate for Africa (FCFA) programme.

Source: Information courtesy of John Marsham, HyCRISTAL Principal Investigator, and Gabriela Lopez-Gonzalez, water@leeds coordinator.

between land, water, climate, society, politics and economics. These integrated solutions need to be considered at the river basin scale and, increasingly, as we consider trade and sustainable consumption and production, at a global scale. They also require interdisciplinary approaches because there are scientific and technological issues to be addressed, there are economic and political issues to be addressed and there are social and cultural issues to be addressed and these all interconnect. Solving a technological problem on water supply may be futile if there is no political or social will or the economic means to utilise that technological advance.

I hope that you have found it useful to read all of the chapters in this book, as they cover physical, environmental, technological, social, political and economic aspects of water resources. The purpose of the book has been to bring together these subject areas related to water in one volume as a learning and reference aid. I hope you agree that it has been highly beneficial to do so and now feel willing, able and inspired to learn more about the interdisciplinary subject of water.

FURTHER READING

Ramirez, R. and Wilkinson, A. 2016. *Strategic reframing: the Oxford scenario planning approach.* Oxford University Press; Oxford.

The book is not about water, but outlines the TUNA issues discussed in Section 13.3 and ways of thinking about future scenarios.

Sharp, E. 2017. *Reconnecting people and water: public engagement and sustainable urban water management.* Routledge; New York.

A book which shows the importance of social science perspectives to deliver informed water management.

Smith, L., Porter, K., Hiscock, K., Porter, M.J. and Benson, D. (eds) 2015. *Catchment and river basin management: integrating science and governance.* Earthscan, Routledge; New York.

A book with examples from around the world showing the importance of bringing disciplines together and the wide range of institutions needed to provide integrated catchment management approaches.

Maia, R. and Pereira, L. (eds) 2015. Special issue: water resources management in an interdisciplinary and changing context. *Water Resources Management* 29: 211–661.

A series of 27 papers on water resource management, providing studies from all over the world highlighting the need for an interdisciplinary approach in the context of a rapidly changing world.

United Nations. 2015. *Transforming our world: the 2030 agenda for sustainable development.* United Nations; New York.

The original declaration on the SDGs which sets out the ambition, principles and approach.

Classic papers

Burt, T.P. 2001. Integrated management of sensitive catchment systems. *Catena* 42: 275–290.

A classic paper highlighting the importance of spatial variability for integrated catchment management approaches.

Kuylenstierna, J.L., Björklund, G. and Najlis, P. 1997. Sustainable water future with global implications: everyone's responsibility. *Natural Resources Forum* 21: 181–190.

It is interesting to go back to the 1990s and evaluate what was being forecast and what were seen as key issues for water resources for the 2020s and 2030s.

Scheffer, M., Hosper, S.H., Meijer, M.L., Moss, B. and Jeppersen, E. 1993. Alternative equilibria in shallow lakes. *Trends in Ecology and Evolution* 8: 275–279.

A paper that illustrates non-linear change in water systems studying shallow lakes as an example.

Stafford-Smith, M., Griggs, D., Gaffney, O., Ullah, F., Reyers, B., Kanie, N., Stigson, B., Shrivastava, P., Leach, M. and O'Connell, D. 2017. Integration: the key to implementing the Sustainable Development Goals. *Sustainability Science* 12: 911–919.

A paper that points out how there are links across the SDGs that need to be worked on so that there is overall success rather than sector-specific success.

PROJECT IDEAS

- Draw a diagram that shows links between the SDGs (this may get rather complicated and end up looking like spaghetti).

- For each SDG undertake some research to show what water-related progress has been made to date and what further actions are proposed by governments, businesses and NGOs.

- For a case study river basin, report on what integrated catchment management activity has taken place, what groups are collaborating and what datasets are available to evaluate success. Suggest what further opportunities there are for further integrated catchment management work in the basin.

REFERENCES

Burt, T.P. 2001. Integrated management of sensitive catchment systems. *Catena* 42: 275–290.

Challinor, A., Adger, W.N., Benton, T.G., Conway, D., Joshi, M. and Frame, D. 2018. Transmission of climate risks across sectors and borders. *Philosophical Transactions of the Royal Society A: Mathematical, Physical and Engineering Sciences* 376: doi: 10.1098/rsta.2017.0301.

Church, J.A., Clark, P.U., Cazenave, A., Gregory, J.M., Jevrejeva, S., Levermann, A., Merrifield, M.A., Milne, G.A., Nerem, R.S., Nunn, P.D., Payne, A.J., Pfeffer, W.T., Stammer, D. and Unnikrishnan, A.S. 2013. Sea level change. *In:* Stocker, T.F., Qin, D.,

Plattner, G.K. *et al.* (eds), *Climate change 2013: the physical science basis. Contribution of Working Group I to the Fifth Assessment Report of the Intergovernmental Panel on Climate Change.* Cambridge University Press; Cambridge; New York, 1137–1216.

Cody, R., Harmouche, J. and Narasimhan, S. 2018. Leak detection in water distribution pipes using singular spectrum analysis. *Urban Water Journal* 15: 636–644.

de Villiers, M. 1999. *Water wars. Is the world's water running out?* Phoenix Press; London.

Higgins, S.L., Thomas, F., Goldsmith, B., Brooks, S.J., Hassall, C., Harlow, J., Stone, D., Völker, S. and White, P. 2019. Urban freshwaters, biodiversity, and human health and well-being: setting an interdisciplinary research agenda. *WIREs Water* 6: doi: 10.1002/wat2.1339.

Kragt, M.E., Newham, L.T.H., Bennett, J. and Jakeman, A.J. 2011. An integrated approach to linking economic valuation and catchment modelling. *Environmental Modelling and Software* 26: 92–102.

Kuylenstierna, J.L., Björklund, G. and Najlis, P. 1997. Sustainable water future with global implications: everyone's responsibility. *Natural Resources Forum* 21: 181–190.

Maia, R. and Pereira, L. (eds) 2015. Special issue: water resources management in an interdisciplinary and changing context. *Water Resources Management* 29: 211–661.

Ramirez, R. and Wilkinson, A. 2016. *Strategic reframing: the Oxford scenario planning approach.* Oxford University Press; Oxford.

Rogger, M., Alaoui, A., Agnoletti, M., Bathurst, J., Bodner, G., Borga, M., Chaplot, V., Gaál, L., Gallart, F., Glatzel, G., Hall, J., Holden, J., Holko, L., Horn, R., Kiss, A., Kohnova, S., Leitinger, G., Lennartz, B., Parajka, J., Peth, S., Perdigão, R., Plavcová, L., Robinson, M., Salinas, J.L.,

Santoro, A., Szolgay, J., Tron, S., van den Akker, J., Viglione, A. and Blöschl, A. 2017. Land-use change impacts on floods at the catchment scale – challenges and opportunities for future research. *Water Resources Research* 53: 5209–5219.

Rowley, R.J., Kostelnick, J.C., Braaten, D., Li, X. and Meisel, J. 2007. Risk of rising sea level to population and land area. *EOS Transactions of the American Geophysical Union* 88: 105–116. http://dx.doi.org/10.1029/2007EO090001.

Scheffer, M., Hosper, S.H., Meijer, M.L., Moss, B. and Jeppersen, E. 1993. Alternative equilibria in shallow lakes. *Trends in Ecology and Evolution* 8: 275–279.

Stafford-Smith, M., Griggs, D., Gaffney, O., Ullah, F., Reyers, B., Kanie, N., Stigson, B., Shrivastava, P., Leach, M. and O'Connell, D. 2017. Integration: the key to implementing the Sustainable Development Goals. *Sustainability Science* 12: 911–919.

Steffen, W., Broadgate, W., Deutsch, L., Gaffney, O. and Ludwig, C. 2015. The trajectory of the Anthropocene: the great acceleration. *The Anthropocene Review* 2: 81–98.

Thomas, F. 2015. The role of natural environments within women's everyday health and wellbeing in Copenhagen, Denmark. *Health & Place* 35: 187–195.

United Nations. 2015. *Transforming our world: the 2030 agenda for sustainable development.* United Nations; New York.

Völker, S. and Kistemann, T. 2013. 'I'm always entirely happy when I'm here!' Urban blue enhancing human health and well-being in Cologne and Düsseldorf, Germany. *Social Science and Medicine* 78: 113–124.

WHO and UNICEF. 2017. *Progress on drinking water, sanitation and hygiene: 2017 update and SDG baselines.* World Health Organization; Geneva.

Glossary

Absolute water scarcity According to the United Nations' use of the term, this occurs when the renewable freshwater available is less than 500 m^3 per person per year.

Acid neutralising capacity The ability of a substance to change an acidic substance into a neutral one, as well as buffering an acidic solution.

Adiabatic cooling This is the process whereby a rising parcel of air cools, due to the expansion of the parcel.

Adit A passage into a mine which is nearly horizontal and may be used for access or drainage.

Adsorption–desorption Adsorption is the uptake or attracting of a substance to the surface of another material, while desorption is the release of a substance from the surface of another material. This may be through the surface or from the surface itself.

Advect To transfer by the flow of a fluid, often moving a body of mass horizontally.

Aerosols Microscopic particles contained within the atmosphere that interfere with the Earth's incoming energy from the Sun and the outgoing energy being emitted from the Earth's surface. Aerosols can result in either cooling or warming of the Earth's climate.

Albedo The proportion of incoming energy from the Sun that is reflected by a surface. Snow has a high albedo, whereas tarmac has a low albedo.

Allochthonous Something which originates a large distance away from where it is found.

Alluvial fan A depositional landform which develops when a fast-moving stream is given the opportunity to slow its speed and become more spread out in its arrangement. They often occur as a river exits a constrained geomorphology such as a canyon and enters a flatter, wider expanse of land.

Anions *Ions* formed when an atom or group of atoms has gained one or more negatively charged electrons. The result is that the atoms have a greater number of electrons (negatively charged) than protons (positively charged) and so anions are negatively charged.

Aquifer   Rock that is porous enough to take in and retain water and permeable enough to allow groundwater to pass through it freely.

Aquitard   A layer of slightly permeable rock or other material that is adjacent to a more permeable layer of rock, the latter of which readily holds and transmits water. The aquitard restricts flow of water in or out of the more free-flowing *aquifer* layer. If the rock layer completely stops water flow to and from the adjacent layers then it is often known as an aquiclude.

Artesian well   A well drilled into an aquifer which is trapped between two layers of impermeable rock. The aquifer receives its water from a higher altitude than the one at which it is situated and this creates pressure which forces the water to rise upwards in the well. Owing to this natural pressure there is no need for a pump in an artesian well with the water rising above the level of the water table. Dependent upon the pressure which exists, the water level can reach the same level as the ground surface.

Asymptote   A straight line which approaches a curve but never quite meets it. The distance between the curve and the straight line approaches zero as they tend to infinity.

Autochthonous   Something which originates from the location in which it is found.

Avulsion   With reference to fluvial geomorphology, this is the action whereby a river channel rapidly abandons its current location in favour of a new channel, often as a result of becoming choked with sediment.

Baseflow   The stable portion of a river's discharge that is fed by groundwater.

Bases   Substances that have a pH higher than 7 and release hydroxide ions (OH^-).

Bedload transport   The movement of material in a river which has partial contact with the riverbed. Movement may be through the processes of rolling, sliding and saltating.

Benthic layer (aka Benthic zone)   Ecological zone found at the deepest point of a body of water such as the sea floor or a lake bottom.

Benthos   The community of organisms that live around the bed of a pond, lake, river or sea.

Bioaccumulate   The build-up of toxic substances, such as heavy metals and pesticides, in the tissue of living things.

Bioavailable (Bioavailability)   The potential for adsorption and use of a nutrient by an organism.

Biochemical oxygen demand (BOD)   A measure of the amount of oxygen required in a given set of conditions for aerobic organisms in water to break down the organic matter present.

Biofilm   A build-up of organisms, including bacteria, algae and fungi, in a slimy, thin layer on solid surfaces in watery environments.

Biofuel   Fuel which is the product of recent living matter, providing what is sometimes considered to be a renewable energy source.

Boundary layer   The section of the atmosphere just above the Earth's surface. Here there is sharing of heat and moisture from the Earth's surface into the atmosphere.

Brownian motion   A very irregular movement of small particles in a fluid. The movement occurs due to the particles being hit by the molecules in the fluid.

Buffer zone   A section of land which is permanently vegetated and often used to retain pollutants in an effort to improve water, soil and air quality.

Buffering capacity   This is the ability of a substance to resist pH changes and in the context of water is related to the ability of the water to maintain a steady pH.

C_3 plants   Plants which directly fix carbon dioxide. The carbon is fixed into a compound containing three carbon atoms. Eighty-five per cent of plants are C_3, including most temperate zone plants and broadleaf plants. C_3 plants are disadvantaged in hot, dry conditions.

Capillary action   The process whereby a liquid can flow in a narrow space without the need for the influence of other forces or in opposition to other forces, such as gravity.

Capillary water   The water which is retained in soil pore spaces and in a continuous layer around soil particles. The water is held so strongly that gravity cannot drain it. The tension at which the capillary water is held in the soil may be greater than 30 times the atmospheric pressure.

Carbon footprint   The amount of carbon that an individual or an organisation uses over a given time period. It is often expressed in terms of the equivalent amount of carbon dioxide.

Cardinal scale   A method of measurement using numbers that denotes a quantity such as one, two, three.

Catchment   Hydrological unit which may be outlined by topography and which is the natural drainage area that is drained by a stream or river.

Cations   *Ions* formed when an atom or group of atoms has lost one or more negatively charged electrons. The result is that the atoms have a greater number of protons (positively charged) than electrons (negatively charged) and so cations are positively charged.

Centrifuge   A piece of apparatus which spins samples around, applying centrifugal forces to them to separate fluids.

Chemical oxygen demand (**COD**)   A test used in environmental sciences to indirectly determine the amount of organic compounds in the water. This method is often used to determine the amount of organic pollution which is found in surface waters.

Cloud seeding   The process whereby ice crystals or silver iodide crystals are spread in the upper layers of clouds to act as cloud condensation nuclei and stimulate the creation of precipitation.

Coagulation   The process in water treatment to remove impurities whereby chemicals are used to aid the clumping together of suspended solids so that they can be filtered out.

Colloids   A mix of substances where one substance of tiny insoluble particles is suspended throughout another substance. The two cannot normally be separated by simple filtering and the substances do not settle out from suspension over time.

Common property resource   A natural resource which is owned by the general public. The use of the resource is often regulated by a country's government and no individual has exclusive rights to its use. Common property resources include those from public waters, land and air.

Complexes   Molecules which are made up of two or more component molecules and are joined with a weak bond in place of a covalent bond.

Condensation   The change in state from a gaseous vapour to a liquid. It is the process of becoming more dense. Pertaining to water, it occurs when humid air comes into contact with a cold surface.

Conglomerate   In geological terms this is the term used to describe a rock which contains large individual clasts in a matrix of finer-grained particles which have been cemented together in formation.

Conjunctive use   The joint use of surface and groundwater supplies to service a need and sharing of water storage between groundwater and surface water zones.

Connate water In geological and sedimentological terms this is water which has been trapped in the pores of rock during the formation process.

Consumers In the ecological food chain these organisms gain their energy by eating other organisms.

Consumptive water use The withdrawal and use of water where the water is not returned to the system that provided the source of the water.

Convective The molecular motion responsible for the transfer of energy, such as heat, through a fluid.

Coriolis effect The rotation of the Earth on its axis causes a moving object or fluid to be apparently deflected. Deflection occurs to the right in the northern hemisphere and to the left in the southern hemisphere and is stronger at higher latitudes.

Covalent bond A chemical bond formed between two non-metal atoms. It involves the sharing of a pair of electrons between the atoms.

Critical load A threshold in environmental pollution. When it is exceeded a system will begin to fail, due to an excessive level of pollution being in the system, which affects its functioning.

Cyclonic The atmospheric characteristics found in a low-pressure weather system where the air movement occurs over a large-scale circular formation. Movement is clockwise in the southern hemisphere and anticlockwise in the northern hemisphere.

Delphi survey A method used for decision-making whereby a series of questionnaires are completed on a particular subject by experts in that field to result in an informed outcome.

Delta The depositional landform produced at the mouth of a river where it meets another water body such as the ocean or a lake. It is usually triangular in shape.

Desalination The process whereby salt and dissolved solids are removed from water, particularly seawater. This process is often undertaken to allow otherwise unsuitable water sources to be used for domestic purposes.

Desertification The occurrence whereby semi-arid land which was once habitable and productive becomes desert. Prolonged drought, possibly due to a changing climate and also to the overuse of the land by human activity, results in the top layers of soil becoming dry and therefore more easily eroded, especially by the wind.

Detergent A water-soluble chemical which is used for the purposes of cleaning through the action of combining with particles of dirt and making them more soluble.

Detritivores Organisms which gain their energy by feeding on dead and decaying organic matter, including plant litter.

Dew-point temperature The temperature at which a cooling air parcel becomes saturated with water vapour, and at which water vapour condenses to form liquid water.

Diagenesis The physical, chemical and biological changes which sediment undergoes after its initial deposition and when it is in the process of becoming sedimentary rock.

Diffuse pollution Pollution which comes from sources which are difficult to precisely locate. The pollution is likely to have come from a number of locations over a wide area. An example of diffuse pollution is nitrate leaching from agricultural land.

Displacement flow The process whereby water infiltrating at the top of a slope rapidly pushes out water in the soil at the bottom of the slope. This has an influence over storm hydrographs.

Domestication The influence of humans over the evolution of a plant or animal species.

Dose–response relationship An expression of the amount of toxin that can enter an environmental system and the extent of the response of the system to that amount of toxin.

Drainage basin The area of land drained by a single river system.

Drainage density The measure of the total length of all the rivers and streams in a drainage basin divided by the total drainage basin area.

Ecological footprint A measure of the demands that humanity places on the Earth's natural resources and ecosystems. It is a standardised measure which determines the area of land and water required to sustain a human population of a given size.

Ecosystem services Services provided by ecosystems that support human life in some way (e.g. food, medicine and clean water). Evaluating these services helps to focus public attention on environmental issues that could result in the loss of these services.

Effective rainfall The rainfall that is not stored on the land surface nor infiltrates into the soil. It produces the overland flow which generates quick-responding stormflow in the river.

Electrical conductivity The ability of material to transport electrical charge. In water the electrical conductivity will typically increase as the concentration of dissolved material increases.

Endemic In ecological terms this refers to species of plants or animals which belong to a specific location due to their evolution in biogeographically isolated areas, such as the Galapagos Islands.

Endorheic basin An internal drainage basin which is a closed hydrological system, retaining water rather than allowing it to flow to other water bodies such as other rivers or the ocean. They are most likely to be found in desert locations, although they can form in other climatic zones, and can be seasonal or permanent.

Ensemble When related to environmental modelling this refers to a group of models which represent an environmental system being considered all at once. Each model has a slight difference in its representation of the system, but comparison allows users to see the spread of possible outcomes and predictions from the group of models.

Enteric pathogens Bacteria which occur in the intestines of humans and animals and cause disease.

Epidemiology The study of health and disease patterns and causes and application of this study to control health and disease problems.

Epilimnion Within a water body this is the layer at the surface which is warmer and less dense than the layer which is trapped below it and is cooler and more dense.

Eutrophication The enrichment of plant nutrients in freshwater and marine water bodies; this results in the accelerated bloom of plants in the water, leading to a detrimental effect on other organisms in the water.

Evapotranspiration The process whereby water from the Earth's surface is lost to the atmosphere as water vapour through evaporation and transpiration by plants.

Exorheic basin The dominant type of drainage basin in the world, these systems are open and drain to other rivers and, ultimately, the seas and oceans.

Field capacity The term used to describe the maximum amount of water which can be held as capillary water in soil once all gravitational water has drained away.

Flocculation The following stage after coagulation; this is the process whereby a liquid is slowly mixed so that smaller particles agglomerate to form larger, suspended particles which are visible.

Folded mountains Mountain ranges which are formed through the coming together of two continental tectonic plates. As they both have the same density the collision results in the rock layers being pushed inwards and upwards.

Food web A representation of the connectivity of food chains and trophic levels in an ecological community.

Genotype The entire genetic make-up of an individual organism.

Germ theory of disease A theory developed in the 1800s stating that the presence and actions of microorganisms in the body are the cause of many diseases. The research of Louis Pasteur and Robert Koch provided scientific evidence for the theory.

Glacial mass balance The difference over time between the total accumulation and melt of a glacier. The mass balance can be either positive or negative, depending upon whether accumulation exceeds loss or vice versa.

Greenhouse gases These gases are present in the Earth's atmosphere and can absorb and emit infrared radiation and therefore aid in the regulation of the Earth's temperature. Water vapour, carbon dioxide, methane, nitrous oxide and ozone are some of the most abundant greenhouse gases present in the atmosphere.

Gross primary productivity The total amount of biomass formed during photosynthesis, within a specific time period.

Head loss This is the measure of the reduction in the total pressure of a fluid between one point and another. It is the result of friction between adjacent particles in the fluid moving past each other, the friction between the fluid and the substance which it is moving through, and also the turbulence which occurs as a result of the redirection of fluids.

Headroom With respect to water availability this term refers to the difference which exists between supply and demand. Headroom can be classified as target headroom or available headroom, each representing the supply–demand balance of water utility.

Helicoidal flow The flow of water in a river meander which is corkscrew in pattern. As a result of this the water level on the outside of the meander bend becomes super-elevated, and the return flow of the water is along the channel bed.

Hermaphrodite An organism which has both female and male reproductive organs.

Hydraulic conductivity A measure of the transmission of water through soil at a specific hydraulic gradient. It is denoted by the letter K.

Hydraulic gradient The change in hydraulic pressure per unit distance in a specific direction.

Hydraulic head A term representing the energy of water at a given point in an aquifer. It is often measured via the level at which water stands in an unconfined or confined well. It is the sum of three component parts: the pressure head, the elevation head and the velocity head at a specific point.

Hydrological pathways The generic term used to describe the ways in which water can flow through a catchment. These pathways can be categorised into three main groups: above the ground, including evaporation, interception, stemflow and throughfall; overland flow; and below the ground, including infiltration, macropore flow and subsurface matrix flow.

Hygroscopic aerosols A specific type of aerosol which is water soluble and has the ability to attract water vapour from the air.

Hygroscopic water   Water which exists as a film around soil particles. In this form it is not available to plants, as a result of the strong attraction between the water molecules and the soil particles.

Hypolimnion   Within a water body this represents a lower layer which is cooler and denser than the water found above it. There is little mixing between this layer and the one above it, due to its characteristics.

Hyporheic zone   A zone found below a river channel where there is a mixing of groundwater and channel water in the substrate.

Hypoxia   The condition whereby the dissolved oxygen available is below the required level to support most life forms. It is the result of the oxygen being used in the decomposition of organic matter exceeding oxygen replenishment through photosynthesis and from exchange with the atmosphere. The concentration of oxygen in hypoxic waters is usually less than 2–3 parts per million.

Igneous rock   Rock which is formed through the solidifying of magma once it is has escaped a volcano.

Infiltration   The process whereby water is absorbed into the soil and then transported downwards through the soil profile.

Infiltration capacity   The upper rate at which water can flow into the soil from the surface. This rate can change through time, depending on how wet the soil is and the surface conditions.

Infiltration-excess overland flow - When the rate of rainfall or irrigation water supply to the soil surface exceeds the infiltration capacity, excess water will flow across the surface. This is also known as Hortonian overland flow.

Infiltration rate   The time it takes for a unit of water added to a surface to enter a unit of soil.

Infrared radiation   Generally invisible to humans, a form of electromagnetic radiation that is released by all solids, liquids and gases.

Interception   The process whereby precipitation is captured by plants as it falls and, instead of reaching the ground, is lost back to the atmosphere through *evapotranspiration*.

Interflow   The movement of water which has infiltrated the surface layers of the soil and then moves laterally in the unsaturated zone of the soil above the water table; often also known as *throughflow*.

Intertropical Convergence Zone (ITCZ)   The region where the trade winds from the northern and southern hemispheres converge. Conditions are favourable for cloudiness and heavy rainfall.

Ion   A charged atom or molecule. The charge results from an imbalance in the number of electrons, which are negatively charged, as compared to the number of protons, which are positively charged. If there is one more proton than electrons, then the atom or molecule is said to be positively charged and it will be attracted to an atom or molecule that is negatively charged.

Irrigation   The process whereby water is artificially applied to the land, often used to aid crop production in arid climates or dry periods elsewhere.

Jet stream   Narrow, high-speed winds caused by sharp temperature gradients, located within Rossby waves in the upper atmosphere. They can be thousands of kilometres long and hundreds of kilometres wide.

Karst   A distinctive landscape created by the dissolving of limestone rocks. Such landscapes often include sinkholes and underground caverns.

Latent heat of vaporisation   The amount of energy which is required to change a specific mass of a liquid into a gas without a change in temperature.

Lentic A word used to describe a feature or organism associated with slow-moving or static water bodies such as lakes.

Levees Raised embankments which are constructed to protect land from rising flood waters.

Littoral zone The zone in a lake closest to the shore and where light can reach the bottom, allowing for a diverse range of plants and algae to exist.

Long profile A graphical representation of the change in altitude of a river channel along its longitudinal course from head to mouth.

Lotic A feature or organism associated with fast-moving water bodies such as rivers.

Lysimeter An instrument used to measure the water which flows through or is lost from the soil and which can also be used to measure the chemical composition of the water as it flows through.

Macropore flow The transfer of water through the soil between large pores greater than 0.1 mm in diameter.

Macropores Openings within the soil which are greater than 0.1 mm in diameter and can result in rapid flow. They can be produced as a result of structural failings in the soil or biological activity such as the burrowing of animals, earthworms and plant roots.

Marginal costs The increase or decrease in the total cost of production, should one extra unit of a product be produced.

Matrix flow The transfer of water through the soil between pores smaller than 0.1 mm in diameter.

Meiofauna Benthic organisms which range in size from approximately 0.1 mm to 1 mm.

Mesohabitats These units combine to make up a reach of a river and have similar characteristics, such as deep, slow-flowing pools.

Metabolism The various chemical transformations that take place in living organisms to maintain life.

Metalloids Elements whose properties are considered to be a mix of metal and non-metal and which as a result are difficult to classify. Boron is considered a metalloid.

Metamorphic See *Metamorphic rocks*.

Metamorphic rocks Rocks which have undergone considerable structural and mineralogical alterations due to heat and pressure underground.

Miasmatic theory of disease A historic theory of disease which suggested that diseases were the result of miasma, a form of 'bad air'. The spread of cholera and the Black Death were historically attributed to this theory.

Monomers A molecule that can bond with other molecules to produce very long chains of molecules.

Nanomaterials This term is used to broadly refer to materials which have a structural component which is less than 100 nanometres. The properties of these materials differ from others in the sense that they have an increased relative surface area and quantum effects which can result in changes in other properties.

Natural greenhouse effect Atmospheric greenhouse gases, such as carbon dioxide and water vapour, absorb 90% of the long-wave radiation emitted from Earth, resulting in an average global temperature approximately 35°C warmer than would be experienced without the natural levels of greenhouse gases present.

Nektonic (nekton) The term used to refer to organisms which move around in the water column, rather than being restricted to one region within it.

Net primary production This is the difference between the biomass which is fixed by plants during photosynthesis (gross primary production) minus the energy which is used by plants during respiration.

Neuston   The term used to refer to organisms which exist on or just below the surface of a water body.

Non-point-source pollution   See *Diffuse pollution*.

Nosocomial   Originating from or occurring within a hospital. The term is usually used with reference to an infection.

Orographic   The term used when considering things in relation to mountains or their influence.

Osmoregulation   The process whereby a balance is kept by cells and simple organisms between the fluid and the electrolytes they contain and their surrounding conditions.

Overland flow   The process whereby water runs over the surface of the land. Overland flow is often classified into two types: *infiltration-excess overland flow* and *saturation-excess overland flow*.

Oxbow lake   A body of water produced when a meander becomes cut off from the rest of the river channel as a result of a change in the river's course.

Palaeolimnological   Pertaining to ancient lakes through a study of their sediments and fossils.

Parasites   Organisms which exist by living on or inside another organism, taking nourishment from the host organism but not giving anything back to it.

Parent rock   The original rock from which another material may be formed. It is usually referred to in relation to soil formation and the parent rock which was weathered to produce the initial soil particles.

Partial contributing area concept   With respect to the process of *infiltration-excess overland flow* this is the notion that this will occur only in localised areas, rather than over a whole catchment.

Partial pressure   The pressure that would be exerted by a single gas if that gas were the only one present in a specific volume.

Patches   Also known as microhabitats in river systems, these are discrete areas used by a species for breeding or obtaining resources.

Pathogen   The term used to describe organisms which can cause disease.

Pelagic photic zone   This is the zone at the top of a body of water where the abundance of light allows for *photosynthesis* to take place.

Percolates   The ability of a liquid such as water to move through small holes and pores in a substance such as soil.

Permafrost   The term used to describe soil and bedrock materials which are frozen below 0°C for a minimum of two years. Permafrost is predominantly found in high latitudes.

Permeability   The ability of a substance to transmit fluids through it. The permeability of soil can be influenced by its composition; for example, a sandy soil is more permeable than a clayey soil.

Photolysis   The action of the breakdown and separation of molecules by light.

Photosynthesis   The process by which organisms such as plants and algae (autotrophs) create carbohydrates and release oxygen using light energy, carbon dioxide and water.

Phreatic zone   The zone within an aquifer which is filled with water.

Piezometric   A term related to the measurement of water pressure inside an aquifer.

Planktonic   The name given to small organisms which are suspended in the water column and move by the action of drifting and floating.

Plutonic   The term used to refer to igneous rocks which have solidified at a great depth below the Earth's surface.

Point-source pollution Contamination of a water body which has originated from a specific location such as a pipe from an industrial plant.

Pore spaces The gaps which exist between soil particles.

Pore water pressure The pressure of water which is held in soil pores or rock aquifers under saturated conditions. Pore water pressure is measured relative to atmospheric pressure.

Potable water Water of a high enough quality that it can be safely consumed by humans.

Precautionary principle A decision-making approach which believes that lack of scientific evidence for warnings about future threats of serious damage should not be used as an excuse to avoid action in order to prevent damage from happening; action should be taken as early as possible.

Precipitation The condensation of water vapour to form water droplets in the atmosphere, which are then deposited on the Earth's surface in a liquid (e.g. rain) or solid form (e.g. snow, hail).

Producers In ecological terms this refers to organisms which are able to produce their own energy and biomass through processes such as *photosynthesis*.

Profundal zone A deep zone of water found in oceans and lakes. Due to the depth, light penetration is limited.

Property rights Rights pertaining to the authority that determines how a resource can be used.

Q-methodology A research method which takes into consideration the individual's viewpoint.

Rating equation The equation for a curve of best fit derived from the plotting of known river discharge and water level values. From this it is possible to determine the river discharge value at a particular point along a river if the water level is at a particular height.

Recharge The process whereby water moves downwards from the surface to groundwater aquifers to replenish the water stored there.

Recovery efficiency This is the ratio between usable water and the injection of water through aquifer storage and recovery. It indicates the level of mixing which occurs between stored water and native water in an aquifer. Efficiency may begin low but improve with subsequent flushes of water.

Recurrence interval The period of time in which a flood of a particular magnitude is likely to return. For example, a 1-in-100-year flood event would happen, on average, once every 100 years at a particular location.

Respiration A chemical process undertaken by organisms to produce energy from glucose, normally involving the uptake of oxygen into cells and the release of carbon dioxide.

Riffles The accumulation of coarse sediment which forms bar deposits across a river, which tend to be spaced apart by between five and seven times the channel width.

Riming The process of cloud growth whereby supercooled water droplets are deposited on the ice crystal surfaces in the clouds. This method leads to the rapid growth of clouds.

Riparian Pertaining to the banks of a river or stream.

River basin See *Catchment*.

River discharge The volume of water flowing through a given point on a river at a given time.

River reach A length of a river which shares similar physical characteristics.

River regime This term refers to the annual variation in river discharge.

Rossby waves Large upper-atmosphere undulations which disturb the belt of prevailing westerly winds associated with the Ferrel cell. They contain jet streams.

Runoff See *Overland flow*. This term sometimes also includes *throughflow* water that makes it to river systems.

Salinity The concentration of all of the salts dissolved in water.

Saltating A method of sediment transport whereby sediment grains are bounced along a bed surface.

Saturation-excess overland flow A flow that occurs when all of the pore spaces within the soil become filled with water, and therefore saturated, forcing the excess water to move across the surface.

Saturation vapour pressure In a closed container evaporation will occur until there are the same number of molecules returning to the liquid as there are escaping from it. This is the point where the vapour is said to be saturated. The vapour pressure differs according to temperature.

Schmutzdecke A sticky, red-brown biological layer that naturally forms on the top of a slow sand filter.

Sea-level rise Relative to a fixed datum mean sea levels can rise due to thermal expansion of water and *salinity* changes and due to increased volumes of water in oceans due to melting ice sheets on land. This global phenomenon is known as eustatic sea-level rise. Mean sea levels may also appear to rise locally with respect to the surrounding land if the land surface elevation falls due to, for example, groundwater abstraction, mining or compaction.

Sedimentary rocks These rocks are formed through a build-up of sediments, whether that be through transport and deposition in water, precipitates from a solution or secretions from organisms. These sediments build up over time and are subject to forces to become consolidated and solidified.

Sensible heat This refers to the energy which can increase the temperature of a substance, but its phase remains unaltered.

Shear stress A stress acting upon a particle in the same direction as the surface it is resting upon. In rivers the shear stress is the velocity of flowing water. When a sediment particle can be lifted from the riverbed then the critical flow velocity (critical shear stress) is reached.

Single mass curve A line on a graph showing the cumulative total of a substance that accrues over time. In water resources, the cumulative river discharge over a year could be drawn as a single mass curve.

Soil water tension The force required to remove from the soil the water which is held in the *pore spaces* around the soil particles.

Solute load The amount of dissolved material which is carried in stream or river water at a given point in space for a set period of time. It is usually measured in units of mass (e.g. kilograms).

Solutes Substances which are dissolved within another substance to form a solution.

Sorption The process whereby one substance is taken up by another substance which is in a different phase state.

Specific capacity The term used to measure the performance of a well. It is derived by dividing the yield of the well by the amount of drawdown.

Specific discharge A measure of the amount of groundwater discharge across a cross-sectional area of a porous medium. It is also known as the Darcy velocity.

Specific heat The amount of heat which is required to increase the temperature of a given unit mass of a substance by 1°C.

Stage   The height of the water level in a river channel at a given point in time.

Stakeholders   Those who have an interest in a resource which is available. It may be an individual or a group which have a claim on a resource.

Stomata   Openings in a plant through the surface of the leaves. Through these openings oxygen can be taken in and water and carbon dioxide can be expelled.

Stratification   The development of layers in a body of water, due to different densities in the water.

Stream order   The dimensionless numbering method used to identify how tributaries have merged into a stream and hence its relative position in the stream network.

Sublimation   The process by which a solid changes directly into a gas.

Suspended sediment   Particles larger than 0.45 μm which are suspended in water.

Sustainable Development Goals (SDGs)   A set of 17 goals with underlying targets agreed by the United Nations in 2015. All countries have a responsibility to deliver actions to achieve the global targets.

Sustainable urban drainage systems   A technique used within urban environments to mitigate against the issues of flooding and surface water quality which can occur through the traditional approach of routing runoff through pipe networks directly to a watercourse, as well as that of water harvesting. Methods used include the creation of permeable surfaces, underground storage areas and wetlands.

Symbiosis   The biological process whereby two organisms can exist in the same geographic location without causing harm to the survival and productivity of the other.

Thermocline   A sharp temperature gradient which exists in a body of water (ocean or lake) where there is a dramatic difference between the water temperature above and the water temperature below.

Throughflow   The movement of water draining through the soil in a downslope direction.

Total dissolved solids   The total mass of inorganic and organic substances which are dissolved in water.

Total economic value   An approach in environmental economics which does not restrict the value placed on a natural resource to its monetary worth. Instead the total economic value considers the sum of all the values that a resource can have, those which come from its direct and indirect use, and those which come from the mere existence of the resource.

Total hydraulic potential   See *Hydraulic head*.

Toxicity   A measure of how toxic a substance is; the degree to which a substance can harm an organism.

Transboundary   Describes something which moves across boundaries or borders; it may be used to refer to freshwater resources or pollution.

Transboundary pollution   Pollution which originates in one jurisdiction but can be transported across a boundary or border between countries.

Transmissivity   The volume of horizontal water flow through an aquifer.

Transpiration   The evaporation of water which is released into the atmosphere through the pores of plant leaves.

Tropopause   The boundary in the Earth's atmosphere which separates the *troposphere* from the stratosphere. Typically air temperature falls with rising altitude in the troposphere. However, once the troposphere is reached, air temperature rises with altitude.

Troposphere The lower layer of the atmosphere that extends between 6 and 15 km in altitude above the Earth's surface.

Tsunami An energetic sea wave triggered by an earthquake, landslide or meteor impact in the ocean, which can become very large once it reaches shallow water and cause devastation in coastal zones.

Turbidity A measure of the clarity of water. Fine particles suspended within the water reduce the penetration of light into the water body.

Ultraviolet (UV) On the spectrum of electromagnetic radiation this form has a shorter wavelength than visible light but a longer wavelength than X-rays. Too much exposure can damage living tissue.

Vadose zone The zone of water which is below the Earth's surface but remains above the water table and is therefore aerated.

Variable source area concept The theory put forward that the areas in a catchment where overland flow is generated will vary in space (location and size) and time. The areas where overland flow occurs are influenced by a number of factors, including rainfall, temperature, topography and vegetation.

Vesicle A cavity filled with fluid or air.

Volatilisation (also volatilise) The process whereby a substance is evaporated and becomes a vapour.

Water scarcity According to the United Nations' use of the term, this occurs when the renewable freshwater available is less than 1000 m^3 per person per year. However, other definitions suggest that water scarcity occurs below 500 m^3 per person per year. The UN defines the latter as *absolute water scarcity*.

Water softening The process of reducing the levels of metal ions in the water, predominantly calcium and magnesium, which otherwise mean that it is hard.

Water stress According to the United Nations' use of the term, this occurs when the renewable freshwater available is less than 1700 m^3 per person per year.

Water table The upper limit of the saturated zone of the soil or rock.

Watershed In some regions this term has the same meaning as river *catchment*, while in other regions it refers to the high point that forms the divide between two adjacent catchments; thus, care over interpreting the meaning is required.

Weathering The process whereby the physical and chemical erosion of rock causes it to break down into smaller particles.

Index